Optimizing and Measuring Smart Grid Operation and Control

Abdelmadjid Recioui
Université M'hamed Bougara de Boumerdes, Algeria

Hamid Bentarzi
Université M'hamed Bougara de Boumerdes, Algeria

A volume in the Advances in Computer and Electrical Engineering (ACEE) Book Series

IGI Global
PUBLISHER of TIMELY KNOWLEDGE

Published in the United States of America by
IGI Global
Engineering Science Reference (an imprint of IGI Global)
701 E. Chocolate Avenue
Hershey PA, USA 17033
Tel: 717-533-8845
Fax: 717-533-8661
E-mail: cust@igi-global.com
Web site: http://www.igi-global.com

Library of Congress Cataloging-in-Publication Data

Names: Recioui, Abdelmadjid, 1979- editor. | Bentarzi, Hamid, 1965- editor.

Title: Optimizing and measuring smart grid operation and control / Abdelmadjid Recioui and Hamid Bentarzi, editors.
Description: Hershey, PA : Engineering Science Reference, 2021. | Includes bibliographical references and index. | Summary: "This book presents the research activities and trends on the topic of smart grids, focusing the smart grid main components, its operation, and control and optimization aspects"-- Provided by publisher.
Identifiers: LCCN 2020029727 (print) | LCCN 2020029728 (ebook) | ISBN 9781799840275 (hard cover) | ISBN 9781799869214 (soft cover) | ISBN 9781799840282 (ebook)
Subjects: LCSH: Smart power grids.
Classification: LCC TK3105 .O68 2021 (print) | LCC TK3105 (ebook) | DDC 621.31--dc23
LC record available at https://lccn.loc.gov/2020029727
LC ebook record available at https://lccn.loc.gov/2020029728

This book is published in the IGI Global book series Advances in Computer and Electrical Engineering (ACEE) (ISSN: 2327-039X; eISSN: 2327-0403)

British Cataloguing in Publication Data
A Cataloguing in Publication record for this book is available from the British Library.

All work contributed to this book is new, previously-unpublished material. The views expressed in this book are those of the authors, but not necessarily of the publisher.

For electronic access to this publication, please contact: eresources@igi-global.com.

Advances in Computer and Electrical Engineering (ACEE) Book Series

Srikanta Patnaik
SOA University, India

ISSN:2327-039X
EISSN:2327-0403

Mission

The fields of computer engineering and electrical engineering encompass a broad range of interdisciplinary topics allowing for expansive research developments across multiple fields. Research in these areas continues to develop and become increasingly important as computer and electrical systems have become an integral part of everyday life.

The **Advances in Computer and Electrical Engineering (ACEE) Book Series** aims to publish research on diverse topics pertaining to computer engineering and electrical engineering. **ACEE** encourages scholarly discourse on the latest applications, tools, and methodologies being implemented in the field for the design and development of computer and electrical systems.

Coverage

- Analog Electronics
- Electrical Power Conversion
- Computer Architecture
- Qualitative Methods
- Chip Design
- VLSI Fabrication
- Algorithms
- Applied Electromagnetics
- Power Electronics
- Computer Hardware

IGI Global is currently accepting manuscripts for publication within this series. To submit a proposal for a volume in this series, please contact our Acquisition Editors at Acquisitions@igi-global.com or visit: http://www.igi-global.com/publish/.

The Advances in Computer and Electrical Engineering (ACEE) Book Series (ISSN 2327-039X) is published by IGI Global, 701 E. Chocolate Avenue, Hershey, PA 17033-1240, USA, www.igi-global.com. This series is composed of titles available for purchase individually; each title is edited to be contextually exclusive from any other title within the series. For pricing and ordering information please visit http://www.igi-global.com/book-series/advances-computer-electrical-engineering/73675. Postmaster: Send all address changes to above address.

Titles in this Series

For a list of additional titles in this series, please visit: www.igi-global.com/book-series

New Methods and Paradigms for Modeling Dynamic Processes Based on Cellular Automata
Stepan Mykolayovych Bilan (State University of Infrastructure and Technology, Ukraine) Mykola Mykolayovych Bilan (Mayakskaya Secondary School, Moldova) and Ruslan Leonidovich Motornyuk (Main Information and Computing Center, Ukraine)
Engineering Science Reference • © 2021 • 326pp • H/C (ISBN: 9781799826491) • US $195.00

Examining Quantum Algorithms for Quantum Image Processing
HaiSheng Li (Guangxi Normal University, China)
Engineering Science Reference • © 2021 • 336pp • H/C (ISBN: 9781799837992) • US $195.00

Design and Investment of High Voltage NanoDielectrics
Ahmed Thabet Mohamed (Aswan University, Egypt & Qassim University, Saudi Arabia)
Engineering Science Reference • © 2021 • 363pp • H/C (ISBN: 9781799838296) • US $195.00

Research Advancements in Smart Technology, Optimization, and Renewable Energy
Pandian Vasant (University of Technology Petronas, Malaysia) Gerhard Weber (Poznan University of Technology, Poland) and Wonsiri Punurai (Mahidol University, Thailand)
Engineering Science Reference • © 2021 • 407pp • H/C (ISBN: 9781799839705) • US $225.00

HCI Solutions for Achieving Sustainable Development Goals
Fariza Hanis Abdul Razak (Universiti Teknologi MARA (UiTM), Malaysia) Masitah Ghazali (Universiti Teknologi Malaysia (UTM), Malaysia) Murni Mahmud (International Islamic University Malaysia, Malaysia) Chui Yin Wong (Multimedia University, Malaysia) and Muhammad Haziq Lim Abdullah (Universiti Teknikal Malaysia, Melaka, Malaysia)
Engineering Science Reference • © 2021 • 300pp • H/C (ISBN: 9781799849360) • US $195.00

Innovations in the Industrial Internet of Things (IIoT) and Smart Factory
Sam Goundar (The University of the South Pacific, Fiji) J. Avanija (Sree Vidyanikethan Engineering College, India) Gurram Sunitha (Sree Vidyanikethan Engineering College, India) K Reddy Madhavi (Sree Vidyanikethan Engineering College, India) and S. Bharath Bhushan (Sree Vidyanikethan Engineering College, India)
Engineering Science Reference • © 2020 • 300pp • H/C (ISBN: 9781799833758) • US $225.00

IGI Global
PUBLISHER of TIMELY KNOWLEDGE

701 East Chocolate Avenue, Hershey, PA 17033, USA
Tel: 717-533-8845 x100 • Fax: 717-533-8661
E-Mail: cust@igi-global.com • www.igi-global.com

Table of Contents

Detailed Table of Contents

Pricing policy is one of the tools allowing the involvement of customers in the balance between the supply and the demand in smart grids. The present chapter aims at presenting the smart metering action including the bidirectional measurement of energy for smart houses equipped with renewable energies as well as the way a smart meter communicates data at the required timing to and from the control center. A typical bill establishment explaining how the net billing is produced along with a discussion about different pricing policies that the utility may adopt to reduce the peak load demand is also presented. The work is concluded by a typical simulation of a smart city modeled in LABVIEW software.

Many models of phasor measurement units (PMU) have been implemented; however, few dynamic models have been developed when the power system parameters change. It is necessary to use a method that can somehow estimate the frequency and correct the phasors. The conventional way to determine frequency is to detect zero crossings per unit time. However, this method has many drawbacks such as high cost and low accuracy. Also, after the frequency determination, the phasor should be corrected by suitably modifying the algorithm without omitting any data. This chapter presents different estimation techniques such as discrete Fourier transform (DFT), smart discrete Fourier transform (SDFT) that may be used to estimate the phasors. These estimated values would be incorrect if the input signals are at an off-nominal frequency and the phase angles would drift away from the true values. To correct this issue, first of all, the off-nominal frequency has been estimated using different techniques such as least error squares and phasor measurement angle changing, and then it is used to correct the phasors.

During a transient operation condition of power smart grid, line current may include unwanted components that may cause unnecessary tripping of protection system. The disturbance mainly appears in a form of harmonics and sub-harmonics. In this case of signal waveforms including harmonics, the low pass filter may be used. However, this type of filter does not provide the ability to reject sub-harmonics. This chapter presents the digital filtering design issue based on optimization approach for removing sub-harmonics and hence improving the measurement. The first point of view is to reach an unified accurate phasor measurement algorithm that is immune to nearly all disturbances (sub-harmonics) in power grid including FACT devices and renewable energy sources, simultaneously with required speed of convergence. The second point focuses on reducing the computational requirement and algorithm complexity through designing recursive digital filter with reduced order.

Section 2
Management and Control in Smart Grids

The conventional energy meters are not suitable for long operating purposes as they spend much human and material resources. Smart meters, on the other hand, are devices that perform advanced functions including electrical energy consumption recording of residential/industrial users, billing, real-time monitoring, and load balancing. In this chapter, a smart home prototype is designed and implemented. Appliances are powered by the grid during daytime, and a photovoltaic panel stored power during the night or in case of an electricity outage. Second, consumed power from both sources is sensed and further processed for cumulative energy, cost calculations and bill establishment for different proposed scenarios using LABVIEW software. Data are communicated using a USB data acquisition card (DAQ-USB 6008). Finally, a simulation framework using LABVIEW software models four houses each equipped with various appliances. The simulator predicts different power consumption profiles to seek of peak-demand reduction through a load control process.

In these last decades, electrical power grids become more intelligent. Hence, sophisticated software and hardware were introduced to the power grid, which makes it a smart grid. This chapter is an introduction on smart-grid technology; thus, microgrids are explained, and the use of multiagent system in centralized/decentralized energy management systems are discussed and compared. Smart agents are an emerging technology for decentralized computation and data storage. Hence, in this chapter, decentralized energy management system is created basing on multi-agent system technique where sources and loads are considered as separated agents each of them. After that, these sources and load create a microgrid and each microgrid can be considered as an agent. The work proposes an approach for load supplying optimization to decrease the microgrid cost and enhance its efficiency.

It is very important that the installed renewable energy system should produce the maximum power outputs with minimum costs, and that can only be achieved with the selection of the best optimization technique applied for the best control strategies along with the introduction of the hybrid energy storage systems (HESS). This chapter presents some optimization techniques applied in control strategies for hybrid energy storage systems in distributed renewable energy systems. The integration of energy production and consumption component through the smart grid concept enables increased demand response and energy efficiency. Hybrid energy storage systems and their applications in the renewable energy systems are extensively discussed besides control strategies involved. The storages systems will play an important role in future related to smart grid.

This chapter presents different techniques for obtaining the optimal number of the phasor measurement units (PMUs) that may be installed in a smart power grid to achieve full network observability under fault conditions. These optimization techniques such as binary teaching learning based optimization (BTLBO) technique, particle swarm optimization, the grey wolf optimizer (GWO), the moth-flame optimization (MFO), the cuckoo search (CS), and the wind-driven optimization (WDO) have been developed for the objective function and constraints alike. The IEEE 14-bus benchmark power system has been used for testing these optimization techniques by simulation. A comparative study of the obtained results of previous works in the literature has been conducted taking into count the simplicity of the model and the accuracy of characteristics.

A microgrid is a representation of controllable electrical power network, which comprises the distributed generation (DG) such as micro-hydro, wind turbines, photovoltaic arrays, bio-mass generation, battery energy storage systems (BESSs), and a group of interconnected load. A typical microgrid can be operated in either grid connected or island mode. The issue of unintentional islanding in microgrid is a major challenge for power engineers as it may harm the service personnel working during maintenance or restoration of the main grid. In literature, various passive, active, and hybrid islanding detection methods (IDMs) have been developed and implemented by researchers/engineers. In this chapter, the proposed IDMs from the recent literature have been analyzed and briefly discussed. The primary requirement of IDMs is fast and accurate detection of main grid failure and should not deteriorate the power quality of distribution networks.

In recent years, power quality (PQ) has become an increasingly major concern for both electric utilities and the end users. Accordingly, the electrical engineering community has to deal with the analysis, diagnosis, and solution of PQ issues using system approach rather than handling these issues as individual problems.

This chapter describes the analysis of PQ using advanced signal processing tools represented in Hilbert and wavelet transforms (HT-WT) and artificial intelligence tools represented in artificial neural network and support vector machine (ANN-SVM) for detection and classification of power quality disturbances, respectively. These techniques were successfully simulated using LABVIEW software capabilities. The results of simulation indicate that the signal processing techniques are effective mechanisms to detect and classify power quality disturbances. At the end, the combination of WT as a tool of detection and features extraction with SVM as a classifier tool resulted as the best combination for PQ monitoring system.

This chapter presents how root cause analysis based on fault tree technique can be used to study mal-operation of protective relay in a smart power grid. This approach is used to identify disturbances first and root cause of mal operation of protective relays which may lead to unnecessary tripping and hence to blackout. Once the critical root causes are identified, mitigation measures first such as blocking protection functions and digital filters may be used in view of increasing the reliability of the considered protection system. The proposed approach has many advantages as it allows obtaining an important quantitative figure (security). These permits to strongly strengthen the elements of the protecting system which are most likely appropriate to failure and hence the impact on the overall system's cost is significant. Another main advantage is that it takes into account the reliability of the software part of the system which is considered to have a significant contribution in the overall reliability of the protecting system.

The testing approach is facing many difficulties regarding the actual implementations in the modern smart power grids. One of these challenges is the testing of hardware devices such as protective relays, PMUs, and smart meters before its final deployment to the power grid. One way to overcome this is the real-time simulation of power grid. The hardware-under-test (HuT) is plugged to a real-time simulator via signal conditioning circuit (SCC). SCC is an interface circuit involving power amplifier and measurement sub-circuit between the real-time power grid simulator and the HuT. In this chapter, some advanced developed techniques and approaches will be presented.

Preface

The energy sector constitutes a vital element for the development of any society. In particular, the society has become dependent on electricity at every level as the electric grids are providing electricity to residential areas, businesses and industries. The ever increasing energy demand and climate change issues in the 21st century made it a priority for many nations to find new energy resources and enhance energy efficiency (Chellali et al., 2011). Traditional power grids face some challenges that limit their efficiency. Among these deficiencies are the non-optimal dimensioning and usage of grid resources. In order to meet customers' demand of electric energy, power capacity must be able to meet the worst case scenario that is the peak of demand (Recioui et al., 2016). Moreover, additional capacity must be available to deal with the uncertainty in generation and consumption. Based on that, the grid resources are, for most of the time, underutilized (Barbato and Capone, 2014). Furthermore, power failure problems such as voltage sags, blackouts, and overloads have occurred are frequent (Dekhandji et al., 2019; Dekhandji et al., 2020). Also, the integration of medium- to small-sized renewable energy source (RES) plants in the current power grids may lead to instability in the power grid (Recioui and Dassa, 2017). From economic and environmental point of view, changes must be implemented to the current unstable and inefficient grid. This requires it to be reliable, scalable, manageable, and extensible. On the other hand, the grid should also be interoperable, secure, and cost-effective. The right mix of these requirements is found in the modern electric infrastructure called ''Smart Grid''.

A smart grid is an electricity network that uses digital and other advanced technologies to monitor and manage the transport of electricity from all generation sources to meet the varying electricity demands of end-users. It offers a lot of valuable technologies that are already in use today or that can be used within the near future. Smart grids co-ordinate the needs and capabilities of all generators, grid operators, end-users and electricity market stakeholders to operate all parts of the system as efficiently as possible i.e. minimizing costs and environmental impacts while maximizing system reliability, and stability. Smart Grid includes electric network, digital control appliance, and intelligent monitoring system (Zhao et al., 2013). All of these can:

- Deliver electricity from producers to consumer.
- Control energy flow.
- Reduce the loss of energy.

Make the performance of the electric network more reliable and controllable.

In the short term, a smarter grid will function more efficiently, enabling it to deliver a specific level of service at an expected less costs while offering great societal benefits such as less impact on the environment. We can expect the Smart Grid to support the kind of transformation that the internet has already brought to our daily life (European Smart Grids Technology Platform, 2006).

There are many smart grid technology areas (each consisting of sets of individual technologies) spanning the entire grid, from generation through transmission and distribution to various types of electricity consumers. Some of the technologies are actively being deployed and are considered mature in both their development and application, while others require further development and demonstration. A fully optimized electricity system will deploy all the technology areas. However, not all technology areas need to be installed to increase the "smartness" of the grid (Technology Roadmap Smart Grids, 2011). The key technologies are highlighted in the following sections.

- Wide-area monitoring and control: Real-time monitoring and display of power system components and performance across interconnections and over large geographic areas help system operators to understand and optimize power system components, behavior and performance. Advanced system operation tools avoid blackouts and facilitate the integration of variable renewable energy resources (Technology Roadmap Smart Grids, 2011; Recioui et al., 2015; Recioui, 2018).
- Information and communications technology integration: The underlying communication infrastructure, whether using private utility communication networks (radio networks, meter mesh networks) or public carriers and networks (Internet, cellular, cable or telephone), supports data transmission for deferred and real-time operation, and during outages (Technology Roadmap Smart Grids, 2011; Saoud and Recioui, 2017; Recioui and Grainat, 2020).
- Renewable and distributed generation integration: Integration of renewable and distributed energy resources can present challenges for the dispatch ability and controllability of these resources and for operation of the electricity system. Energy storage systems can alleviate such problems by decoupling the production and delivery of energy. Smart grids can help through automation of control of generation and to ensure supply and demand balance (Technology Roadmap Smart Grids, 201; Azzougui and Recioui, 2018).
- Transmission enhancement applications: There is a number of technologies and applications for the transmission system:
    - Flexible AC transmission systems (FACTS) are used to enhance the controllability of transmission networks and maximize power transfer capability.
    - High voltage DC (HVDC) technologies are used to connect wind and solar farms to large power areas, with decreased system losses and enhanced system controllability.
    - Dynamic line rating (DLR), which uses sensors to identify the current carrying capability of a section of network in real time, can optimize utilization of existing transmission assets, without the risk of causing overloads.
    - High-temperature superconductors (HTS) can significantly reduce transmission losses and enable economical fault-current limiting with higher performance (Technology Roadmap Smart Grids, 2011; Bouaraki and Recioui, 2017).
- Distribution grid management: Distribution and sub-station sensing and automation can reduce outage and repair time, maintain voltage level and improve asset management. Advanced distribution automation processes real-time information from sensors and meters for fault location, automatic reconfiguration of feeders, voltage and reactive power optimization, or to control distributed generation (Technology Roadmap Smart Grids, 2011).

- Advanced metering infrastructure: Advanced metering infrastructure (AMI) involves the deployment of a number of technologies providing customers and utilities with data on electricity price and consumption, including the time and amount of electricity consumed (Technology Roadmap Smart Grids, 2011; Recioui et al.,2019; Recioui, 2019). AMI will provide a wide range of functionalities:
  - Remote consumer price signals, which can provide time-of-use pricing information.
  - Ability to collect, store and report customer energy consumption data for any required time intervals or near real time.
  - Improved energy diagnostics from more detailed load profiles.
  - Ability to identify location and extent of outages remotely via a metering function that sends a signal when the meter goes out and when power is restored.
  - Remote connection and disconnection.
  - Losses and theft detection.
- Electric vehicle charging infrastructure: Electric vehicle charging infrastructure handles billing, scheduling and other intelligent features for smart charging during low energy demand.
- Customer-side systems: Customer-side systems, which are used to help manage electricity consumption at the industrial service and residential levels, include energy management systems, energy storage devices, smart appliances and distributed generation.

FOCUS OF THE BOOK

A better understanding of the control of power grid; which no longer focuses only on power flow but also takes the measurement and communication into account, led to further development of smart grid supporting the integration of renewable energy sources and allowing better reliability and efficiency.

With the fast technical development of PMU technology, integrated system of monitoring, control and protection that avoids the power system instability and the blackout became possible.

The future vision of this kind of development is the integration of an Internet of Things (IoT) technology in smart grid, which can also be used in monitoring, control and protection system. This may take time; however, 30 years ago nobody even considered the possibility of the PMU development.

Therefore, the purpose of this book is twofold. First, to be used by the students in order to introduce some topics related to smart-grids, where the fundamentals as well as details in the many facets of this domain are presented. Second, it can be used as reference for researchers in the field.

This book addresses to many actors in the university and R&D sectors. As its title indicates, the book is devoted to present research findings and works relative to modern smart grid operation and control. So, mainly researchers of the different aspects of smart grid will be the main target audience of this book. However, the book will include also some theoretical background and state of art items that will be beneficial for both undergraduate and post-graduate students at different aspects of engineering. The book may be used as a textbook for some parts relevant to the topic presented by the teachers. Thus, the main audience targeted by this book includes researchers in research laboratories and universities along with students and faculty members in some rudimentary or even advanced courses.

ORGANIZATION OF THE BOOK

To better focus of the previously traced objectives, the book can be divided into three main parts.

In Section 1 of the book, a number of chapters that discuss fundamental components for realizing smart grids and its measuring system is included. These chapters can be used as an introductory course in this domain either at the undergraduate or graduate level. Relative information is often summarized here in order to make each chapter as self-contained as possible. At the same time, after the introduction to the fundamentals, the advances in the area are summarized in a survey manner with appropriate references, so that the student can immediately build upon the fundamentals, while the practicing researcher can easily find relative information.

In Section 2 of the book, more precisely, we highlight how it is possible to employ PMUs in order to control and protect employing the least number of PMUs. Additionally, we also describe in detail s the main features provided by a PMU in the domain of monitoring and protection of large-scale smart grids.

Finally, Section 3 focuses on reliability and quality issues in smart grids. A set of chapters is included to highlight how important is this to a smart grid implementation campaign. It is worth to mention that some chapters in this book provide a survey that summarizes the optimization techniques which can be used in the domain of smart grids. A detailed description of each chapter is provided in what follows.

Chapter 1

The smart grid can be regarded as a layered user-involved platform, comprising the power system layer, a control section, a communication link, a security section and an application layer. The communication link ensures a two-way communication in the smart grid system. It is one of the most crucial agents that enable smart grid applications. The smart grid can be cast as an involved collection of networks, involving both high voltage and communication infrastructures along with various intelligent electronic devices (IEDs) [7-8]. Communication networks constitute the required infrastructure enabling a utility to monitor and control these devices from a remote location. In the smart grid system, diverse communication technologies and architectures are brought together. Communication networks should satisfy some desired properties such as reliability, latency, bandwidth and security, corresponding to each smart grid application. The involvedness of the smart grid may result in complications in selecting right communication networks because of the large number of parameters and diverse requirements to be taken into account in relation to the applications and utility expectations. This chapter presents the key technologies and requirements of a smart grid communication system with clear state of art descriptions.

Chapter 2

Smart meters play a major role in the smart grid at the customer side. Sensors and sensor networks are the main parts of smart devices and smart metering. Generally, sensors are located at different parts of the grid. Sensors play an important role in remote monitoring and demand side management. As the sensors are located at all the parts of the grid, it is able to measure/monitor temperatures, operation of grid devices, power quality disturbances and detect outages. This process helps the control centers to receive accurate and actual condition of the grid. This chapter focuses on the importance smart meter deployment in smart grid system. It also gives emphasis to the evolution of smart meters from the conventional meters and the hardware components employed in smart meter and its functions. The significance of communication networks and the protocols used for the proper implementation of the smart meter has been deliberated. The responsibility of smart meters in the implementation of DSI has been enlightened. The advancements in smart metering technology lead the world to new dimensions even though there are lot of implementation issues and challenges. The various kinds of solutions for the issues has been discoursed in this chapter. It also discussed about the present and future market developments of smart grid.

Chapter 3

The main goal of smart grid is to promote active customer participation and decision making as well as to create the operation environment in which both utilities and electricity users influence each other. New technologies are being deployed, including advanced meters, controllable appliances, distributed energy generation and storage systems, i.e., plug-in hybrid electric vehicle batteries, stand-alone storage systems, and communications capabilities. New laws are being proposed to allow electricity consumers to access pricing information. New dynamic pricing policies are likely to be implemented at the retail level over the next years. Energy management controllers are primarily designed to control load within a single home. They often take into account the utility data like load forecasts or ToU pricing for scheduling the household appliances. At the customer side, customers have the incentive to shift their electricity usage from high peak hours to low peak hours so that their electricity bills can be reduced . In this chapter, the crucial significance of pricing policies as well as Net metering have been discussed. The latter allows an owner of rooftop solar panels to participate in the electric grid as a distributed energy producer. Solar panels first supply power to the home, and at times when the solar panel produces more power than the home is using the power is sent back onto the grid to be used by other customers.

Chapter 4

An electric power system is a large interconnected system that produces, transmits and distributes an electric energy to different consumers. Stability of the smart power system is of a great concern, since it is subjected to different disturbances that may cause false measurement and hence mal-operation of protective and control systems which may lead to complete system collapse. Therefore, many techniques have been developed to mitigate the disturbance effects by introducing the digital filtering for computing the fundamental phasors of line voltages and currents. This chapter discusses phasor and frequency measurements under transient system conditions. Both electromagnetic and electromechanical transients are considered. After explaining the basics of phasors, synchrophasors, and frequency and system transients, suggestions have been made for improving the interoperability of PMUs under transient conditions.

Chapter 5

Many techniques have been developed to mitigate the disturbance effects by introducing the digital filtering for computing the fundamental phasors of line voltages and currents. Several digital algorithms have been proposed in the literature for estimating the fundamental phasors, such as the discrete Fourier transform, cosine algorithm, least square algorithm, Kalman filter and wavelet transform. The discrete Fourier transform (DFT) filter which has proven reliability and robustness in relaying application was considered as the most widely used and most popular algorithm for estimating phasors. However, during a transient time of smart grid condition, line fault current may include unwanted components that are associated to the fault impedance level as well as to the power transmission line type. A considerable exponentially decaying sub-harmonic or sub-synchronous resonance is associated with fundamental and integer harmonic of the line current waveform. This type of fault is more severe since it reduces the speed of convergence of the DFT algorithm to correctly estimate the fundamental phasors. In this chapter, a filtering approach is used to completely remove or significantly mitigate the effect of fault disturbances on phasor measurement using digital band-pass filter. The digital filter has been specified to meet several objectives as constant group delay, minimum phase and sharp narrow band-width gain frequency response. A recursive digital filter is then designed using a multi-objective optimization technique based on the spiral optimization algorithm.

Chapter 6

Smart meters (SM) are a key component of SG, which can be used for reducing peak demand and adapting flexible demand to fluctuating generations. This device enables two-way communication between suppliers and consumers in real time. Moreover, it has the ability to remotely control electrical devices to manage loads and demands. In a SG, Demand Side Management (DSM) has been developed since early 1980s to balance the time varying demand load of consumers and maximum power generation capacity in the power system. In DSM, the pricing mechanisms and direct control strategies are employed by the energy suppliers to affect consumers' consumption behaviors and reshape the total load. This chapter focused on describing strategies for decreasing the load demand and power consumption costs, the active role of the customer in the smart grid is better defined and clarified. A simulation in LABVIEW software and a hardware application are performed to demonstrate the smart metering services in decreasing the pressure on the local electricity grid by taking power from RES and also managing customer energy use and save money by controlling the loads consumption. A simulation in LABVIEW software is performed to visualize some peak demand mitigation techniques. As an application, a hardware circuit is designed to demonstrate an automatic and a manual load control technique performed on a house in case of an increase in power demand. Simulation results based on four houses design has been presented, the work concentrated mainly on the power consumed by the four houses, different cases were analyzed and some solutions were proposed to manage the energy consumption of the system.

Chapter 7

Micro-grid energy management approaches are often classified as centralized and decentralized approaches. In centralized ones, all control processing occurs in a single agent with lack of coordination between the micro-grid performers. Their advantage is that Off-the-shelf energy management algorithms can be directly applied; however, if the dimensionality of configuration space increases, it leads to an increasing running time. Whereas, in decentralized approaches Control processing is distributed among the agents and each one computes its solution independently; then, a strategy for coordinating the micro-grid participants is implemented; where each participant evaluates its own optimal settings. The advantage of these approaches is that dimensionality of configuration space does not increase running time since each agent is independent from others; but in the other hand, coordination is not always possible, which does not complete the decoupled management. This chapter proposes an approach based on a decentralized energy management system to enhance the performance of a smart-grid. This approach is build upon Multiagent system where the coordinator commands are merged to solve the problem of microgrid management. Each agent calculates its coefficients and needs and exchanges the data with the other data to improve the performance of the overall microgrid.

Chapter 8

In the case of micro grids with integration of photovoltaic and wind power systems, the power variations are basically due to the irregular power generations from renewable energy sources and improbability of load consumption. The grid connected microgrids have essentially all the support from high inertia power generators whereas off grid micro grids mainly depends on Energy Storage Systems(ESS) in order to balance the mismatch between the power generated by the renewable source and the power being consumed by the user. The main focus of this chapter is to review various optimization techniques that can be used to develop the adaptive controls strategies for controlling the power production from renewable energy sources mainly solar photovoltaic and the charging of HESS. The overall effect on performances of the Renewable Energy Power System (REPS) with different optimization techniques is discussed.

Chapter 9

State estimation of power network may allow the scheduling generation and interchange; monitoring outages and scheduling alternatives; supervising scheduled outages; scheduling frequency and time corrections; coordinating bias settings; and emergency restoration of system. This can be achieved either by state estimation algorithms or by means of PMUs with extreme precision, time synchronization, and excellent performance. Measuring state estimation is achieved through complex bus voltages that enable the estimation of the neighborhood bus voltage magnitudes and angles by using the line flow measurement (both real and reactive power). The static and dynamic behaviors of the power grid may be obtained from the information gathered by PMUs. In this chapter, a binary teaching learning based optimization (BTLBO) algorithm for the optimal placement of phasor measurement units (PMUs) is proposed. The optimal PMU placement problem is formulated to minimize the number of PMUs installation subject to full network observability at the power system buses during the fault. The effectiveness of the proposed method is verified by the simulation of IEEE14-bus.

Chapter 10

A Microgrid is a group of small, interconnected electrical network consisting of distributed energy resources (DERs), energy storage devices, load and have the capability to act as a single controllable power network. A Microgrid can be interconnected at low (400 V) and medium voltage (69 KV) distribution networks. Also, it can be operated in both grid connected or island mode. The term islanding is defined as the phenomena in which both generation and connected load of mirogrid becomes isolated from the main utility grid.The issue of unintentional islanding in microgrid is a major challenge for power engineers as it may harm the service personnel working during maintenance or restoration of the main grid. In literature, various passive, active and hybrid islanding detection methods (IDMs) have been developed and implemented by researchers/engineers.In this chapter, the proposed IDMs from the recent literature have been analyzed and briefly discussed. The primary requirement of IDMs is fast and accurate detection of main grid failure and should not deteriorate the power quality of distribution networks.

Chapter 11

Power quality is generally meant to express the quality of voltage and/or the quality of current which can be defined as: the measure, analysis, and improvement of the bus voltage to maintain a sinusoidal waveform at rated voltage and frequency; this definition includes all momentary and steady-state phenomena. The objectives of this chapter were to develop a PQM system that had the ability to detect power quality disturbances at the minimum possible of time and to correctly identify them with high precision, which is the major area of research in power system field. To do that, WT and HT were introduced as signal processing tools used for detecting and characterizing PQDs and they have proved their efficiency and flexibility in detecting different disturbances, either the amplitude variation disturbances or the frequency variation ones. Also the machine learning tools ANN and SVM were proposed in the literature to classify the different PQDs.

Chapter 12

A protection system is a vital part of any electric power system and plays an incredible role in maintaining high availability required in present day. The mal-operation (false trip) of one relay of the protective scheme to operate as intended may place at risk the stability of the whole power grid and hence may lead to blackout. In this chapter, Root Cause Analysis (RCA) based on Fault Tree Analysis (FTA) is used to identify disturbances first and root cause of false trips of protective relays which may lead to unnecessary tripping in cascading and then to the bulk blackout. Once the critical root causes are identified for this mal-operation of these relays, mitigation measures such as blocking functions and digital filters may be used in view of increasing the reliability of the considered protection system.

Chapter 13

Appropriate relay testing provides first guard against relay false-tripping. Relay testing can aid to confirm the design of relay, compare the performance of different relays, verify relay settings, classify system conditions that might cause unplanned relay operation, and carry out post-event analysis to understand the causes of unintentional or incorrect relay actions. Relay tester enhancements need to continue because of the use of relays in smart power grids where the conditions that are not the same as in the conventional one. This leads to new relay technologies. This chapter focuses mainly on the design and the implementation of the Class D amplifier which is the amplification part of the testing system. The Class D amplifier has been used in this work to amplify the simulated disturbances that are generated by SIMULINK power system model to be then injected to the protective relay and monitor its response. LABVIEW has been used for developing the graphical user interface and controlling NI 7851Board

Chapter 14

The protection of power transformers is crucial for the continuity of the power supply. A power transformer is mostly protected against internal faults using a differential protection which is sensitive and a fast clearing technique. Differential relays are prone to mal-operation in the presence of transformer inrush currents. Inrush currents are produced from transformer magnetic flux. The first solution to this problem has been investigated by introducing an intentional time delay in the differential relay. Another proposal has been performed by desensitizing the relay for a given time, to overcome the inrush condition. Others have suggested adding a voltage signal to restrain. This chapter aims at presenting the recent developed blocking techniques of differential protection during inrush conditions. This is following a number of questions that has been arisen while applying differential relays for transformer protection.

Chapter 15

As the penetration of renewable generation increases, the impact on power system dynamics is becoming increasingly apparent, and will become a more integral part of system planning and renewables integration studies. Historically, power systems have been based around large synchronous generators connected to a strongly meshed transmission network, with the dynamic characteristics of such systems being well understood. Phasor Measurement Units (PMUs) for data measurement has opened a vast number of potential applications in the power system world, ranging from monitoring, protection, and control to enhance the smartness and efficiency of power systems. This chapter presents the application of PMUs to Grid connected PV system synchronization as well as to solve islanding problem in a network including high penetrations of solar energy generation. In the first application, a PMU is used to provide data that can be used to generate pulses which in turn they are used to generate an equivalent reference signal i_{ref} that will be used in the hysteresis controller. However, in second application, the solution of the use of synchrophasor measurements makes the islanding detection easier basing on the phasor of the PV system generator with respect to a reference from the grid. This method is performing a continuous monitoring in order to disconnect the islanded generator, and leave it connected if it is not islanded.

Abdelmadjid Recioui
University M'hamed Bougara of Boumerdes, Algeria

Hamid Bentarzi
University M'hamed Bougara of Boumerdes, Algeria

REFERENCES

Azzougui, Y., & Recioui, A. (2018). Hardware and software load power control in smart home applications based on Taguchi optimisation technique. *International Journal of Intelligent Systems Design and Computing.*, *2*(3-4), 203–223. doi:10.1504/IJISDC.2018.097453

Barbato, A., Capone, A., Carello, G., Delfanti, M., Merlo, M., & Zaminga, A. (2011). House energy demand optimization in single and multi-user scenarios. *Proceedings of the IEEE International Conference on Smart Grid Communications (SmartGridComm),* 345–350.

Bouaraki, M., & Recioui, A. (2017). Optimal placement of power factor correction capacitors in power systems using Teaching Learning Based Optimization. *Algerian Journal of Signals and Systems.*, *2*(2), 102–109.

Chellali, F., Khellaf, A., Belouchrani, A., & Recioui, A. (2011). A Contribution to the actualization of the wind map of Algeria. *Renewable and Sustainable Energy Reviews, Elsevier.*, *15*(2), 993–1002. doi:10.1016/j.rser.2010.11.025

Dekhandji, F. Z., Kassah, M. T., & Bensalah, A. (2020). Harmonic Reduction and Elimination in Three Phase PWM Inverters using a Spiral-Inspired Optimization Technique, Algerian J. Env. *Sc. Technology*, *6*(3), 1436–1447.

Dekhandji, F. Z., Talhaoui, S., & Arkab, Y. (2019, December). Power Quality Detection, Classification and Monitoring Using LABVIEW. *Algerian Journal of Signals and Systems*, *4*(2), 101–111.

European Smart Grids Technology Platform. (2006). European Commission Directorate-General for Research Information and Communication Unit.

Recioui, A. (2018). Application of Teaching Learning-Based Optimization to the Optimal Placement of Phasor Measurement Units. In Handbook of Research on Emergent Applications of Optimization Algorithms. IGI Global. doi:10.4018/978-1-5225-2990-3.ch018

Recioui, A. (2019). *Home Load-Side Management in Smart Grids Using Global Optimization in Handbook of Research on Smart Power System Operation and Control*. IGI Global. doi:10.4018/978-1-5225-8030-0.ch005

Recioui, A., Bentarzi, H., & Ouadi, A. (2015). Application of a binary teaching learning-based algorithm to the optimal placement of phasor measurement units. *Progress in Clean Energy*, *1*, 817–830. doi:10.1007/978-3-319-16709-1_60

Recioui, A., & Dassa, K. (2017). Design of Standalone Micro-Grid Systems Using Teaching Learning Based Optimization. *Algerian Journal of Signals and Systems.*, *2*(2), 75–85.

Recioui, A., Djehaiche, M., & Boumezrag, A. (2016). *Load Side Management in Smart Grids using a Global Optimizer*. Presented at the 8th International Exergy, Energy and Environment Symposium, Antalya, Turkey.

Recioui, A., Djemai, H., & Boucenna, F. (2019). A Smart Metering Simulation in LABVIEW. *Algerian Journal of Signals and Systems.*, *4*(2), 89–100.

Recioui, A., & Grainat, Y. (2020). Application and Optimization of MIMO Communication in Wide Area Monitoring Systems. *International Journal on Data Science.*, *1*(2), 82–98. doi:10.18517/ijods.1.2.82-98.2020

Saoud, A., & Recioui, A. (2017, September). A review on Data communication in smart grids. *Algerian Journal of Signals and Systems*, *2*(3), 162–179.

Technology Roadmap Smart Grids, International Energy Agency – IEA. (2011). *Executive Director, IEA: Nobuo Tanaka, section: Smart Grid Deployment*. IEA.

Zhao, Z., Lee, W. C., Shin, Y., & Song, K. B. (2013). An optimal power scheduling method for demand response in home energy management system. *IEEE Transactions on Smart Grid*, *4*(3), 1391–1400. doi:10.1109/TSG.2013.2251018

Acknowledgment

The editors of this book entitled *Optimizing and Measuring Smart Grid Operation and Control* would like to express their gratitude to the laboratory of signals and systems at the University M'hamed Bougara of Boumerdes, Algeria in which the idea behind this book originated. We would like to thank all the members of this laboratory who made this idea a reality and contributed effectively to the content of this book. We are also very grateful to the authors from other affiliations and countries whose contribution has been an added value to make this book a global success. Many thanks also go to all the reviewers who well contributed to the improvement of the individual chapters and hence the quality of the book.

Section 1
Smart Grid and Measurements

Chapter 1
Data and Communication Infrastructure in Smart Grids:
State of the Art and Trends

Abdelmadjid Recioui
https://orcid.org/0000-0001-9028-3910
Université M'hamed Bougara de Boumerdes, Algeria

Youcef Grainat
Université M'hamed Bougara de Boumerdes, Algeria

ABSTRACT

The communication infrastructure constitutes the key element in smart grids. There have been great advances to enhance the way data is communicated among the different smart grid applications. The aim of this chapter is to present the data communication part of the smart grid with some pioneering developments in this topic. A succinct review of the state of art projects to improve the communication link is presented. An illustrative simulation using LABVIEW is included with a proposed idea of introducing some newly technologies involved in the current and future generations of wireless communication systems.

INTRODUCTION

Smart grids incorporate with the electrical grid a data communication infrastructure that gathers and evaluates data in either real- time or offline about power transmission, distribution and consumption (Locke and Gallagher, 2010). The power grid involves four elements: electricity generation plants, transmission substations, distribution substations and customers (Güngör et al., 2011). Power generated from plants from a variety of sources (including classical fossil fuel, solar, wind and nuclear sources) is attenuated to the voltage fit for residential use. Home appliances power utilization is sensed through their electric smart meters (Güngör et al., 2010; Liserre et al., 2010; Hauser et al., 2005). This blend of power levels has to be monitored through the data gathering system. As an example, the power generation can be continually controlled by using the real-time energy consumption of the end users. Similarly, the end

DOI: 10.4018/978-1-7998-4027-5.ch001

user can monitor the real-time power usage of the home and can obtain the real-time cost of the power supplied from the service provider (Kim et al., 2015). On the other hand, the users can supply electricity in a smart grid. For example, users having homes equipped with photovoltaic (PV) systems are able to generate electricity and return it to the grid to help balancing the loads by "peak clipping", i.e., sending power back to the grid when the demand is high. This information exchange between different elements provides the smart grid with predictive information generate recommendations to utilities, their suppliers and their customers on how to manage power in an optimal manner (Hauser et al., 2005).

The new information and communication technologies (ICTs) are becoming progressively integrated within the society. One key area demonstrating this is the next generation of electricity called Smart Grid (SG) or intelligent grid. The Smart Grid adds a two way flow of information, automation and control to the actual power grid. It optimizes energy efficiency through the mutual real time exchange of information between power supplier and consumers through the integration of developed sensors, smart meters and the information technology (IT) (Kuzlu et al, 2014;Park and Kim, 2014).

Due to the importance of the communication infrastructure in the success of smart grid implementation and operation, there has been a variety of research works in the literature focusing on this topic (Saoud and Recioui, 2017). The authors of (Locke and Gallagher, 2010; Yan et al., 2012; Wang et al., 2011; Khan, 2013) provide surveys about the requirements that the smart grid needs from communication perspectives. Smart grid technologies and standards are discussed to provide an overview of the smart grid paradigm and integration of different communication technologies in (Gungor et al., 2011; Fan et al., 2012). Some studies focus on a particular standard or communication technologies such as power line communication (PLC) (Galli et al., 2010) and wireless communication (Pipattanasomporn et al., 2012; Wietfeld et al., 2011). In (Aravinthan et al., 2011), the authors evaluate the network performance for a long-distance distribution line and propose communication architecture for distribution level applications. Furthermore, the appropriate communication technologies for transmission-level applications have been discussed in (Dong and Kezunovic, 2011).

Wide-area monitoring, wide-area control and wide-area protection use Wide Area Networks and have been identified as the next-generation solution to improve power system planning, operation and protection in the smart grid. These applications involve the use of system wide information and selected local information to counteract the propagation of large disturbances (Kuzlu et al, 2014). Wide-area monitoring, control and protection applications offer higher data resolution and shorter response time than classical supervisory control and data acquisition (SCADA) and energy management (EMS) systems. While SCADA/EMS provides a measurement update interval of several seconds or even minutes, wide-area monitoring, control and protection applications provide high-resolution data.

The IEEE Standard for Synchrophasors for Power Systems (IEEE Std.C37.118), defines measurement and data transmission formats for real-time data reporting in electric power systems (IEEE, 2006). For wide-area monitoring applications, a minimum message size containing measurements made by a PMU is 52 bytes. Similar to those for wide-area protection and control applications, the required response time for wide area monitoring applications is also in the range of milliseconds to minutes, and communication system reliability requirement is very high.

While wide-area protection and control applications offer more advanced protection/control systems as compared with traditional power systems, more stringent performance and availability requirements are needed. For example, the required response time for wide-area protection and control applications should be in the range of milliseconds to minutes, and the communication system reliability requirement should be very high. Their typical message size can vary depending upon communication protocols implemented. Typical protocols used for these applications include MIRRORED BITS, IEC 61850 Generic Object-Oriented Substation Event (GOOSE), and ETHER-CAT communications (Dolezilek et al., 2012).

BACKGROUND

Smart Grids

A Smart Grid is a concept for transforming an electric power grid by using advanced communication, automated controls and other forms of information technology. It integrates new innovative tools and technologies from generation, transmission, and all the way to consumer appliances and equipment (Massoud Amin and Wollenberg, 2005). The smart grid can be viewed as a superposition of communication networks on the electric grids. It can improve efficiency, reliability, safety, and security of electricity supply to the customers, with a seamless integration of renewable and alternative energy sources, through automated control and modern communications technologies. The various components of the electric grid are linked together via a two-way communications and power flows to provide interoperability among them. Thus, consumers can not only draw power but also supply surplus power to the grid using smart meters that enable monitoring and measuring of these bidirectional flows (TrongNghia, 2016).

1. Wide Area Monitoring Systems

Synchronized Phasor Measurement Unit (PMU) is a monitoring device, which was first introduced in mid-1980s. They use synchronization signals from the global positioning system (GPS) satellites and provide the phasors of voltage and currents measured at a given substation. As the PMUs become more and more affordable, their utilization will increase not only for substation applications but also at control centers for Energy Management System (EMS) applications (Recioui, 2018, Recioui et al., 2017).

Phasor measurements are necessary when dealing with wide disturbances that often cause widespread blackouts in power system network. PMUs are important for power system state estimation, network protection and control. The results obtained from PMUs are time-tagged by GPS signal (Azzougui et al., 2019).

Phasor Data Concentrators (PDCs) are devices able to concentrate measurements from different PMUs, time-align them and communicate with them as a single stream to other PDCs or to monitoring and control devices. A synchrophasor network can be developed which can derive measurements from different parts of the power network through PMUs (Recioui et al., 2015).

The phasor data concentrator (PDC) collects synchrophasor data and other quantities (i.e., frequency, rate of change of frequency, nodal injected/absorbed powers, etc.) estimated by the PMUs and transmits the information to other applications such as visualization tools. Among other functionalities, the PDC is able to communicate with the PMUs and encapsulates their datagram. The synchrophasors are aggregated and time-aligned according to their timestamps with minimum time latency. Indeed, for real communication networks and devices, measurements can be delayed due to non-determinism in the data streaming, network traffic, and different reporting rates among the devices. For this reason, an adaptive algorithm was developed that determines, based on an event timeout, the amount of time each dataset has to actively wait for the rest of phasors with the same timestamp. When the event is triggered, the current dataset is forwarded to the application. This solution always ensures that the available measurements will be forwarded in an acceptable time range (~20ms in case of PMUs streaming at 50 fps) thus increases the determinism of the process. Finally, we note that GPS synchronizes the PDC; hence we are able to identify and solve eventual bottlenecks in the whole chain that might result in an increased total latency (Azzougui et al., 2019; Recioui et al., 2015; Recioui, 2016). Fig. 1 shows data paths which will require different data size pipes and characteristics that correspond to different operational levels.

Figure 1. Representation of a WAMS
(Recioui et al., 2015)

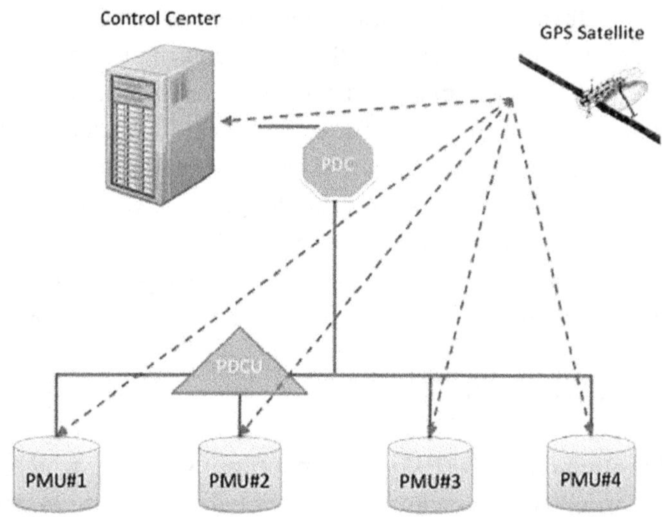

2. Smart Grid Communication

Maintaining communication between the components of the smart grid by the communication infrastructure is very essential. Besides the electricity connection, smart grids permit to have data connection to vehicle information over the grid. Hence, there is a need for a secure communication network with high bandwidth capacity and speed.

The smart grid is an interactive platform, consisting of a power system layer, a control layer, a communication layer, a security layer and an application layer. The communication layer is one of the most critical elements that enables smart grid applications. In the smart grid environment, a communication network can be represented by a hierarchical multi-layer architecture classified by data rate and coverage range. According to the smart grid communication, infrastructure can be generally classified into HAN, NAN and WAN based on the data rate and coverage range. Smart grid applications are grouped by their data rate and coverage range required for their successful deployment. A comparison of various communication technologies that can support smart grid applications in terms of data rate and coverage distance is presented in Table 1.

Generally, there are three types of networks: Home Area Networks (HANs), Business Area Networks (BANs) and Neighborhood Area Networks (NANs) (Kuzlu et al., 2014). In the following, each of these types of networks is detailed.

- HAN: HAN operates within small area (10 meters). It ensures the transmission of information between different electronic devices that are inside or within close vicinity of the house. From the devices included, we have: smart meter, smart appliances, smart sockets, plug in vehicle chargers. The smart meter corresponds with the rest of devices to gather data as power usage information of energy-consuming devices, the status of renewable energy resources and the produced energy by these resources etc in order to be sent to utility company. HAN permits an efficient home energy management. Wireless and power-line communications are the most fitting options to implement HAN (Wang et al., 2011; Jingcheng et al., 2012).

Table 1. Comparison of communication technologies for the smart grid

Technology	Standard/Protocol	Max. Theoretical Data Rate	Coverage Range
Fiber optic	PON	155 Mbps–2.5 Gbps	Up to 60 km
	WDM	40 Gbps	Up to 100 km
	SONET/SDH	10 Gbps	Up to 100 km
DSL	ADSL	1–8 Mbps	Up to 5 km
	HDSL	2 Mbps	Up to 3.6 km
	VDSL	15–100 Mbps	Up to 1.5 km
Coax. Cable	DOCSIS	172 Mbps	Up to 28 km
PLC	HomePlug	14–200 Mbps	Up to 200 m
	Narrowband	10–500 kbps	Up to 3 km
Ethernet	802.3x 10	Mbps–10 Gbps	Up to 100 m
Bluetooth	802.15.1	721 kbps	Up to 100 m
ZigBee	ZigBee	250 kbps	Up to 100 m
	ZigBee Pro	250 kbps	Up to 1600 m
WiFi	802.11x	2–600 Mbps	Up to 100 m
WiMAX	802.16	75 Mbps	Up to 50 km
Wireless Mesh	Various (e.g., RF mesh, 802.11, 802.15,802.16)	Depending on selected Protocols	Depending on deployment
Cellular	2G	14.4 kbps	Up to 50 km
	2.5G	144 kbps	
	3G	2 Mbps	
	3.5G	14 Mbps	
	4G	100 Mbps	
Satellite	Satellite Internet	1 Mbps	100–6000 km

- NAN: Neighborhood Area Network comprises multiple HANs. Smart meters (HAN gateways) transfer data of energy consumed by each house and control data to the NAN network. This last sends this data to Local Data Centers for storage (Bouhafs et al., 2012). NAN operates within area of hundreds meters, i.e, few urban buildings. The NAN can be implemented by PLC, Wi-Fi, and cellular technologies. Since the deployment environment of NAN in an urban area is complex and harsh, wireless mesh network (WMN) is considered as to be the most suitable technology for NAN (Niyato and Ping, 2012.

- WAN : Wide Area Network is a link that ensures communications between all entities of smart grid, i.e. power plants, transmission and distribution subsystems, utility companies and customer premises.It is expanded over a wide geographical area. NANs and HANs are incorporated in the WAN (Tariq and Dooley, 2013;Weixiao et al., 2014).

There are several technologies that can be applied to the smart grid:

1. **ZigBee:** Zigbee is classified as wireless communications technology that is moderately low in power usage, data rate, complexity, and cost of operation. It is an ultimate technology for smart lightning, energy monitoring, home automation, and automatic meter reading, etc. it is preferable to facilitate the communication between smart meters, and among intelligent home appliances and in home displays. Zigbee Smart Energy Profile (SEP) allows utilities to transmit messages to the home owners so that they can get the information of their real-time energy consumption (Güngör et al., 2011) smart grid technologies
    a. *Advantages:* as advantages we can list the low price, small size and it uses relatively small bandwidth. It is ideal to be used in smart grid implementation for mobility and robustness. (Peizhong et al., 2011; Gezer and Buratti, 2011).
    b. *Disadvantages:* some of disadvantages are: the small battery that bound its lifetime, small memory, limited data rate, low processing capability and small delay requirements (Güngör et al., 2011).
2. **Wireless Mesh:** A mesh network is flexible network composed of an assembly of nodes, where new nodes can join the group and every node can operates as an independent router. In PG&E's Smart Meter system, each smart device is supplied with radio module that routes the metering data through close meters. Every meter works as a signal repeater until the gathered data reaches the electric network access point. This data is transmitted to the utility via a communication network (Gungor et al., 2011).
    a. *Advantages:* we can list its cost effectiveness with dynamic self-organization, self-healing, self-configuration, high scalability services, that give many benefits as improving the network performance, balancing the load on the network extending the network coverage range and good coverage (Yarali, 2008). Home energy management are some of the applications that wireless mesh technology can be used for.
    b. *Disadvantages:* as major challenges, we have Network capacity, fading and interference. There is also the coverage challenge as the meter density is not able to provide complete coverage of the communications network. Providing the balance between reliable and flexible routing, a sufficient number of smart nodes, taking into account node cost, are very important for mesh networks.as the metering information passes through each access point, some encryption techniques are applied to the data for security purposes. While data packets travel around many neighbors, there can be loop problems causing additional overheads in the communications channel that would result in a reduction of the available bandwidth (Lewis et al., 2009).
3. **Cellular Network Communication:** Cellular networks are widely operated in many countries. They can be used to allow the communication between distinct components and devices in smart grid. The following characteristics that can be achieved using the telecommunication wireless

network technologies like 2G, 3G, even 4G that are deployed in smart grid networks: application awareness, support for large numbers of simultaneous cell connections, high service coverage, and prioritized routing of data. The WiMAX is one of the 4G solutions that is most preferable for smart grid implementation (Jingcheng et al., 2012).

a. *Advantages*: the already existing infrastructure with wide area of deployment, data transfer high rates, available security algorithms that are already implemented in the cellular communication, Lower cost, better coverage, lower maintenance costs, and fast installation features.
b. *Disadvantages*: cellular networks are joint with other users and are not completely dedicated to the smart grid communications. This may present a serious problem in case of emergency state of the grid (Güngör et al., 2011).

4. **Powerline Communication:** Powerline Communication (PLC) is a technology that makes use of the power lines to transmit data and electrical power (NETL, 2007; Vivekanandan and Wong, 2006). This system sends modulated carrier signals on the power transmission wires. There is a variation of few hundred of bits per second to millions of bits per second for data rates on power lines, inversely proportional to the power line distance. Thus, to give an alternative broadband networking infrastructure (NETL, 2009; Huang et al, 2006) without installing dedicated network wires, power line communication is largely used for in-door environment. a survey of communication network
   a. *Advantages:* the wide-spread infrastructure that decreases installation costs. The standardization efforts on PLC networks, the cost-effective, ubiquitous nature, and widely available infrastructure of PLC make it strong and popular (Paruchuri et al., 2008). PLC technology is appropriate to urban areas for smart metering, monitoring and control applications.
   b. *Disadvantages:* the network topology, the number and type of the devices connected to the powerlines, wiring distance between transmitter and receiver, all of this that have an effect on the quality of signal transmitted over powerlines (Gungor et al., 2011). All this make PLC technology not appropriate for data transmission. To fix that, some hybrid solutions are proposed like the combination of PLC and other technology as GPRS OR GSM.

COMMUNICATION FRAMEWORKS

IEEE C37.118 Communication System

IEEE C37.118 is the improved version of IEEE 1344 which was the first available synchrophasor communication standard. It defines methods for evaluation of Synchrophasor measurements, time synchronization, application of time-tags and format of messages exchanged over the network. It does not put any restriction on the communication mode, protocol or media and messages can be transmitted in unicast, multicast or broadcast fashion over any communication medium and transport protocol. Originally, IEEE C37.118 addressed the performance of synchrophasors only under steady state conditions ignoring system disturbances and noise. However, a revision of IEEE C37.118 in 2011 accounts for more precision and support for dynamic power system conditions.

IEEE C37.118 describes four types of messages: data, configuration, header and command. Data messages are used to send actual real time measurements made by the PMU. Data from multiple PMUs may be transmitted in a single message correlated to a particular time stamp (i.e., PDC functionality).

Header messages contain descriptive information (e.g., filtering, scaling algorithms etc) which is sent by the PMU/PDC but actually provided by the user. Command messages are used to control the operation of device sending synchrophasor measurements. In short, data, configuration and command messages are expressed in machine-readable format while header is descriptive information in human-readable format. Further, data, configuration and header are the message types sent by the data sources whereas command message is received by the data sources (Martin et al., 2008; Khan et al., 2016).

IEC 61850-90-5 Communication System

IEC 61850-90-5 is derived from IEC 61850 which was initially proposed for substation automation. IEC 61850 is a complete communication system that addresses modeling of power system components, abstraction of services and communication protocols and methods (Apostolov, 2013). It was designed with several objectives in mind:

- interoperability and integration between power system components from different vendors,
- device/service modeling,
- self description and object auto-discovery due to structured meta-data,
- reliability mechanism through retransmission,
- reduced substation cost through multicast and replacing of expensive relay-to-relay wiring with wireless communication, and (vi) support for machine to machine sharing of configuration data using Substation Configuration Language (SCL). However, IEC 61850 also has limitations including lack of security mechanism and restricted communication to only the local network. IEC61850-90-5 inherits all the features of IEC 61850 while also overcoming its limitations. IEC 61850- 90-5 includes a security mechanism based on Group Domain of Interpretation (GDOI) and also allows transmission of time-critical protocols over wide-area networks by relying on transport and network layer protocols.

DATA COMMUNICATION

In smart grids, the information can be collected at the data server and reproduced by the service provider in order to develop new services and business models. Hence, all electric grid elements must share the common understanding that the smart grid needs to be integrated with a data-networking infrastructure in order to be categorized them as "smart things" (Recioui et al., 2016).

In smart grids, care must be taken to deal with the different data sets. Because of their nature, distribution and real-time constraints of certain needs, big Data techniques are suitable for advance and efficient data management for this kind of applications. The large volume of data will help utilities to do things they never could do before such as better understanding the customer behaviour, conservation, consumption and demand, keeping track of downtime and power failures etc. At the same time, this will present challenges for utilities that lack the systems and data analysis skills to deal with these data. So, the main goal of utilities now is the ability to manage high volume data and to use advanced analytics to transform data collected to information, then to knowledge and finally to actionable plans (Daki et al., 2017, Saoud and Recioui, 2017).

As data size, latency and reliability requirements for different smart grid applications vary widely and are not easily obtainable to practitioners, this paper presents a comprehensive compilation of information from various use cases and smart grid-related standards on potential smart grid applications and their associated communication network requirements. These are discussed in terms of typical payload, data sampling requirements, as well as latency and reliability requirements for smart grid applications deployed in a Home Area Network (HAN), Neighborhood Area Network (NAN) and Wide-Area Network (WAN). Additionally, the paper also discusses, based on selected smart grid projects around the world, various communication technologies that have been implemented to support real-world smart grid applications.

Data quality is a subjective standard used to determine if a set of data is suitable for a particular business purpose. To quantify the data quality problem further, fifteen data quality aspects have been defined in (Wang et al., 2001). The suitability of PMU data to the intended WAMC application function is measured by a set of metrics that are referred to as the PMU data quality measures. Three critical PMU data quality attributes including data accuracy, data latency, and data completeness are of great importance.

Data Communication at the Grid Level

The WAMC system is proposed to enhance the power system operation and control by aggregating the fine-grained and timely information that was originally used for power system protection purposes at higher levels in the system hierarchy (Phadke et al., 1983). The data accuracy is a relevant quality attribute for all of the computer systems providing services based on measurements. This quality attribute is measured by the discrepancy between the PMU measurements and the corresponding power system states. A set of metrics has been proposed by the IEEE standard C37.118.1 to determine the performance of the PMU measurements under a variety of conditions. For example, the Total Vector Error (TVE) is introduced as a primary tool to evaluate the synchrophasor measurement accuracy in the power system steady state, and the delay time and response time are proposed to evaluate the PMU response to step changes (IEEE, 2005).

A prominent feature of the synchrophasor technology is that the PMU measurements are provided with time information, represented by their time-stamps, to track the power system states that change with respect to time. Consequently, neither the inaccurate PMU measurements nor the delayed synchrophasors can represent the true states of the power system.

Figure 2. Generic synchrophasor system.

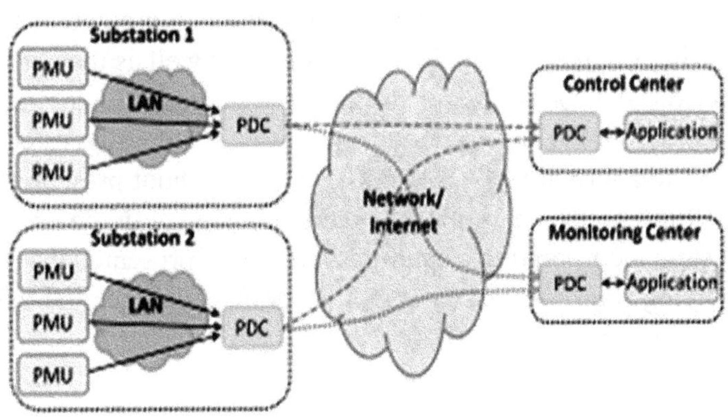

In (Johnston et al., 2006), a set of delay values was presented to prove the feasibility of transferring synchrophasors over a multicast-based communication infrastructure. In another related project (Ragib et al., 2006), the PMU communication delays in IP networks were examined by performing simulations in ns-2. The extra PMU communication delays due to encryption and decryption of the synchrophasor data streams were highlighted by this paper. A similar work with an extended scope can be found in (Kansal and Bose, 2012). Simulations of the IP network was extended by (Chenine and Nordström, 2011) to investigate the PMU data frame losses due to data alignment performed by the PDC. To ensure the timeliness of the PMU measurements, research efforts were also made to implement quality of service mechanisms to prioritize the PMU traffic. Results from simulations of an IP network incorporating the multi-protocol label switching scheme in OMNET++ were reported in (Chenine et al., 2010), and a latter work with a similar concept can be found in (Deng et al., 2012). In (Zhu et al., 2012), the challenges posed by a hierarchical infrastructure with multiple layers of data concentrations (performed by PDCs) were discussed.

The empirical results regarding the PMU communication delays or data frame losses are those of (Chao et al., 2012) from China, (Cyr and Kamwa, 2010) from Canada, and (Decker et al, 2011) from Brazil. The primary contribution of the above works was not to report empirical results. As a result, these works provided neither detailed descriptions of the ICT systems in question nor elaborated discussions on the presented empirical evidence. In (Zhu et al., 2013; Zhu et al., 2013), a comprehensive and detailed description of the utility IP network from where the empirical results were obtained are provided. Additionally, the research also emphasized on the interpretation of the empirical evidence by analyzing the root causes of the patterns of the studied data. Finally, efforts have been made to generalize the PMU communication delays by identifying distributions that offer the best fit to the empirical datasets.

To account for the challenges posed by the delayed PMU input, or the data frame losses, many delay-robust control schemes have been proposed. According to (Chaudhuri et al., 2010), the delayed input essentially create an excessive time-variant phase-lag to the damping controller, which, in turn, run the risk of deteriorating the control functionality. Such a challenge could be overcome by deploying adaptive compensations calculated from phasors extracted from rotating coordinates. An alternative is to propose control strategies against a set of nominal delays that are modeled using Pade approximations. A typical work along this track is the predictor framework presented by (Majumder et al., 2007). Following the same principle, a parameter-dependent H∞ controller coupled with scheduled gains, referred to as GSPOD, was proposed to manage the delayed input and model variations in (Wu et al., 2004). Though promising outcomes have been presented, these Pade approximation based designs tend to yield conservative control outcome with respect to delays that deviate significantly from the nominal delay horizons. Research efforts were also made to apply the generalized predictive control scheme together with an on-line Kalman filter to manage the PMU input delay as well as the changes of the plant model (Yao et al., 2010). Moreover, the robustness of the fuzzy-logic based wide-area damping system was investigated by (Kamwa et al., 2011). The reported simulation results suggested that this design could tolerate PMU input delays to approximately 300 milliseconds without presenting any notable function degradation. To increase the stability margin of the damping control in the presence of input delays, a PMU-based damping system incorporating a band-pass filter was presented in (He et al., 2009). Finally, a more recent work applying trajectory extrapolation to compensate for the communication network impairments can be found in (Wang et al., 2012).

Supervisory Control and Data Acquisition (SCADA)

Communication between multiple automation components is important in any smart grid implementation. SCADA system has a role at the core of smart grid system to maintain the real-time monitoring and control of power distribution (Metke and Ekl, 2010)]. A standard SCADA system comprises the following components (Stouffer et al., 2006):

- Control Servers: hosting control software and accessing subordinate control modules.
- Human-Machine Interface (HMI): it permits operators to monitor the system states; changes control settings, and manually overrides automatic control operations in emergencies.
- Remote Terminal Unit (RTU): with wireless radio, it interfaces to conduct data acquisition and control.
- Programmable Logic Controller (PLC): that performs the logic control functions performed by electrical hardware.
- Intelligent Electronic Devices (IED): a smart sensor and actuator to acquire data, communicate to other devices, and perform local processing and control.

As it is illustrated in Figure 3, the control center holds the control server, the HMI, engineering workstations, and the data historian, which are all connected by a LAN and exposed through a router. It collects measurements and logs information from the field devices, visualizes them to the HMI, and may generate actions based upon detected events. The wide area networks enable the communication protocols between the control center and the field sites, which are typically implemented using power/telephone line, cable, radio microwave and satellite (Tan et al., 2016).

Figure 3. SCADA system general layout
(Tan et al., 2016)

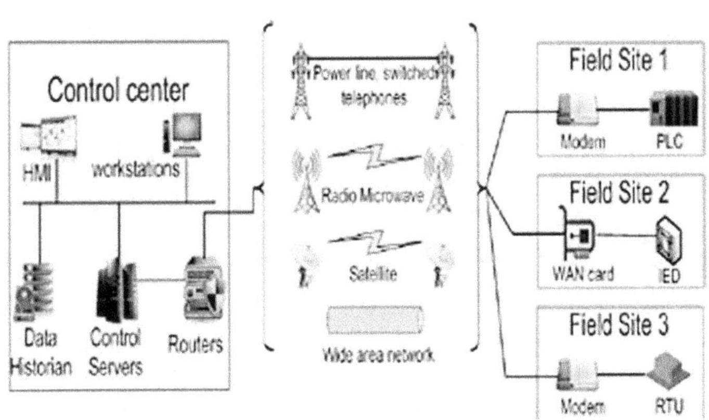

Synchronized Phasor Measurement Unit (PMU)

PMU is a monitoring device that uses synchronization signals from global positioning system (GPS) satellites. It affords the phasors of voltage and currents computed at some substation. PMUs will be used at control centers for energy management system(EMS) not only for substation applications (Marijic et al, 2010).The energy management system (EMS) uses the information received from PMUs at control centers for improving state estimation, monitoring, control a,d protection. When the PMUs are effective utilized, they will be useful in mitigating blackouts and learning the real time behavior of the power system (Wang et al., 2011).

Data Communication at User Level

Smart distribution grids's enablers are automated meters and smart meters. The term smart grid is used broadly to refer to the next generation of electrical energy transmission and distribution infrastructures, which will be characterized by a tight integration with Information and Communication Technologies (ICT). Automated meter term is used for a device that can (i) measure consumption of electric energy with a variable time granularity and (ii) report the measured consumption to a Meter Data Management System (MDMS).

Figure 4. Typical smart metering architecture

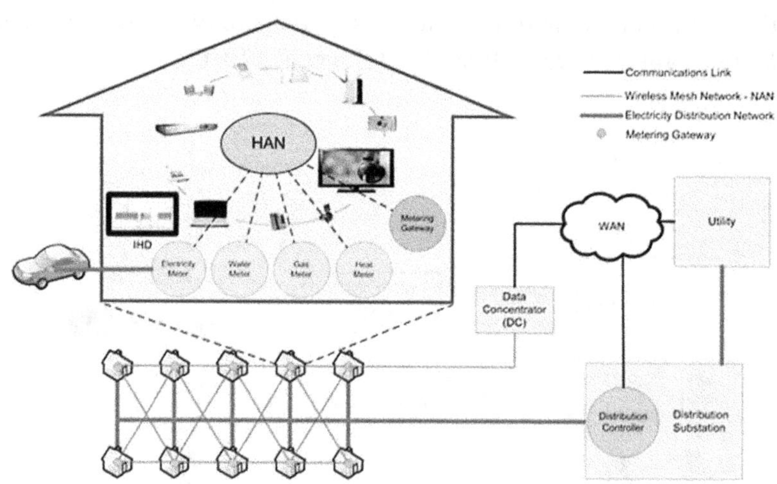

Smart meter are used for an automated meter that can also (iii) receive pricing information or direct load control commands and can (iv) exchange information with smart home appliances, that permits to optimise energy use and to participate in demand response (Asghar et al., 2017).

Unfortunately, the data collected by smart meters may also serve for invading consumers' privacy. Several recent works have pointed out that electricity consumption data may allow one to reveal private information, such as household occupancy or economic status (Wood and Newborough, 2003; Kalogridis et al., 2011). As a consequence, smart meter data are subject to serious privacy and security concerns.

Figure 5. Detailed architecture of an AMI

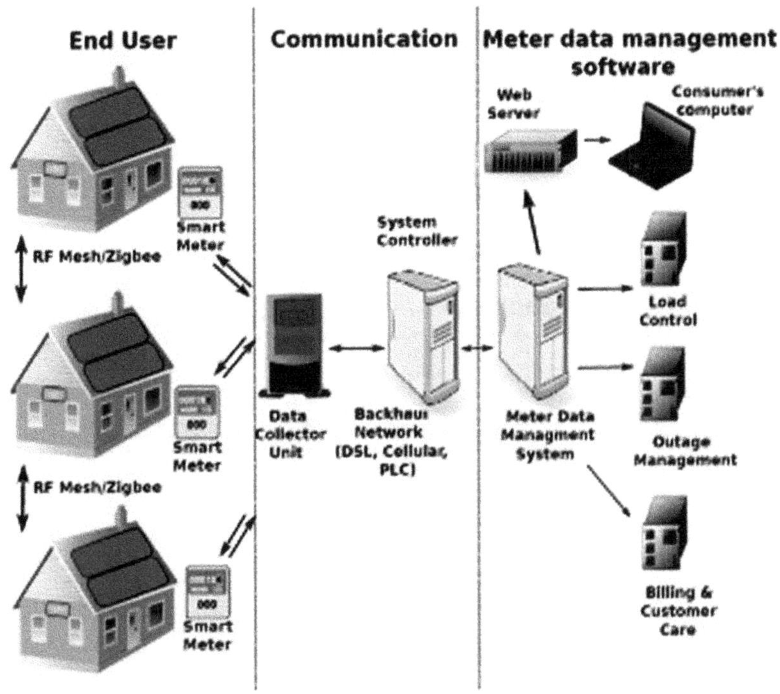

The privacy concerns and their perception within the public have delayed the roll out of smart metering in a number of countries (Cuijpers and Koops, 2013), and call for new technical solutions.

What makes the problem different from standard data security and privacy is the combination of three factors: the legacy of energy technologies that are based on closed systems, the interweaving with legal and regulatory aspects that introduce additional constraints, and the complex structure of the energy sector, with a variety of interconnected stakeholders, thus requiring more standardized solutions.

There has been significant recent works showing that individual appliances (based on their load signatures) can be identified from the detailed analysis of energy consumption traces (Batra et al., 2014; Hart, 1992). Frequent meter readings can also be used to infer the occupancy of a household, and data mining algorithms can also be used to invade the privacy of consumers in more sophisticated ways, e.g., by revealing their life-styles and economic status. Recent work also shows that fine enough measurements could reveal consumers' interests as well, e.g., (Greveler et al., 2012) show that they can estimate the displayed TV channel from the electricity usage profile with a sampling period of 0.5 s. Privacy is thus a serious concern, and calls for a data governance framework tailor-made for the smart grid.

A number of recent articles survey security issues in smart grids (Komninos et al., 2014; Sharma and Saini, 2015). (Lu et al., 2012) provide guidelines on designing security schemes for smart grids. (Baumeister, 2016) reviews and categorises the literature on smart grid security. In (Anderson and Fuloria, 2010), the authors discuss the security economics of electricity metering, while the potential effects of hacking have been reviewed by several security specialists. (Mo et al., 2012) discuss security approaches for smart grids. Privacy challenges in the smart grid have been considered recently in (Sharma and Saini, 2015; Borges, 2017).

Figure 6. Different domains of the smart metering infrastructure
(Asghar et al., 2017)

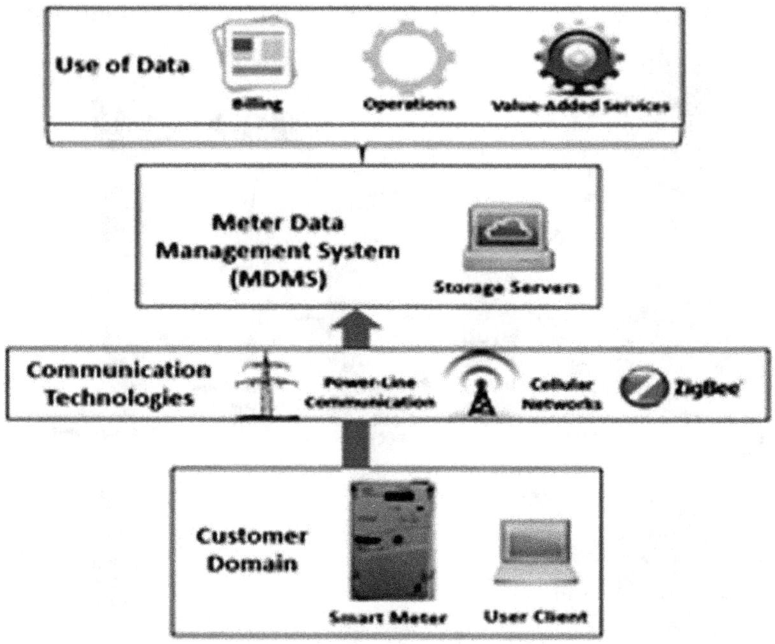

BIG DATA MANAGEMENT

Cloud Computing

Several problems related to big data management for smart grids are solved by cloud computing. It allows utilities to guarantee the flexibility, agility and efficiency in terms of saving cost, energy and resources (NETL, 2007). Using cloud computing in smart grid has huge advantages as a result of redundancy, rollback recovery and multi-location data backup that augment data fault tolerance and security (Jingcheng et al, 2012).

Cloud computing is based on service models that can be proposed in public, private, or hybrid manner: (1) software as a service (SaaS) that gives applications and make them available to customers over the Internet, (2) platform as a service (PaaS) delivers hardware and software tools and gives customers the ability to create their own applications, (3) infrastructure as a service (IaaS) offers hardware, software, servers, and other IT infrastructure components over the Internet, (4) data as a service (DaaS) permits customers in addition to run applications to store data on-line, (5) communication-as-a service (CaaS) is useful for messaging tools including voice over IP (VoIP), instant messaging (IM), and video conferencing, and (6) monitoring as a service (MaaS) is used for security services to ensure a third party security (Güngör et al., 2011).

challenges of using Big Data technologies are satisfied by Cloud computing frameworks. there are several cloud solutions that can hold Big Data such as Amazon Elastic Compute Cloud (Amazon EC2), Google Compute engine, Microsoft Azure Cloud, IBM Docker Cloud, etc.

Figure 7. Cloud computing components. Cloud computing relies on several concepts that are required for Big Data management in smart grid
(NETL, 2007)

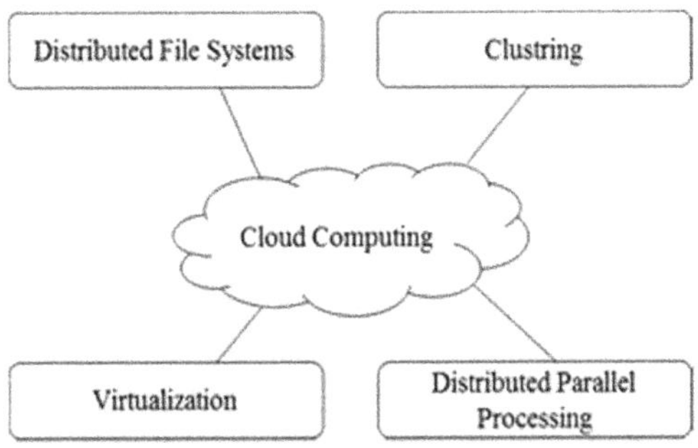

MIMO COMMUNICATION SYSTEMS

MIMO systems promise high efficiency in providing low probability of bit error (i.e. reliability) without increasing the transmitted power or the bandwidth (Shahid, 2005). Therefore, to achieve high capacity with reliability, time-, frequency- and space-diversity could be included, even in dense scattering and multipath environments. The above mentioned advantages of the MIMO systems mean that with the same power and bandwidth, a MIMO system would perform better than its SISO counterpart.

1. Orthogonal Frequency Division Multiplexing (OFDM)

OFDM is a modulation technique that consists in dividing high-data rate stream into N smaller streams with 1/*N*-th of the original data-rate, modulating them onto N narrowband subcarriers, and then, transmitting them simultaneously, at the receiver end, the data can be recovered using filters (Prasad, 2004; Parsons, 2000).

The OFDM subcarriers almost experience flat fading in frequency selective channel as they are narrowbands, which makes OFDM a key technique, overcoming issues related to multipath fading. Besides increasing the channel capacity, Sub-carriers in OFDM overlap but don't interfere with one another as they are orthogonal. OFDM improves the bandwidth efficiency and removes the guard intervals between sub-carriers.

The inverse Fast Fourier Transform (IFFT) and the Fast Fourier Transform (FFT) are used to generate the orthogonal subcarriers. Each OFDM carrier corresponds to one element of this discrete Fourier spectrum. The amplitudes and phases of the carriers depend on the data to be transmitted. The data transitions are synchronized at the carriers, and can be processed together, symbol by symbol. The channel distorts the transmitted OFDM symbols, causing Inter Symbol interference (ISI) and Inter carrier (ICI) which affects the subcarriers orthogonality and makes it impossible to be correctly demodulated at the receiver.

2. Space Time Block Codes (STBC)

Space Time Block Coding (STBC) is a technique to achieve higher diversity at the receiver end to mitigate multipath fading without increasing the transmitted power or bandwidth.

Figure 8. Comparison of the bandwidth utilization. (a) FDM, (b) OFDM.

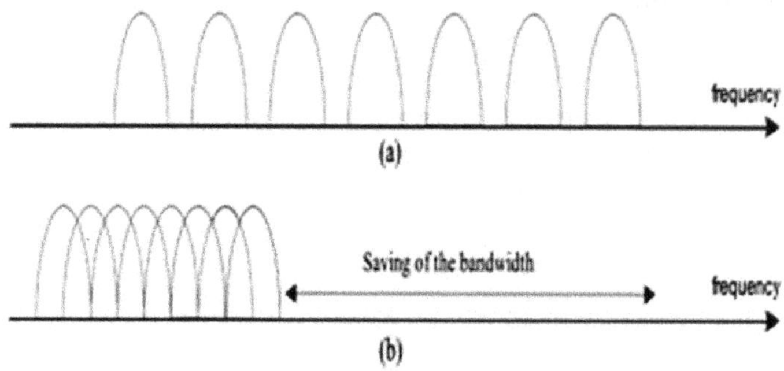

Space time coding promises to maximize the system capacity. The system capacity is defined in (Shahid, 2005), "The maximum possible transmission rate such that the probability of error is arbitrarily small." In 1998, Alamouti discovered a very simple transmitter diversity technique for two Transmit antennas, which provides a full diversity order, and has no loss of capacity (if the number of Rx antennas is equal to one), and possesses a simple and fast maximum likelihood (ML) decoding. Instead of being joined, the transmitted signals are decoded separately at decoders due to the orthogonality between the columns (and rows) of the code. The Alamouti code could be considered as the original Space-Time Block Code (STBC), that is the Space Time Codes (STCs) constructed from orthogonal designs (Wysocki, 2006).

CASE STUDY: OPTIMIZATION OF SMART GRID COMMUNICATION SYSTEMS EMPLOYING MIMO TECHNOLOGIES

1. System Description

The work emulates a small scale wide area monitoring system. The power system considered in this work is the IEEE 14-bus standard system. Previous works have shown that with one zero injection at bus 7, this system can be made completely observable using only 3 PMUs placed at buses 2,6 and 9. The raw data collected from the three PMUs is sent both through the SISO channel and the MIMO channel. In the MIMO system, a 2×2 (2 antennas at the transmitter and two antennas at the receiver) system is considered here. To account for the delays, the three PMUs are assumed to be located at different distances from the PDC. This is translated into different delays introduced into the set data. The three PMUs are synchronized using the GPS clock. In the SISO channel, the data is sent as a stream while in the MIMO channel; data is splitted into two streams (sent at two antennas) and is combined back at the receiver. Due to the fact that data arrives at the receiver from different directions, maximum combining is used to ensure that the received data undergoes the minimum possible error. The MIMO channel is assumed to exhibit a Rayleigh fading which is close to reality. To assess the performance of both systems, two performance indicators are considered: the time of data flow through the channel to account for the data latency and the bit error rate to account for data completeness. It should be noted that the state of the power system state is estimated using the received data from the three PMUs.

2. Results and Discussions

Fig. 9 represents the SISO and MIMO implementations of the previously defined systems. The same amount of data is sent over the SISO and MIMO systems. At the receiver side, the received data is assessed for completeness with reference to the sent data and latency as measured by the SIMULINK counter for both systems. Figures 10, 11 and 12 show a comparison of the data received from the three PMUs using the SISO and MIMO systems in relation to the originally transmitted data.

Figure 9. MIMO and SISO implementations

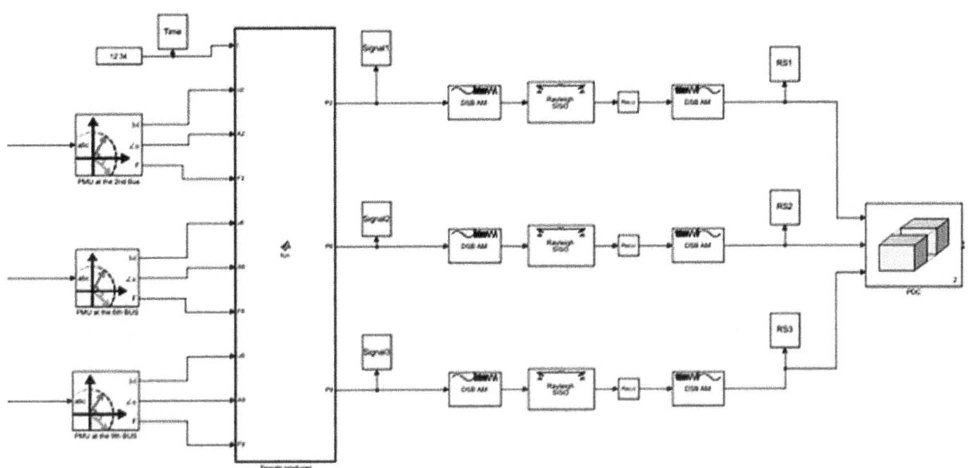

Figure 10. Comparison of the raw data and the data received from SISO and MIMO systems from PMU at bus 2

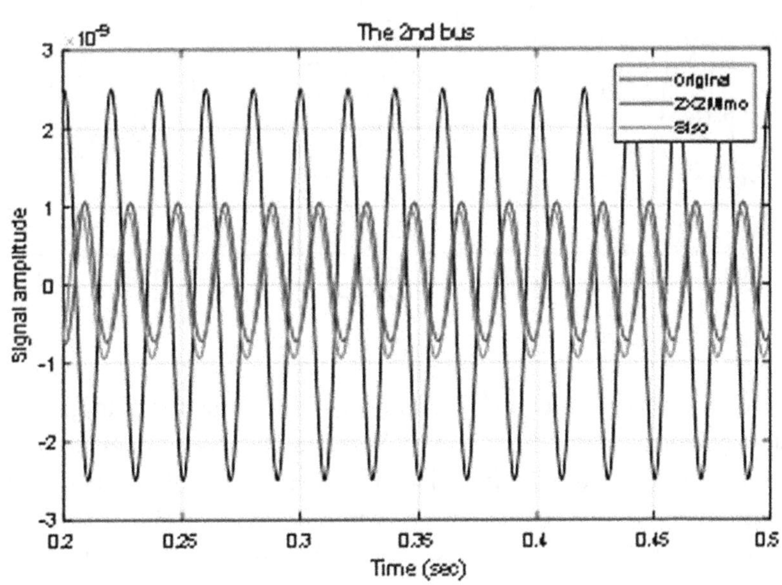

Figure 11. Comparison of the raw data and the data received from SISO and MIMO systems from PMU at bus 6

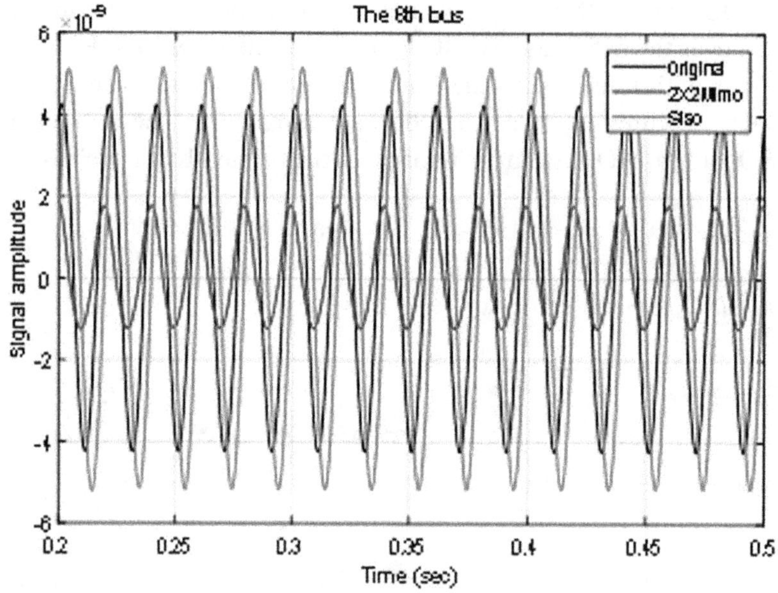

Figure 12. Comparison of the raw data and the data received from SISO and MIMO systems from PMU at bus 9

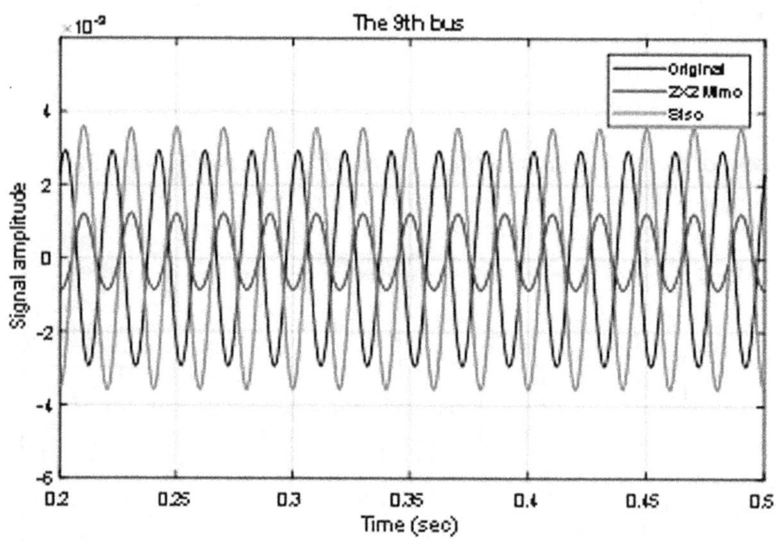

Table 2 summarizes the performance indicators for both systems for comparison. It is clearly seen that the MIMO system performs better than the SISO system in terms of data completeness. This is due to the fact that MIMO systems are mainly introduced to cure the impairments of the channel. Furthermore, the signal reception at the receiver is done as so to minimize the Bit Error Rate (BER). However, the SISO system is better in terms of latency as the data is a single stream and it is traveling a single

Table 2. Summary of The Performance Measures of the MIMO and SISO Systems

Bus N° (with PMU)	System	Latency (Delay) (Sec)	Bit Error Rate (completeness)
2	MIMO	0.0406625	2.07435×10^{-3}
	SISO	0.0193750	2.61042×10^{-3}
6	MIMO	0.0306250	2.70365×10^{-3}
	SISO	0.0153125	3.07656×10^{-3}
9	MIMO	0.0203125	2.09768×10^{-3}
	SISO	0.0103125	3.05326×10^{-3}

channel towards the receiver while in the MIMO system, data is splitted and is scattered in the channel and combined back at the receiver. Hence, the designer has to compromise between the two performance measures when it comes to practical system design.

It should be noted that MIMO systems are introduced mainly to cope with the limited power and bandwidth. So, a MIMO system promises more data to be put onto the communication system and hence the system would use more data and incorporate many applications simultaneously. In a wide area system, data can be used for monitoring, control and protection using the same data package. Furthermore, the study made here aims at highlighting the promising applicability of the MIMO communication systems into mart grid applications. A complete design of the MIMO system itself must be done to improve further the whole system. This includes the transmitter/Receiver architectures, the channel models and the modulation/coding schemes. Hence, the Smart grid application and the MIMO communication system constitute a complex system that needs to be optimized for performance at all levels.

CHALLENGES AND OPPORTUNITIES

The development of the Smart Grid will mean a drastic change in power use and administration. Users will become active participants in energy management and will be able to control their consumption. On the other hand, utilities will be able to control demand peaks and manage the grid efficiently from generation to distribution.

Different types of wireless networks are available, but which one is the suitable for a SG, depends on the system architecture and varieties of communication modules and links. The multihop wireless networking is definitely necessary, as electric equipment out of range of each other need to exchange information. First of all, in order to simplify network organization and maintenance the entire environment needs to be self-organized. Moreover, communications modules may pertain heterogeneous nature in terms of coverage, computing power and power efficiency. Wireless mesh networks (WMNs) are considered as important networking for SG. However, when WMNs are applied in SG wireless infrastructure, a few challenging issues still remains (e.g. security level). Without effective measures to prevent security attacks, the primary of users and confidentiality of grid information cannot be guaranteed. advanced technologies and applications offered by smart grid will shave the energy consumption behaviors of consumers, and hence, a considerable reduction in energy consumption will be achieved. The integration of SG technology into the existing system has been accelerated with the SG standardization efforts of different organizations, government regulations, and SG solutions of technology and solution providers as depicted in Tables I and II. A description of communication requirements, potential applications, and the SG roadmap has been presented in this paper as the keys to SG development.

The challenges include the deployment of large-scale embedded computing, legacy power grids, intelligent appliances, and next-generation communications and collaborations that will provide the foundation for a post-carbon society. In this section, we discuss the context that gives these challenges urgency as well as the technical challenges that need to be addressed by smart grid communication infrastructures.

Modeling, analysis and design a suitable communication infrastructure meet many new challenges. The models to be used must be capable of accounting for uncertainty as a way to simulate emerging behavior. The numerical tools to perform the analysis must be capable of solving very large scale problems. In fact, the power system is tightly coupled and non-linear (Quinn, 2009) and does not benefit from the sparsity that typically characterized this problem. The control system and particularly communication infrastructure must be designed to manage uncertainty and inconsistencies to be resilient or gracefully degrade when necessary. Finally the performance metric must be adjusted to the new nature of the power system.

Communication infrastructure should be capable of providing fail proof and nearly instantaneous bidirectional communications among all devices ranging from individual loads to the grid wide control centers including all important equipment at the electricity distribution and transmission system. This involves processing vast number of data transactions for analysis and automation. It requires a high performance communication infrastructure capable of providing fast intelligent local subsecond responses coordinated with a higher level global analysis in order to prevent or contain rapidly evolving adverse events (Kolter and Johnson,2011).

To address the variability of the net demand, as renewable resources growing over the long run, efficient communication infrastructure for information exchange among demand response, storage devices and utilization of plug-in electric vehicles (PEVs)/plug-in hybrid electric vehicles (PHEV) will complement the remedies (Komninos et al., 2014).

Demand response can be implemented through either automatic or manual response to price signals, or through a bidding process based on direct communications between the consumers and the market/ system operators or through intermediaries such as aggregators or local utilities.

CONCLUSION

Smart Grid is a foundational technology that will provide considerable changes to the existing power grid, consumer lives, the quality of power delivery, and traditional energy resources. Traditional energy sources produce energy with high environmental damage, as opposed to renewable energy sources (solar,wind, waves, and biomass), which do not present this problem, but have remarkable different dynamic characteristics than traditional ones. An integration of these into the existing energy grid is necessary and involves new technologies. The information flow is an important concept to define these technologies.

Information and communication technologies (ICT) represent a fundamental element in the growth and performance of smart grids. A sophisticated, reliable and fast communication infrastructure is, in fact, necessary for the connection among the huge amount of distributed elements, such as generators, substations, energy storage systems and users, enabling a real time ex-change of data and information necessary for the management of the system and for ensuring improvements in terms of efficiency, reliability, flexibility and investment return for all those involved in a smart grid: producers, operators and customers.

The WAMC systems offer many opportunities to improve the real-time situational awareness of the power system. Essentially, these systems are built following a similar principle as the conventional SCADA systems but with continuous streaming of data frames at high resolutions. The functionality of the WAMC applications running on top of them are heavily, but not exclusively, dependent on the capabilities of the underlying ICT systems. The design of the ICT systems supporting the WAMC applications, i.e., selecting the communication protocols, architecture and configurations need to be coordinated with the design and parameterization of the application algorithms.

Smart meters are supposed to help transform the delivery network into a two-way information system which can signal price changes to the customer; the customer in turn will be able to set rules so that heavy-load appliances such as cars and dishwashers work when the electricity is cheapest. This 'demand response' will help cope with a growing number of fluctuating energy sources such as solar and wind. Peak demand shaving will help governments meet energy security goals and will also save utilities money, as customers on day-night tariffs are typically paying less then the wholesale cost of energy during the peak.

IEEE C37.118 is a weak communication framework from a security point of view due to no built-in security mechanism. On the other hand, IEC 61850-90-5 provides strong protection against cyber attacks by relying on GDOI based security mechanism. However, this assumes the KDC and communicating devices secure from unauthorized access. In terms of required network resources, IEEE C37.118 is highly efficient with very small packet size compared to IEC 61850-90-5. This results in much lower bandwidth requirement for IEEE C37.118 compared to IEC 61850-90-5.

As an illustration, the potential benefits that a communication system designer within the framework of smart grid would gain when opting for a MIMO communication link has been demonstrated. It has been shown that for the same application, a MIMO system outperforms its SISO counterpart in terms of data completeness which is a major issue as all decisions and control actions results from correct data reception. Also, It has been that the MIMO system is relatively sluggish compared to the SISO system. This issue can be addressed by considering more details of the MIMO system design. On the other hand, the MIMO system promises higher data capacity to be injected into the system which makes it beneficial as many smart grid applications can be considered simultaneously. It should be noted that the smart grid system using MIMO communication is very complex and opens the window to an optimization task that has to be done on both the application itself and the MIMO communication system.

REFERENCES

Anderson, R., & Fuloria, S. (2010), On the security economics of electricity metering. *Proc. of Workshop on Economics of Internet Security (WEIS).*

Apostolov, A. (2013). Impact of IEC 61850 on the Interoperability and Reliability of Protection schemes. PES-GM.

Aravinthan, V., Karimi, B., Namboodiri, V., & Jewell, W. (2011). Wireless communication for smart grid applications at distribution level -feasibility and requirements. IEEE Power and Energy Society General Meeting, 1–8.

Asghar, M. R., D´an, G., Miorandi, D., & Chlamtac, I. (2017). *Smart Meter Data Privacy: A Survey*. IEEE.

Azzougui, Y., Recioui, A., & Mansouri, A. (2019). PMU Optimal Placement in Wide Area Monitoring Systems using Grey Wolf Optimization technique. *Algerian Journal of Signals and Systems*, *4*(1), 1–7.

Batra, N., Kelly, J., Parson, O., Dutta, H., Knottenbelt, W., Rogers, A., Singh, A., & Srivastava, M. (2014), NILMTK: An Open Source Toolkit for Non-intrusive Load Monitoring. *Proc. of ACM Intl. Conf. on Future Energy Systems (e-Energy).*

Baumeister, T. (2016). *Literature review on smart grid cyber security*. https://csdl-techreports.googlecode.com/svn/trunk/techreports/2010/10-11/10-11.pdf

Borges, F. (2017). *On privacy-preserving protocols for smart metering systems*. Springer. doi:10.1007/978-3-319-40718-0

Bouhafs, F., Mackay, M., & Merabti, M. (2012). Links to the future: Communication requirements and challenges in the smart grid. *Power and Energy Magazine, IEEE*, *10*(1), 24–32. doi:10.1109/MPE.2011.943134

Chao, L., Wu, X., Wu, J., Li, P., Han, Y., & Li, L. (2012), Implementations and Experiences of Wide-area HVDC Damping Control in China Southern Power Grid. *IEEE PES General Meeting*. doi: 10.1109/PESGM.2012.6345363

Chaudhuri, N., Ray, S., Majumder, R., & Chaudhuri, B. (2010), A New Approach to Continuous Latency Compensation With Adaptive Phasor Power Oscillation Damping Controller (POD). *IEEE Transactions on Power Systems*, *25*(2), 939–946. doi:10.1109/TPWRS.2009.2031908

Chenine, M., Al Khatib, I., Ivanovski, J., Maden, V., & Nordström, L. (2010), PMU traffic shaping in IP-based Wide Area communication. *5th International Conference on Critical Infrastructure (CRIS)*, 1–6, 10.1109/CRIS.2010.5617564

Chenine, M., & Nordström, L. (2011), Modeling and Simulation of Wide-Area Communication for Centralized PMU-Based Applications. *IEEE Transactions on Power Delivery*, *26*(3), 1372 –1380. doi:10.1109/TPWRD.2011.2106805

Cuijpers, C., & Koops, B.-J. (2013). Smart metering and privacy in Europe: Lessons from the Dutch case. *In European Data Protection. Eds. Springer, 2013*, 269–293.

Cyr, C., & Kamwa, I. (2010), WACS design at Hydro-Quebec. *PMU Tutorial, IEEE PES General Meeting*.

Daki, H., El Hannani, A., Aqqal, A., Haidine, A., & Dahbi, A. (2017). Big Data management in smart grid: Concepts, requirements and implementation. *Journal of Big Data*, *4*(13), 13. Advance online publication. doi:10.118640537-017-0070-y

Decker, I. C., Silva, A. S., Agostini, M. N., Prioste, F. B., Mayer, B. T., & Dotta, D. (2011). Experience and applications of phasor measurements to the Brazilian interconnected power system. *European Transactions on Electrical Power*, *21*(4), 1557–1573. doi:10.1002/etep.537

Deng, Y., Lin, H., Phadke, A., Sandeep, S., & Thorp, J. M. L. (2012), Communication network modeling and simulation for wide area measurement applications. IEEE PES ISGT. doi:10.1109/ISGT.2012.6175664

Dolezilek, D., Fischer, N., & Schloss, R. (2012). *Improvements in Synchronous Wide-Area Data Acquisition Design and Deployment for Telecontrol and Teleprotection.* https://www.selinc.com/WorkArea/DownloadAsset.aspx?id=99365

Dong, Y., & Kezunovic, M. (2011). Communication infrastructure for emerging transmission-level smart grid applications. IEEE Power and Energy Society General Meeting, 1–7. doi:10.1109/PES.2011.6039640

Fan, Z., Kulkarni, P., Gormus, S., Efthymiou, C., Kalogridis, G., Sooriyabandara, M., Zhu, Z., Lambotharan, S., & Chin, W. (2012). Smart grid communications: Overview of research challenges, solutions, and standardization activities. *IEEE Communications Surveys and Tutorials*, *99*, 1–8.

Galli, S., Scaglione, A., & Wang, Z. (2010). Power line communications and the smart grid. *IEEE International Conference Smart Grid Communications (SmartGridComm)*, 303–308. 10.1109/SMARTGRID.2010.5622060

Gezer, C., & Buratti, C. (2011), A ZigBee smart energy implementation for energy efficient buildings. *Proc. IEEE 73rd Veh. Technol. Conf. (VTC Spring)*, 1–5.

Greveler, U., Justus, B., & Loehr, D. (2012). *Multimedia content identification through smart meter power usage profiles*. Computers, Privacy and Data Protection.

Gungor, V. C., Sahin, D., Kocak, T., & Ergüt, S. (2011). *Smart grid communications and networking.* Türk Telekom, Tech. Rep. 11316-01.

Gungor, V. C., Sahin, D., Kocak, T., Ergut, S., Buccella, C., Cecati, C., & Hancke, G. P. (2011). Smart grid technologies: Communication technologies and standards. *IEEE Transactions on Industrial Informatics*, *7*(4), 529–539. doi:10.1109/TII.2011.2166794

Hart, G. (1992). Nonintrusive appliance load monitoring. *Proceedings of the IEEE*, *80*(12), 1870–1891. doi:10.1109/5.192069

Hauser, C. H., Bakken, D. E., & Bose, A. (2005). A failure to communicate: Next generation communication requirements, technologies, and architecture for the electric power grid. *IEEE Power & Energy Magazine*, *3*(2), 47–55. doi:10.1109/MPAE.2005.1405870

He, J., Lu, C., Wu, X., Li, P., & Wu, J. (2009). Design and experiment of wide area HVDC supplementary damping controller considering time delay in China southern power grid. *Generation, Transmission Distribution, IET*, *3*(1), 17–25. doi:10.1049/iet-gtd:20080129

Huang, S., Jan, R., & Yang, W. (2006). A ring-based information collection architecture in wireless sensor networks. *International Journal of Sensor Networks*, *1*(3/4), 190–199. doi:10.1504/IJSNET.2006.012034

IEEE. (2006). C37.118-2005. *IEEE Standard for Synchrophasors for Power Systems.*

IEEE. (2011). IEEE Standard for Synchrophasor Measurements for Power Systems. *IEEE Std C37.118.1-2011 (Revision of IEEE Std C37.118-2005)*, 1–61. doi:. doi:10.1109/IEEESTD.2011.6111219

Jingcheng, G., Xiao, Y., Liu, J., Liang, W., & Chen, C. L. (2012). A survey of communication/networking in Smart Grids. *Future Generation Computer Systems*, *28*(2), 391–404. doi:10.1016/j.future.2011.04.014

Johnston, R., Hauser, C., Gjermundrod, K., & Bakken, D. (2006). Distributing Time Synchronous Phasor Measurement Data Using the Grid Stat Communication Infrastructure. *In System Sciences, 2006. HICSS, 06*, 245b–245b. doi:10.1109/HICSS.2006.127

Kalogridis, G., Cepeda, R., Denic, S., Lewis, T., & Efthymiou, C. (2011). Elecprivacy: Evaluating the privacy protection of electricity management algorithms. *IEEE Transactions on Smart Grid*, *2*(4), 750–758. doi:10.1109/TSG.2011.2160975

Kamwa, I., Dobrescu, M., Heniche, A., Cyr, C., & Cadieux, P. (2011). A fundamental study of wide-area damping controllers with application to fuzzy-logic based PSS design for dynamic shunt compansators. *Power Systems Computation Conference (PSCC).*

Kansal, P., & Bose, A. (2012). Bandwidth and latency reuirements for smart transmission grid applications. *Smart Grid, IEEE Transaction*, *3*(3), 1344–1352. doi:10.1109/TSG.2012.2197229

Khan, R., McLaughlin, K., Laverty, D., & Sezer, S. (2016). Analysis of IEEE C37.118 and IEC 61850-90-5 Synchrophasor. *Proceedings of Power and Energy Society General Meeting (PESGM).* 10.1109/PESGM.2016.7741343

Khan, R. H., & Khan, J. Y. (2013). A comprehensive review of the application characteristics and traffic requirements of a smart grid communications network. *Computer Networks*, *57*(3), 825–845. doi:10.1016/j.comnet.2012.11.002

Kim, J., Filali, F., & Young-Bae, K. (2015). Trends and Potentials of the Smart Grid Infrastructure: From ICT Sub-System to SDN-Enabled Smart Grid Architecture. *Applied Sciences (Basel, Switzerland)*, *5*(4), 706–727. doi:10.3390/app5040706

Kolter, J. Z., & Johnson, M. J. (2011). Redd A public data set for energy disaggregation research. *Proc. of ACM KDD Workshop on Data Mining Applications in Sustainability (SustKDD).*

Komninos, N., Philippou, E., & Pitsillides, A. (2014). Survey in smart grid and smart home security: Issues, challenges and countermeasures. *IEEE Communications Surveys and Tutorials*, *16*(4), 1933–1954. doi:10.1109/COMST.2014.2320093

Kuzlu, M., Pipattanasomporn, M., & Rahman, S. (2014). *Communication network requirements for major smart grid applications in HAN, NAN and WAN. In Computer Networks* (Vol. 67). Elsevier.

Lewis, R. P., Igic, P., & Zhongfu, Z. (2009). Assessment of communication methods for smart electricity metering in the U.K. *Proc. IEEE PES/IAS Conf. Sustainable Alternative Energy (SAE)*, 1-4. 10.1109/SAE.2009.5534884

Liserre, M., Sauter, T., & Hung, J. Y. (2010). Future energy systems: Integrating renewable energy sources into the smart power grid through industrial electronics. *IEEE Ind. Electron*, *4*(1), 18–37. doi:10.1109/MIE.2010.935861

Locke, G., & Gallagher, P. D. (2010). *Nist Framework and Roadmap for Smart Grid Interoperability Standards, Release 1.0; Smart Grid technologies: Communication technologies and standards*. National Institute of Standards and Technology.

Lu, X., Wang, W., & Ma, J. (2012). Authentication and integrity in the smart grid: An empirical study in substation automation systems. *International Journal of Distributed Sensor Networks, 8*(6), 175262. doi:10.1155/2012/175262

Majumder, R., Chaudhuri, B., & Pal, B. (2007). Implementation and test results of a wide-area measurement-based controller for damping interarea oscillations considering signal-transmission delay. *Generation, Transmission Distribution, IET, 1*(1), 1–7. doi:10.1049/iet-gtd:20050493

Marijic, Z., Ilic, Z., & Bazant, A. (2010). Fixed-data-rate power minimization algorithm for ofdm-based power-line communication networks. *IEEE Transactions on Power Delivery, 25*(1), 141–149. doi:10.1109/TPWRD.2009.2035286

Martin, K. E., Hamai, D., Adamiak, M. G., Anderson, S., Begovic, M., Benmouyal, G., Brunello, G., Burger, J., Cai, J. Y., Dickerson, B., Gharpure, V., Kennedy, B., Karlsson, D., Phadke, A. G., Salj, J., Skendzic, V., Sperr, J., Song, Y., Huntley, C., ... Price, E. (2008). Exploring the IEEE Standard C37.118-2005 Synchrophasors for Power Systems. *IEEE Transactions on Power Delivery, 23*(4), 1805–1811. doi:10.1109/TPWRD.2007.916092

Massoud Amin, S., & Wollenberg, B. F. (2005). Towards a SmartGrid: Power delivery for the 21st Century. *IEEE Power & Energy Magazine, 3*(5), 34–41. doi:10.1109/MPAE.2005.1507024

Metke, A., & Ekl, R. (2010). Smart grid security technology. In *Innovative Smart Grid Technologies* (pp. 1–7). ISGT.

Mo, Y., Kim, T.-H., Brancik, K., Dickinson, D., Lee, H., Perrig, A., & Sinopoli, B. (2012). Cyber-physical security of a smart grid infrastructure. *Proceedings of the IEEE, 100*(1), 195–209. doi:10.1109/JPROC.2011.2161428

NETL. (2007). *A systems view of the modern grid*. NETL.

NETL. (2009). *West Virginia Smart Grid implementation Plan*. NETL.

Niyato, D., & Ping, W. (2012). Cooperative transmission for meter data collection in smart grid. *Communications Magazine, IEEE, 50*(4), 90–97. doi:10.1109/MCOM.2012.6178839

Park, N., & Kim, M. (2014). Implementation of load management application system using smart grid privacy policy in energy management service environment. *Cluster Computing, 17*(3), 653–664. doi:10.100710586-014-0367-y

Parsons, J. D. (2000). Mobile lRadio Propagation Chanel (2nd ed.). John Wiley & Sons, Inc.

Paruchuri, V., Durresi, A., & Ramesh, M. (2008). Securing powerline communications. *Proc. IEEE Int. Symp. Power Line Commun. Appl., (ISPLC)*, 64–69.

Peizhong, Y., Iwayemi, A., & Zhou, C. (2011). Developing ZigBee deployment guideline under WiFi interference for smart grid applications. *IEEE Transactions on Smart Grid, 2*(1), 110–120. doi:10.1109/TSG.2010.2091655

Phadke, A. G., Thorp, J. S., & Adamiak, M. G. (1983). A New Measurement Technique for Tracking Voltage Phasors, Local System Frequency, and Rate of Change Frequency. *IEEE Transactions on Power Apparatus and Systems*, *102*(5), 1025–1038. doi:10.1109/TPAS.1983.318043

Pipattanasomporn, M., Kuzlu, M., & Rahman, S. (2012). Demand response implementation in a home area network: a conceptual hardware architecture. IEEE Innovative Smart Grid Technologies (ISGT) Conference, 1–8. doi:10.1109/ISGT.2012.6175675

Prasad, R. (2004). *OFDM For Wireless Communications Systems*. Artech House.

Quinn, E. (2009). *Privacy and the new energy infrastructure*. Available at SSRN http://ssrn.com/abstract=1370731

Ragib, H., Rakesh, B., & Himanshu, K. (2006). Analysis NASPInet Data Flows. IEEE PES Power System Conference and Exhibition. doi:10.1109/PSCE.2009.4840100

Recioui, A. (2018). *Application of Teaching Learning-Based Optimization to the Optimal Placement of Phasor Measurement Units. In Handbook of Research on Emergent Applications of Optimization Algorithms*. IGI Global.

Recioui, A., Bentarzi, H., & Ouadi, A. (2015). Application of a binary teaching learning-based algorithm to the optimal placement of phasor measurement units. *Progress in Clean Energy*, *1*, 817–830.

Recioui, A., Bentarzi, H., & Tsebia, M. (2016). PMU Deployment in Power System Oscillation Monitoring. *Sustainable Aviation*, 312-322.

Recioui, A., Tazibt, M., & Benhenni, H. (2017, September). Fault Detection in Wide Area Monitoring Systems via PMU Optimal Placement. *Algerian Journal of Signals and Systems*, *2*(3), 149–161.

Saoud, A., & Recioui, A. (2017). A review on Data communication in smart grids. *Algerian Journal of Signals and Systems*, *2*(3), 162–179.

Shahid, M. (2005). *(MIMO) Systems with Multicarrier Time Delay Diversity Modulation*. Naval Postgraduate School.

Sharma, K., & Saini, L. M. (2017). Performance analysis of smart metering for smart grid: An overview. *Renewable & Sustainable Energy Reviews*, *49*, 720–735. doi:10.1016/j.rser.2015.04.170

Stouffer, K., Falco, J., Scarfone, K., Stouffer, K., Falco, J., & Scarfone, K. (2006). Guide to supervisory control and data acquisition (scada) and industrial control systems security. *SPIN*.

Tan, S., Debraj, D., Wen-Zhan, S., Yang, J., & Das Sajal, K. (2016). *Survey of Security Advances in Smart Grid: A Data Driven Approach*. IEEE COMST.

Tariq, F., & Dooley, L. S. (2013). Smart Grid Communication and Networking Technologies: Recent Developments and Future Challenges. In Smart Grids (pp. 199–213). Springer London.

TrongNghia, L., Wen-Long, C., Dang Khoa, T., & Tran Hiep, N. (2016). Advanced Metering Infrastructure Based on Smart Meters in Smart Grid. *Intechopen*, *2016*. Advance online publication. doi:10.5772/63631

Vivekanandan, V., & Wong, V. W. S. (2006). Ordinal MDS-based localisation for wireless sensor networks. *International Journal of Sensor Networks*, *1*(3/4), 169–178. doi:10.1504/IJSNET.2006.012032

Wang, R., Ziad, M., & Lee, Y. (2001). Data Quality. The Kluwer International Series on Advances in Database Systems. Springer.

Wang, S., Gao, W., Wang, J., & Lin, J. (2012). Synchronized Sampling Technology-Based Compensation for Network Effects in WAMS Communication. *IEEE Transactions on Smart Grid*, *3*(2), 837–845. doi:10.1109/TSG.2012.2183902

Wang, W., Xu, Y., & Khanna, M. (2011). A survey on the communication architectures in smart grid. *Computer Networks*, *55*(15), 3604–3629. doi:10.1016/j.comnet.2011.07.010

Weixiao, M., Ruofei, M., & Hsiao-Hwa, C. (2014). Smart Grid Neighborhood Area Networks: A Survey. *IEEE Network*, *28*(1), 24–32. doi:10.1109/MNET.2014.6724103

Wietfeld, C., Georg, H., Groening, S., Lewandowski, C., Mueller, C., & Schmutzler, J. (2011). Wireless M2M communication networks for smart grid applications. Wireless Conference 2011 – Sustainable Wireless Technologies (European Wireless), 1–7.

Wood, G., & Newborough, M. (2003). Dynamic energy-consumption indicators for domestic appliances: Environment, behaviour and design. *Energy and Building*, *35*(8), 821–841. doi:10.1016/S0378-7788(02)00241-4

Wu, H., Tsakalis, K., & Heydt, G. (2004). Evaluation of time delay effects to wide-area power system stabilizer design. Power Systems. *IEEE Transaction*, *19*(4), 1935–1940. doi:10.1109/TPWRS.2004.836272

Wysocki, L. C. (2006). Complex Orthogonal Space Time Processing in Wireless Communications. Academic Press.

Yan, Y., Qian, Y., Sharif, H., & Tipper, D. (2013). A Survey on Smart Grid Communication Infrastructures: Motivations, Requirements and Challenges. IEEE Communications Surveys & Tutorials, 15(1), 5 – 20.

Yan, Y., Qian, Y., Sharif, H., & Tipper, D. (2013). A Survey on Smart Grid Communication Infrastructures: Motivations, Requirements and Challenges. *IEEE Communications Surveys and Tutorials*, *15*(1), 5–20. doi:10.1109/SURV.2012.021312.00034

Yao, W., Jiang, L., Wu, Q., Wen, J., & Cheng, S. J. (2010). Design of wide-area damping controllers based on networked predictive control considering communication delays. *IEEE PES General Meeting*, 1–8. doi: 10.1109/PES.2010.5589556

Yarali, A. (2008). Wireless mesh networking technology for commercial and industrial customers. Proc. Elect. Comput. Eng., CCECE, 47–52.

Zhu, K., Al-Hammouri, A., & Nordström, L. (2012). To concentrate or not to concentrate, Performance Analysis of ICT system with Data Concentrations for Wide-area Monitoring and Control Systems. *IEEE PES General Meeting*. doi: 10.1109/PESGM.2012.6344977

Zhu, K., Chenine, M., Nordström, L., Holmström, S., & Ericsson, G. (2013). An Empirical Study of Synchrophasor Communication Delay in a Utility TCP/IP Network. *International Journal of Emerging Electric Power Systems*, *14*(4), 341–350. doi:10.1515/ijeeps-2013-0062

Zhu, K., Chenine, M., Nordström, L., Holmström, S., & Ericsson, G. (2013). Design Requirements for Wide-Area Damping Systems - Using Empirical Data from a Utility IP Network. *IEEE Transactions on Smart Grid Revision*.

KEY TERMS AND DEFINITIONS

Data Communication: Data transmission (also data communication or digital communications) is the transfer of data (a digital bitstream or a digitized analog signal over a point-to-point or point-to-multipoint communication channel. Examples of such channels are copper wires, optical fibers, wireless communication channels, storage media and computer buses. The data are represented as an electromagnetic signal, such as an electrical voltage, radiowave, microwave, or infrared signal.

MIMO Systems: Multiple-input and multiple-output, or MIMO is a method for multiplying the capacity of a radio link using multiple transmission and receiving antennas to exploit multipath propagation.

Phasor Measurement Unit: A phasor measurement unit (PMU) is a device used to estimate the magnitude and phase angle of an electrical phasor quantity (such as voltage or current) in the electricity grid using a common time source for synchronization. Time synchronization is usually provided by GPS and allows synchronized real-time measurements of multiple remote points on the grid. PMUs are capable of capturing samples from a waveform in quick succession and reconstructing the phasor quantity, made up of an angle measurement and a magnitude measurement. The resulting measurement is known as a synchrophasor. These time synchronized measurements are important because if the grid's supply and demand are not perfectly matched, frequency imbalances can cause stress on the grid, which is a potential cause for power outages.

Protocols: A communication protocol is a system of rules that allow two or more entities of a communications system to transmit information via any kind of variation of a physical quantity. The protocol defines the rules, syntax, semantics and synchronization of communication and possible error recovery methods. Protocols may be implemented by hardware, software, or a combination of both

Chapter 2
Role of Smart Metering and Implementation Issues in Smart Grid

Maheswari Maruthakutti
Nizwa College of Technology, Oman

Loganathan Nachimuthu
Nizwa College of Technology, Oman

Suthanthira Vanitha N.
Muthayammal Engineering College, India

ABSTRACT

Smart grid provides the digital technology that allows for two-way communication between the utility and the customers. The smart grid consists of controls, computers, automation, and new technologies and equipment working together. The smart grid will move the energy industry into a new era of reliability, availability, and efficiency with economic and environmental health. A smart meter plays a major role in the smart grid and it is an electronic device that measures and records the energy consumption. It enables two-way communication between the meter and the supplier through advanced metering infrastructure (AMI). The mode of communication is enabled by either wireless or wired. The wireless communication includes Wi-Fi, wireless mesh networks, Zig Bee, cellular communications, and low power range Wi-Fi. This chapter deliberates about the evolution of electricity metering, major components of smart meter, communication infrastructure and protocols for smart metering, demand-side integration, recent developments, issues faced and solutions, merits and demerits.

DOI: 10.4018/978-1-7998-4027-5.ch002

1. INTRODUCTION

Smart grid provides the digital technology that allows for two-way communication between the utility and the customers. The definition for smart grid as per the European technology platform is "A Smart Grid is an electricity network that can intelligently integrate the actions of all users connected to it – generators, consumers and those that do both – in order to efficiently deliver sustainable, economic and secure electricity supplies" as discussed in European Commission (2006). For the implementation of the smart grid, it requires information and communication technologies, sensing, measurement, control and automation technologies, power electronics and storage devices deliberated by Ekanayake et al (2012). Information and communication technologies are used to provide two-way communication between the different components of the power system and the load. Intelligent Electronic Devices (IED), Phasor Measurement Unit (PMU), Wide Area Monitoring, Protection and Control (WAMPC) and smart meters are the most important components used in sensing, measurement, control and automation technologies respectively. It provides advanced protection, high security and less energy consumption. Power electronics and storage devices part deals about the different types of high voltage transmission and energy storage devices for reliable and flexible operation of the power system reflected by Ekanayake et al (2012).

Smart meters play a major role in the smart grid at the customer side. Sensors and sensor networks are the main parts of smart devices and smart metering. Generally, sensors are located at different parts of the grid. Sensors play an important role in remote monitoring and demand side management. As the sensors are located at all the parts of the grid, it is able to measure/monitor temperatures, operation of grid devices, power quality disturbances and detect outages. This process helps the control centers to receive accurate and actual condition of the grid.

A smart meter is an electronic device that measures and records the energy consumption. It also communicates the measured data into the central system frequently. It enables two-way communication between the meter and the supplier through Advanced Metering Infrastructure (AMI). The mode of communication is enabled by either wireless or wired. The wireless communication includes Wi-Fi, wireless mesh networks, Zig Bee, cellular communications and low power range Wi-Fi.

This chapter is about the evolution of electricity metering, major components of smart meter, communication infrastructure and protocols for smart metering, demand side integration, recent developments, real time implementation, issues faced and solutions, merits and demerits.

2. EVOLUTION OF ELECTRICITY METERING

Electricity meters named as energy meters are used to record the energy supplied to the customer for over a duration and most of these meters are accumulation type. Accumulation meters located in the customer side are read manually to calculate the bill for energy consumed over a period. Nowadays the customers are started to use more advanced meters due to the advancements in the technology. These advanced meters are very much useful to identify the energy consumed by a particular customer for a specific period and the pattern of energy consumption. Based on the pattern the energy suppliers will design the tariff amount for a customer. It will in turn help the customer to analyze the energy consumption pattern and utilize the energy accordingly. Smart meters are still having more features like two-way communication, real time energy usage, bill price and control over the electrical appliances.

The evolution of electrical metering from accumulation meter to smart meter is depicted in Figure 1. It is clear that manual reading type meters were widely used before the year 2000. Automatic Meter Reading (AMR) techniques have been developed around the year 2000 which communicates in one –way from the energy meters to the energy supplier through low speed communication networks. Thereafter one – way communication and AMR leads path to tremendous development in the meter infrastructure. These developments are the main resources for the implementation of smart concept in the power grid and enables two-way communication, tariff change, demand side bidding, remote connection, distribution automation and power quality monitoring.

Figure 1. Evolution of Electricity Metering [2]

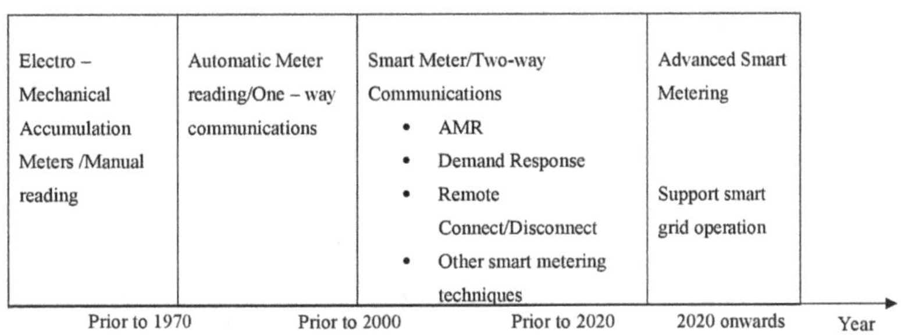

Smart meters and conventional meters are mainly differing in the way of communication and medium of communication as shown in Figure 2. The early one uses two-way communication through high speed communication networks whereas the later one employs only one –way communication through low speed communication networks. Smart meter communicates through gateway and Home Area Network (HAN) controller. Gateway permits the transfer of smart meter data from energy suppliers, distribution network operators and other services. It may receive data either directly from smart meters or through data management company.

Figure 2. Comparison between Conventional and Smart Meters [2]

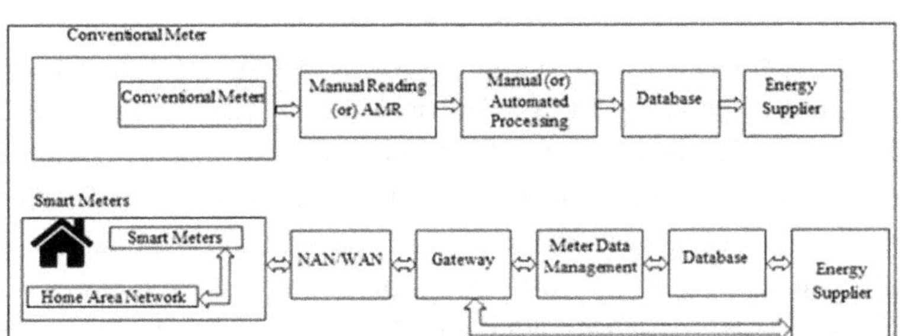

The usage of smart meters provide the following benefits compared to Automatic Meter Reading (AMR) to the customers

- More reliable energy supply and reduced customer complaints
- Easy payments for Distributed Generation (DG) output
- Remote controlling of DG becomes simple and utility cost reduces
- Home area automation implemented easily
- Facilitating the users to adopt electric vehicles or heat pumps during peak demand

3. FUNCTIONAL BLOCK DIAGRAM AND HARDWARE COMPONENTS OF SMART METER

The four major components of smart metering are smart meters, a two-way communication network, a Meter Data Management (MDS) system and HAN. The electro-mechanical meters record the energy consumption over a period of time. But electronics meters measure the various parameters like voltage, frequency, reactive power and power factor apart from electrical energy. It is more accurate than conventional meters. The energy consumption can be read from the display of electronic meters over a period of time. But nowadays the two-way communication electronic meters namely smart meters are designed and installed in many places. The basic block diagram of smart meter is shown in Figure 3. The block diagram has five major divisions such as signal acquisition, signal conditioning, analog to digital converters (ADC), computation and communication. The following section covers the detailed explanation about the function and type of hardware used in each block of the smart meter.

Figure 3. Functional Block Diagram of Smart Meter

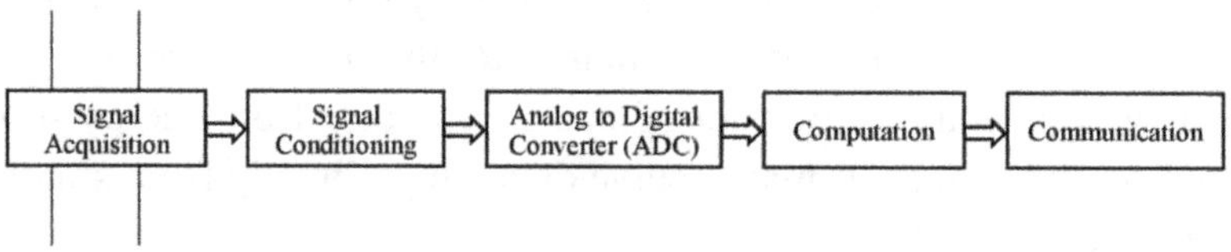

3.1 Signal Acquisition

The primary function of the smart meter is to obtain system parameters precisely and continuously for the further computation and communication. In electrical system, the voltage magnitude, its frequency, current magnitude and its displacement are the fundamental quantities. The other parameters such as power, power factor and Total Harmonic Distortion (THD) can be estimated from these fundamental quantities. In general, the current sensors measure the current at the load side and the voltage sensor measures at the supply side. In case of low cost meters, the current sensing has done by current sensing shunt resistors and the voltage by resistive voltage divider as shown in Figure 4.

The value of resistance used in current sensing shunt varies from 100 μΩ to 500 mΩ. The amount of voltage drop across these resistors are proportional to the current flowing through it. The self-heating parameter decides the current rating of this shunt resistors. The voltage drop across the current sensing resistor will be small and it is matched with ADC through a Programmable Gain Amplifier (PGA). In case of voltage sensor, resistor divider provides the voltage between the phase conductor and the neutral.

Manganin alloy is used for the resistive divider due to its constant impedance over typical operating temperature ranges.

For high power applications, Current Transformer (CT) can be used to sense the current. CT handles high power and consumes less power than shunt resistors. The nonlinear phase response of CT may cause measurement errors at low currents and large power factors. Whenever smart meters require high precision over a wide operating range, the sophisticated sensing techniques such as Rogowski coils, optical methods and Hall effect sensors are used.

Figure 4. Voltage and Current Sensing Circuit

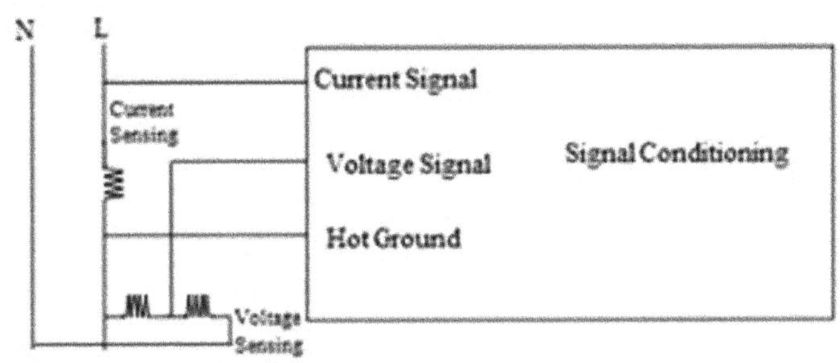

3.2 Signal Conditioning

The input signals are prepared for ADC process in the signal conditioning stage. The process may be addition or subtraction, amplification or attenuation and filtering. These stages are realized through discrete elements or combined with ADC as a part of integrated circuit or built in function of 'System on Chip' architecture. To maintain accuracy of the input signal, it is necessary to remove components of the input signal above the Nyquist frequency i.e., half the sampling rate of the ADC. Hence a low pass filter is placed before the ADC stage. The sampling frequency is determined by the functions of the meter. The sampling frequency will be chosen as very high value if the meter measures harmonic contents apart from fundamental components as explained by Millman, J. and Grabel, A. (1999) and Allen, P.E. and Holberg (2002).

3.3 Analog to Digital Conversion

The signal obtained from the sensors need to be sampled and digitized for the further processing by metering software. In electrical system, generally voltage and current are the two fundamental parameters measured. Hence there are two signals in a single phase meter, a multiplexer is used to send signals to the ADC. ADC converts analog signals into digital signal. For better understanding the concept of ADC, sampling of 3 bit ADC is shown in Figure 5. ADC uses $8(=2^3)$ levels so any voltage between 0.8 and 0.6V is represented by 111 and voltage between 0.6 and 0.4 V is represented by 110 and so on. The performance of ADC is determined by its resolution. It is defined as the ratio between the voltage range and total number of levels (2^n) i.e.,

$$Resolution = \frac{Voltage\ Range}{2^n},$$

where n is the number of bits in the ADC

For a 3-bit ADC as shown in Figure 5, the voltage range is 1.6V. Hence the resolution is 1.6/8 = 0.2 V. The resolution will be low when high bits are used in ADC. If 8-bit ADC is used then the resolution is $1.6/2^8 = 6.25$mV. There are many methods are available for the conversion of analog signal to digital signal like flash type, successive approximation type, integrated or dual slope type and sigma – delta type. Successive approximation type and Sigma-delta type are widely used methods for ADC in smart meters as conversed by Roy Choudhury, D. and Shail B.Jain (2010) and Kester,W. (2009).

Figure 5. Simplified Successive Approximation ADC

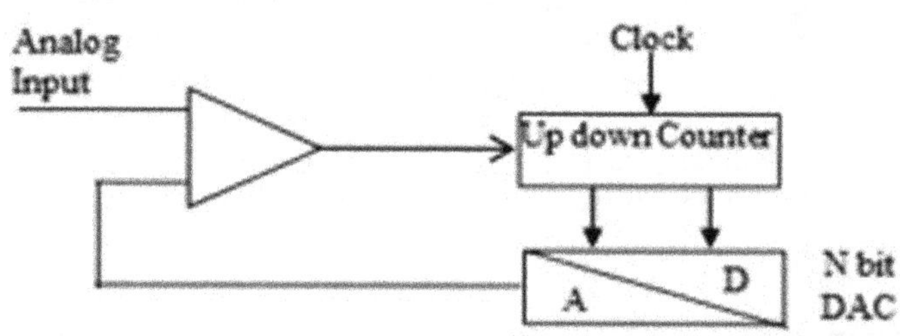

3.4 Computation

Computation is the major function block of a smart meter which carries all kind of arithmetic and logical operations, communication and storage functions as shown in Figure 6. The computation process includes the following

- Arithmetic operations on input signals
- Timestamping of data
- Preparation of data for communication
- Handling of routines with regular input such as payment, tamper detection
- Storage of data

Figure 6. Block Diagram of Computation

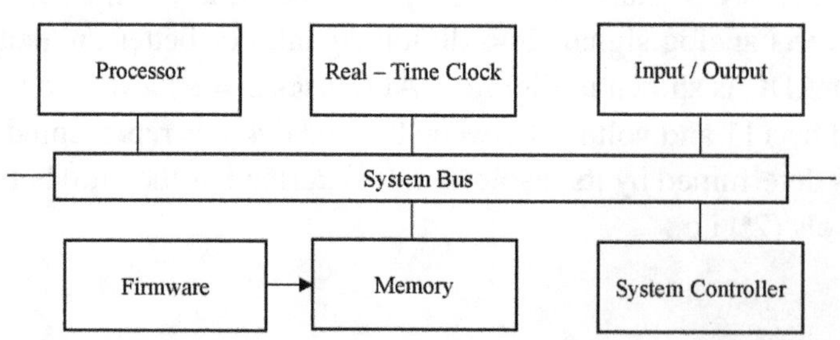

- System updates and
- Co-ordinating different functions

Large number of arithmetic operations are required to derive the various parameters, hence Digital Signal Processor (DSP) is used. Apart from arithmetic operations, the meter has to perform routine tasks and large number of other operations like payment, tamper detection and system updates. Therefore, the processor should be capable of performing more parallelism or buffering actions. The required arithmetic operations to derive different parameters are listed in Table 1.

Table 1. Arithmetic Operations Required for Different Parameters

Parameter Required	Arithmetic Operation
Instantaneous Voltage/Current	Multiplication
Peak Voltage/Current	Comparison
RMS Voltage/Current	Multiplication
System frequency	Zero detection, Comparison
Phase Displacement	Zero detection, Comparison
Power Factor	Trigonometric Function
Instantaneous Apparent Power	Multiplication
Instantaneous Real Power	Multiplication
Instantaneous Reactive Power	Multiplication
Energy Use/ Protection	Integration
Harmonic Voltage Distortion	Fourier Series
Total Harmonic Distortion	Multiplication and addition

In computation process, memory unit plays a major role. It requires both volatile and non-volatile memory for its operation as discussed by Millman, J. and Grabel, A. (1999) and Webster (1998). To store the data temporarily during the operations, volatile memory is required. The size of these memory depends on the quantity, rate, complexity and rate of communication to/from ports. Similarly, non-volatile memory is also required to information like serial number and maintenance access codes. Additionally, the details of energy consumption also need to be stored until it is communicated to the billing company.

3.5 Input / Output

The measured parameters in smart meter is displayed in the form of text and graphs for the customer. The devices used as display unit are Liquid crystal displays(LCD) and the light emitting diodes(LED) in seven segment alphanumeric and matrix format. LEDs produce light when polarized directly at low voltage and few milli ampere of current.

A small key pad or touch screen is used as input device for human interface. It facilitates the selection of smart appliance to be controlled and the payment options. Apart from this, calibration input is also provided to calibrate the smart meter when variations occur in voltage references, sensor tolerances or other system gain errors. Remote calibration and control capability through communication lines is also provided in some models of smart meter.

To create the awareness among the customer about the energy consumption and tariff, a separate display unit may be located in the visible places in the residence like kitchen or living room. These kind of approach will encourage the customers to reduce their energy use either throughout the year.

3.6 Communication

Smart meters stand different from conventional meters by the way of communicate the data to the energy suppliers. It employs many types of network adapters for communication purpose. It may be wired or wireless communication. Public Switched Telephone Network (PSTN), power line carrier, cable modems and Ethernet are different types of wired communication. The wireless communication includes ZigBee, infrared and GSM/GPRS/CDMA.

4. COMMUNICATION TECHNOLOGIES USED IN SMART METER

Communication techniques plays a vital role in smart grid technology. The major role of this is to act as a medium to transfer the collected information from one place to another. For the successful operation of a smart meter, the three communication interfaces required are

- Wide Area Network (WAN)
- Neighborhood Area Network (NAN)
- Home Area Network (HAN)
- A typical smart metering communication architecture is shown in Figure 7.

Figure 7. Communication Interfaces in Smart Meter

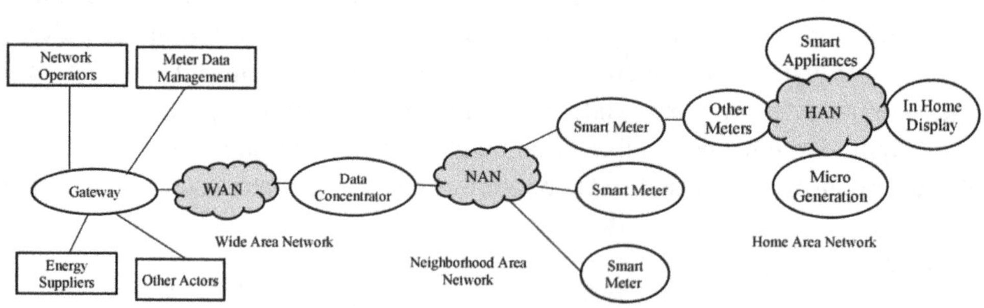

4.1 Home Area Network

The integration of smart meter, microgeneration, in home display, smart appliances, smart socket, plug-in hybrid electric vehicles and HVAC (Heating, Ventilation, Air Conditioning) is called as Home Area Network (HAN). In general, it uses wired or wireless communications and networking protocols to ensure the operation of all connected appliances and the interfacing of smart meter. HAN includes security mechanisms as an important criterion to protect the consumer data and the metering system. HAN also providing the other services like centralized energy management and services for the better performance and user convenient. Energy monitoring and display, controlling of smart appliances, smart plugs and controlling of HVAC, scheduling and remote operation of house hold appliances are the add-on services provided by energy management function. Also HAN is expected to provide aids to the utilities through the following

- Demand Response and Management
- Management of Micro-Generation
- Charging of electric vehicles

To provide the above said functions, smart meters are following two methods in general. As deliberated by Kushiro, et al. (2003) and Young, S. and Stanic, R. (2009), either smart meters may be used as the interface to the suppliers, network operators and the other actors or a separate control box which is directly interfaced to the outside world through the NAN and WAN as shown in Figure 8 & 9.

Figure 8. Smart Meter Act as an Interface

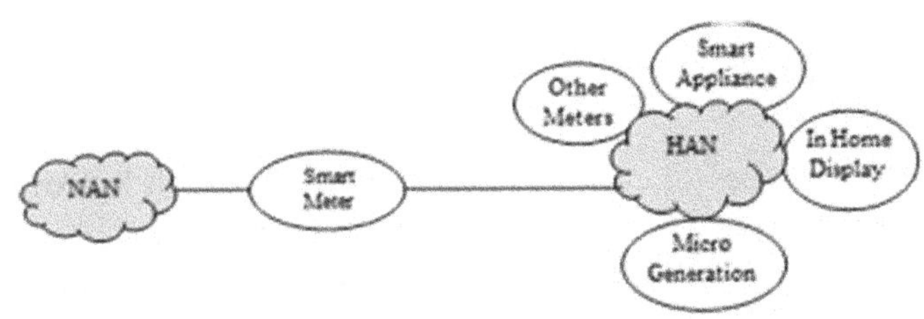

Figure 9. Controller Act as an Interface

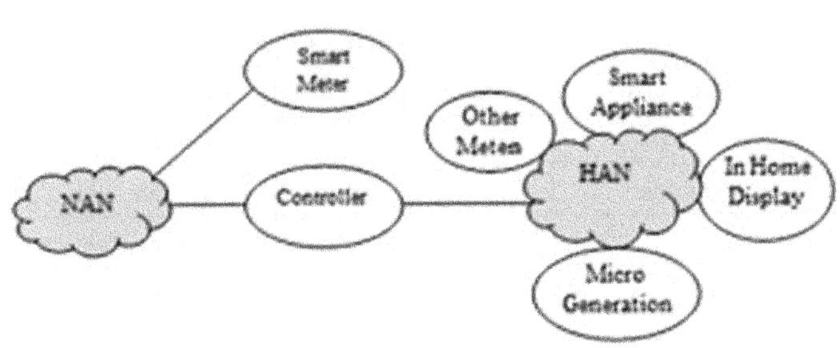

4.2 Neighborhood Area Network (NAN)

The transfer of consumption reading from smart meters is the major function of the Neighborhood Area Network (NAN). Diagnostic messages, real-time messages and firmware upgrades are the other supports provided by NAN to the power system. The expected data volume transferred from a simple meter is 100 kB per day and for firmware upgrades may require 400 kB of data to be transferred as specified in Smart Metering Implementation Programme (2010). The data transfer value may increase rapidly based on the real-time functions added to the smart metering infrastructure. NAN selects the communication technology based on its data transfer capacity such as Zig Bee technology is used to transfer energy consumption data from each household to data concentrator. Zig Bee has 250kB/s data transfer capacity.

4.3 Data Concentrator

The data concentrator acts as a communicator between the smart meters and the gateway. It copes up the meters automatically detecting them, creates and optimizes repeating chains, coordinate the bidirectional delivery of data and monitors the conditions of the meters.

4.4 Meter Data Management System

Database is the heart of the meter data management system. It has capable of data acquisition, validation, adjustment, storage and calculation in order to provide better service to the customers. The critical issue associated with meter data management system is to make it open and flexible with the existing system. Apart from this, meter data management system also provides functions such as remote meter connection/disconnection, power status verification, supply restoration verification and on-demand reading of remote smart meters.

4.5 Protocols for Communications

Smart meter employs different types of communication protocols such as Power Line Carrier (PLC), GPRS and combination of both. The communication between the meter and data concentrator PLC is used and GPRS is employed in between the data concentrator and the gateway to the data management

Table 2. Different Types of Protocols for Demand Side Applications

Types of Protocol	Different Kinds of Demand Side Applications			
	Local AMR	Remote AMR	Smart Metering	HAN
TCP/IP		Applicable	Applicable	Applicable
IEC 62056	Applicable	Applicable	Applicable	Applicable
SML	Applicable	Applicable	Applicable	Applicable
IEC 61334 PLC		Applicable	Applicable	
EN 13757 M-Bus	Applicable	Applicable	Applicable	Applicable
SITRED	Applicable	Applicable	Applicable	
PRIME	Applicable	Applicable	Applicable	
Zigbee Smart Energy			Applicable	Applicable
EverBlu	Applicable	Applicable	Applicable	
OPERA/UPA		Applicable	Applicable	
IEC 62056-21 'FLAG'	Applicable	Applicable		
IEC 62056-21 'Euridis'	Applicable	Applicable		
ANSI C12.22		Applicable	Applicable	

system. The following table summarizes the protocols which are applicable for different demand side applications.

5. DEMAND SIDE INTEGRATION

Demand side integration (DSI) is a set of measures to use loads and local generation to support network operation/management and improve the quality of power supply. DSI potential depends on

- Availability and timing of information provided to consumers
- Duration and timing of their demand response
- Performance of the ICT infrastructure
- Metering
- Automation of end use equipment and pricing

The most important terminologies and definitions used in DSI are

- *Demand-Side Management (DSM)* is the utility activity which controls the customer electricity usage. It encourages the customer to change their electricity usage patterns through proper planning, implementation and monitoring activities.
- *Demand Response (DR)* is the mechanism to manage the demand in response according to the supply conditions.
- Demand Side Participation is a set of schemes used in an electricity market by end-use customers to contribute to economic, system security and environmental benefits.

5.1 Services Provided by DSI

Flexible loads, distributed generation and storage are the demand side resources provide various services to the power system by modifying the load consumption patterns. The services are load shifting, valley filling, peal clipping, dynamic energy management, energy efficiency improvement and strategic load growth as presented in IEA Demand Side Management Programme . The movement of load between times of day i.e., from on peak to off peak or seasons is called as load shifting. For example, the loads such as washing machine consumes 1 kW for 2 hours is shifted to off peak loads.

The increase of off peak demand through storing of energy is called as valley filling, for example charging the battery of an electric vehicles and thermal storage in an electric heater. Load shifting only shifts the load from one time to another but valley filling add the new loads during the off peak demand. The total energy consumption remains same in case of load shifting but increases in case of valley filling. Peak clipping means reducing the peak load demand when demand reaches the thermal limits of feeders/transformers. It is done through switching off the domestic appliances like control of electric water heaters or air conditioners.

The development of smart metering and home automation system, makes possible the intelligent control of all domestic appliances. Hence the load curve becomes flexible and can be controlled to meet the system needs. For the effective implementation of DSI, the utility should know not only about the loads are installed but also about their usage. Hence it requires two-way communication between the smart meter and the network operators.

5.2 Implementations of DSI

Price-based schemes and incentive based schemes are the two methods in which DSI can be implemented in the customer service as deliberated by Gellings, C.W. (2009) . The customer changes its load according to the electricity price then it is called as price based DSI. Incentives are given to the customer for load modification apart from their electricity rates is known as incentive based DSI. There are various kinds of DSI programs are developed and implemented in the power system as shown in Figure 10.

5.2.1 Price-Based DSI Implementations

The customers are easily attracted by the effective tariff and pricing mechanism in the electricity markets. These schemes include the prices like time of use rates, real-time pricing and critical peak pricing. The definition for these terms are as follows,

- Time of use (ToU): The prices are predefined for 24 hours and use different unit price for different time blocks. It represents the cost of generation, transmission and distribution of power in different time periods.
- Real Time Pricing (RTP): According to the fluctuation in the whole sale market, the fluctuation in RTP also occurs. The changes in the RTP will be intimated to the customers either a day ahead or an hour ahead.
- Critical Peak Pricing (CPP): It is the combination of both ToU and RTP. The basic price will be ToU but the peak price is replaced by CPP when the system is suffering from some operational problems.

Figure 10. DSI Implementation

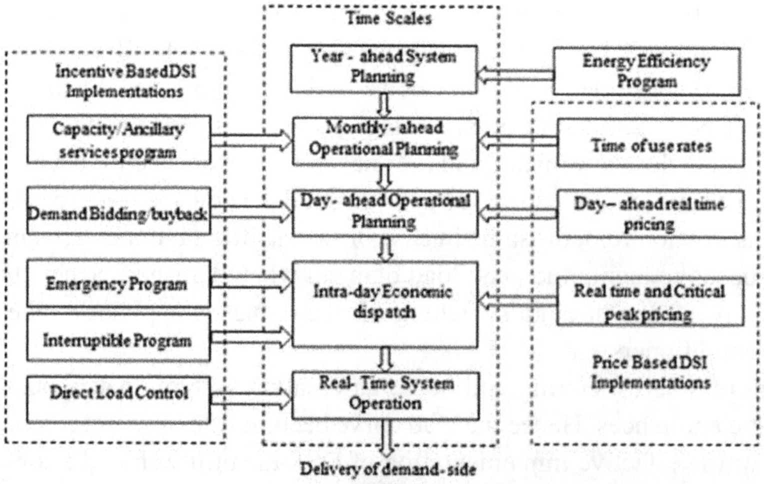

DSI performance is estimated by the response for demand to price signal. It is estimated by the price elasticity. The percentage change in quantity demanded in response to a 11 percent change in price is called as price elasticity. There are different types of elasticities used to evaluate the available amount of the demand side resources.

- Price elasticity of demand: The ratio between percentage change in demand and percentage change in price is called as price elasticity of demand. Generally, it should be a negative number.

$$Price\ Elasticity\ of\ Demand = \frac{\Delta D / D}{\Delta P / P}$$

Where D is the demand, ΔD is the change in demand, P is the price and ΔP is the change in price.

- Elasticity of substitution: It is a measure of ration between the percentage change in peak to off peak demand and percentage change in peak to off peak price.
- Long term price elasticity: It is the ratio between the annual energy consumption to average change in energy price.

5.2.2 Incentive Based DSI Implementation

The various kinds of incentive based DSI are direct load control, interruptible/curtailable service, demand-side bidding/ buy-back programs, emergency demand response programs, capacity market programs and ancillary services market programs.

- Direct load Control: The controlling of customer appliances like air conditioners, water heater etc., by the program operator based on short notice. It is offered to residential and small commercial customers.
- Interruptible/curtailable service: The services may be interrupted during contingency period. If the customer accepts for these condition, then there will be a discount or reduction in their tariff. These facilities are offered only to industrial customers.
- Demand-side bidding/ buy-back programs: The bids offered for the curtailment based on the electricity market by the customers. It is applicable only to industrial customers.
- Emergency demand response programs: Incentives has to be provided to the customers for the load reduction during the shortage of system reserve.
- Capacity market programs: Customer offers load curtailment as system capacity to replace conventional generation.

6. STANDARDS FOR INFORMATION EXCHANGE IN SMART METERING

The major function of smart meter is to communicate information from meter to data concentrator and receives back the control signals from it. Hence smart meter requires standards for these communications. AMR sends the measured data to the data concentrator once in a while for example once in a month.

But AMI does bidirectional communication frequently. Communication system in smart meters plays a crucial role in smart grid. There are two standards widely used for smart meter communication namely, IEC 62056 and ANSI C12.22. IEC 620556 defines the transport and application layers for smart metering under a set of specifications called Companion Specification for Energy Metering (COSEM). The sending and receiving of meter data to and from external systems is specified by ANSIC 12.22. It is suitable for any communication network. Figure 11 shows the steps followed in ANSI 12.222 to transport messages.

Figure 11. Basic Smart Meter Architecture in ANSI C12.22

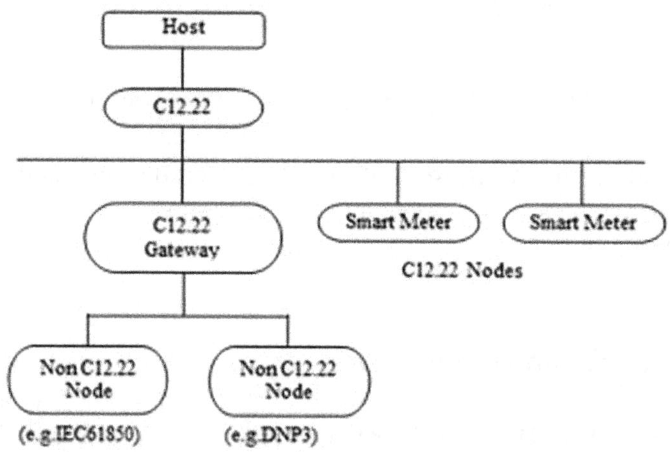

7. ISSUES AND CHALLENGES

The existing grid can be efficiently managed by means of integration of the grid using technical advancements. But there are lot of difficulties and challenges are encountered in the smart meter implementation. The implementation and maintenance of smart meter in the distribution side requires billion dollars of investment. The replacement of existing meters with smart meters is the biggest challenge for the utility companies. The existing meters does not have the required infrastructure to synchronize with the new technology. When the smart meters are implemented, the other devices need to be integrated with the smart meter for the complete operation of the smart grid. The integration of all devices will create issues when the number of customers are increased. The installation of communication networks becomes very difficult in terrestrial areas.

The energy consumption data is collected and transmitted continuously by the smart meter. It is very difficult and expensive job. This process will create the impression of privacy and security risks about the data and signals being transmitted among customers. Also this data reveals the information about the presence of customers in their houses and the details of appliances that are used. Hence some customers are not willing to transmit these data to their neighbors. Hence there should be a clarity about the parameters to be transmitted and the authentication to access the data as presented by C. Bennett and D. Highfill (2008).

The huge amount of data has to be transferred between the smart meter and the server in the base station. The storage, maintenance and management of these huge data will be tedious work in smart grid system. Hence for the speedy transfer of data and security, the communication networks need to be selected properly. If the bandwidth of the communication network is small then there will be more traffic in the data transmission and the quantity of data transferred also will be low. There are lot of devices need to be integrated for modulation, demodulation and storage along with the smart meter. This will increase the deployment cost of the smart meter. If the data transfer is done through the public communication networks like mobile networks, it may lead to many security issues, weak authentication, error handling, weak protocols and improper session management as discussed by Cleveland (2008) and Ramakrishna Kappagantu (2018) . Apart from these security issues, utility companies encounter some difficulties in facing the problems like network coverage, data capacity and propagation issues as deliberated by Robinson and Stuber. If wired communication is employed instead of wireless, then the physical damage of cables will lead to discontinuity in data transfer as deliberated by Depuru et al (2011) and Silva . The Figure 12, 13 and 14 shows the challenges in design, implementation, utilization and maintenance of smart meter system.

Apart from utility companies, there are other people are interested to collect and analyses the energy consumption data of a customer. It may lead to revengeful ex-spouses like civil accusers, blackmailers, terrorists and political leaders etc as discussed by Ericsson (2010). The future of the smart grid depends on the policies decided by the utility companies and respective governments. Even though the customer gateways are intelligent and are compatible with other devices they are prone to physical as well as security risks.

Figure 12. Design Issues for a Smart Meter System

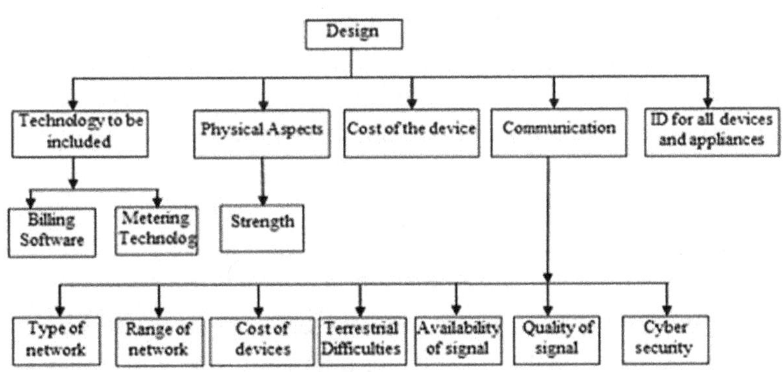

Figure 13. Maintenance Issues for a Smart Meter System

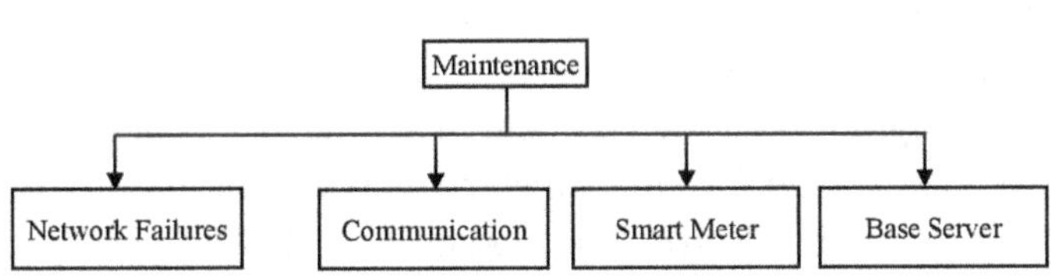

Figure 14. Challenges with Data Transfer for a Smart Meter System

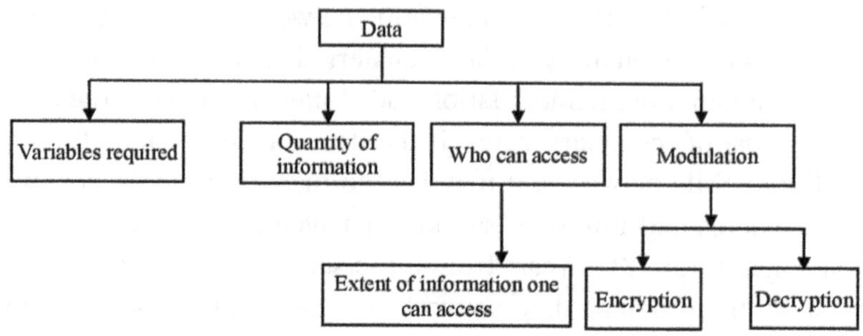

8. ADVANTAGES OF SMART METERING SYSTEM

In smart grid system, it is required to find the details of daily workflow, workforce management, asset management, call center philosophy and billing system etc. The operation of SCADA system is augmented by smart meters. Smart meters can provide numerous assistances like efficient power system control and monitoring, operational decisions those are taken to reduce outages and losses as presented by Mahmood et al (2008) . Smart meters are able to accomplish energy cost allocation, fault analysis, demand control and power quality analysis. Apart from this, smart meters can also plan the maintenance schedule and provide accurate billing. It also identifies the presence of harmonic component in the current supplied and it will lead to find the problems associated in the sources as explained by Lee and Lai (2009). Smart metering system can also be engaged with pattern recognition techniques to determine the operation of the information devices and financial incentives to the customer as elucidated by Boait (2009) . It also urges the customers to conserve energy by helping them maintain the quantity and cost of the energy consumption. It also enhances the power system to implement new and dynamic tariff system for the benefits of the customer. This will help the utility companies to implement pre-paid smart card system.

9. STATUS OF SMART METERING

In view of the advantages and applications, smart meter systems are being under large scale deployment worldwide. The smart meter deployment has grown dramatically since 2011 as shown in Figure 15. In U.S. households, nearly 70% conventional meters i.e., 88 million are replaced by smart meters as per the status of 2018 as presented by Adam Cooper and Mike Shuster (2019). According to the survey and plan, it is expected to reach 107 million meters at the end of year 2020.

Worldwide the application of smart meters has been increased in the sectors like commercial, industrial and residential. Among all these, the residential segment is expected to grow very fast in the forthcoming years. The market region for smart meter has been analyzed with respect to six regions such as North America, Europe, Asia- Pacific, South America, Middle East and Africa. Out of all these regions, Asia-Pacific is expected to dominate the market due to its widespread use of these solutions in the commercial, residential and industrial sector as shown in Figure 16. It is projected that the Asia-Pacific market will grow at the highest CAGR from 2017 to 2022. It will act as a largest market for infrastructure and industrial sector because it takes several steps like implementing smart grid, smart city and gas pipeline, repair and upgrading of aging infrastructure and managing growing electricity and water scarcity. Hence from the analysis report it is evident that the smart meter deployment will play a major role in energy industries in the fore coming years.

10. CONCLUSION

This chapter deliberated about the importance smart meter deployment in smart grid system. It also gives emphasis to the evolution of smart meters from the conventional meters and the hardware components employed in smart meter and its functions. The significance of communication networks and the protocols used for the proper implementation of the smart meter has been deliberated. The responsibility of smart meters in the implementation of DSI has been enlightened. The advancements in smart metering technology leads the world to new dimensions even though there are lot of implementation issues and challenges. The various kinds of solutions for the issues has been discoursed in this chapter. It also discussed about the present and future market developments of smart grid.

Figure 15. U.S. Projected Smart Meter Installation (15)

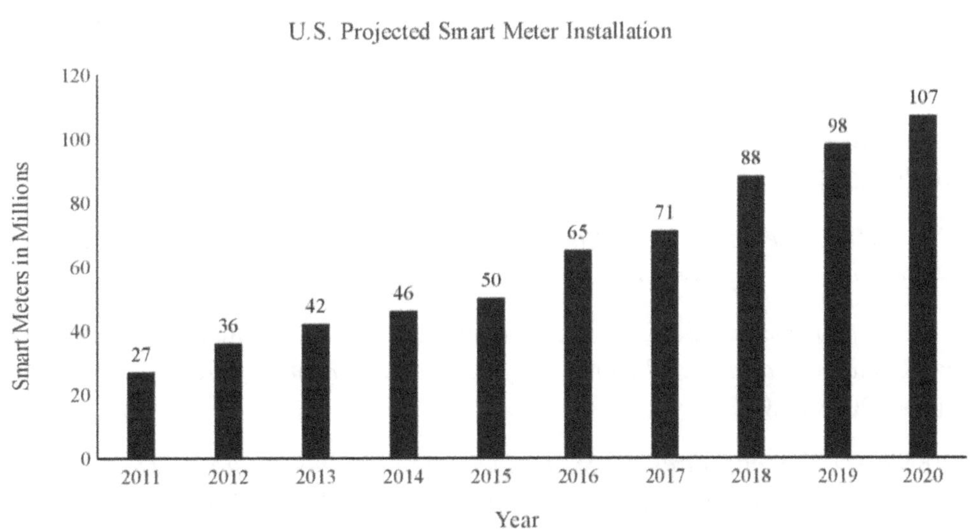

Figure 16. Smart Meter Market, By Region,2022 (USD Billion)
(Source: Markets and markets analysis)

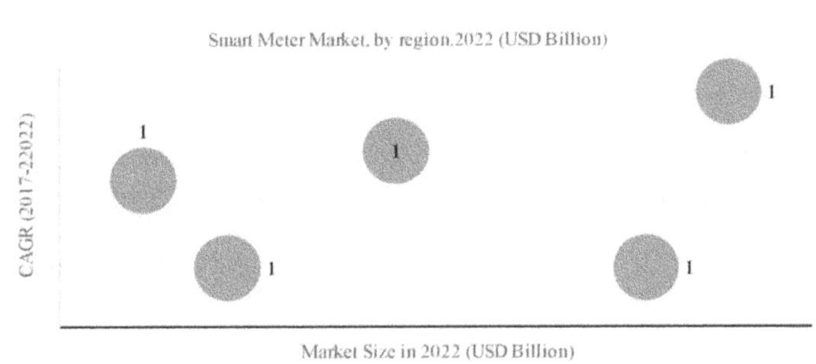

REFERENCES

Allen, P. E., & Holberg, D. R. (2002). *CMOS Analog Circuit Design.* Oxford UniversityPress.

Bennett, C., & Highfill, D. (2008). Networking AMI smart meters. *Proc. IEEE Energy 2030 Conference*, 1 – 8.

Boait, P. (2009). Smart metering of renewable micro generators by output pattern recognition. *Proc. International Conference and Exhibition on Electricity Distribution*, 1 – 4. 10.1049/cp.2009.0541

Cleveland, F. M. (2008). Cyber security issues for advanced metering infrastructure. *Proc. IEEE Power and Energy Society General Meeting - Conversion and Delivery of Electrical Energy*, 1 – 5.

Cooper, A., & Shuster, M. (2019). *Report on Electric Company Smart Meter Deployments: Foundation for a Smart Grid.* https://www.edisonfoundation.net/iei/publications/Documents/IEI_Smart%20Meter%20Report_2019_FINAL.pdf

Depuru, S. S. S. R., Wang, L., Devabhaktuni, V., & Gudi, N. (2011, March). *Smart European Commission (2006) European SmartGrids Technology Platform: Vision and Strategy for Europe's Electricity.* http://ec.europa.eu/research/energy/pdf/smartgrids_en.pdf

Ekanayake, J., Jenkins, N., Liyanage, K., Wu, J., & Yokoyama, A. (2012). *Smart grid: technology and applications.* Wiley.

Ericsson, G. N. (2010). Cyber Security and Power System Communication - Essential Parts of a Smart Grid Infrastructure. *IEEE Transactions on Power Delivery*, *25*(3), 1501–1507. doi:10.1109/TPWRD.2010.2046654

Gellings, C. W. (2009). *The Smart Grid: Enabling Energy Efficiency and Demand IEA Demand Side Management Programme.* http://www.ieadsm.org

Kappagantu, R., & Arul Daniel, S. (2018). Challenges and Issues of Smart Grid Implementation: A Case of Indian Scenario. *Journal of Electrical Systems and Information Technology*, *5*(3), 453–467. doi:10.1016/j.jesit.2018.01.002

Kester, W. (2009). *ADC Architectures III: Sigma-Delta ADC Basics, Analogue Devices, MT-022 Tutorial, 2009.* http://www.analog.com/static/imported-files/tutorials/MT-022.pdf

Kushiro, N., Suzuki, S., Nakata, M., Takahara, H., & Inoue, M. (2003). Integrated residential gateway controller for home energy management system. *IEEE Transactions on Consumer Electronics*, *49*(3), 629–636. doi:10.1109/TCE.2003.1233787

Lee, P. K., & Lai, L. L. (2009). Smart Metering in Micro-Grid Applications. *Proc. IEEE Power & Energy Society General Meeting*, 1 – 5.

Mahmood, A., Aamir, M., & Anis, M. I. (2008). Design and implementation of AMR smart grid system. *Proc. IEEE Canada Electric Power Conference*, 6 - 7. 10.1109/EPC.2008.4763340

Millman, J., & Grabel, A. (1999). *Microelectronics* (2nd ed.). McGraw-Hill Book Company.

Robinson, R. E., & Stuber, M. G. (n.d.). *Itron white paper: advanced metering infrastructure risk analysis for advanced metering*. www.itron.com/asset.asp?path=support/whitepaper/pdf/itr_ 016898.pdf

Roy Choudhury, D., & Shail, B. J. (2010). Linear Integrated Circuits. New Age International Publishers.

Silva, D. (n.d.). *New 'smart' electrical meters raise privacy issues*. Available: http://www.physorg.com/news176703307 .html

Smart Metering Implementation Programme: Statement of Design Requirements, Ofgem E-Serve. (2010). http://www.decc.gov.uk/assets/decc/consultations/smartmeter-imp-prospectus/225-smart-metering-imp-programme-design.pdf

Webster, J. G. (1998). *The Measurement, Instrumentation and Sensors Handbook*. CRC Press.

Young, S., & Stanic, R. (2009). *Smart Meter to HAN Communications*. Smart Grid Australia Intelligent Networking Working Group. http://smartgridaustralia.com.au/SGA/Documents/IN_Work_Group_Smart-Meter_HAN_Comms.pdf

Chapter 3
Smart Metering and Pricing Policy in Smart Grids

Fatma Zohra Dekhandji
Université M'hamed Bougara de Boumerdes, Algeria

ABSTRACT

Pricing policy is one of the tools allowing the involvement of customers in the balance between the supply and the demand in smart grids. The present chapter aims at presenting the smart metering action including the bidirectional measurement of energy for smart houses equipped with renewable energies as well as the way a smart meter communicates data at the required timing to and from the control center. A typical bill establishment explaining how the net billing is produced along with a discussion about different pricing policies that the utility may adopt to reduce the peak load demand is also presented. The work is concluded by a typical simulation of a smart city modeled in LABVIEW software.

INTRODUCTION

Electricity networks are undergoing significant transformation. Clean, small, distributed energy and demand-side resources are challenging the traditional axiom of electricity from large, remote power generation facilities delivered over extensive transmission and distribution (T&D) infrastructure to consumers. This transformation is the result of a number of diverse and disruptive technology innovations. These innovations are changing the design of the electricity network, the flow of electricity in the system, and are driving utilities to forge a complex set of new relationships with stakeholders (e.g., end users, energy services companies, generators).

Smart grid is a network of technologies that delivers electricity from power plants to the end user and connects all supply,grid and demand elements via an effective communication system. The system is the amalgamation of engineering, information and communication technologies, and management of the power grid. The improvements in these technologies can be applied to enhance automation, foster integration of distributed generation from renewables, secure the power system architecture, and enable efðcient demand-side energy management (Recioui and Dassa, 2017;Chellali et al.,2014).

DOI: 10.4018/978-1-7998-4027-5.ch003

A modern electricity distribution system is more complex, has greater redundancy, and allows for greater choice over the manner in which users generate, deliver, and consume electricity. The information and communication technology in the smart grids enables it to make envisioned benefits a reality. These technologies encompass a wide range of operations, such as detecting and identifying faults and a quick response to power outages; providing consumers with near real-time information on the amount and cost of the power they use; improving the security of the system; and linking all elements of the grid to enable better decision making on resource use. With continual up-gradation and modifications these technologies will produce more and better quality data that will give the utilities more flexibility and new opportunities to improve their analysis in areas, such as customer load patterns and tariffs, and thus offer better services to their customers. The technology makes more efficient use of the grid to identify and correct supply and demand imbalances instantaneously and detect faults in a self-healing process that improves service quality, enhances reliability, and reduces costs (Gurlin, 2011; Chafai, 2013; Ouadi, 2013).

The main goal of smart grid is to promote active customer participation and decision making as well as to create the operation environment in which both utilities and electricity users influence each other. New technologies are being deployed, including advanced meters, controllable appliances (Graditi et. al.,2016), distributed energy generation and storage systems, i.e., plug-in hybrid electric vehicle batteries, stand-alone storage systems, and communications capabilities. New laws are being proposed to allow electricity consumers to access pricing information. New dynamic pricing policies are likely to be implemented at the retail level over the next years (Hubert and Grijalva, 2012; Yalcintas et. al., 2015). Energy management controllers (Hu and Li, 2013) are primarily designed to control load within a single home. They often take into account the utility data like load forecasts or ToU pricing for scheduling the household appliances. At the customer side, customers have the incentive to shift their electricity usage from high peak hours to low peak hours so that their electricity bills can be reduced (Jazayeri et al, 2005; Rahimi and Ipakchi, 2010).

Demand Response (DR) is a term to mean ''changes in electricity usage by end customers from their normal consumption patterns in responseto changes in the price of electricity over time''. Price based DR programs consider ñattening demand ñuctuations as an objective. Both the customer and the utility will get beneðts from DR. It encourages the customer to reduce the peak demand in response to the incentives (Ozturk et.al., 2013). A DR strategy coordinates the requirements between the energy provider and the customer (Medina et. al., 2010). At the utility side, by reducing high peaks, DR programs are helpful in protecting grid from the risk of outages, reduce the usage of spinning reserves during peak load periods, balance the supply demand ratio, and improve the grid reliability (Mahmood et al., 2016; Rasheed et al., 2016; Rasheed et al., 2015). Further DR beneðts include: (i) lower electricity price in wholesale market, (ii) adequacy saving and operational security, (iii) integrated resource planning studies, and (iv) improved choice for using DR (Qdr, 2006; Cappers et. al., 2016). Contrary to DR programs, the integration of renewable energy into residential units provides reliable, efðcient and most attractive solution nowadays. It can curtail electricity cost at residential premises and ñatten the peaks at utility premises.

Smart grid gives opportunity to the end users to bi-directionally communicate with the utility in real time, so consumers can tailor their energy consumption based on individual preferences like price concern, user comfort, etc. Based on different usage patterns of energy, the smart grid offers differential pricing scheme in order to avoid different risk factors like blackout or load shedding, thus allows the user to curtail energy consumption during peak demand. The objective of appliance scheduling in differential pricing environment is to optimally schedule the ON-OFF cycles of appliances subject to end user electricity cost minimization.

The main objectives of this chapter are to present a software implementation of a smart metering system using LABVIEW. The system is to be used in a smart city where houses are equipped with renewable energies and at the same time are connected to the main grid. The smart meter, hence, possesses a bidirectional measurement capability. Real scenarios are simulated in the form of load profile variations. The aim of the measurement is to establish the electricity bill. As a next step, direct load control is tested and is embedded within the LABVIEW program. Finally, we will talk about the control center or Data Management System (DMS) which is in charge of receiving and storing the metering data for processing purposes.

BACKGROUND

Climate change, awareness of energy efficiency, new trends in electricity markets, the obsolescence of the actual electricity model, the urgent need to decarbonize electricity supply and swap aging resources and to make effective application of swiftly evolving information and communication technologies (Hossain et. al., 2016), and the gradual conversion of consumers to consumer profiles are the main agents of progressive change in electricity systems towards the Smart Grid paradigm. The introduction of multiple distributed generation and storage resources, with a strong involvement of renewable energies, exposes the necessity of advanced metering or Smart Metering systems, able to manage and control those distributed resources (Noelia, 2016).

Smart Grid and Smart Metering

1. Smart Grid

A Smart Grid is a concept for transforming an electric power grid by using advanced communication, automated controls and other forms of information technology. It integrates new innovative tools and technologies from generation, transmission, and all the way to consumer appliances and equipment. (Massoud and Wollenberg, 2005). The smart grid can be viewed as a superposition of communication networks on the electric grids. Hence, it can improve efficiency, reliability, safety, and security of electricity supply to the customers, with a seamless integration of renewable and alternative energy sources, through automated control and modern communications technologies .where various components of the electric grid are linked together via two-way communications and power flows to provide interoperability among them. Thus, consumers can not only draw power but also supply surplus power to the grid using smart meters that enable monitoring and measuring of these bidirectional flows (TrongNghia et. al., 2016).

2. Smart Metering

Smart metering has been recognized as a major part of the smart grid system, where it is built with smart meters, control devices and a communication link. The smart meter is the key element of this system which is the combination of all energy metering and intelligence, Smart meters are capable of communicating with each other and executing command signals remotely and locally. They provide good solutions in overcoming the problems that are faced with the old grid system.

Smart meters are embedded with different technologies and services. Therefore, implementation of smart meters appears to be multi-standard and incompatible but the meter developers ought to include the consideration of basic abilities such as:

- Remote provision of metering data and related information to the utility;
- Two-way communications between the meter and the utility;
- Remote operation for disabling and enabling supply;
- Provision of information to home and networks;
- Automatic processing, transfer, management and utilisation of metering data;
- Security tampers detections and remote configurations (Koponen et. al., 2008).

3. Home Automation

With the increase in consumption of energy and population, there is a grave need to conserve energy in every way possible. The inability to access and control the appliances from remote locations is one of the major reasons for energy loss (Palaniappan et. al., 2015).

Home automation is building mechanization for a home, called a smart home. It includes the control and computerization of lighting, warming, (for example, brilliant indoor regulators), ventilation, aerating and cooling (HVAC), and security, and also home machines, for example, washer/dryers, stoves or fridges/cooler. And it does not only refer to reducing human efforts but also energy efficiency and time saving. Wi-Fi is regularly utilized for remote checking and control (Katre1 and Rojatkar, 2017). This system can also make use of a host of communication methods such as Z-wave, GSM, Bluetooth, ZigBee and PLC. Different controlling devices and configurations can be found in existing systems. Such systems have been found already in many places for a wide variety of applications.

4. Home Control

In the actual era of smart homes and smart grids, advanced technological systems that allow the control of domestic tasks are developing rapidly. There are numerous technologies and applications that can be installed in smart homes today. They enable communication between home appliances and users, and remote control capabilities. Home Control is as an application of the smart home that is able to automate or assist the users. And provide to them high comfort levels, improve security, facilitate energy management, reduce environmental emissions and save energy. Furthermore, the users can control home electricity usage through a direct and bidirectional communication with home appliances (Lobaccaro et. al., 2016).

5. Evolution of Electricity Meter

In recent years, domestic and industrial users have shifted from traditional meters to smart meters. Electromechanical meters were a dominant part of electricity measurement before 1970. They could only measure the electrical energy. However, it had been identified that the requirement of a meter which could communicate and measure the electrical energy along with other electrical parameters was necessary. Therefore, solid state electronic meters were introduced to measure the overall electrical parameters.

Eventually, automatic meter reading was added to electronic meters and it was a great achievement since it could send the data in near time. However, it could only provide the one-way communication. This limitation was overcome by the introduction of smart meters which can provide two-way communication. Smart meters can measure all the electrical parameters like electronic meters and communicate data in a meaningful way. The consumer is updated with electricity usage, cost, tariffs and other notifications sent by the utility. Smart meters have different functionality to manage the end user loads and run them in an optimal way to reduce the electricity bill as well as to conserve the energy (Weranga et.al, 2014).

6. Smart Meter

Smart meters are different from electronic meters because of their additional functionalities and features. Apart from electricity measurements and automatic meter reading (AMR), they allow two-way communication between the meter and the base station. Load profiling, prepayment, remote disconnection and reconnection, power outage notification, tamper detection, and multi-tariffing are also possible with smart meters. Therefore AMR system is introduced by combining the communication infrastructure to electronic meters. Meanwhile more features and functions are added to AMR system. Ultimately Advanced Metering Infrastructure has been developed to today's technology with two-way communication and data management system (Saoud and Recioui, 2017)). Fig. 1 shows the evolution of smart meter technology from electronic meters to AMI (Weranga et.al, 2014).

Figure 1. Evolution of the Smart meter technology

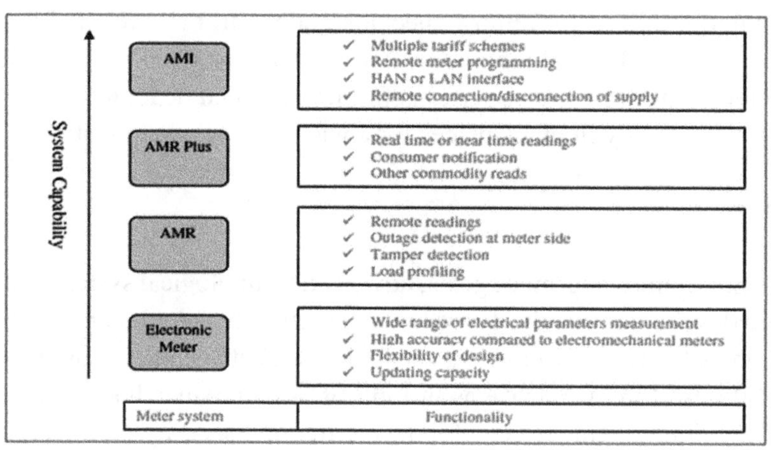

7. The Hardware Structure of a Smart Meter

Figure 2 shows the functional bock diagram of a smart meter. It includes signal acquisition, signal conditioning, Analogue to Digital Conversation (ADC), computation and communication. Smart meters use voltage and current sensors to get the input signals. Signal conditioning, ADC, and computations are done inside the micro controller unit (MCU). Additional hardware components are required for other operations like communication, time and date measurements, and data backup and storage. A smart meter is typically composed of following hardware components:

- Voltage and current sensing unit
- Power supply
- Energy measurement unit (metering IC)
- Microcontroller
- Real time cock
- Communicating system (Weranga et.al, 2014).

Figure 2. The functional bock diagram of a smart meter

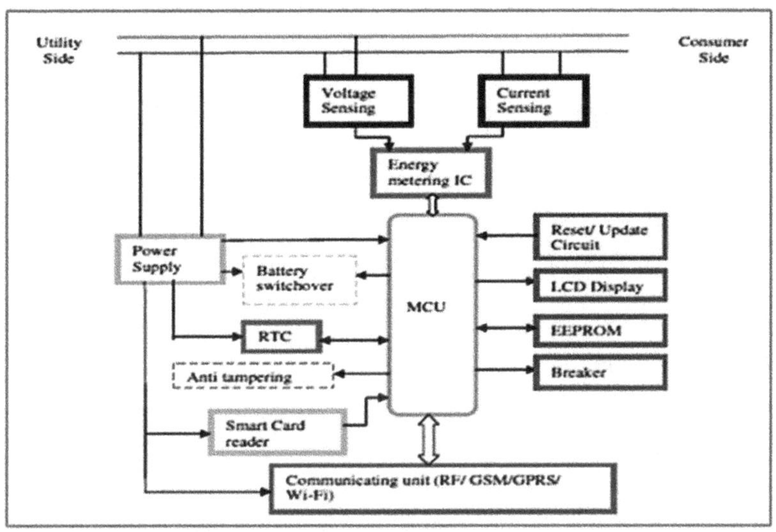

TWO-WAY POWER MEASUREMENT

Smart meters are an advanced form of meters. They differ from electronic meters because of added functionalities. Apart from electricity measurements and automatic meter reading (AMR), they allow two-way communication between the meter and the utility.

Smart Metering System Simulation Using LABVIEW

1. Smart Meter Design

The presentation of the smart meter design is divided into two sections:

- Customer energy consumption
- Photovoltaic panels energy production.

Fig. 3 shows the front panel of a typical house.

Fig. 4 shows the total block diagram of one house where both energy consumption and production are included. This will be illustrated in details below.

Figure 3. A typical house measurement front panel.

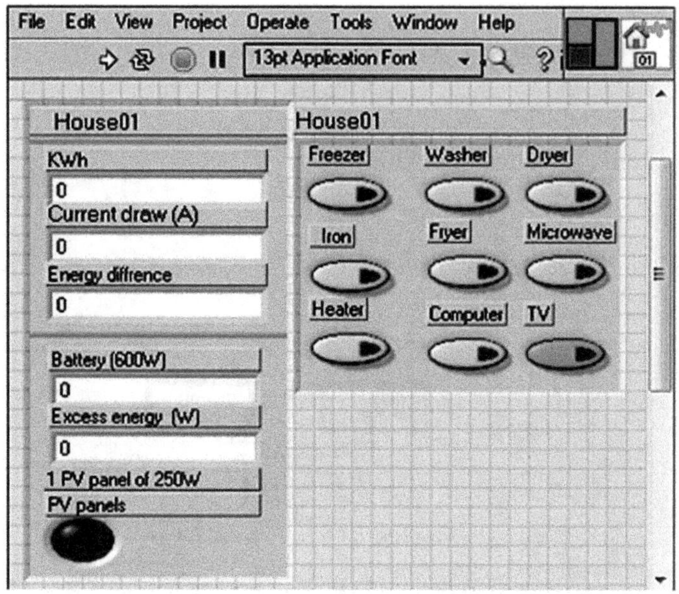

Figure 4. Typical house measurement block diagram.

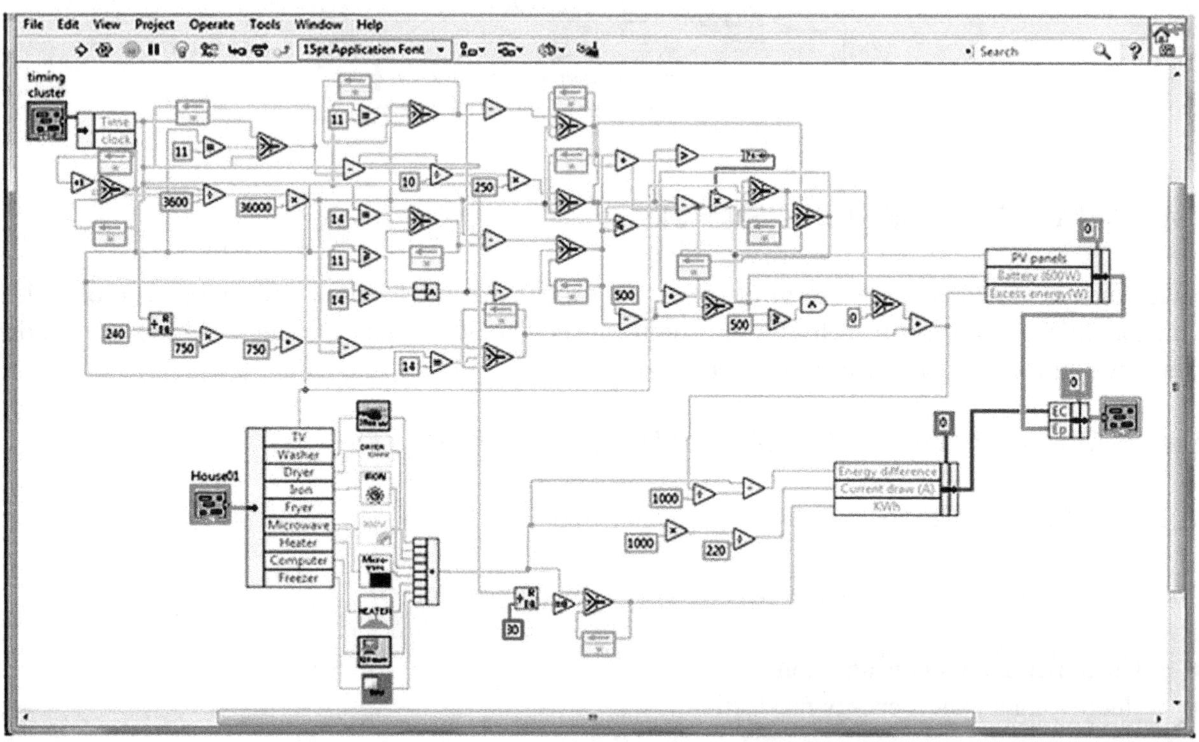

2. The Biderectional (Two-Way Flow Of Electricity) meter reading

- Measuring Customer Energy consumption:
    - Block diagram A: represents 8 house appliances, each with an on/off button where all are grouped in a cluster, the Kwh consumed by the 8 appliances are added in order to calculate the totale Kwh consumed of the house. An example of microwave subVI is shown in Fig. 5.
    - Block diagram B: A cluster is used to represent all smart meter output: current draw, total Kwh with an update every 30 sec, energy difference between Kwh and excess energy.
    - Block diagram C: represents subVI of the microwave which is a function that calculate the Kwh consumed by counting the ON time of the microwave accumulatively, a conversion into hour is done then a multiplication with wattage of the microwave, and dividing by 1000 to get energy consumed in kilowatt. All of the other house appliances are of this same subVI but with the right wattage.

Figure 5. Energy consumption Block diagram and Microwave sub-VI.

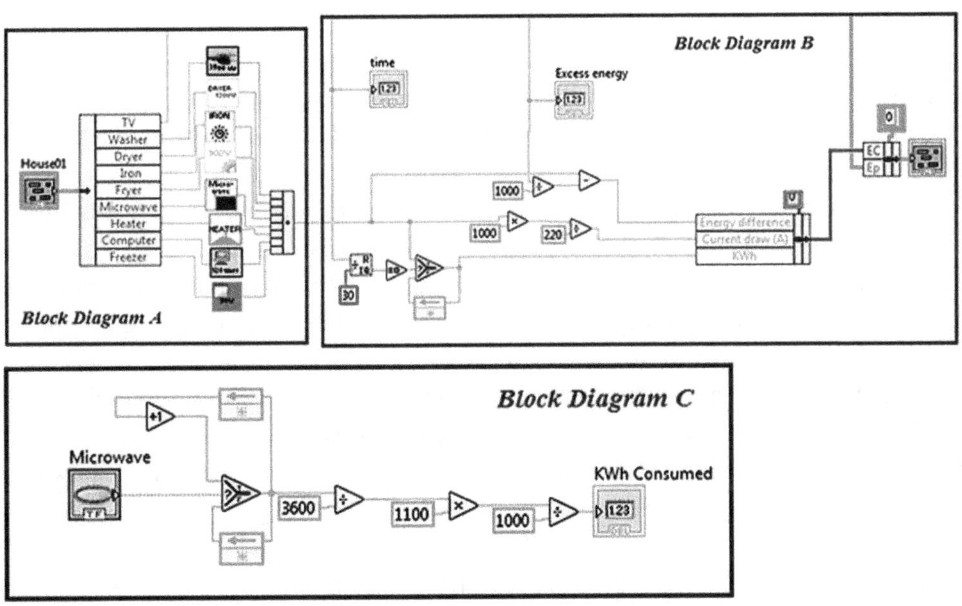

- Measuring photovoltaic panels Energy production:
    - Photovoltaic solar panel (PV Panels): Photovoltaic (PV) technologies are an expanding market segment in modern power markets. Photovoltaic use large areas of photovoltaic cells, known as PV or solar cells, to directly convert sunlight into usable electricity. These cells are usually made from silicon alloys, solar panels energy depend on the size of the panel, the efficiency at which the panel converts sunlight to electricity, the amount of sunlight the panel receives, the orientation of the panel relative to the sun, and a host of other variables.

3. Grid Fall Back

With a grid fallback system, the solar array generates power, which in turn charges a battery bank. Energy is taken from the battery and run through an inverter to power one or more circuits from the distribution panel in the house.

When the batteries run flat, the system automatically switches back to the grid power supply. The solar array then recharges the batteries and the system switches back to solar power. With a grid fallback system, you do not sell electricity back to the electricity companies, until your batteries have been charged first. Once your batteries have been fully charged, any excess power generated will be exported to the grid.

Grid fallback systems provide most of the benefits of a grid interactive system, with the additional benefit that you use your own power when you need it, rather than when the sun is shining. This reduces your reliance on external electricity supplies during peak load periods, which ensures that your system has an overall environmental benefit. Solar panels rarely power electrical equipment directly. This is because the amount of power the solar array collects varies depending on the strength of the sunlight (irradiance). This makes the power source too variable for most electrical equipment to cope with (Boxwell, 2017).

4. Measuring the Production of the PV Panels

The requirements for solar electricity in the ten houses that are to run a few lights, or some relatively low-power electrical equipment such as a laptop computer, a small TV, a compact fridge or other appliances with small wattages.

The solar panels generate electricity which is stored in battery bank. The PV module charges the battery bank during the 3hours of full sun starting at 11AM, where systems are designed to always have enough power to charge the battery bank. No excess energy will be fed into the grid unless the batteries are fully charged and the home appliance connected to the grid is turned off. When PV modules are off, the appliance will draw energy from the battery bank.

The battery banks installed are assumed to be enough for energy consumption when PV panels are off and the PV module is 250 watt under standard test conditions.

The Block Diagram D in fig. 6 represents a complex block diagram for energy production system where it performs the following functions:

- Turns on a LED when the PV panels start producing energy at 11:00 till 14.00.
- When PV panels goes ON the batteries get charged immediately, after charging any excess of energy will be sent to the grid if the appliance connected to the PV module is off.

5. The Real Time Simulation

- Load profile: The term load profile describes the pattern of electricity usage for a customer or a group of customer over a given period (Lo et. al., 2005).

A load profile will vary according to customer type (typical examples include residential, commercial and industrial), temperature and holiday seasons. It enhances our knowledge of electrical consumption patterns. Besides, it has many applications including demand response, load forecasting, and non-

Figure 6. Energy production block diagram

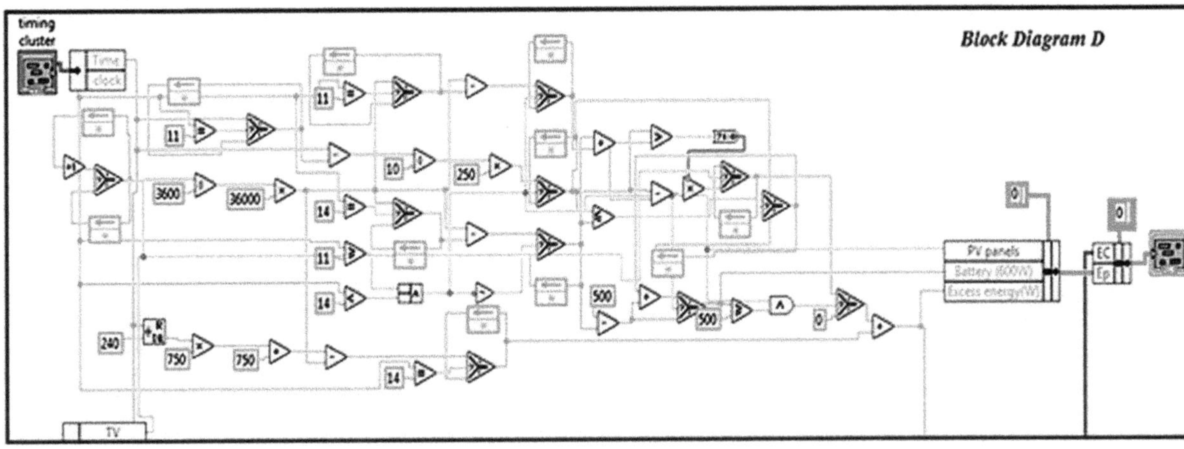

technical loss detecting, etc. With widespread use of AMI equipment, a lot of information can be mined from electrical consumption data of individual customers through load profiling (Wang et.al., 2015).

Using waveform chart, the load profile of the ten houses is drawn for duration of 3days as shown in fig. 7. Fig. 7 also represents the cumulative kWh consumed by the ten houses. Fig. 8 represents the instantaneous kWh consumed by the ten houses .the energy provided by the utility (10kw has been taken as an example) and the limit of energy provided from the utility plus the Solar Panels (12kw). If the limit of 12Kw is exceeded some loads would be turned off automatically this is known as direct load control as will be seen in the next section.

For instance, taking only energy consumption, the customers will benefit from metering solutions through greater understanding of their own energy consumption, allowing them to better manage costs of their usage. It would also allow them to understand which energy plan may suit their consumption more effectively. Suppliers benefit by meeting energy demand, improving efficiencies in load management and smoothing out demand by offering customers more tailored tariff plans.

In general, by load profiling everyone can benefit by better understanding and managing energy demand. Additionally, a potential value added maybe also achieved by forecasting future demand to reduce the strain on the system during heavy usage periods.

- Direct Load control: Load control system can be operated by an electricity supplier or network operator, a market or system operator, a demand side response service provider or end user. Special tariff rates such as real time pricing, day ahead pricing, time of use pricing, and critical peak pricing should be introduced with load control, aimed at compensating the reduced autonomy of the consumer. This allows the operator the right to control load demand. Where modern smart metering systems support information display devices, appliance and equipment controllers, and integrated load control systems. Smart appliances individually measure the energy consumption of each appliance at the consumer end. The data from these devices are collected at the control centre in near real time. The control signals are sent back to the appliance controllers from the control centre. Therefore two-way communication link should be established to communicate between the load manager and the controlled loads.

Figure 7. Load Profile (Cumulative)

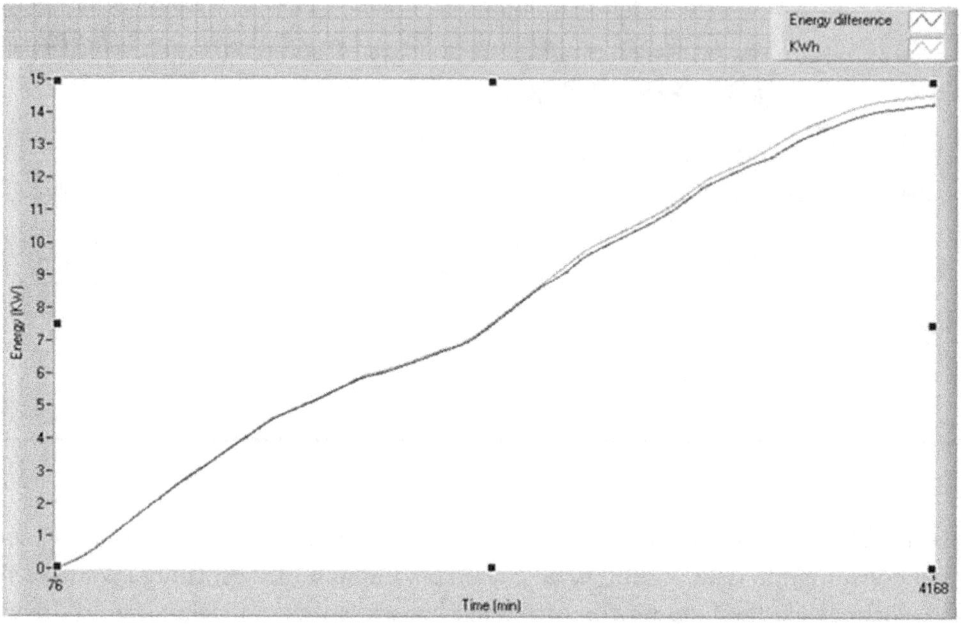

Figure 8. Load Profile (instantaneous)

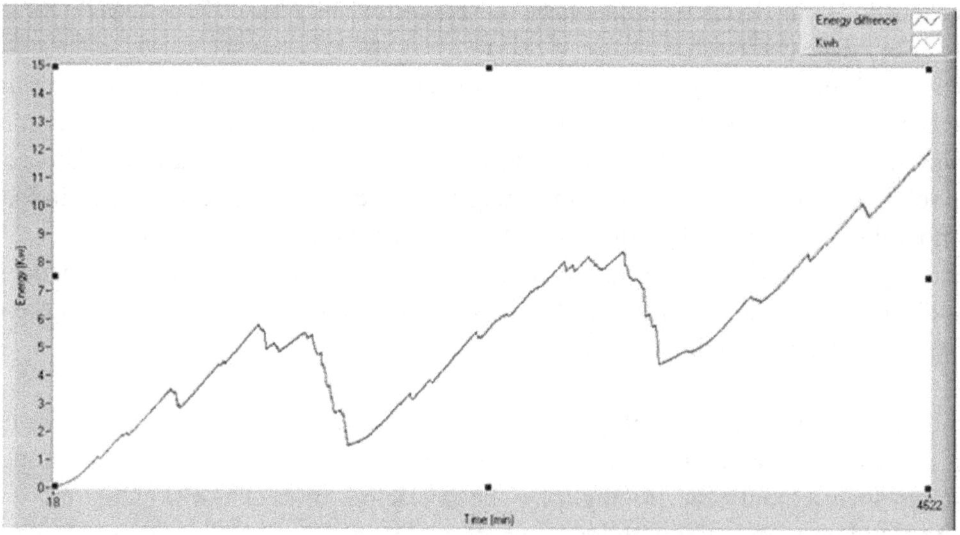

At the operating centre, the load manager sends signals to initiate or terminate a load control program. The signal may also be generated automatically when the demand exceeds the pre-set load levels. Instructions are given to reduce the demand to pre-set levels, cycle the loads on and off in regular intervals, or to switch off the load completely (Weranga et.al, 2014).

In our work we proposed a LABVIEW program where the consumer will be notified by the excess of energy that he is using (practically, this could be done by sending him a text message to his mobile phone) and ask him to turn off the loads that consumes a big amount of electricity such as an air conditioner after having them turned off automatically by the utility for a short period of time.

Fig. 9 shows a dialog box that will pump out to alert the user and stop the execution of the program until clicking on the Ok button and turning off the loads.

Fig. 10 shows the Block diagram of the dialog box, where it is connected to the microwave and iron.

Fig. 11 shows the front panel of the dialog box with the limit of 15 KWh.

Figure 9. Window of the Dialog Box

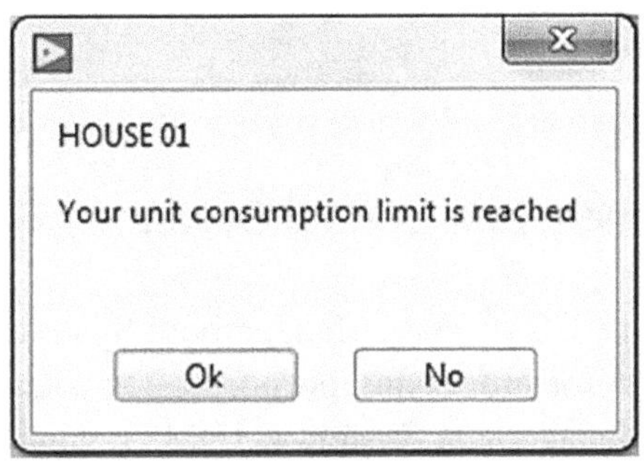

Figure 10. Dialog box block diagram

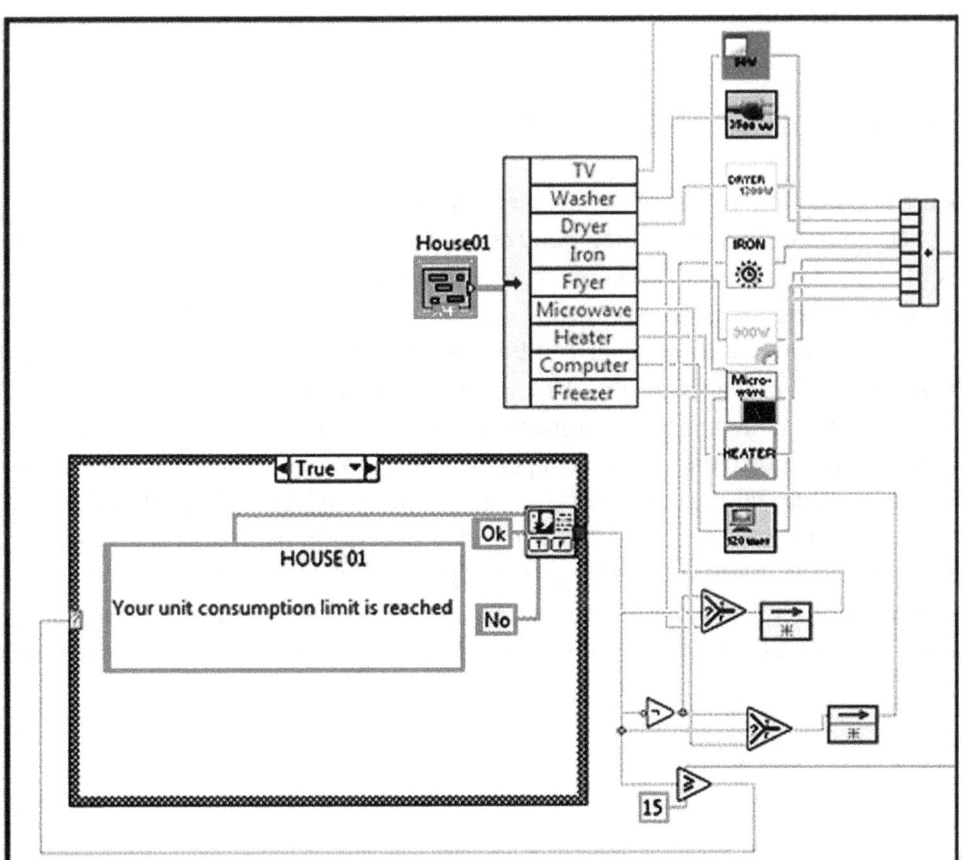

Figure 11. Dialog Box front panel (case of 15 kWh)

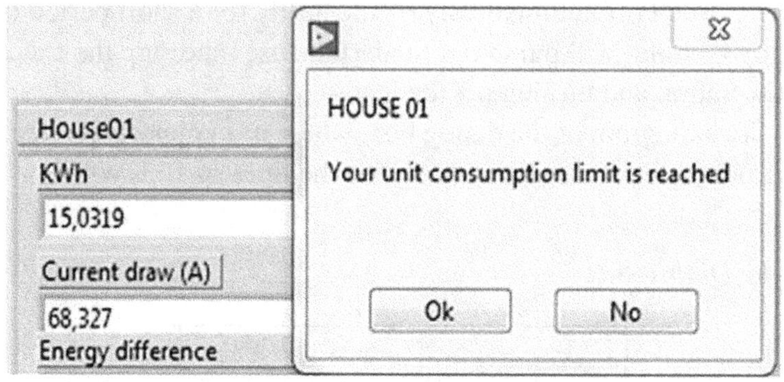

6. The Centralised Management and Control System

- The Control Center (CC): The CC or Data Management System (DMS) is in charge of receiving and storing the metering data for processing purposes. The CC can be seen as a modular system formed by the Meter Data Management System (MDMS), which manages the metering data and additional secondary modules in charge of end-users applications, weather forecasting systems, geographical information systems, control applications, and load management. Fig. 12 represents the control center front panel where the data and parameters of ten houses are displayed and Fig. 13 is its block diagram.

BILL ESTABLISHEMENT AND PRICING POLICY

This part focuses on the importance of the net metering and adopting correct power pricing policies to maximize the net economic benefits of electricity consumption to society. An example of how net metering works, is illustrated by the simulation of Three scenarios using LABVIEW software that shows the different cases of homes energy consumption that may occur and its associated billing system description. At the end of a given month, the customer is billed only for the net electricity used. If more energy is produced than consumed, producers receive benefit for this negative balance, such as, renewable energy credits (REC), which is credited on the customer's account toward the next billing cycle. If at the end of the year a surplus remains, then the customer depending on the utility policy may (a) be paid for the total REC collected at avoidance cost rate or retail cost rate, (b) the total REC collected can be transferred and could be used as a compensation for a possible negative balance in the following years, or (c) the total REC collected are granted back to the utility. (Poullikkas et. al., 2013). This section goes into details about the different form of pricing policies that exists since they are a valuable "soft" tool for managing demand and reducing electricity consumption and peak demand (Munasinghe and Warfor, 1982).

Figure 12. Control Center front panel

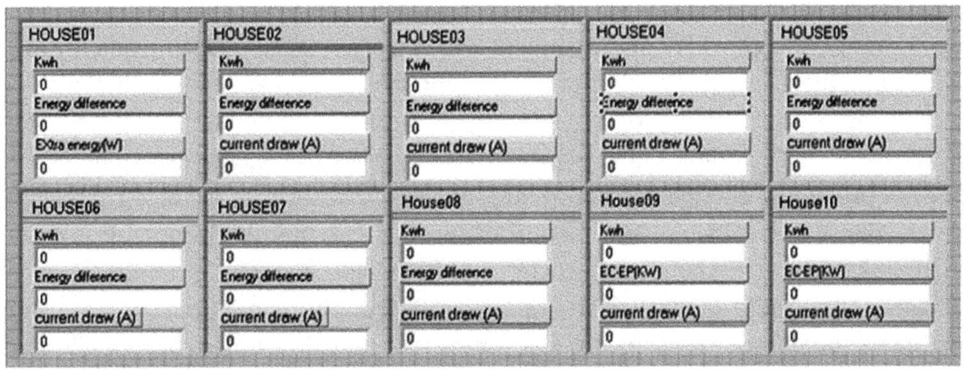

Figure 13. Control center block Diagram

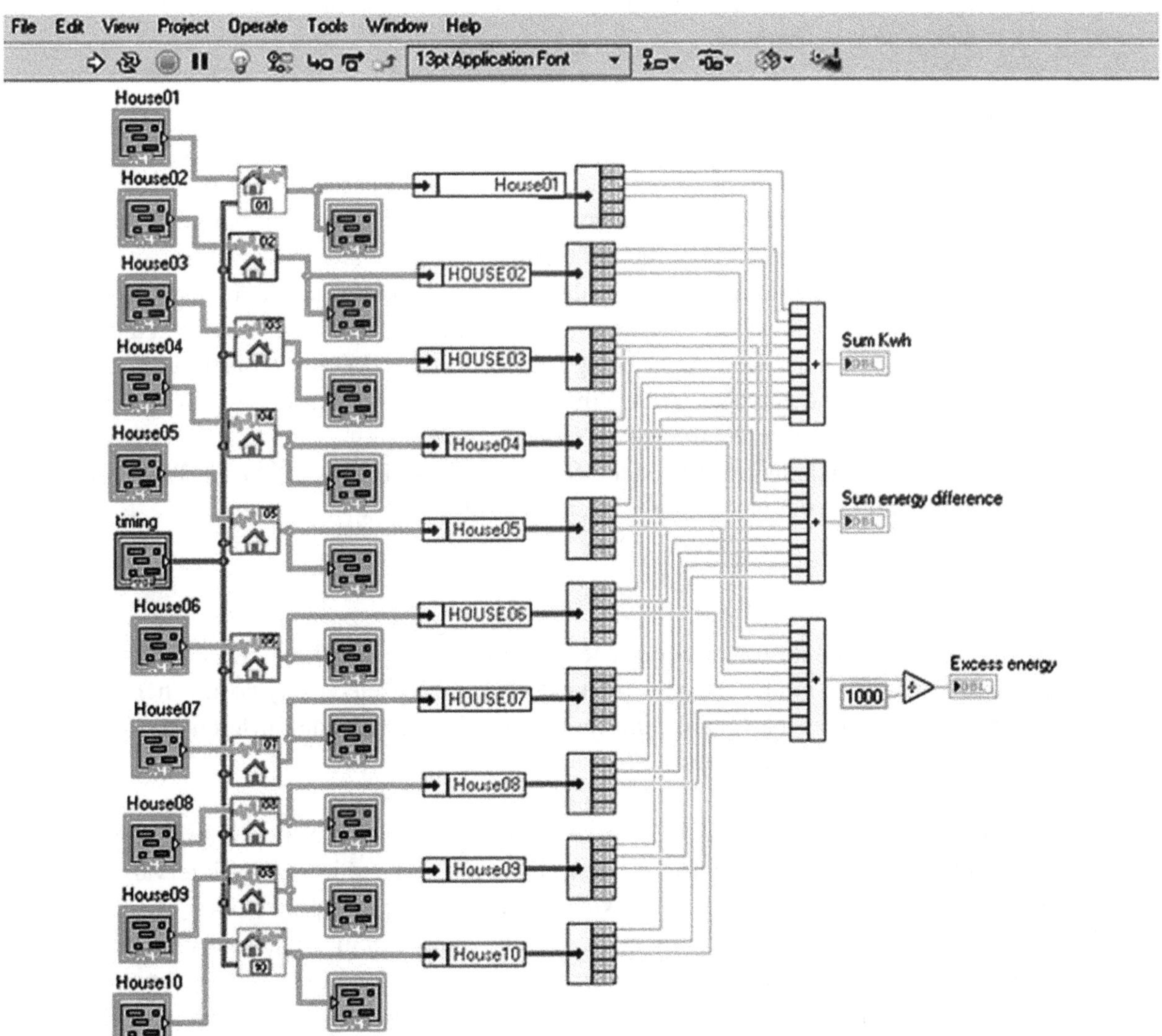

1. Net Metering

Net metering is a new concept where an instrument which has a special metering and billing agreement between utilities and their customers, facilitates the connection of small, renewable energy-generating systems to the power grid. This new program is being developed to encourage small scale renewable energy systems to ensure that customers always have a reliable source of energy even when their renewable generators are not producing energy, and to provide substantial benefits to the electric power-generating system as well as the environment. When a net metering client's renewable generator is producing more power than is being consumed, the electric meter runs backward generating credits. Whenever the net metering customer uses more power than is being produced, the meter runs forward normally. Net metering customers are charged only for the net power that they consume from the electricity service provider that has accumulated over a specific period. In other words, if their renewable energy generating systems make more electricity than is consumed, they may be credited or paid for the excess electricity contributed to the grid over that same period. Net metering is also a way to increase the energy in the power grid to keep up with increase in demand during peak power use times, and this is of particular interest to states facing power shortages.

2. Final Billing

The power produced by the solar system is consumed by the house, and the surplus power is sent to the Grid. We have designed an Energy Billing Statement with details included for explaining how the statement is calculated. Three examples on how the statement may appear are shown below; the first is an example of a normal solar billing statement (Illustration A). The second is an example of a solar billing statement for someone whose solar system produced more energy than they received from the electrical grid, resulting in an Excess Generation credit being applied to their account (Illustration B). The last one is very similar to the second one but this time the consumer has been absent for a long period, thus resulting in a considerable excess generation credit (Illustration C).

Fig. 14 represents the simulation results of the first scenario in House 01. From this figure, some useful information to calculate the bill can be extracted as follows:

3. Pricing Policy

The Distribution companies are starting to adopt a new pricing strategy also known as Load management and demand control in order to reduce the peak demand, and save the electricity cost. When it comes to electricity savings the consumers can change the consumption patterns under different tariff schemes.

Modern smart metering systems support following tariff schemes

- Real time pricing (RTP): Price signals are generated in hourly basis or half hourly basis, in most cases reflecting the true cost of electricity. Consumers are updated in regular intervals (hourly of half hourly) by sending price signals to the smart meters or home display units.
- Time of use pricing (TOU): It utilizes time-varying electricity pricing signals to stimulate customers to shift their power consumption from peak hours to off-peak hours, which eventually reduces the peak load of the grid. By setting high electricity prices in peak-demand hours and low prices in off-peak-demand hours, the user is encouraged to shift the use of some appliances such as

washing machines to low price hours, to save the household's electricity cost. The most widely adopted time-varying electricity tariff is Time-of-Use (TOU) which is widely employed in many countries. In TOU, one day is divided into multiple periods and different electricity prices are applied in each period.

- Critical peak pricing: The tariff is same as the TOU but higher rates are applied at higher demands. Prior to applying the higher rates, the consumers are notified.
- Day ahead pricing: This is similar to RTP, but the price schedule is sent to the consumer 24h before the rates are applied. Generally, different prices are defined for different time locations (on hourly basis most of the time) of the next day.

Figure 14. Scenario 01

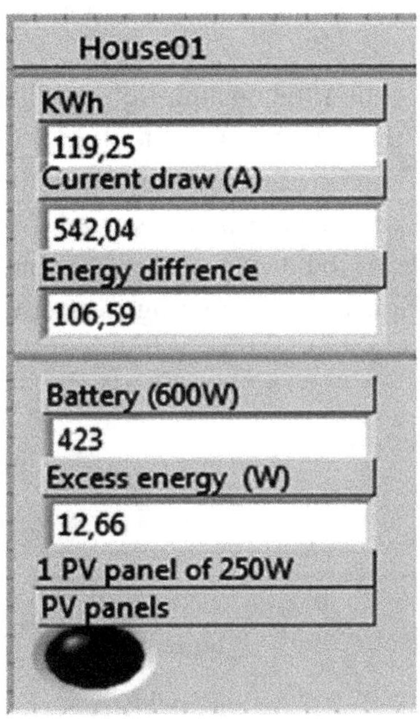

Figure 15. Scenario 02

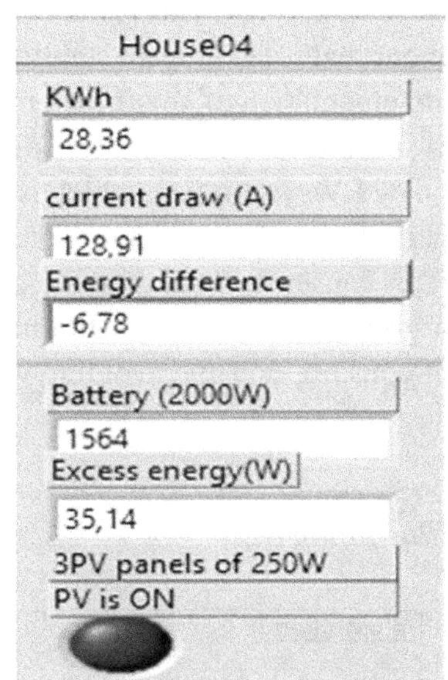

CONCLUSION

Electric power is a crucial form of energy in the world today. A well maintained electric network is required to provide accuracy, reliability and sufficiency to consumers. The current electric infrastructure is aging and it is being pushed to do more than it was originally designed to do. Modernizing the grid to make it "smarter" and more resilient through the use of technologies that communicate and work together to deliver electricity efficiently and allow the consumers to better manage their own energy consumption and costs because they have easier access to their data.

"Smart grid" technologies are made possible by two-way communication technologies, control systems, and computer processing, and the integration of Smart Metering. Smart metering generally concerns the usage of a *smart* meter at customers place and the regular process of reading, processing and giving the feedback of consumption data to the customer. These *smart* meters have been called smart because of the fact that they included at least one microprocessor. The future of smart metering solutions with the strong involvement of renewable energies seems to be bright as they are increasingly perceived as essential elements of more environmental friendly energy systems.

Consequently, the present work aimed to provide a well defined view of the internal architecture of the smart meter, its working which includes the bidirectional measurement of energy as well as the way it communicates and send data at the required timing to the control centre, at the end we tried to design a typical bill statement which explains how the net billing is done and discussed different pricing policies that the utility may adopt to reduce the peak hours demand; the work is modeled and simulated with the graphical programming language LABVIEW software.

Due to the feedback and communication that Smart Meters provides, they are able to offer possibilities to introduce demand response approach which is important on electricity market dealing with peak loads, where this work aimed at generating Synthesized Load Profiles for the 10 houses modeled with smart meters. Therefore, smart meters may become a gateway into the home of the customer, and provide the data about real usage to gain stability in the power system.

The implementation of smart meters in the power grid allows solving consumption predictions and the lack of mechanisms to make consumers as an active part of the network. Grid fallback system is a relatively new PV system type which is very advantageous for low-power residential PV systems.

Electricity cannot be stored economically in large quantities, and has to be consumed instantly on demand. This causes the load duration curve for most utility systems to be very peaky as a consequence it led to a growing awareness of the importance of pricing policies in maximizing the net economic benefits of consumption and avoiding waste.

Table 1. Illustration A

Illustration A				
Customer Number:	**##########**			**##########**
Previous Bill				DA 00.00
Meter Reading Details				
ElectricMeter	***Previous***	***Current***	***Consumption***	
Use (kWh)	000	119.25	119.25	
Generation (kWh)	000	12.66	12.66	
Total Net Billing Consumption			**106.59**	
Type of charge	***How we calculate this charge***		***Amount***	
Delivery Tax				DA020.25
Energy Charge	106.59 kWh x DA4.42			DA 471.13
NEt Charge	OkWh x DA15.94			DA 491.38
				DA000.00
Total Electric Bill				**DA491.38**
Your next scheduled meter read date is July 23, 2019				
Billing Period May 24, 2019 - June 23, 2019				

Table 2. Illustration B

Illustration B					
Customer Number:	**##########**				**##########**
Previous Bill					DA 00.00
Meter Read Detail					
ElectricMeter	***Previous***	***Current***	***Consumption***		
Use (kWh)	000	28.36	28.36		
Generation (kWh)	000	35.14	35.14		
Total Net Billing Consumption			**-6.78**		
Type of charge	***How we calculate this charge***			***Amount***	
Delivery Tax					DA 020.25
Energy Charge	00.00 kWh x DA4.42				DA000.00
NEt Charge	6.78kWh x DA15.94				DA020.25
					DA108.07
Total Electric Bill					**DA -87.82**
Your next scheduled meter read date is July 23, 2019					
Billing Period May 24, 2019 - June 23, 2019					

Table 3. Illustration C

Illustration C					
Customer Number:	**##########**				**##########**
Previous Bill					DA 00.00
Meter Read Detail					
ElectricMeter	***Previous***	***Current***	***Consumption***		
Use (kWh)	000	10.32	10.32		
Generation (kWh)	000	67.50 67.50			
Total Net Billing Consumption			**-57.18**		
Type of charge	***How we calculate this charge***			***Amount***	
Delivery Tax					DA 020.25
Energy Charge	00.00 kWh x DA4.42				DA000.00
NEt Charge	57.18 kWh x DA15.94				DA020.25
					DA911.45
Total Electric Bill					**DA-891.20**
Your next scheduled meter read date is July 23, 2019					
Billing Period May 24, 2019 - June 23, 2019					

Figure 16. Scenario 03

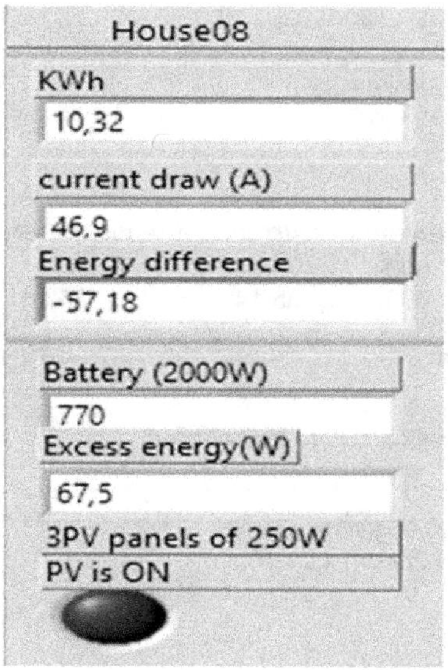

In this chapter, the crucial significance of pricing policies as well as Net metering have been discussed. The latter allows an owner of rooftop solar panels to participate in the electric grid as a distributed energy producer. Solar panels first supply power to the home, and at times when the solar panel produces more power than the home is using the power is sent back onto the grid to be used by other customers.

REFERENCES

Boxwell, M. (2017). *Solar electricity Handbook*. Academic Press.

Cappers, P., MacDonald, J., Page, J., Potter, J., & Stewart, E. (2016). *Future opportunities and challenges with using demand response as a resource in distribution system operation and planning activities.* Lawrence Berkeley Nat. Lab., Berkeley, CA, USA, Tech. Rep. LBNL-1003951.

Chafai, M., Bentarzi, H., Ouadi, A., Zitouni, A., & Recioui, A. (2013). *PC based testing system for protective relay*. 4th International Conference on Power Engineering, Energy and Electrical Drives, POWENG"2013, Istanbul, Turkey.

Chellali, F. B. M., Recioui, A., Yaiche, M. R., & Bentarzi, H. (2014). A hybrid wind/solar/diesel stand alone system optimisation for remote areas in Algeria. *International Journal of Renewable Energy Technology*, *5*(1), 12–24. doi:10.1504/IJRET.2014.059658

Graditi, G., Ippolito, M. G., Telaretti, E., & Zizzo, G. (2016). Technical and economical assessment of distributed electrochemical storages for load shifting applications: An Italian case study. *Renewable & Sustainable Energy Reviews*, *57*, 515–523. doi:10.1016/j.rser.2015.12.195

Gurlin, S. L. (2011). Smart Grid and its Development Prospects In the Asia-Pacific Region. *Journal of Emerging Trends in Computing and Information Sciences.*

Hossain, M., Amanullah, M. T. O., & Shawkat Ali, A. B. M. (2013). Smart Grid: Opportunities, developments and trends. Springer-Verlag.

Hu, Q., & Li, F. (2013). Hardware design of smart home energy management system with dynamic price response. *IEEE Transactions on Smart Grid, 4*(4), 1878–1887. doi:10.1109/TSG.2013.2258181

Hubert, T., & Grijalva, S. (2012). Modeling for residential electricity optimization in dynamic pricing environments. *IEEE Transactions on Smart Grid, 3*(4), 2224–2231. doi:10.1109/TSG.2012.2220385

Jazayeri, P., Schellenberg, A., Rosehart, W. D., Doudna, J., Widergren, S., Lawrence, D., Mickey, J., & Jones, S. (2005). A survey of load control programs for price and system stability. *IEEE Transactions on Power Systems, 20*(3), 1504–1509. doi:10.1109/TPWRS.2005.852147

Katrel S. R., & Rojatkar D. V. (2017). Home Automation: Past, Present And Future. *International Research Journal of Engineering and Technology, 4*(10).

Koponen, P. (2008). *Definition of Smart Metering and Applications and Identification of Benefits Project: European Smart Metering Alliance*. VTT Technical Research Centre of Finland.

Lo, K. L., Zakaria, Z., & Sohod, M. H. (2005). Determination of Consumers' Load Profiles based on Two-stage Fuzzy C-Means, Power System Research Group, Department of Electronics & Engineering. *Proceedings of the 5th WSEAS Int. Conf. on Power Systems and Electromagnetic Compatibility*, 212-217.

Lobaccaro, G., Carlucci, S., & Löfström, E. (2016). A Review of Systems and Technologies for Smart Homes and Smart Grids. *Energies, 9*(5), 348. doi:10.3390/en9050348

Mahmood, D., Javaid, N., Alrajeh, N., Khan, Z., Qasim, U., Ahmed, I., & Ilahi, M. (2016, March 15). Realistic scheduling mechanism for smart homes. *Energies*, *9*(3), 202. doi:10.3390/en9030202

Massoud Amin, S., & Wollenberg, B. F. (2005). Towards a SmartGrid: Power delivery for the 21st Century. *IEEE Power & Energy Magazine, 3*(5), 34–41. doi:10.1109/MPAE.2005.1507024

Medina, J., Muller, N., & Roytelman, I. (2016). Demand response and distribution grid operations: Opportunities and challenges. IEEE Trans. Smart Grid, 1(2), 193–198.

Munasinghe, M., & Warfor, J. J. (1982). Electricity Pricing Theory and Case Studies. Academic Press.

Noelia, U. P. (2016). State of the Art and Trends Review of Smart Metering in Electricity Grids. *Applied Sciences, MDPI, 6*(68). Advance online publication. doi:10.3390/app6030068

Ouadi, A., Bentarzi, H., & Recioui, A. (2013). Optimal multiobjective design of digital filters using spiral optimization technique. *SpringerPlus, 2*(1), 461. doi:10.1186/2193-1801-2-461 PMID:24083108

Ozturk, Y., Senthilkumar, D., Kumar, S., & Lee, G. (2013). An intelligent home energy management system to improve demand response. *IEEE Transactions on Smart Grid, 4*(2), 694–701. doi:10.1109/TSG.2012.2235088

Palaniappan, S. (2015). Home Automation Systems - A Study. *International Journal of Computer Applications, 116*(11).

Poullikkas, A., Kourtis, G., & Hadjipaschalis, I. (2013). *A review of net metering mechanism for electricity renewable energy sources*. Electricity Authority of Cyprus.

Qdr, Q. (2006). *Beneðts of demand response in electricity markets and recommendations for achieving them. United States Dept. Energy, Washington*. Tech. Rep.

Rahimi, F., & Ipakchi, A. (2010). Demand response as a market resource under the smart grid paradigm. *IEEE Transactions on Smart Grid*, *1*(1), 82–88. doi:10.1109/TSG.2010.2045906

Rasheed, M. B., Javaid, N., Ahmad, A., Awais, M., Khan, Z. A., Qasim, U., & Alrajeh, N. (2016). Priority and delay constrained demand side management in real-time price environment with renewable energy source. *International Journal of Energy Research*, *40*(14), 2002–2021. doi:10.1002/er.3588

Rasheed, M. B., Javaid, N., Ahmad, A., Khan, Z. A., Qasim, U., & Alrajeh, N. (2015). An efðcient power scheduling scheme for residential load management in smart homes. *Applied Sciences (Basel, Switzerland)*, *5*(4), 1134–1163. doi:10.3390/app5041134

Recioui, A., & Dassa, K. (2017, June). Design of Standalone Micro-Grid Systems Using Teaching Learning Based Optimization. *Algerian Journal of Signals and Systems*, *2*(2), 75–85.

Saoud, A., & Recioui, A. (2017, September). A review on Data communication in smart grids. *Algerian Journal of Signals and Systems*, *2*(3), 162–179.

TrongNghia, L., WenLong, C., DangKhoa, T., & Tran, H. N. (2016). *Advanced Metering Infrastructure Based on Smart Meters in Smart Grid*. Intech Open.

Wang, Y., Chen, Q., Kang, C., Zhang, M., Wang, K., & Zhao, Y. (2015). Load Profiling and Its Application to Demand Response: A Review. *Science and Technology*, *20*(2), 117-129.

Weranga, K. S. K., Kumarawadu, S., & Chandima, D. P. (2014). *Smart Metering Design and Applications*. Springer Singapore. doi:10.1007/978-981-4451-82-6

Yalcintas, M., Hagen, W. T., & Kaya, A. (2015). An analysis of load reduction and load shifting techniques in commercial and industrial buildings under dynamic electricity pricing schedules. *Energy and Building*, *88*, 15–24. doi:10.1016/j.enbuild.2014.11.069

KEY TERMS AND DEFINITIONS

Demand Response: Peak demand on an electrical grid is simply the highest electrical power demand that has occurred over a specified time period. Peak demand is typically characterized as annual, daily or seasonal and has the unit of power. Peak demand, peak load or on-peak are terms used in energy demand management describing a period in which electrical power is expected to be provided for a sustained period at a significantly higher than average supply level.

Load Control: Load control, also known as demand-side management (DSM), is the process of balancing the supply of electricity on the network with the electrical load by adjusting or controlling the load rather than the power station output. This can be achieved by direct intervention of the utility in real time, by the use of frequency sensitive relays triggering the circuit breakers (ripple control), by time clocks, or by using special tariffs to influence consumer behavior.

Smart Meters: A smart meter is an electronic device that records consumption of electric energy and communicates the information to the electricity supplier for monitoring and billing. Smart meters typically record energy hourly or more frequently, and report at least daily.[7] Smart meters enable two-way communication between the meter and the central system. Such an advanced metering infrastructure (AMI) differs from automatic meter reading (AMR) in that it enables two-way communication between the meter and the supplier.

Chapter 4
PMU Phasor Estimation Using Different Techniques

Hamid Bentarzi
Laboratory Signals and Systems, IGEE, University M'hamed Bougara of Boumerdes, Algeria

Abderrahmane Ouadi
Laboratory Signals and Systems, University M'hamed Bougara of Boumerdes, Algeria

ABSTRACT

Many models of phasor measurement units (PMU) have been implemented; however, few dynamic models have been developed when the power system parameters change. It is necessary to use a method that can somehow estimate the frequency and correct the phasors. The conventional way to determine frequency is to detect zero crossings per unit time. However, this method has many drawbacks such as high cost and low accuracy. Also, after the frequency determination, the phasor should be corrected by suitably modifying the algorithm without omitting any data. This chapter presents different estimation techniques such as discrete Fourier transform (DFT), smart discrete Fourier transform (SDFT) that may be used to estimate the phasors. These estimated values would be incorrect if the input signals are at an off-nominal frequency and the phase angles would drift away from the true values. To correct this issue, first of all, the off-nominal frequency has been estimated using different techniques such as least error squares and phasor measurement angle changing, and then it is used to correct the phasors.

INTRODUCTION

Power systems are the largest manmade dynamic system. They are highly nonlinear and involve complicated electromagnetic and electromechanical phenomena. To operate the system reliably and economically, system conditions may be monitored using voltages and currents phasors. Conventionally, static state estimation has been used to estimate the steady-state power and voltage magnitudes. However, for dynamic applications, measurements are time-skewed without a time reference.

DOI: 10.4018/978-1-7998-4027-5.ch004

Digital computer based measurements; protection and control have become common features of electric power substations. These systems use sampled data to compute various quantities such as voltage and current phasors.

The advent of satellite-based time-keeping system, such as high accurate atomic clocks used in global position system (GPS), with lowest cost GPS receiver providing high precision timing sources, measurements are made with a common time reference, usually GPS clock. The advances of computer, synchronization and communication technologies have made synchronized phasor measurement in power system applications such as protective relay and phasor measurement unit (PMU) within 1μs. Even the measurements accomplished at various sites, are synchronized to each other using the same common time base. Synchronized phasors measurement set provides a vastly improved method for tracking power system dynamic phenomena. They facilitate a number of wide area applications including wide area measurement systems. (WAMS), a power system protection, monitoring and control applications such as wide-area measurement protection and control system (WAMPAC) utilize these measurements.

Stand alone phasor measurement units (PMU) have been used on critical systems, for providing synchronized phasor measurement data. They are only in developed countries because they are relatively expensive, this has conveyed some manufacturers to design modern protective relay includes phasor measurement function and capabilities. The addition of synchrophasor measurement in a protective relay results in increased power system reliability and provides easier disturbance analysis, protection, and control capabilities.

LITERATURE REVIEW

In the power system phasor, amplitude, phase angle and frequency are variables which may be critical and used by many power system control and protection applications. The fast and accurate measurements of these variables are still considered in a contemporary topic of research interest. Many phasor estimation algorithms as the Newton method, Kalman filtering and level tracking, least square algorithm (Giray, M.M., & Sachdev, M.S.,1989; Girgis, A.A., & Brown, R.G., 1981) techniques have also been proposed for synchrophasor computations in (Phadke, A. G., & Thorp, J. S., 1988), a raised cosine filter (RCF) is used to compute phasors during power oscillations and dynamics. The RCF filter needs a comparatively large (4cycles) computational time windows. This algorithm has a very slow time response.

Discrete Fourier Transform (DFT) is widely used as phasor estimation algorithm of fundamental frequency (Nguyen, C.T., & Srinivasan, K., 1984; Sachdev, M.S., & Giray, M.M., 1985; Ouadi, A., Bentarzi H., 2009). Due to their good harmonic rejection property conventional DFT algorithm achieve excellent performance when the signal contains only fundamental frequency and integer harmonic frequency components.

Performance of the DFT algorithms deteriorates at off nominal frequencies and in the presence of DC decaying components as will be exposed in chapter 3. The smart DFT algorithm (Gurusinghe, D.R., Rajapakse, A. D., & Narendra, K., 2012, Ouadi, A. et al., 2010), where the sampling approach is maintained at fixed frequency and fixed window length. The power system frequency is estimated then the estimated phasors are compensated as a function of the estimated frequency. The used technique can be extended for accurate phasor of any harmonic (Girgis, A.A., & Brown, R.G., 1981). The lower frequency components, as the DC decaying which may introduce fairly large error in the phasor estimation (Johns, A.T., & Salman S.K., 1995; Phadke, A. G., & Thorp, J. S., 1988). For improving the DFT phasor accuracy, these unwanted components must be filtered using digital band pass filter and then conventional DFT is performed. This will be the issue of chapter 5 for designing the digital band pass filters algorithms able to remove unwanted components.

An electric power system is a large interconnected system that produces, transmits and distributes an electric energy to different consumers. Stability of the smart power system is of a great concern, since it is subjected to different disturbances that may cause false measurement and hence mal-operation of protective and control systems which may lead to complete system collapse. Therefore, many techniques have been developed to mitigate the disturbance effects by introducing the digital filtering for computing the fundamental phasors of line voltages and currents. Several digital algorithms have been proposed in the literature for estimating the fundamental phasors, such as the discrete Fourier transform, cosine algorithm, least square algorithm, Kalman filter and wavelet transform (Yalla, M. V. V. S., 1992). The discrete Fourier transform (DFT) filter which has proven reliability and robustness in relaying application was considered as the most widely used and most popular algorithm for estimating phasors. However, during a transient time of smart grid condition, line fault current may include unwanted components that are associated to the fault impedance level as well as to the power transmission line type. The first type of disturbance mainly appears in a form of dc exponentially decaying component associated with the line fault current signal waveforms. In this case of signal waveforms including dc components, phasors estimated through application of the DFT filter may have large errors during the first cycles. This problem stems from the fact that the signal containing dc exponential decaying components has non integer harmonics (using the Fourier transform of this dc component) while the DFT filter does not provide the ability to reject non integer type of harmonics. The second kind of disturbance that may occur in power transmission line including series compensation devices in presence of relatively high fault impedance, where a significant distortion is introduced in the faulted line current waveforms. This is due to the behavior of non-linear components integrated in series compensated devices. A considerable exponentially decaying sub-harmonic or sub-synchronous resonance is associated with fundamental and integer harmonic of the line current waveform. This type of fault is more severe since it reduces the speed of convergence of the DFT algorithm to correctly estimate the fundamental phasors.

SYNCHRONIZED PHASORS FOR MEASUREMENT

During steady state in an electric power system, the voltage and current signals are virtually sinusoidal waveforms. A phasor is a vector consisting of magnitude and angle that corresponds to a sinusoidal waveform as illustrated in Fig.1. The phasor of a signal can be derived using Fourier transforms utilizing the data samples of the signal within a selected time window. For a steady state signal, the magnitude is a constant, while the value of the angle depends on the starting point of samples. The angle is a relative quantity and a reference has to be selected.

Measurement devices are placed at different locations in a power grid to capture voltage and current waveforms, from which phasors can be calculated. If the samples obtained by the measurement devices are not synchronized to a common timing reference, the angles of the phasors computed at different locations will not be comparable. This may hinder the understanding and analysis of certain power system phenomena and the development of certain advanced applications. To remove this barrier, phasors measured across the power grid should have a common timing reference such that direct comparison is feasible.

Figure 1. Phasor representation for a sinusoidal waveform

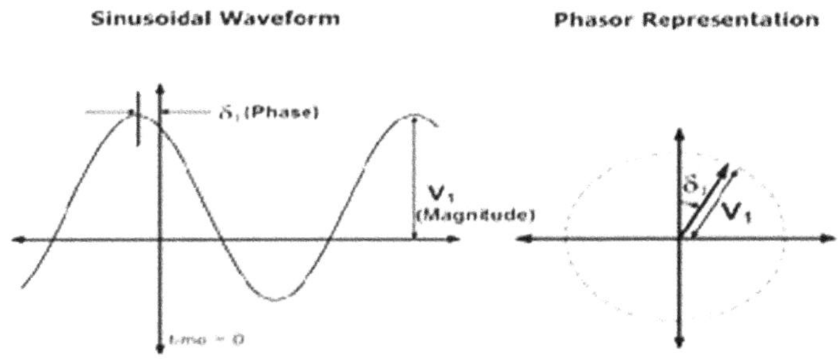

The IEEE standard, C37.118 defines the term and the requirements for synchronized measurement of phasors, or Synchrophasor. A synchrophasor is defined as "A phasor calculated from data samples using a standard time signal as the reference for the measurement". Synchronized phasors from remote sites have a defined common phase relationship." As a result, synchrophasors measured across an interconnected power grid will have a common timing reference and thus can be compared directly as illustrated in Fig.2.

According to the standard, a synchronizing source that provides the common timing reference may be local or global. The synchronizing signal may be distributed by broadcast or direct connection, and shall be referenced to Coordinated Universal Time (UTC). One commonly utilized synchronizing signal is the satellite signal broadcast from the Global Positioning System (GPS) (IEEE Std.C37.118-2005, 2006).

Figure 2. Phasor measurements at remote locations

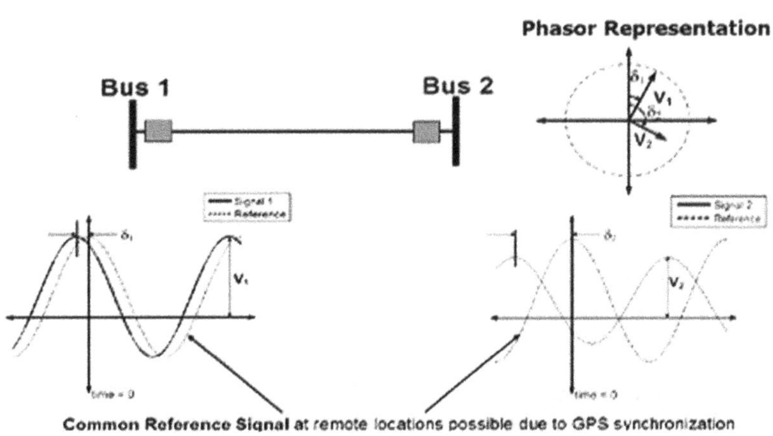

Phasor Measurement Unit Architecture

The general hardware architecture of a synchronized phasor measurement based electronic device is shown in Fig.3 which may include both digital protective devices and phasor measurement unit, generally is called intelligent electronic device (IED). The device includes analogue input block, digital input/output block subsystem, a data acquisition block, microcomputer and communication block port interfaces.

Figure 3. Phasor measurement and protection architecture

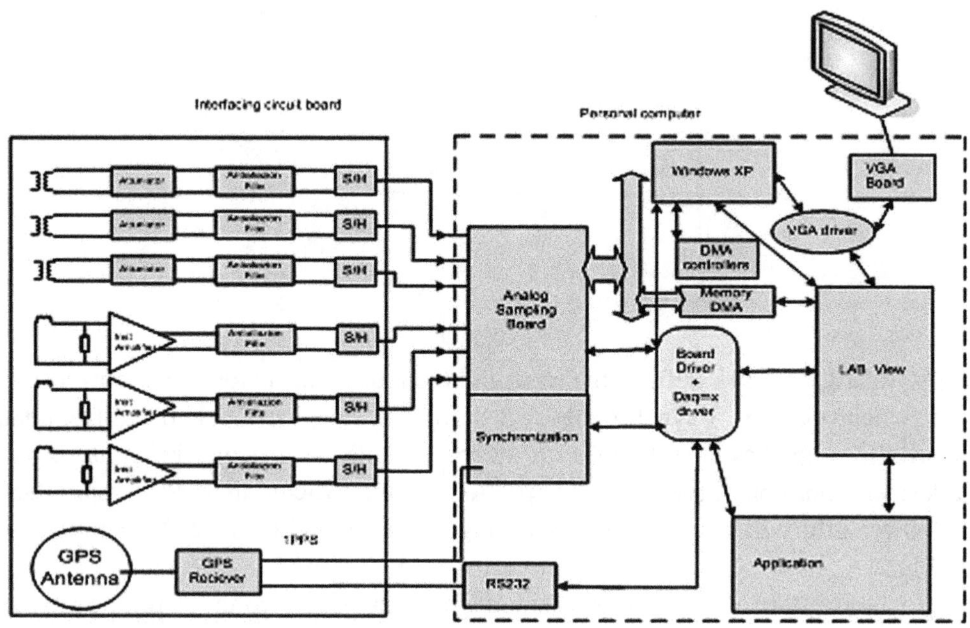

PHASOR MEASUREMENT ALGORITHMS

DFT Estimator

A sinusoidal input signal of frequency f_o is given by:

$$x(t) = X_m \cos(2\pi f_o + \varphi_i) \tag{1}$$

Where,

X_m: maximum value of the input signal,
f_o: the nominal frequency,
Δf: the frequency offset,
φ: the initial phase angle of the input signal.

The signal is conventionally represented by a phasor

$$\overline{X}\overline{X} = \frac{X_m}{\sqrt{2}} e^{j\phi} = Xe^{j\phi} = X\cos\phi + jX\sin\phi \tag{2}$$

Assuming that the signal $x(t)$ is sampled N times per fundamental period (50Hz or 60Hz) waveform to generate the sample set:

$$x_k = X_m \cos\left(\frac{2\pi}{Nf_o}k + \varphi_i\right) \tag{3}$$

The phasor $\bar{X}$ is given by:

$$\bar{X} = \frac{\sqrt{2}}{N}\left(X_c - jX_s\right) \tag{4}$$

Where:

$$X_c = \sum_{k=1}^{N} x_k \cos\left(\frac{2\pi}{N}k\right) \tag{5}$$

And

$$X_s = \sum_{k=1}^{N} x_k \sin\left(\frac{2\pi}{N}k\right) \tag{6}$$

Figure 4. PMU Phasor signal processing model.

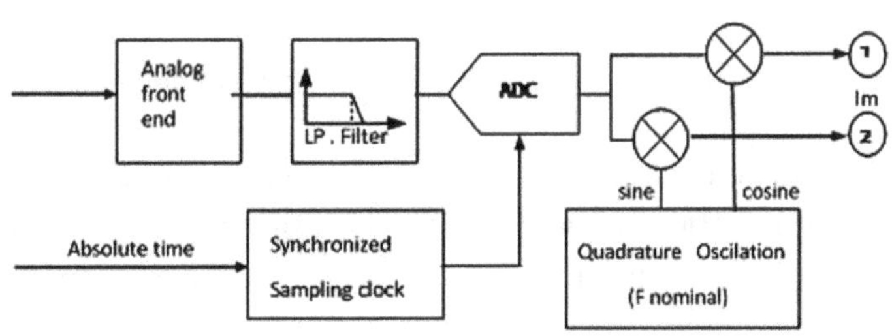

Figure 4 shows a signal processing steps performed for generating the synchronized phasors provided that the power system frequency is operating at nominal frequency. This is based on the fixed frequency sampling synchronized to an absolute time reference, followed by multiplication of the cosine and sine nominal frequency carrier. The conventional phasor representation of a sinusoidal signal is related to the fundamental component by its DFT as follows,

$$\bar{X} = \frac{1}{\sqrt{2}}\left(X_s - jX_c\right) \tag{7}$$

Equation (4) represents a phasor measurement using the DFT algorithm for a period of a fixed data window (a set of N samples). This DFT procedure provides the phasor computation for a fixed-length data window, but Eq.(4) represents a non recursive calculations which are repeated for each data window as shown in Fig.5. Evidently the procedure described by Eqs.(5)-(7) is non-recursive, and requires 2N multiplications and (N-l) additions to produce the phasor X. It may be noted that in progressing from one data window to the next, only one sample (x) is discarded and only one sample (xN) is added to the data set. Therefore, it is advantageous to develop a technique which retains 2(N-1) multiplications and 2(N-1) sums corresponding tothat portion of the data which is common to the old and new data windows.

Let $\overline{X}_r$ be the phasor corresponding to the data set

$x_k\{k=r, r+1,\ldots,N+r-1\}$

and let a new data sample be obtained toproduce a new data set

$x_k\ \{k= r+1, r+2,\ldots,N+r\}$.

The Phasor corresponding to the new data window $\overline{X}_{r+1}$ is then given by:

$$\hat{x}_{r+1} = \hat{x}_r + \frac{\sqrt{2}}{N}\left(x_{N+r} - x_r\right)e^{-j\frac{2\pi}{N}r} \tag{8}$$

The difference between the recursive and non-recursive Phasor calculations is illustrated in Fig.5. The non-recursive Phasor rotates in counter-clockwise direction by an angle $\frac{2\pi}{N}$ as the sample time advances, whereas the recursive phasor remains stationary. More importantly, the computations implied in the recursive formula (8) involves only two samples: x_{N+r} and x_r, whereas the non-recursive formula (5) implies computations with N samples. Now, all phasor measurement systems use the recursive form of phasor calculations.

Unlike Eq.(5), Eq.(8) with the recursive procedure only two multiplications are neededat each new sample time; making this a very efficient computational algorithm. It is interesting that when the input signal is a pure sine wave of fundamental frequency $\hat{x}_{N+r=\hat{x}_r}$ for all r as shown in Fig.5 (c); and consequently for this case, Eq.(8) becomes,

$$\hat{x}_{r+1} = \hat{x}_r \tag{9}$$

Equation (9) shows that when a recursive computation is used to calculate phasors, it leads to stationary phasors in the complex plane when the input signal is a pure sine wave of fundamental frequency. Recall that non-recursive phasor computation leads to phasors which rotate in the complex plane with an angular velocity $\omega 0$.

The phasor computation can be performed using FFT (Fast Fourier Transform) algorithm implemented on Digital Signal Processing (DSP) chips which are custom built for this function. A common form of the FFT requires that there will be 2n samples in the data window, e.g. 8, 16, 32 etc. whereas, the recursive DFT algorithms can be used for any number of samples in a window. The FFT computes harmonic components up to the Nyquist limit. If the window is taken as (60 Hz) cycle, then, 32 samples per cycle allows calculation up to the 15th harmonic for each variable. It is unlikely that any relay application

Figure 5. Phasors from sampled data. (a) Moving window. (b) Non-recursive. and (c) Recursive computation.

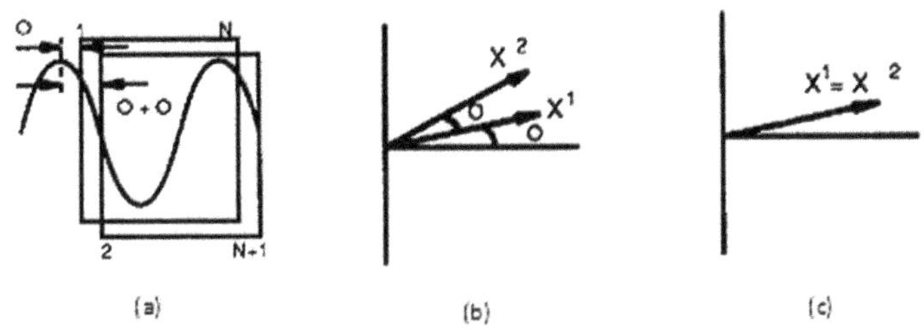

will require higher harmonics than this harmonic order. The recent developed DSPs can calculate an FFT for ten input variables in a sampling interval of (1/32x60) second (Ouadi A., Bentarzi, H., 2010).

If we concentrate on the fundamental components coming from, say, one set of three-phase variables as shown in the functional diagram of Fig.2.1, the microprocessor can calculate the sequence components and make them available for use in a number of functions internal to the station. If the sampling frequency and the frequency of the measurands are not synchronously related, then, they will produce errors in the phasor calculations and the positive sequence voltage will not remain stationary. The latter fact is used to measure the line frequency and to adjust the sampling frequency for making the phasor stationary (phase locked loop) in some relays. The errors in the phasors are small providing the frequency variation is small. By combining the three phase voltages, any oscillating component in the individual phase voltage phasors tends to be canceling out (IEEE Std.C37.118-2005, 2006). The window length of one60Hz (50Hz) cycle is about optimum for relaying purposes, however, consecutive Phasor estimates can be averaged over several m cycles if higher precision is required for metering or control purposes, and a slow response is acceptable. Whether one uses the recursive DFT, or the angle-compensated FFT described previously, this leads to the measurement of a stationary phasor at steady state network conditions if the sampling rate is fixed to N times the network frequency.

If the network frequency undergoes a change characterized by a frequency offset M, it has been shown in (Girgis, A.A., & Hwang, T.L., 1984; Phadke, A. G., Thorp, J. S., & Adamiak, M.G., 1983) that for a small if the positive sequence phasor will undergo the following change, at each r'th sampling interval,

$$X^r\left(f_0+\Delta f\right)=Xe^{-j(N-1)\pi\Delta f\Delta t}\left\{\frac{\sin N\Delta f\Delta t}{N\,\sin\Delta f\Delta t}\right\}e^{j2\pi\Delta f\Delta t} \tag{10}$$

Where f_0 is the power system nominal frequency.

To simplify the computing algorithm, an efficient recursive DFT algorithm can be used as follows:

$$\hat{x}_r=\hat{x}_{r-1}+\frac{\sqrt{2}}{N}\left(x_{N+r-1}-x_r\right)e^{-j\frac{2\pi}{N}(r-1)} \tag{11}$$

Phasor Sequence Quantities

Even though a three-phase system is balanced most of time, it is important to develop an estimator that can accomplish accurate estimations under unbalanced condition. In practice, machine-oscillating modes can be determined by measuring the phase angle of the positive-sequence voltage phasor at the machine terminals. Fortescue has shown that any three phase asymmetric (unbalanced) system can be decomposed into a set of a balanced three phase direct component (1: positive sequence), a balanced three phase inverse component (2: negative sequence) and a balanced three phase homopolar (0: zero sequence) component. The abc three phase (X_a, X_b, X_c) sinusoidal system can be expressed in terms of the symmetrical components (X_0, X_1, X_3) as:

$$\begin{bmatrix} X_a \\ X_b \\ X_c \end{bmatrix} = \frac{1}{3} \begin{bmatrix} 1 & 1 & 1 \\ 1 & \alpha^2 & \alpha \\ 1 & \alpha & \alpha^2 \end{bmatrix} \begin{bmatrix} X_0 \\ X_1 \\ X_2 \end{bmatrix} \tag{12}$$

Or the symmetrical components in terms of the phase sequences can be expressed as:

$$\begin{bmatrix} X_0 \\ X_1 \\ X_2 \end{bmatrix} = \frac{1}{3} \begin{bmatrix} 1 & 1 & 1 \\ 1 & \alpha & \alpha^2 \\ 1 & \alpha^2 & \alpha \end{bmatrix} \begin{bmatrix} X_a \\ X_b \\ X_c \end{bmatrix} \tag{13}$$

Where $\alpha = e^{j\frac{2\pi}{3}}$ and X may be a voltage or current variable

The recursive relation for the positive sequence voltage is similar to Eq.(3),

$$\hat{x}_r = \hat{x}_{r-1} + \frac{\sqrt{2}}{N}\frac{1}{3}\begin{bmatrix} \left(x_{a,N+r-1} - x_{a,r}\right)e^{-j\frac{2\pi}{N}(r-1)} + \alpha.\left(x_{b,N+r-1} - x_{b,r}\right)e^{-j\frac{2\pi}{N}(r-1)} \\ +\alpha^2.\left(x_{c,N+r-1} - x_{c,r}\right)e^{-j\frac{2\pi}{N}(r-1)} \end{bmatrix} \tag{14}$$

Other symmetrical components of voltages and currents are calculated in a similar manner (Phadke, A. G. et al., 1977; Phadke, A. G. et al., 1979; Phadke, A. G. et al., 1981). Measurements of system frequency and time deviation of frequency can be done using the positive sequence voltage. Since the phase angle of this voltage, used here, is measured using narrow-band filtering performed by the DFT algorithm. These measurements provide excellent rejection of the effects of harmonics, noise and dc offsets. Using the positive-sequence voltage corresponds closely to the actual state of the system, as compared to measurements made on a single phase only. Of course, single-phase measurement of frequency and time of variation is also possible where system requirements dictate its use.

Frequency Tracking Algorithm

As reported in the literature (Phadke, A. G., Thorp, J. S., & Adamiak, M.G., 1983), the DFT phasor calculation has an error when the power system operates at off-nominal frequencies. Figure 5 shows the deviation from an ideal of the phasor calculation at off-nominal frequency for angles vary from 0° to 360°.

Figure 6. Phasor calculation errors at off-nominal frequency.

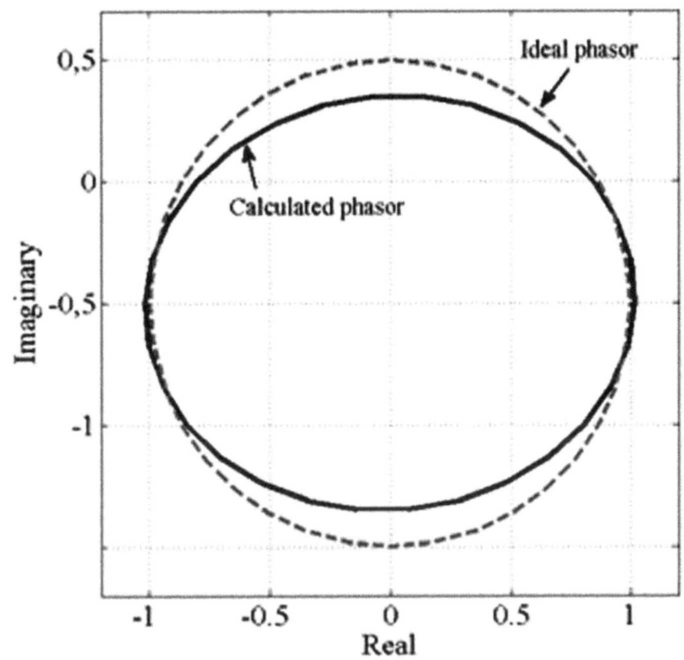

To measure the power system frequency, many techniques have been developed as a modified zero crossing technique (Begovic, M.M., 1993) and level crossing technique (Nguyen, C.T., & Srinivasan, K., 1984), both are abandoned due to their sensitivity to noise, presence of DC components in the signal, and harmonics. Other techniques as the least square technique (Sachdev, M.S., 1985; Giray, M.M., 1989), Kalman filter technique (Girgis, A.A., 1981), leakage effect technique (Girgis, A.A. et al., 1982), and phasor-based technique (Phadke, A.G. et al., 1983; Chen, 1994; Phadke, A.G. et al., 1988) have been proposed.

These developed techniques are mainly used to measure steady state frequency based on the assumption of fixed frequency model. For use in wide area control application, dynamic frequency or instantaneous frequency is more useful to describe the behavior of the system (Bentarzi, H., 2010).

The Smart DFT algorithm developed by (Phadke, A. G., et al., 1977) is used to estimate the instantaneous frequency and then the phasor amplitude and phase angle are corrected.

During power system disturbances, where the sinusoidal frequency f(t) is no longer fixed and is a function of time t. The sinusoidal signal waveform can be represented by its nominal frequency f_0 varying with a frequency deviation $\Delta f(t)$, as follows:

$$x(t) = X_m \cos\left(2\pi\left(f_o + \Delta f\right) + \varphi_i\right) \tag{15}$$

Where,

X_m: maximum value of the input signal,
f_o: the nominal frequency,
Δf: the frequency offset,
φ_i: the initial phase of the input signal .

It is conventionally represented by a phasor:

$$\overline{X} = \frac{X_m}{\sqrt{2}} e^{j(2\pi\Delta ft + \varphi_i)} = Xe^{j\phi} = X\cos\phi + jX\sin\phi \tag{16}$$

Assuming that the signal $x(t)$ is sampled N times per period of nominal frequency fo (50Hz or 60 Hz) to generate the sample set:

$$x_k = X_m \cos\left(\frac{2\pi}{N} k\left(1 + \frac{\Delta f}{f_o}\right) + \varphi_i\right) \tag{17}$$

The phasor tracking algorithm (SDFT) may be applied by defining θ as

$$\theta = \frac{2\pi}{f_o}\frac{\Delta f}{N} \tag{18}$$

And assuming that the sampling rate is m times of the correcting computation frequency, the exact solution of phasor at off nominal frequency can be obtained by the following equations.

Using the sampling set given by Eq.(1), the fundamental frequency component of DFT of{ $\tilde{x}(k)$ } can be obtained:

$$\overline{x}(r) = \frac{\sqrt{2}}{N}\sum_{k=0}^{N-1}\tilde{x}(k+r)je^{-\frac{2\pi}{N}k} \tag{19}$$

Where $\overline{x}(r)$ is a rotating phasorDefining the synchronous phasor as the following:

$$\hat{x}(r) = \overline{x}(r)e^{-j\frac{2\pi}{N}r} = \frac{\sqrt{2}}{N}j\sum_{k=r}^{N+r-1}\tilde{x}(k)e^{-j\frac{2\pi}{N}k} \tag{20}$$

And expressing (11) in the exponential form as:

$$\tilde{x}(k) = \sqrt{2}X \frac{e^{j\left[2\pi\left(1+\Delta f/f_0\right)\frac{k}{N}+\varphi\right]} - e^{-j\left[2\pi\left(1+\Delta f/f_0\right)\frac{k}{N}+\varphi\right]}}{2j} \tag{21}$$

Equation (20) can be rearranged as:

$$\hat{x}(r) = \frac{X}{N}\left[e^{j\varphi}\sum_{k=r}^{N+r-1} e^{j\left(\frac{2\pi\Delta f}{f_0 N}\right)k} - e^{-j\varphi}\sum_{k=r}^{N+r-1} e^{-j\left(\frac{2\pi\Delta f}{f_0 N}+\frac{4\pi}{N}\right)k}\right] \tag{22}$$

To define θ as (14) and use the following expression,

$$\sum_{k=r}^{N+r-1} e^{j\theta k} = \frac{e^{j\theta r} - e^{-j\theta(r+N)}}{1-e^{j\theta}} = \frac{\sin(\theta N/2)}{\sin(\theta/2)} e^{j\theta[r+(N-1)/2]} \tag{23}$$

Then, Eq.(23) can be expressed as:

$$\hat{x}(r) = \frac{X}{N}\left\{e^{j\varphi}\frac{\sin(\theta N/2)}{\sin(\theta/2)} e^{j\theta[r+(N-1)/2]} - e^{-j\varphi}\frac{\sin(\theta N/2)}{\sin\left[\left(\theta+\frac{4\pi}{N}\right)/2\right]} e^{-j\left(\theta+\frac{4\pi}{N}\right)[r+(N-1)/2]}\right\} \tag{24}$$

and the exact phasor can be expressed in a discrete form as:

$$\vec{x}(r) = Xe^{j(r\theta+\varphi)} \tag{25}$$

Rearranging Eq.(24) further, we obtain,

$$\hat{x}(r) = \vec{x}(r)\frac{\sin(\theta N/2)}{N\sin(\theta/2)} e^{j\theta(N-1)/2} - \vec{x}(r)^{*}\frac{\sin(\theta N/2)}{N\sin\left[\left(\theta+\frac{4\pi}{N}\right)/2\right]} e^{-j\left\{\frac{4\pi}{N}[r+(N-1)/2]+\theta(N-1)/2\right\}} \tag{26}$$

where * denotes complex conjugate.

By defining:

$$c1(r) = \vec{x}(r)\frac{\sin(\theta N/2)}{N\sin(\theta/2)} e^{j\theta(N-1)/2} \tag{27}$$

and

$$c2\left(r\right)=\vec{x}\left(r\right)^{*}\frac{\sin\left(\theta N/2\right)}{N\sin\left[\left(\theta+\frac{4\pi}{N}\right)/2\right]}e^{-j\left\{\frac{4\pi}{N}\left[r+\left(N-1\right)/2\right]+\theta\left(N-1\right)/2\right\}} \tag{28}$$

$$\alpha\left(\theta\right)=e^{j\theta} \tag{29}$$

We can obtain the following equations from Eq.(26):

$$\hat{x}\left(r\right)=c_{1}\left(r\right)-c_{2}\left(r\right) \tag{30}$$

$$\hat{x}\left(r-m\right)=c_{1}\left(r\right)\propto^{-m}\left(\theta\right)-c_{2}\left(r\right)\propto^{m}\left(\theta\right)e^{j\frac{4\pi}{N}m} \tag{31}$$

$$\hat{x}\left(r-2m\right)=c_{1}\left(r\right)\propto^{-2m}\left(\theta\right)-c_{2}\left(r\right)\propto^{2m}\left(\theta\right)e^{j\frac{8\pi}{N}m} \tag{32}$$

Solution of Eqs (33-35) gives,

$$\propto^{-m}\left(\theta\right)=\frac{\left[\hat{x}\left(r-2m\right)+\hat{x}\left(r\right)e^{j\frac{4\pi}{N}m}\right]}{2\hat{x}\left(r-m\right)}\pm\frac{\sqrt{\left[\hat{x}\left(r-2m\right)+\hat{x}\left(r\right)e^{j\frac{4\pi}{N}m}\right]^{2}-4.\widehat{x}\left(r-2m\right)^{2}e^{j\frac{4\pi}{N}m}}}{2.\hat{x}\left(r-2m\right)}. \tag{33}$$

According to physical meaning, we take '+"and simplify Eq.(33) as Eqs.(17) and (18). By combining Eqs.(19), (30) and (33),therefore the exact solution of phasor at off nominal frequency can be obtained by

$$\hat{x}\left(r\right)=\left[\vec{x}\left(r\right)+C_{2}\left(r\right)\right]\frac{N\sin\left(\theta/2\right)}{\sin\left(N\theta/2\right)}e^{-j\theta\left(N-1\right)/2} \tag{34}$$

Where:

$$C_{2}\left(r\right)=\frac{\hat{x}\left(r-m\right)-\hat{x}\left(r\right)\alpha^{-m}\left(\theta\right)}{\alpha^{-m}\left(\theta\right)-\alpha^{m}\left(\theta\right)e^{j\frac{4\pi}{N}m}},\ \eta\left(r-2m\right)=\frac{\hat{x}\left(r-2m\right)/\hat{x}\left(r-2m\right)+e^{j\frac{4\pi}{N}m}\hat{x}\left(r\right)/\hat{x}\left(r-m\right)}{2},$$

And,

$$\alpha^{-m}\left(\theta\right) = \eta\left(r-2m\right) + \sqrt{\left[\eta\left(r-2m\right)\right]^2 - e^{j\frac{4\pi}{N}m}}.$$

The instantaneous frequency can then be calculated from equations (22) and (27) and given by the following relation:

$$f(t) = f_o + \Delta f. \tag{35}$$

However since the standard defines the phasors at the nominal frequency (50Hz or 60Hz), the instantaneous phase of the phasor is obtained by the following relation:

$$\phi(t) = \varphi_i + k\theta(t). \tag{36}$$

Power System Frequency and Rate of Change (ROCOF)

Power networks have recently been more complex due to many reasons. The balance between load and generation is a prerequisite for stable power system during its dynamic operation. It has become more difficult to maintain it stable because the expansion of the transmission network does not follow the growth of the power system. The direct consequence is that the security margins are narrow, and power systems often operate at the edge of their instability, resulting in a blackout. Generation-load mismatches cause the system frequency to deviate from its nominal value. Therefore, frequency is an important operating parameter of a power system. Frequencies lower than the nominal value; indicate that the system is overloaded. The system state of the power system can be evaluated as the Direct and accurate frequency measurement as well as its ROCOF of the power system.

One of the different techniques used to measure frequency is the phasor-based technique which gives much closer to the instantaneous frequency allowing tracking dynamics faster. During the electromechanical disturbances, the power system can be evaluated from accurate frequency measurements as well as the rate of change of frequency (ROCOF). A longer time may be needed for accurate measurements of these parameters.

The basic idea mentioned in references (Phadke, A. G., et al., 1977; Phadke, A. G., et al., 1988) is to calculate the positive sequence voltage phasor for a balanced power system operating at off-nominal frequency, while the nominal frequency based DFT and samples are still used. In this work, a technique which is based on the positive sequence voltage phasors obtained by DFT (inherently filtering harmonics) and further smoothed via minimum least square polynomial fitting of the quasi-stationary phasor data has been used to obtain both the operating frequency and the ROCOF parameter.

The sinusoid signal is considered using the parametric equation,

$$x\left(t\right) = X_m \cos\left(\omega_0 t + \phi\right) = X_m \cos\left(2\pi f_0 t + \phi\right). \tag{37}$$

Where f_0 is the nominal angular system frequency (50 Hz or 60 Hz). In the general case where the amplitude is a function of time $X_m(t)$ and the sinusoid frequency is also a function of time $f(t)$, the function which is the difference between the actual and nominal frequencies can be defined as $g=f-f_0$ or also be a function of time, e.g., $g(t)=f(t)-f_0$. The sinusoid can then be written as:

$$x(t) = X_m \cos\left(2\pi \int f\,dt + \phi\right)$$

or,

$$x(t) = X_m \cos\left(2\pi \int \left(f_0 + g\right)dt + \phi\right) \tag{38}$$

or,

$$x(t) = X_m \cos\left(2\pi f_0 t + \left(2\pi \int g\,dt + \phi\right)\right)$$

The synchrophasor representation for this waveform is given by:

$$X(t) = \frac{X_m(t)}{\sqrt{2}} e^{\left(2\pi \int f\,dt + \phi\right)} \tag{39}$$

For the special case when $X_m(t)=X_m$ is constant and $g=\Delta f$ is a constant offset from the nominal frequency,

$$\int g(t)\,dt = \int \Delta f dt = \Delta ft$$

so the synchrophasor is simply written as:

$$X(t) = \frac{X_m(t)}{\sqrt{2}} e^{j(2\pi \Delta ft + \phi)} \tag{40}$$

that will rotate at the uniform rate Δf, the difference between the actual and off-nominal frequency. This concept is illustrated in Fig. 7. It can be considered that a sinusoid at off-nominal system frequency is observed at intervals $\{0, T_0, 2T_0, \ldots, nT_0, \ldots\}$ where $T_0 = \frac{1}{f_0}$ (the nominal power system period) and the corresponding phasor representations are $\{X_0, X_1, X_2, \ldots, X_n, \ldots\}$. If the sinusoid frequency $f \neq f_0$ and $f<2f_0$, the observed phasor will have a constant magnitude, but the phase angles of the sequence of phasors $\{X_0, X_1, X_2, \ldots, X_n, \ldots\}$ will change uniformly at a rate of $2\pi(f-f_0)T_0$, as illustrated in Fig.6. If these values are reported over time, they would continuously increase until they reach 180° where they would wrap around to –180° and continue to increase as shown in Fig.8 (synchrophasors are commonly reported in angles –180° to +180° rather than 0 to 360°).

Figure 7. A sinusoid with a frequency f > f0 is observed at instants that are T0 seconds apart—the phase angle Φ increases uniformly in relation to the frequency difference, f–f_0.

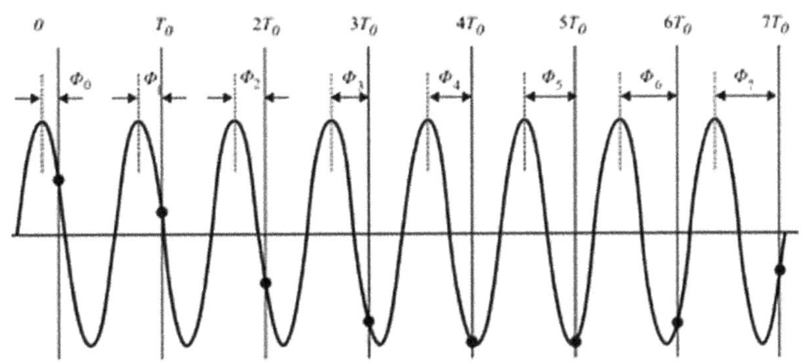

Let $[\Phi_k]$ $\{k=0,1,2,\ldots,n-1\}$ be the vector of "n" samples of the phase angles of the positive sequence measurement. The vector $[\Phi_k]$ is assumed to be monotonically changing over the window of "n" samples. "Δt" is the time interval between samples.

Figure 8. Sampling a power frequency sinusoid at off-nominal frequency

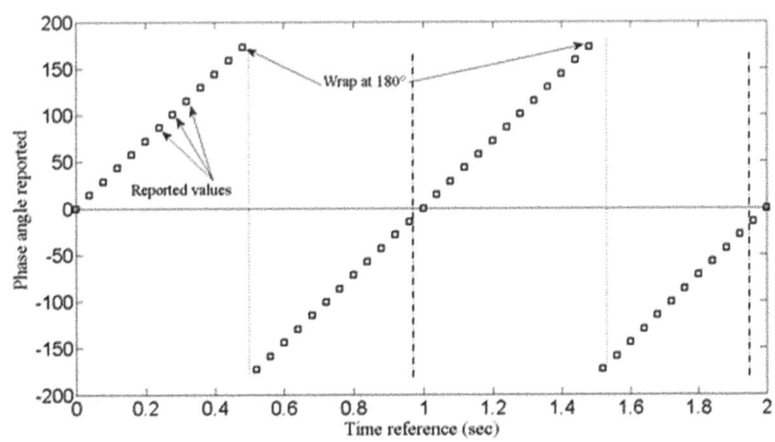

As the phase angles of the phasor estimated may be restricted to a range of 0–2π, it is necessary to make the angles monotonically changing over the entire spanning period. If the frequency deviation from the nominal value, and the rate of change of frequency at t=0 are $\omega 0$ and ω', respectively, the frequency at any time "t" is given by

$$\omega(t) = \omega_0 + t\omega' \tag{41}$$

The phase angle is the integral:

$$\phi(t) = \int \omega dt = \int (\omega_0 + t\omega')\,dt = \phi_0 + t\omega_0 + \frac{1}{2}t^2\omega' \tag{42}$$

If $\phi(t)$ is assumed to be a second degree polynomial,

$$\phi(t) = a_0 + a_1 t + a_2 t^2 \tag{43}$$

at t=0: $\omega_0 = a_1$ and $\omega' = 2a_2$

Or, in terms of Hertz and Hz/s,

$$f_0 = \frac{a_1}{2\pi} \tag{44a}$$

and,

$$f' = \frac{a_2}{\pi} \tag{44b}$$

The vector of "n" angle measurements is given by

$$\begin{bmatrix} \phi_0 \\ \phi_1 \\ \phi_2 \\ \vdots \\ \phi_{n-1} \end{bmatrix} = \begin{bmatrix} 1 & 0 & 0 \\ 1 & \Delta t & \Delta t^2 \\ 1 & 2\Delta t & 2^2\Delta t^2 \\ \vdots & \vdots & \vdots \\ 1 & (n-1)\Delta t & (n-1)^2\Delta t^2 \end{bmatrix} \begin{bmatrix} a_0 \\ a_1 \\ a_2 \end{bmatrix} \tag{45}$$

Or in matrix notation,

$$[\Phi] = [B][A] \tag{46}$$

where $[B]$ is the coefficient matrix. The unknown vector $[A]$ is calculated by the weighted least squares (WLS) technique (IEEE Std.C37.118-2005, 2006) and is given by:

$$[A] = [B^T B]^{-1} B^T [\Phi] = [G][\Phi] \tag{47}$$

Where,

$$[G] = [B^T B]^{-1} B^T \tag{48}$$

The matrix $[G]$ is pre-calculated and stored for use in real time. It has "n" rows and three columns. In real time, $[G]$ is multiplied by $[\Phi]$ to obtain the vector $[A]$, then, the frequency and rate of change of frequency at any time t (which is a multiple of Δt) can be calculated. This time is usually associated with the time tag for which the measurement is posted.

Positive Sequence, Frequency, and ROCOF

The normal positive sequence is calculated using the symmetrical component transformation. Then, frequency is calculated from the rate of change of phase angle. Since phase angle changes relative to the difference between the actual frequency and the nominal frequency, this approach yields the offset from nominal. This algorithm is set up to smooth the estimate, particularly for the case of limited resolution as found in the real PMU (see Fig. 9).

Figure 9. Complete PMU signal processing model—all processing is at the A/D sampling rate; reporting rate is produced by resampling at the system output (decimator stage).

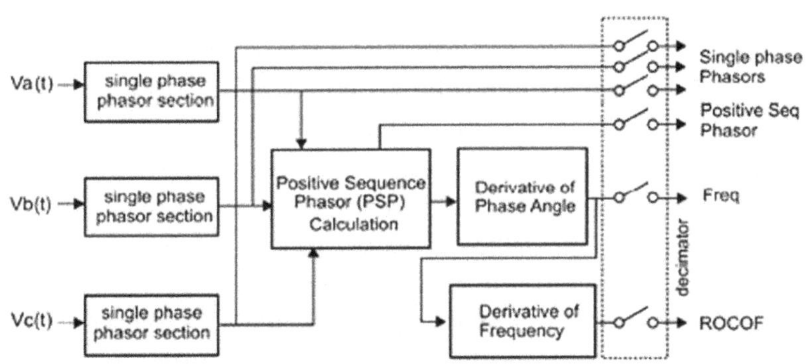

SYNCHROPHASOR MEASUREMENT EVALUATION

The PMU calculated measurements are evaluated against the actual test signals generated from the mathematical equations where theoretical Phasor are analytically calculated. Total vector error (TVE) and frequency error (FE) are determined according to Eqs. (49) and (50) respectively (Gurusinghe, D.R., Rajapakse, A. D., & Narendra, K., 2012).

$$TVE(n) = \sqrt{\frac{x_a(n) - x_m(n)}{x_a(n)}} = \frac{|x_a(n) - x_m(n)|}{|x_a(n)|}. \tag{49}$$

where $x_a(n)$ is actual synchrophasor and $x_m(n)$ is measured Synchrophasor

Frequency and ROCOF Measurement Evaluation

The frequency and ROCOF measurements are evaluated as the absolute value of the difference between the theoretical values and the estimated values given in Hz and Hz/s respectively defined by the two following equations.

Frequency measurement error:

$$FE(n) = |f_a(n) - f_m(n)| \tag{50}$$

ROCOF measurement error:

$$RFE(n) = \left| \left(\frac{df}{f} \right)_a - \left(\frac{df}{f} \right)_m \right| \tag{51}$$

where $f_a(n)$ is actual frequency and $f_m(n)$ is measured frequency. The actual and the measured values are for the same instant of time.

Performance tests need to be performed over the entire ranges of interest and include a range of operating conditions. The PMU algorithm test results illustrate performances of the highest reporting rate of 50 fps (frame per second). It is obvious that if a PMU satisfies performances at the highest reporting rate it also fulfills requested performances at lower reporting rates. However, it is required to evaluate PMU performances at each reporting rate according to (IEEE Std.C37.118-2005, 2006).

Steady-state tests measurement confirm the standard PMU requirements in constant conditions where magnitude, phase angle and frequency of test signal, and all other influence quantities are fixed for the period of measurement. The tests comprise signal frequency, magnitude, phase angle, harmonic distortion and out-of-band interference tests. Each steady-state test is performed over a continuous 5 seconds of test duration. Maximum and mean TVE measurement criterion as specified in (IEEE Std.C37.118-2005, 2006) are used to evaluate the dynamic PMU performances that must satisfies the required guidelines (IEEE Std.C37.118-2005, 2006). Since, only the PMU algorithm is tested, the test of analog input signal waveform is sampled with 36 per cycle (50 Hz) and quantized with a 16 bits A/D resolution to introduce quantization error when performing the PMU measurement algorithms.

1. Steady-State Signal Frequency

Under the signal frequency test, the frequency, f is varied from 45 Hz to 55 Hz with a step resolution of 1 Hz while all other quantities are kept constant. The maximum and mean values of percentage TVE are shown in Fig.10.

The frequency signal deviates from the nominal frequency. This phasor rotation (variable phase) should be considered when evaluating TVE; otherwise calculations may show abnormal TVE values. Frequency of the signal is measured in each case while the FEs can also be determined. The PMU signal frequency TVE compliance is satisfied as the maximum TVE is less than 1% for frequencies between 45 Hz to 55 Hz.

Figure 10. Steady-state signal frequency response at frequency reporting rate of 50 fps.

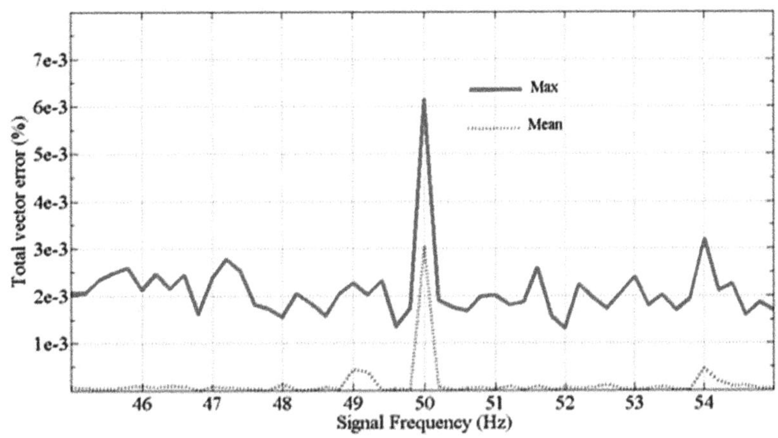

2. Steady-State Signal Voltage Magnitude

The voltage signal magnitude per unit (pu), Xm in Eq.(3) is varied from 0.1 to 1.2 with a step resolution of 0.05 pu while all other quantities are kept constant. Fig. 11 shows the maximum and mean values of percentage TVE. The PMU voltage signal magnitude TVE compliance is satisfied as the maximum TVE is less than 1% for all magnitude variations ranging between 0.1 pu and 1.2 pu.

Figure 11. Steady-state signal magnitude response for voltage at frequency reporting rate of 50 fps

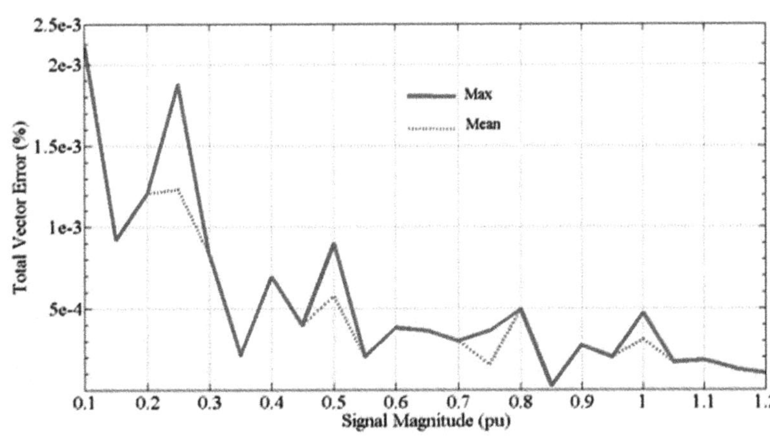

3. Steady-State Signal Current Magnitude

The current signal magnitude per unit (pu), Xm in Eq.(6)is varied from 0.1 to 2.0 with a step resolution of 0.025 pu while all other quantities are kept constant. Fig. 12 shows the maximum and mean values of percentage TVE. The PMU current signal magnitude TVE compliance is satisfied as the maximum TVE is less than 1% for magnitudes variations ranging between 0.1 pu and 2.0 pu.

Figure 12. Steady-state signal magnitude response for current at frequency reporting rate of 50 fps.

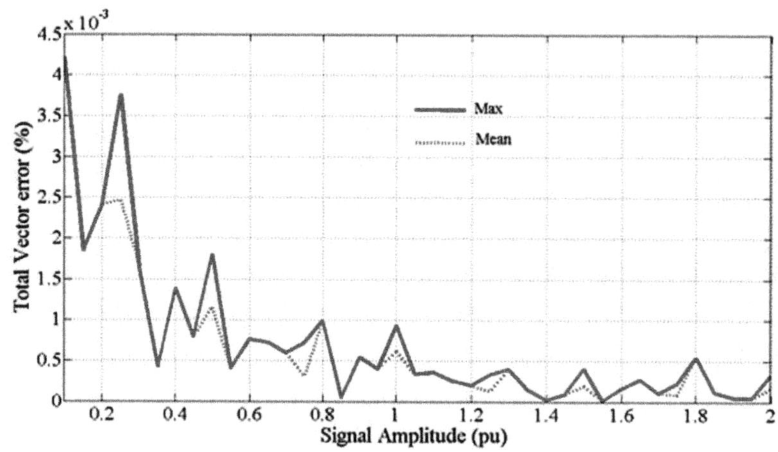

4. Steady-State Phase Angle

The phase angle, in Eq.(6)is varied from –180° to +180° with a step resolution of 18° while all other quantities (magnitude, frequency …) are kept constant. Figure 13 shows the maximum and mean values of percentage TVE. The PMU phase angle TVE compliance is satisfied as the maximum TVE is less than 1% for phase angles between –180° and +180°.

Figure 13. Steady-state phase angle response at frequency reporting rate of 50 fps.

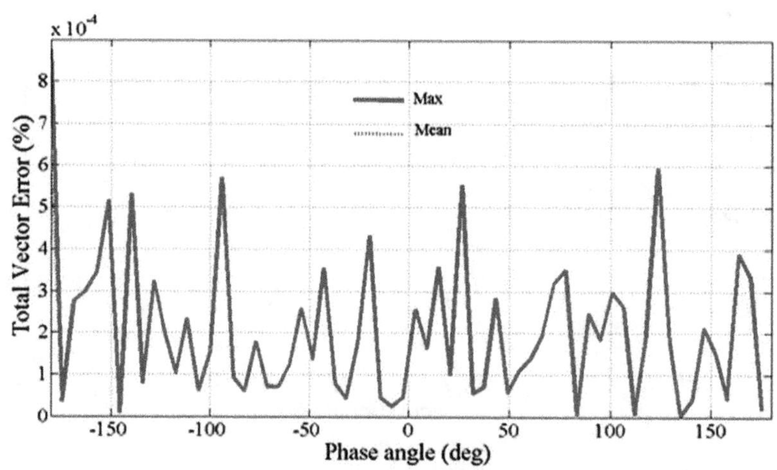

5. Steady-State Harmonic Distortion

A 10% harmonic is introduced into the test signal where Eq.(6) is modified as,

$$x(t) = X_m \left\{\cos\left(2\pi f_0 t + \phi\right) + 0.1\cos\left(2\pi k f_0 t + \phi_1\right)\right\} \tag{52}$$

The k parameter designates the harmonic number, which is varied from 2 to 50 (integer values only) while all other quantities (magnitude, fundamental frequency ...) are kept constant. The harmonic component can either be in phase or out of phase with the fundamental component. When determining TVE, the harmonic component is ignored where only the fundamental component is used for calculating the theoretical phasor. Figure 14 shows the maximum and mean values of percentage TVE. The PMU harmonic distortion TVE compliance is satisfied as the maximum TVE is less than 1% for a 10% harmonic for signals variation ranging between the 2nd and the 50th harmonic.

Figure 14. Steady-state 10% harmonic distortion responses at frequency reporting rate of 50 fps.

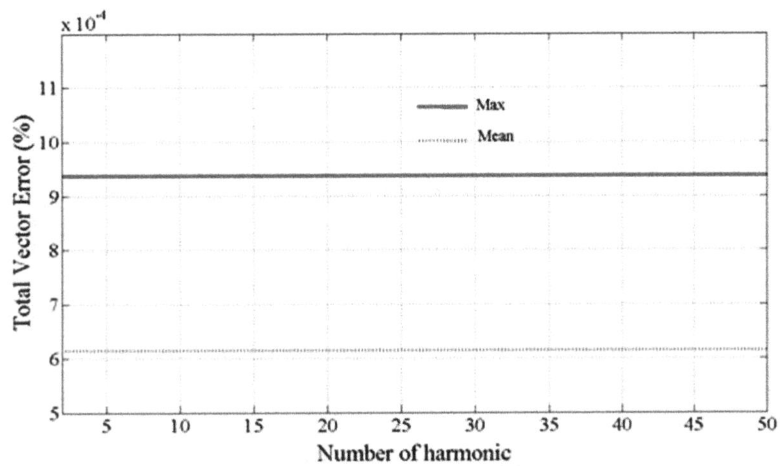

6. Steady-State Out-of-Band Interference

A 10% out-of-band interference is introduced into the test signal where Eq.(6) is modified as:

$$x(t) = X_m \left\{\cos\left(2\pi f_0 t + \phi\right) + 0.1\cos\left(2\pi f_{0b} t + \phi\right)\right\}. \tag{53}$$

The frequency of an interference (out of band) signal, fob is varied from 10 Hz to 100 Hz (second harmonic) while all other quantities (magnitude, fundamental frequency ...) are kept constant. The interference signal should be in phase with the fundamental component. For a 50 frame per second reporting rate the frequency band requirement is from 30 Hz to 90 Hz. The test frequencies are 10 Hz to 25 Hz and 75 Hz to 100 Hz. The interference component is ignored when determining TVE where only the fundamental component is used for calculations. Figure 15 illustrates the maximum, minimum and mean values of percentage TVE. The PMU does not satisfy out of band interference TVE compliance as the maximum TVE is greater than 1.3% for a 10% out of band interference signal.

The PMU satisfies both P class and M class TVE compliance of steady-state except out-of-band interference. This can be explained by the fact that the absence of potential transducer which are considered as pass band filter, which when introduced may enhance significantly the PMU out of band interference performance and may satisfy its percentage TVE compliance.

Dynamic compliances include measurement bandwidth, linear system frequency ramp and step response evaluations.

Figure 15. Steady-state 10% out of band interference response at frequency reporting rate of 50 fps.

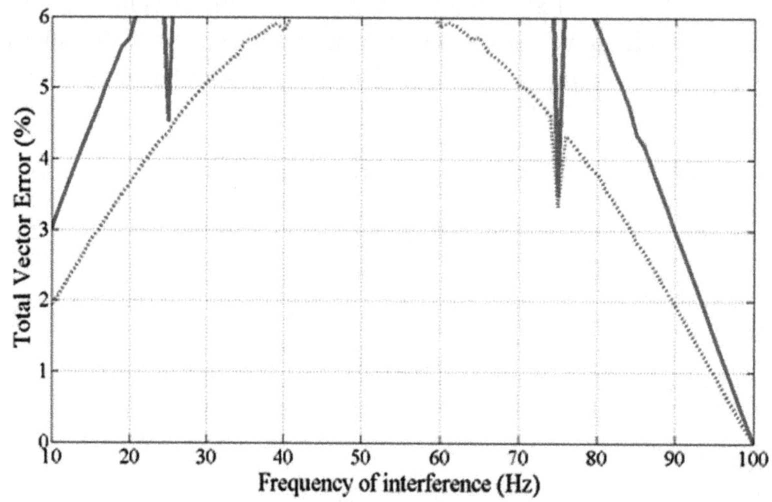

7. Measurement Bandwidth

The synchrophasor measurement bandwidth shall be determined by sweeping the input with sinusoidal amplitude and phase modulation. This shall be done by modulating balanced three-phase input signals (voltages and currents) with sinusoidal signals applied to signal amplitudes and phase angles simultaneously. Mathematically the input signals may be represented by

$$X_a = X_m\left[1 + k_x \cos\left(\omega t\right)\right].\cos\left[\omega_0 t + k_a \cos\left(\omega t - \pi\right)\right] \tag{54a}$$

$$X_b = X_m\left[1 + k_x \cos\left(\omega t\right)\right].\cos\left[\omega_0 t - 2\pi / 3 + k_a \cos\left(\omega t - \pi\right)\right] \tag{54b}$$

$$X_c = X_m\left[1 + k_x \cos\left(\omega t\right)\right].\cos\left[\omega_0 t + 2\pi / 3 + k_a \cos\left(\omega t - \pi\right)\right] \tag{54c}$$

where X_m is the amplitude of the input signal, $\omega 0$ is the nominal power system frequency, ω is the modulation frequency in radians/s, $f_m = \dfrac{\omega}{2\pi}$ is the modulation frequency in Hz, kx is the amplitude modulation factor, and ka is the phase angle modulation factor.

The positive sequence signal corresponding to the above three-phase inputs is given by:

$$X_1 = X_m\left[1 + k_x \cos\left(\omega t\right)\right] \times \cos\left[\omega_0 t + k_a \cos\left(\omega t - \pi\right)\right] \tag{55}$$

At reporting time tags $t=nT$ (where n is an integer and T is the phasor reporting interval) the PMU shall produce a positive sequence measurement of

$$X\left(nT\right) = \left\{X_m / \sqrt{2}\right\}\left[1 + k_x \cos \omega nT\right] \angle \left\{k_a \cos\left(\omega nT - \pi\right)\right\} \tag{56}$$

The modulation tests shall be performed with ω, $k_x\omega$, and k_a over the frequency ranges specified in (IEEE std.1344-2005, 2006).The modulation frequency shall be varied in steps of 0.2 Hz or smaller over the range specified in (IEEE std.1344-2005, 2006).The TVE, FE, and RFE shall be measured over at least two full cycles of modulation. The maximum is the highest value observed at the given reporting rate over the full test interval. This maximum shall be within the specified limits for P class and M class compliance at the given reporting rate. An adequate settling time shall be allowed for each test signal change to prevent parameter change transient effects from distorting the measurement.

Test signals for measurement bandwidth are primarily 50 Hz waveforms that are amplitude or/and phase angle modulated with a sine wave. They could, however, be any kind of signal modulation or other combination that could be input to a PMU to determine a specific type of response characteristic. Modulation level is kept at 10% while modulation frequency is varied over a range that will demonstrate the PMU response characteristics. The tests at each modulation frequency step continue for at least two full cycles of modulation. The input test signal is represented,

$$x\left(t\right) = X_m\left[1 + k_x \cos\left(\omega t\right)\right].\cos\left[\omega_0 t + k_a \cos\left(\omega t - \pi\right)\right] \tag{57}$$

The magnitude modulation level, kx and phase angle modulation level, ka are kept for 10% while the modulation frequency fm is varied from 0.1 to 5 Hz. The maximum, minimum and mean values of percentage TVE are shown in Fig.16. Frequency of the signal is measured in each case and FEs are determined.

Figure 16. Magnitude and phase angle modulation response at reporting rate of 50 fps.

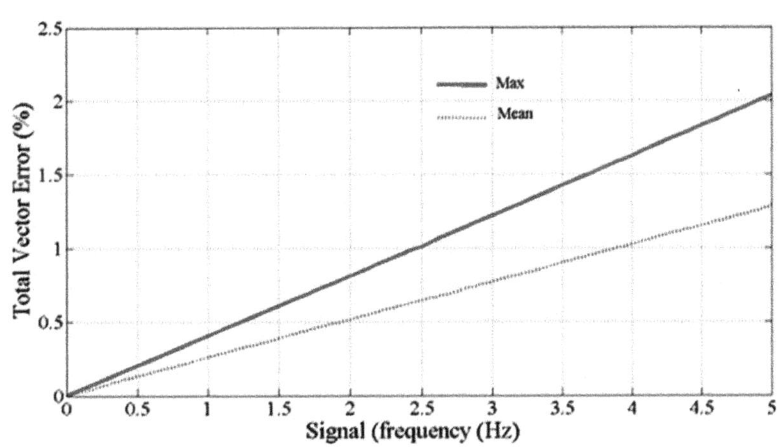

It is necessary to repeat the same test with only the phase angle modulated signal where kx = 0 and ka = 0.1. It is important to allow an adequate settling time to prevent parameter change transient effects from distorting the measurement. The PMU satisfies measurement bandwidth TVE compliance up to modulation frequency of 2 Hz as maximum TVE is less than 3% but it does not comply TVE requirement beyond the modulation frequency of 2 Hz. Thus, the PMU satisfies P class compliance but violates M class compliance.

8. Ramp of System Frequency

The input signal frequency is linearly ramped to test performances during power system frequency changes. The input test signal is represented as,

$$x\left(t\right) = X_m \cos\left(2\pi f_0 t + \pi R_f t^2\right) \tag{58}$$

Measurement performance during system frequency change shall be tested with linear ramp of the system frequency applied as balanced three-phase input signals (voltages and currents). Mathematically the input signals may be represented by

$$X_a = X_m \cos\left(2\pi f_0 t + \pi R_f t^2\right) \tag{59a}$$

$$X_b = X_m \cos\left(2\pi f_0 t - 2\pi / 3 + \pi R_f t^2\right) \tag{59b}$$

$$X_c = X_m \cos\left(2\pi f_0 t + 2\pi / 3 + \pi R_f t^2\right) \tag{59c}$$

where X_m is the amplitude of the input signal, f_0 is the nominal power system frequency, and $R_f \left(= \frac{df}{dt}\right)$ is the frequency ramp rate in Hz/s of fixed value in present equations.

The positive sequence X_1 signal corresponding to the above three-phase inputs is given by

$$X_1 = X_m \cos\left(2\pi f_0 t + \pi R_f t^2\right) \tag{60}$$

At reporting time tags *t=nT* (where *n* is an integer and *T* is the phasor reporting interval) the PMU shall produce a positive sequence measurement as

$$X\left(nT\right) = \left\{X_m / \sqrt{2}\right\} \angle \left\{\pi R_f \left(nT\right)^2\right\} \tag{61}$$

During ramp tests, the true values of frequency, frequency deviation, and ROCOF for the specified test signals at reporting time tags t = nT are given, respectively, by

$$f\left(nT\right)=\frac{\omega_0}{2\pi}+R_f nT \tag{62}$$

$$\Delta f(nT)=R_f nT \tag{63}$$

$$\frac{d}{dt}\left[f\left(nT\right)\right]=R_f \tag{64}$$

The signal frequency ramp rate, R_f is varied from negative ramp (−1.0 Hz/s) to positive ramp (+1.0 Hz/s) while the ramp range is from 45 Hz to 55 Hz. The maximum and mean values of percentage TVE are shown in Fig. 17.

Figure 17. Linear frequency ramp response at reporting rate of 50 fps.

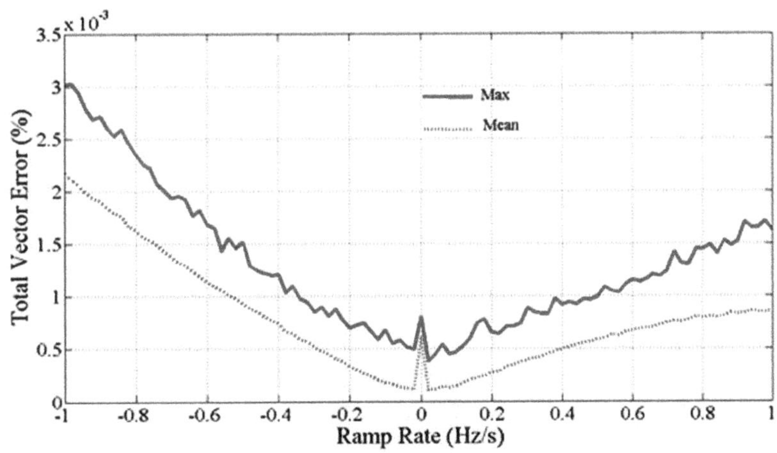

The PMU satisfies frequency ramp TVE compliance of both P class and M class as the maximum TVE is less than1% for ramp rate between −1.0 Hz/s to +1.0 Hz/s while the ramp range is from 45 Hz to 55 Hz.

9. Step Response

Step responses provide a method of comparing the PMU response to a sudden input change. which may be observed during load switching and faults. Step responses include magnitude and phase angle steps. Steps include positive and negative steps of 10% magnitude and 10% phase angle. The step is initiated by a signal at a precise time, which allows determining the response time, delay time and maximum overshoot/undershoot. As PMU response time and delay time are small compared to the PMU reporting interval it is difficult to characterize the response of a single step. Therefore, the equivalent sampling approach explained in (Phadke, A. G., & Thorp, J. S., 1988; Gurusinghe, D.R., et al., 2012) should be used to achieve the required measurement resolution.

A unit step function u(t) is applied to the input signal magnitude and phase angle and it is represented by,

$$x(t) = X_m\left[1 + k_x.u(t)\right].\cos\left[\omega_0 t + \phi_0 + k_a u(t)\right] \tag{65}$$

The magnitude step size, km and the phase angle step size, ka are taken as −0.1 and +0.1 for negative and positive steps respectively. Figure 18 represents magnitude positive step waveforms of input test signal, actual and measured magnitude response, and TVE response in the same timeline. Phase angle positive step response waveforms of input test signal, actual and measured phase angle response, and TVE response are illustrated in Fig.19.

Figure 18. Waveforms of magnitude positive step response at reporting rate of 50 fps.

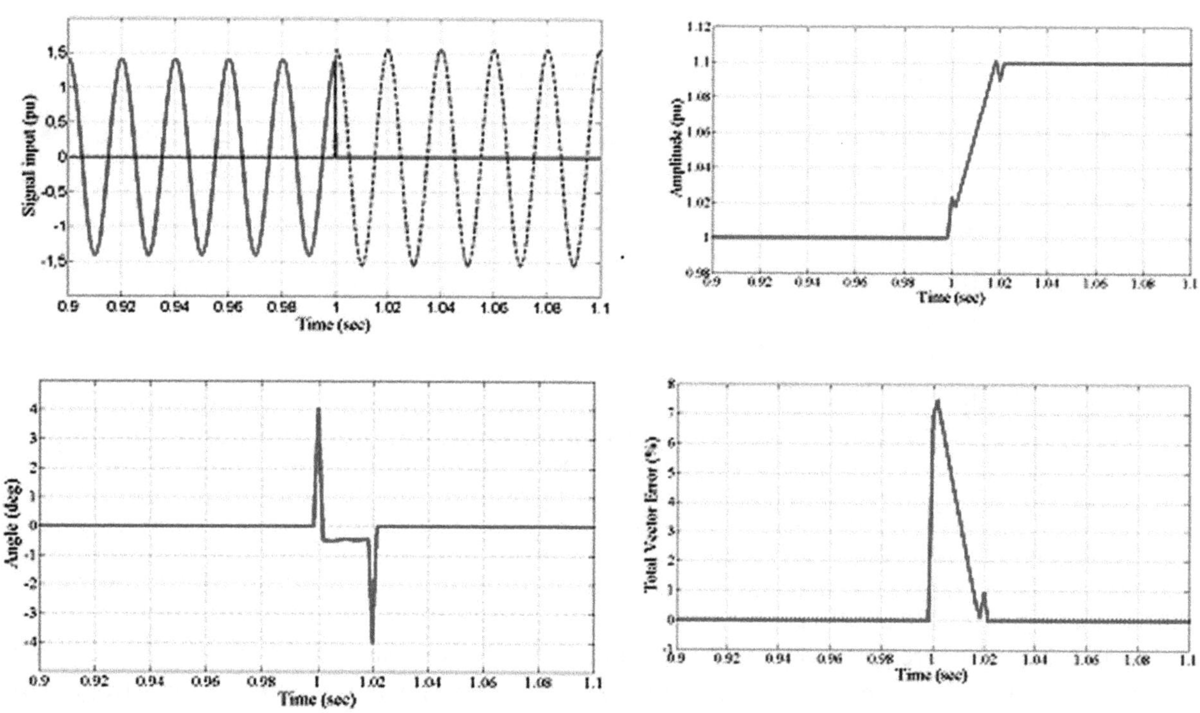

Reported values of a single step are represented by the dots on the continuous response curve, which is obtained by the equivalent time sampling approach. In magnitude step, the PMU fails to comply P class TVE demand as response time exceeds 1.7/f0 (0.028) seconds but it complies M class as the response time is less than 0.079 seconds. Furthermore, maximum overshoots/undershoots of the PMU is zero but the delay time is higher than 0.004 seconds at 60 fps reporting rate.

Figure 19. Waveforms of phase angle positive step response at reporting rate of 50 fps.

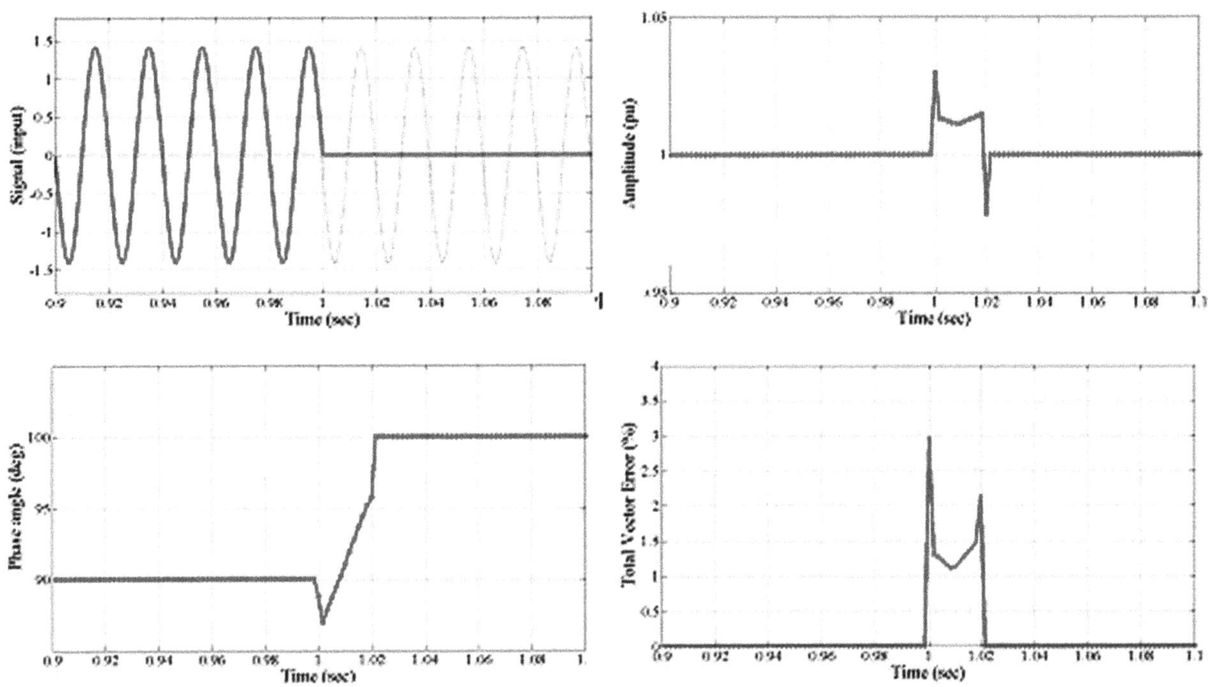

In phase angle step, the PMU fails to comply both P class and M class TVE demand as response time of TVE exceeds P class requirement of 1.7/f0 (0.028) seconds and M class requirement of 0.079 seconds. Furthermore, maximum overshoots/undershoots of the PMU exceeds 10% of step magnitude and the delay time is also higher than 0.004 seconds at 60 fps reporting rate. Table III is summarized step performances of the actual PMU.

CONCLUSION

This chapter discusses phasor and frequency measurements under transient system conditions. Both electromagnetic and electromechanical transients are considered. After explaining the basics of phasors, synchrophasors, and frequency and system transients, suggestions have been made for improving the interoperability of PMUs under transient conditions.

This chapter reviewed the PMU performance tests specified in (IEEE Std. C37.118-2005, 2006), developed a test methodology, and discussed some practical issues in the test environment. The PMU evaluation method used in this paper is simple, repeatable, and can be performed at any facility with commonly available standard signal playback equipment. The approach is based on the mathematically generated signals played back into the PMU using playback equipment with precise GPS synchronization. An actual PMU was tested using the proposed method and some sample test results were presented.

In dynamic tests, the signal is not purely sinusoidal and undergoes changes in its amplitude, phase angle, and frequency over a given interval. Even during steady-state tests, harmonics and noise superimposed on the signal are changing. Therefore, it is necessary to continue steady-state tests for over 5 seconds and the modulation tests over at least two full cycles of modulation. In the case of step response evaluation, proper resolution should be maintained to enable accurate determination of the response and delay times.

The actual PMU tested in the work satisfied steady-state TVE compliance tests of both P and M classes except the out of band interference test. In dynamic tests, the PMU satisfied the measurement bandwidth test of P class, and the linear frequency ramp test and the overshoot/undershoot requirements of the magnitude step response of both performance classes.

REFERENCES

Begovic, M. M., Djuric, P. M., Dunlap, S., & Phadke, A. G. (1993). Frequency tracking in power networks in the presence of harmonics. *IEEE Transactions on Power Delivery*, *8*(2), 480–486. doi:10.1109/61.216849

Bentarzi, H. (2010). Improving Monitoring, Control and Protection of Power Grid Using Wide Area Synchro-Phasor Measurements. *The 12th WSEAS Int. Conf. on Automatic control, modeling and simulation.*

Chen, J. (1994). *Accurate frequency estimation with phasor angles* (Master thesis). Virginia Polytechnic Institute and State University.

Giray, M. M., & Sachdev, M. S. (1989). Off-normal frequency measurements in electric power systems. *IEEE Transactions on Power Delivery*, *4*(3), 1573–1578. doi:10.1109/61.32645

Girgis, A.A., & Brown, R.G. (1981). Application of Kalman filter in computer relaying. *IEEE Trans. on PAS*, *100*(7), 3387-3395.

Girgis, A. A., & Ham, F. M. (1982). A new FFT-based digital frequency relay for load shedding. *IEEE Trans. on PAS*, *101*(2), 433–439. doi:10.1109/TPAS.1982.317125

Girgis, A. A., & Hwang, T. L. (1984). Optimal estimation of voltage phasors and frequency deviation using linear and non-linear Kalman filtering: Theory and limitations. *IEEE Trans. on PAS*, *103*(10), 2943–2951. doi:10.1109/TPAS.1984.318297

Gurusinghe, D. R., Rajapakse, A. D., & Narendra, K. (2012). Valuation of Steady-State and Dynamic Performance of a Synchronized Phasor Measurement Unit. *IEEE elect. Pow. And Ene. Conf. 2012.*

IEEE Std. C37.118-2005. (2006). *IEEE Standard for Synchrophasors for Power Systems*, IEEE Publication.

Johns, A. T., & Salman, S. K. (1995). Digital protection for power system. IEE power series N°15.

Nguyen, C. T., & Srinivasan, K. (1984). A new technique for rapid tracking of frequency deviations based on level crossing. *IEEE Trans. on PAS*, *103*(8), 2230–2236. doi:10.1109/TPAS.1984.318536

Ouadi, A., Bentarzi, H., & Maun, J. C. (2009). A new computer based phasor measurement unit framework. *Proc. 6th Int. Conf. SSD'09*, 1-6. 10.1109/SSD.2009.4956773

Ouadi, A., Bentarzi, H., & Maun, J. C. (2010). Improvement of Phasor Measurement Unit Performance. *Proc. 9th WSEAS International Conference on CIRCUITS, SYSTEMS, ELECTRONICS, CONTROL & SIGNAL PROCESSING (CSECS '10)*, 40-44.

Phadke, A. G., Hlibka, T., Adamiak, M. G., Ibrahim, M., & Thorp, J. S. (1981). A Microcomputer Based Ultra High Speed Distance Relay: Field Tests. *IEEE Trans. On PAS*, *100*(4), 2026–2036. doi:10.1109/TPAS.1981.316480

Phadke, A. G., Hlibka, T., Ibrahim, M., & Adamiak, M. G. (1979). A Microcomputer Based Symmetrical Component Distance Relay. *Proceedings of PICA*, 47-55. 10.1109/PICA.1979.720045

Phadke, A. G., Ibrahim, M., & Hlibka, T. (1977). Fundamental basis for Distance Relaying with Symmetrical Components. *IEEE Trans. on PAS*, *96*(2), 635–646. doi:10.1109/T-PAS.1977.32375

Phadke, A. G., Ibrahim, M., & Hlibka, T. (1977). Fundamental basis for Distance Relaying with Symmetrical Components. *IEEE Trans. on PAS*, *96*(2), 635–646. doi:10.1109/T-PAS.1977.32375

Phadke, A. G., & Thorp, J. S. (1988). *Computer Relaying for Power Systems*. John Wiley and Sons.

Phadke, A. G., Thorp, J. S., & Adamiak, M. G. (1983). A New Measurement Technique for Tracking Voltage Phasors, Local System Frequency, and Rate of Change of Frequency. *IEEE Transactions on Power Apparatus and Systems*, *102*(5), 1025–1038. doi:10.1109/TPAS.1983.318043

Phadke, A. G., Thorp, J. S., & Adamiak, M. G. (1983). A new measurement technique for tracking voltage phasors, local frequency, and rate of change of frequency. *IEEE Trans. on PAS*, *102*(5), 1025–1038. doi:10.1109/TPAS.1983.318043

Sachdev, M. S., & Giray, M. M. (1985). A least error squares technique for determining power system frequency. *IEEE Trans. on PAS*, *104*(2), 437–443. doi:10.1109/TPAS.1985.319059

Yalla, M. V. V. S. (1992). A digital multifunction protective relays. *IEEE Transactions on Power Delivery*, *7*(1), 193–201. doi:10.1109/61.108907

KEY TERMS AND DEFINITIONS

Discrete Fourier Transform (DFT): Is the algorithm that transforms the time domain signals to the frequency domain components. The DFT phasor calculation has an error when the power system operates at off-nominal frequencies.

Phasor Measurement Unit (PMU): Has been defined by the IEEE as "a device that produces Synchronized Phasor, Frequency, and Rate of Change of Frequency (ROCOF) estimates from voltage and/or current signals and a time synchronizing signal". A common time source (GPS) may be used for synchronization. The resulting measurement is known as a synchrophasor.

Phasors Estimation: Phasors that consists of the magnitude and phase angle of an electrical phasor quantities such as voltage or current of the power grid can be estimated using a different techniques such as the Newton method, Kalman filtering and level tracking, least square Discrete Fourier Transform, Smart Discrete Fourier Transform. PMU can capture samples from a signal waveform in quick succession and reconstructing the phasor quantity in the form of an angle and a magnitude measurement.

Smart Discrete Fourier Transform (SDFT): The Smart DFT algorithm developed by (Phadke, A. G., et al., 1977) is used to estimate correctly the instantaneous frequency and then the phasor amplitude and phase angle.

Chapter 5
Phasor Measurement Improvement Using Digital Filter in a Smart Grid

Abderrahmane Ouadi
Institute of Electrical and Electronic Engineering, University M'hamed Bougara of Boumerdes, Algeria

Abdelkader Zitouni
University M'hamed Bougara of Boumerdes, Algeria

ABSTRACT

During a transient operation condition of power smart grid, line current may include unwanted components that may cause unnecessary tripping of protection system. The disturbance mainly appears in a form of harmonics and sub-harmonics. In this case of signal waveforms including harmonics, the low pass filter may be used. However, this type of filter does not provide the ability to reject sub-harmonics. This chapter presents the digital filtering design issue based on optimization approach for removing sub-harmonics and hence improving the measurement. The first point of view is to reach an unified accurate phasor measurement algorithm that is immune to nearly all disturbances (sub-harmonics) in power grid including FACT devices and renewable energy sources, simultaneously with required speed of convergence. The second point focuses on reducing the computational requirement and algorithm complexity through designing recursive digital filter with reduced order.

INTRODUCTION

An electric power system is a large interconnected system that produces, transmits and distributes an electric energy to different consumers. Stability of the smart power system is of a great concern, since it is subjected to different disturbances that may cause false measurement and hence mal-operation of protective and control systems which may lead to complete system collapse. Therefore, many techniques have been developed to mitigate the disturbance effects by introducing the digital filtering for computing

DOI: 10.4018/978-1-7998-4027-5.ch005

the fundamental phasors of line voltages and currents. Several digital algorithms have been proposed in the literature for estimating the fundamental phasors, such as the discrete Fourier transform, cosine algorithm, least square algorithm, Kalman filter and wavelet transform. The discrete Fourier transform (DFT) filter which has proven reliability and robustness in relaying application was considered as the most widely used and most popular algorithm for estimating phasors. However, during a transient time of smart grid condition, line fault current may include unwanted components that are associated to the fault impedance level as well as to the power transmission line type. The first type of disturbance mainly appears in a form of dc exponentially decaying component associated with the line fault current signal waveforms (Boutora, S., Bentarzi, H., & Ouadi, A., 2011). In this case of signal waveforms including dc components, phasors estimated through application of the DFT filter may have large errors during the first cycles. This problem stems from the fact that the signal containing dc exponential decaying components has non integer harmonics (using the Fourier transform of this dc component) while the DFT filter does not provide the ability to reject non integer type of harmonics. The second kind of disturbance that may occur in power transmission line including series compensation devices in presence of relatively high fault impedance, where a significant distortion is introduced in the faulted line current waveforms. This is due to the behavior of non-linear components integrated in series compensated devices. A considerable exponentially decaying sub-harmonic or sub-synchronous resonance is associated with fundamental and integer harmonic of the line current waveform. This type of fault is more severe since it reduces the speed of convergence of the DFT algorithm to correctly estimate the fundamental phasors.

The digital filtering design issue based on optimization approach is aimed to achieve different perspectives. The first point of view is to reach a unified accurate phasor measurement algorithm that is immune to nearly all disturbances in power grid, (including other type of FACT devices) introducing considerable and different kind of waveform distortions affecting both line voltage and current signals, simultaneously with required speed of convergence. Although, only the line current waveform distortion filtering is of concern and considered in this work. The second point, focus on reducing the computational requirement and algorithm complexity through designing recursive digital filter with reduced order.

Digital filters exist in two types: Finite impulse response (FIR) and Infinite impulse response (IIR) or recursive. FIR filters suffer from the problem of high order (hence implementation and performance issues) if strict requirements are imposed at the design stage. Furthermore, IIR filters can have smaller group delay than its equivalent FIR filters (Ouadi, A., Bentarzi, H.; & Recioui, A., 2014; Ouadi, A., Bentarzi, H., & Zitouni A.,2018). The optimal design of an infinite impulse response (IIR) filter consists in choosing a set of coefficients of the filter to have a frequency response that optimally approximates the desired response (Ouadi, A., Bentarzi, H.; & Recioui, A., 2013; Kaur, R. Dhillon, J.S. Singh, D., 2010; Kezunovic, M., & Kasztenny, B., 2000).

Different techniques exist for the design of digital filters. Windowing method; in which the ideal impulse response is multiplied by a window function, is the most popular. There are various kinds of window functions (Butterworth, Chebyshev, Kaiser etc.), depending on the requirements on ripples in the passband and stopband, stopband attenuation and the transition width. These various windows limit the infinite length impulse response of ideal filter into a finite window to design an actual response. Furthermore, windowing methods do not allow sufficient control of the frequency response in the various frequency bands and other filter parameters such as transition width. The designer always has to compromise between the design specifications (Ouadi, A., Bentarzi, H.; & Recioui, A., 2014; Ouadi, A., Bentarzi, H., & Zitouni A.,2018).

Due to the presence of the denominator of the transfer function, the stability condition of the filter should be taken into account in the optimal design (Ouadi, A., et al., 2014; Ouadi, A., Bentarzi, H., & Zitouni A.,2018; IEEE Standard C37.118.1-2011., 2011; Kaur, R. Dhillon, J.S. Singh, D., 2010; Yu, C. –S., 2006), resulting in a constrained optimization problem. Several sufficient conditions (Ouadi, A., Bentarzi, H., & Zitouni A.,2018; Kaur, R. Dhillon, J.S. Singh, D., 2010; Tamura, K., & Yasuda, K., 2011) have been established for the parameterization that represents the filter's denominator by a single polynomial. The triangle-based stability conditions (Ouadi, A., Bentarzi, H.; & Recioui, A., 2014) are necessary and sufficient and have been incorporated into several design procedures (Phadke, A.G., Thorp, J.S., 1988; Math Works Inc., 2002; Tamura, K., & Yasuda, K., 2011) that formularize the filter's denominator by cascaded second-order sections (SOSs). In (Widraw B. et al., 1975), variable transformation is used to convert the finite stability region into the entire coefficient space, such that the original constrained design problem becomes an unconstrained one in the transformed space. However, the transformation increases the nonlinearity of the objective function, which makes it hard to find good (global optimum) solutions in general. In (Tamura, K., & Yasuda, K., 2011), a perturbed stability triangle is proposed to guarantee the SOS to have its zeroes inside a circle of given radius. It is combined with the Gauss–Newton strategy, resulting in an improved design. In (Antoniou, A., & Lu, W.-S., 2007), the conditions presented for the SOS with zeroes inside a circle of given radius enclose a triangular stability domain and can be easily incorporated into any constrained optimization formulations based on the SOS parameterization. A method that divides the overall design of an IIR filter into successive designs of its second-order sections is presented in (Nagaraj, S. B. et al, 2010), where one section is first designed, and then, another section is appended until all sections are designed.

Because of finite word length effects occurring in practical implementations of the designed filters, not only stability of the filter is of great importance but a stability margin is necessary as well. The poles of the transfer function should not lie too close to the unit circle. The sensitivity of pole locations to coefficient quantization increases with decreasing distance from the unit circle. Poles close to the unit circle may considerably enhance quantization noise and increase the maximum amplitude of small scale limit cycle. Consequently, it is desirable to have control over the maximum radius when designing IIR filters.

Linear-phase filters are usually designed as non-recursive (FIR) filters which can have constant group delay over the entire baseband. However, when highly selective filters are required, a very high filter order is needed which makes these filters uneconomical or impractical. To eliminate this problem, attempts have been made to develop methods to design recursive (IIR) filters whose delay characteristics approximate a constant value in the pass-band. This includes IIR filter design approach that can satisfy both magnitude and phase characteristics simultaneously (Gu, J., Shen, K., Yu S., & Yu, C., 2006; Guo, Y. Kezunovic M., & Chen, D., 2003; Girgis, A.A., & Brown, R. G., 1982). The design of IIR filters with constant group delay in the pass-band is also carried out by using all-pass structures through evaluation of phase response instead of approximating the group delay directly (Benmouyal, G., 1995; Osman, A., & Malik, O., 2004; Gu, J. –C., Yu, S. –Li., 2003). Some other methods used an indirect approach based on model reduction techniques where a linear-phase FIR filter that meets the required specifications is first designed and then a lower order IIR filter that meets the original amplitude specifications while maintaining a linear-phase response in the pass-band is obtained (Isaksson, A., 1988; Yu, S.L., & Gu J.C., 2001; Phadke, A.G., Hlibka, T., & Ibrahim, M., 1976). (Coretlazzo, G. et al. 1984) have achieved the simultaneous design in both magnitude and group delay of IIR and FIR filters based on multiple criterion optimizations. (Lutovac M.D., 1997) has developed a new design method for elliptic IIR filters that provide the implementation of half of the multiplication constants with few shifters and adders. (Sullivan, J. L., & John W. A., 1998) have proposed the algorithm based on the peak–constrained least–squares optimality criterion for cascaded IIR filters, which can design a filter that has an equalized group delay without the use of all pass filters, and it can simultaneously meet the frequency response magnitude specifications by using all of the filter coefficients available to optimize the filter. Lang, M.C. (2000) has used least square method for designing IIR filter with prescribed magnitude and phase response . This parameterization of the transfer function has been used for designing IIR filters. Gordana Jovanovic-Dolecek (2006) has proposed a method for the design of IIR notch filters with desired magnitude characteristic, which can be either maximally flat or equi-ripple. Xi Zhang, (2008) has proposed a novel method for designing maximally flat IIR filters with flat group delay responses in the passband.

Under these circumstances, evolutionary and metaheuristic optimization methods find their place. These are referred to as global optimizers while the more familiar, traditional techniques such as conjugate gradient and the quasi-Newtonian methods are classified as local optimizers. The distinction between local and global search of optimization techniques is that the local techniques produce results that are highly dependent on the starting point or initial guess, while the global methods are totally independent of the initial conditions (Nagaraj, S. B., et al., 2010). Though they possess the characteristic of being fast in convergence, local techniques, in particular the quasi-Newtonian techniques have direct dependence on the existence of at least the first derivative. In addition, they place constraints on the solution space such as differentiability and continuity; conditions that are hard or even impossible to satisfy in some situations (Nagaraj, S. B., et al., 2010).

Previously, global optimization techniques have been implemented in the design of digital filters. One such approach using neural networks has been described in (Kaur, R., Dhillon, J.S., Singh, D., 2010). Also, use of PSO in the design of frequency sampling finite impulse response (FIR) filter has been described in (Nongpiur, R.C., Shpak, D.J., & Antonious, A., 2013; Antoniou, A., 1993; Antoniou, A., 2006). Differential evolution has been used in the design of digital filters has been implemented in Storn's work (Dumitrescu, B., & Niemisto, R., 2004; C. Y.-F. et al., 2008; Jiang, A., & Kwan, H. K., 2009; Jiang, A., & Kwan, H. K., 2010). Tang et al. (1998); and Nurhan Karaboga (2004) have used Hierarchical Genetic Algorithms to design and optimization of IIR filter structures. Use of Particle Swarm Optimization (PSO) and Genetic Algorithms (GA) in the design of digital filters is described in (Ababneh, J. I., Bataineh M. H., 2007).

Compared with traditional optimization techniques and other global optimizers, the spiral optimization method is easy to implement and very efficient in reaching optimum solutions. Spiral optimization method has been recently developed based the analogy to spiral phenomena (Ouadi, A., Bentarzi, H.; & Recioui, A., 2013; Ouadi, A., Bentarzi, H., & Zitouni A.,2018). The spiral phenomena occurring in nature are approximated to logarithmic spirals as in Fig. 1. Examples of natural spiral dynamics include whirling currents, low pressure fonts, nautilus shells and arms of spiral galaxies. Logarithmic spirals discrete processes to generate spirals that can form an effective behavior in metaheuristics. A two-dimensional algorithm has been first proposed (Ouadi, A., Bentarzi, H.; & Recioui, A., 2013), and then, a more generalized n-dimensional version has been recently suggested (Ouadi, A., Bentarzi, H., & Zitouni A.,2018).

In the present work, the use of the spiral optimization technique is presented and used to solve the multiobjective IIR filter design. First, based on practical requirements involved in power system application, a detailed mathematical IIR filter design formulation is presented. Next, the spiral optimization algorithm detailed and implemented to solve the optimization task. A first optimization example aims at matching a desired magnitude response only. Later, the design is improved by considering the other requirements including minimum and constant group delay.

DIGITAL FILTERING APPROACH

A digital filter based solution is proposed to remove unwanted disturbances using digital filter design techniques. The filter time response must be included in the requirements. The present filtering application imposes different kind of specifications. On one hand, the time domain requirement where both a high speed and accurate system response are needed. On the other hand, the frequency domain requirements (DC, sub-synchronous and harmonic components elimination) which are the magnitude response within small bandwidth including sharp frequency edges as well as an approximately constant group delay in this band are required too. Usually the best optimum value of all the objective functions of this filter design can be obtained for some values of design variables. A compromise or a trade-off between the objective functions must be made to achieve a satisfactory filter design.

The considered recursive digital filter must satisfy three multi-objective functions. These functions are: 1) meet a specified or a desired magnitude response specification; 2) an approximately constant group delay; and 3) a minimum time response or settling time which involves a minimum phase or a group delay. The optimization approach considers the discrete-time transfer function which is formulated on the basis of some desired amplitude response and a stability margin δ parameter. A norm of the weighted error function is then minimized with respect to the transfer-function coefficients with a prescribed maximum pole radius referred to as stability margin. The stability margin parameter is varied to optimize the filter coefficients which minimizes mainly the magnitude response, satisfies the best approximately constant group delay and the lowest group delay that leads to minimum settling time or time delay of the system dynamic response.

Filter Transfer Functions

In the general case an IIR filter can be described by its discrete-time difference equation,

$$y[n]+\sum_{i=1}^{2N-1} c_i y[n-1]=\sum_{j=0}^{2M} d_j x[n-j] \tag{1}$$

Where x(n) and y(n) are discrete-time input and output signals. Equation (1) can be transformed into the Z-domain and assuming c_i and d_i are real coefficients a second order form transfer function can be obtained, having 2M conjugate zeros and 2N conjugate poles; called hereafter second order sections (SOS), as:

$$H(z)=H_0\frac{\prod_{j=1}^{2M} a_{0j}+a_{1j}z+z^2}{\prod_{j=1}^{2N} b_{0j}+b_{1j}z+z^2} \tag{2}$$

Where a_{ii} and b_{jj} are real coefficients and H_0 is a positive multiplier constant. The polar formulation is also useful and is written as:

$$H(z) = H_0 \frac{\prod_{j=1}^{2M}\left(z - r_{aj}e^{j\theta_{aj}}\right)\left(z - r_{aj}e^{-j\theta_{aj}}\right)}{\prod_{j=1}^{2N}\left(z - r_{bj}e^{j\theta_{bj}}\right)\left(z - r_{bj}e^{-j\theta_{bj}}\right)} \tag{3}$$

Where r_{aj}, θ_{aj} and r_{bj}, θ_{bj} are the radii and angles of the zeros and poles, respectively.

Filter Stability Margin

The system stability margin parameter increases and the system settling time decreases as poles are moved toward the origin (hence decreasing the pole radius). This action, in fact, brings two required and important properties to the designed system. First, the system time or dynamic response is enhanced as settling time is decreased. Second, the system stability becomes more robust which is a very useful property, particularly in practical implementation. Indeed, the rounding or truncation of the filter coefficients may lead to an unstable implementation if the stability margin is too small. It is therefore desirable to approximate a given response by a transfer function with a prescribed maximum pole radius named stability margin as shown in Fig.1.

Figure 1. Stability region and stability margin parameter δ.

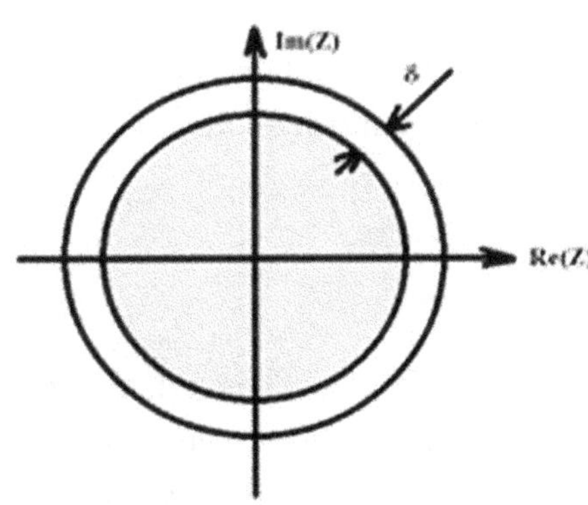

Magnitude Response Objective Function

The amplitude and the phase responses of a recursive filter is given by

$$H(z) = \left|H\left(e^{j\omega t}\right)\right|,\ \varphi\left(x,\omega\right) = \arg\left(H\left(e^{j\omega t}\right)\right) \tag{4}$$

Where ω is the frequency and x is a column vector with 2M+2N+1 elements and depend on the used formulation, that is a Cartesian form as

$$x = \left[a_{01}a_{11}a_{21}b_{01}b_{11}b_{21}\ldots b_{2M}H_0\right] \tag{5-a}$$

Or in polar form as

$$x = \left[\rho_a \rho_b H_0\right] \tag{5-b}$$

Where

$$\rho_a = \left[r_{a1}, \theta_{a1}, \ldots, r_{aM}, \theta_{aM}\right] \tag{5-c}$$

and

$$\rho_b = \left[r_{b1}, \theta_{b1}, \ldots, r_{bN}, \theta_{bM}\right] \tag{5-d}$$

The superscript T denotes the transpose operation. An approximation error can be formulated as the difference between the actual amplitude response $M(x,\omega)$ and the desired amplitude or magnitude response $M_a(\omega)$ as:

$$e(x) = M(x,\omega) - M_d(\omega) \tag{6}$$

By sampling the error function e(x,ωi), the actual and the desired amplitude responses M(x,ωi) and Md(ω) at frequencies: ω1, ω2, …, ωk, the column error vector is constructed as:

$$\left[E(x)\right] = \left[e_1(x), e_2(x), \cdots, e_k(x)\right]^T \tag{7}$$

Where $e_i(x) = M(x,\omega_i) - M_d(\omega_i)$ and $\{\omega_i$: 1,2,…,$k\}$ is a dense set of frequencies which are distributed over in the pass-band and stop-band of the filter. A weighting or penalty error is included to control portions of the actual filter response curve that are most important to the filter response. This involves modifying the error to the form:

$$\overline{e}_i(x) = W(\omega)\left(M(x,\omega_i) - M_d(\omega_i)\right) \tag{8}$$

Where $W(\omega)$ is a weighting piece wise constant function over all frequency space, which is assigned a positive value greater than one or less than one to increase or decrease the magnitude approximation in a given band.

A recursive filter can be designed by finding a point $x = \hat{x}$ in (8) such that

$\bar{e}_i(x) \approx 0$ for $i=1,2,\ldots,k$.

Such a point can be obtained when solving the optimization problem by minimizing the error function $\bar{e}_i(x)$. The design of a recursive filter that approaches a specified response $M_d(\omega_i)$, can be performed by minimizing the error objective function in terms of Lp norm error which is formulated as:

$$\min_x \mu\left(x\right) \text{ Subject to } r_{aj} \leq 1-\delta \tag{9-a}$$

Where,

$$\mu\left(x\right) = e_{\max}\left(x\right)\sum_{i=1}^{l} W_i\left(\omega\right)\left\{\left[\frac{\left|e_i\left(x\right)\right|}{e_{\max}\left(x\right)}\right]^p\right\}^{\frac{1}{p}} \tag{9-b}$$

And $e_{\max}\left(x\right) = \max_{1 \leq i \leq l}\left(e_i\left(x\right)\right)$

The p is a positive integer. The spiral technique is used to minimize $\mu(x)$ for increasing values of p.

Group-Delay Objective Function

The group delay is derived from the phase relation and defined as:

$$\tau\left(x,\omega\right) = -\frac{d\varphi\left(x,\omega\right)}{d\omega} \tag{10-a}$$

Where $\varphi(x,\omega)$ is the phase response of the filter, and stated as

$$\varphi\left(x,\omega\right) = \arg\left(H\left(e^{j\omega T}\right)\right) \tag{10-b}$$

The group delay for a synthesized recursive filter is desired to be unchanged in the considered region one hand, in the other hand a minimum group delay is another required property needed in certain applications. In the present design, a constant group delay property is considered which is defined as:

$$\tau(x,\omega_i) = \tau_c \text{ for } \omega_i \in \omega_p \tag{11}$$

Where τ_c is constant, and ω_p is the pass-band regions of the filter. In the present application τc is an unknown but can be considered as the mean value over the pass-band region, which can be determined as:

$$\tau_{c,m} = \frac{1}{n_k} \sum_{i=1}^{n_k} \tau\left(\bar{x}_m, \omega_i\right) \text{ for } \omega_i \in \omega_p \tag{12}$$

Where $\bar{x}_m$ is the optimal filter coefficient determined by minimizing the magnitude objective function for an mth stability margin parameter.

The stability margin parameter is varied for discrete values, from which an optimal constant group delay is determined by minimizing the following objective error function:

$$\min_m E_\tau\left(m\right) = \frac{1}{\tau_{c,m}} \sum_{i=1}^{n_k} \left(\tau\left(\bar{x}_m, \omega_i\right) - \tau_{c,m}\right) \text{ for } \omega_i \in \omega_p \tag{13}$$

The multi-objective optimization problem is solved by discretizing the stability margin δm pa,ameter, the magnitude optimization algorithm is used to generate the corresponding filter coefficients' described by (10), in which basis the group delay is synthesized where a feasible and optimal solution can be obtained by minimizing the objective function (13).

DIGITAL FILTER IMPLEMENTATION AND TESTS RESULTS

The optimization algorithm was implemented using MATLAB in which spiral optimization technique is used. The digital filter to be optimized is to be used to eliminate harmonics and sub-harmonics in a power network with a fundamental frequency of 50 Hz. The filter is a band-pass type and is desired to satisfy the magnitude response to ideally pass only frequencies confined in the interval [45Hz, 55Hz] while rejecting all other frequency content. The sampling frequency is taken to be 1800 Hz. The filter is of order 10 and is hence composed of 5 cascaded SOS's. At start up, the filter is optimized to match the magnitude response specifications only. Next, more constraints are added to the optimization process including minimum and linear phase and constant group delay to enhance the designed filter performance.

Figure 2. Magnitude filter response (SOOF)

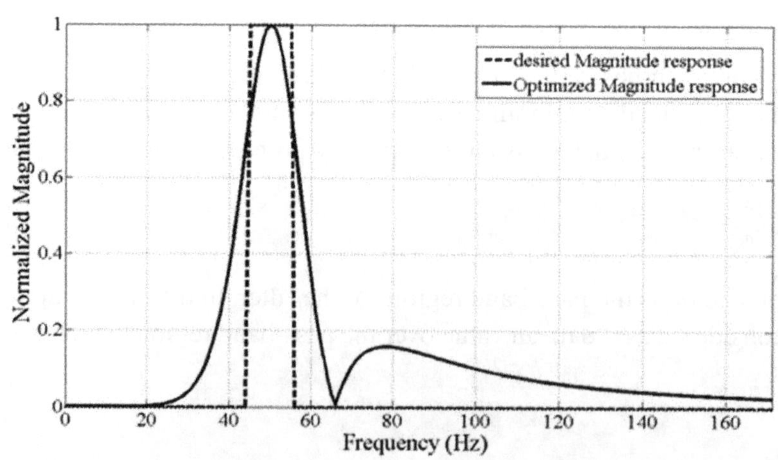

Figure 3. Step response of the filter and time delay (about 100 ms) (SOOF).

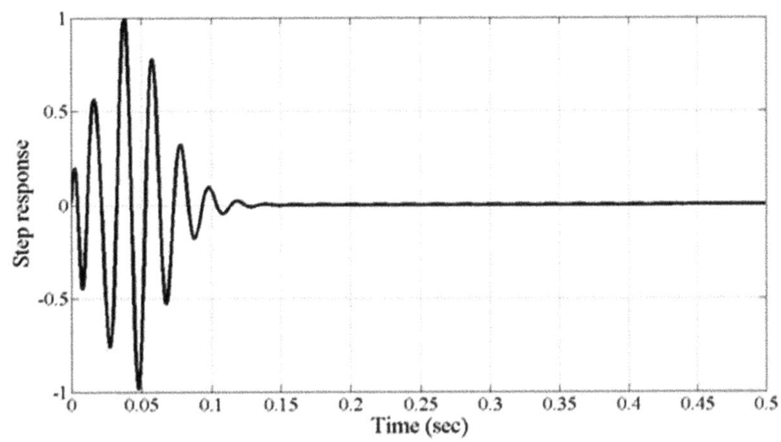

Single Objective Design

The purpose of this part is to design a filter which only satisfies the magnitude response described earlier without considering any other performance criteria. The filter is designated single objective optimized filter (SOOF) hereafter. Fig. 5 shows both the desired and the optimized magnitude response of the digital filter. It can be noted that the filter has good performance in the sense that it fulfills the requirements of magnitude response. Indeed, the filter response falls exactly within the desired response and it attenuates all other frequencies as the overall Side-lobe level does not exceed -16dB.

However, the optimized filter is not of practical use as it suffers from drawbacks in the dynamic properties. First, as shown in figure 3, the time delay is in the order of 100 ms which is not suitable in the present applications as the requirements specify that the time delay should not exceed one cycle (20 ms). In addition, the group delay is not constant as shown in figure 4. Furthermore, the phase delay is nonlinear as shown in figure 5 and accordingly a non constant group delay. As a result, it is necessary to include all the preceding performance criteria into the design process and hence, the problem becomes

Figure 4. Group delay response of single objective optimized filter (SOOF).

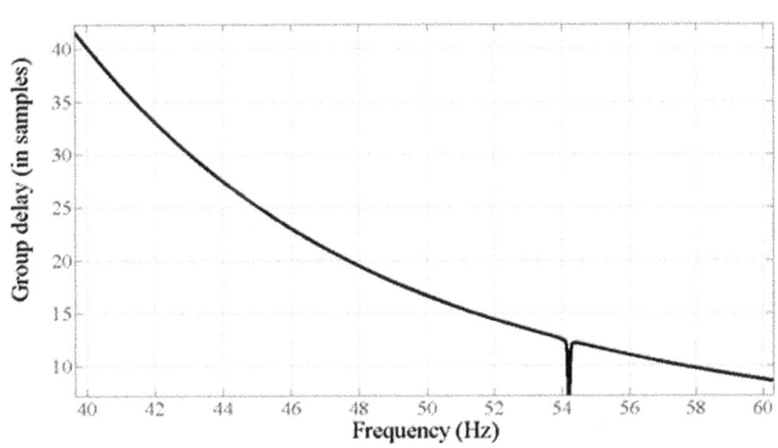

Figure 5. Phase response of filter with non linear phase or non constant group delay (SOOF).

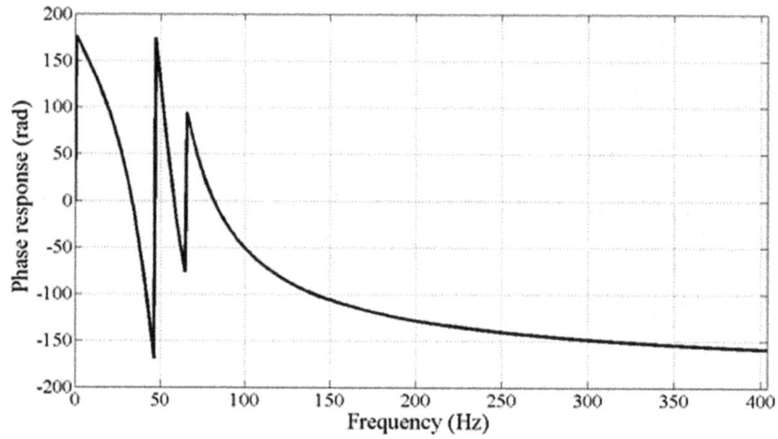

a multi-objective optimization task. The results are summarized in table 1 where are presented the filter SOS coefficients and SOS gains.

Multi-Objective Filter Design

The inclusion of the constant and minimum group delay in the optimization task besides magnitude response criterion produced a filter which satisfies almost all requirements. The filter compromises between these criteria to produce the best possible trade-off. The filter is thereafter labeled multi-objective optimized filter (MOOF). The magnitude response of the MOOF is shown in figure 6. The filter does not show good performance in terms of this characteristic compared to the SOOF. Indeed, the SOOF bandwidth is better than the MOOF which makes it having better selectivity.

Yet, the MOOF performs better in terms of the other criteria. In terms of phase and group responses, figures 7 and 8 illustrate the phase and group delay responses. The filter is characterized by a linear phase with an almost constant group delay in the passband. Furthermore, the group delay is minimum producing a small time response as shown in Fig. 9. Indeed, the time delay of the MOOF is about 14.9 ms which conforms to the desired requirements. This is due to the fact that the stability margin or equivalently the pole radii have been taken into account in the optimization process and these latter have been lowered to a value of 0.595 instead of being closer to unity. Table 2 summarizes the SOS coefficients and gains of the MOOF.

Table 1. Single-objective optimized filter (SOOF) SOS coefficients and gains

Section	Numerator			Denominator			Gain
1	1	−1,9520335260	0,9999998000	1	−1,8473812327	0,8836	0,00089772300915
2	1	−1,9999996000	0,99999980000	1	−1,9999996000	0,999999800000010	1
3	1	−1,9999996000	0,99999980000	1	−1,8473812327	0,88360000000	1
4	1	−1,8851094580	0,99999980000	1	−1,8473812327	0,883600000000	1
5	1	1,99999980000	0,99999980000	1	−1,8473812327	0,88360000000	1

Figure 6. Magnitude filter response (MOOF).

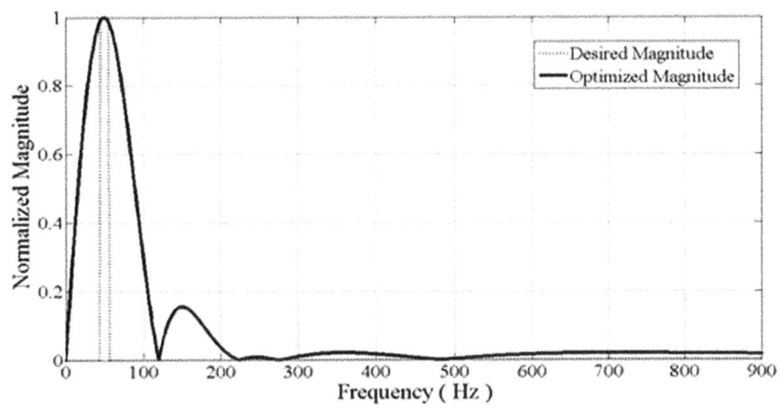

Figure 7. Optimized filter phase response MOOF.

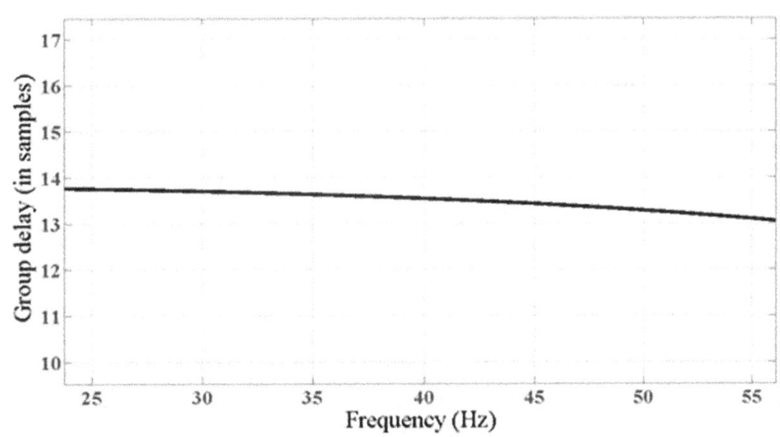

Figure 8. Group delay response of optimized filter MOOF.

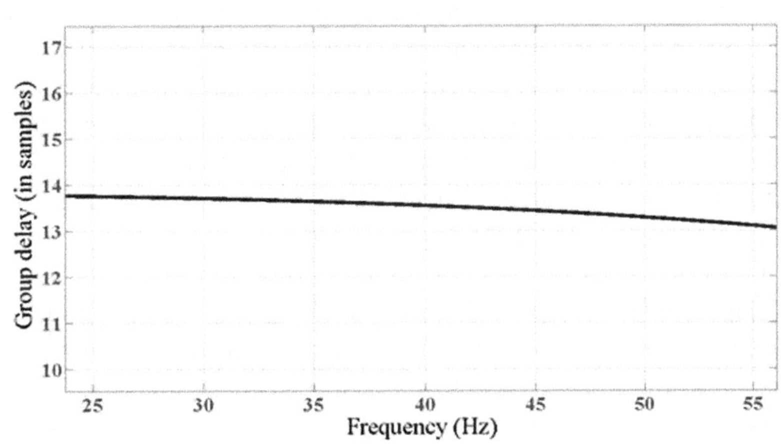

Figure 9. Optimized filter step response and time delay (14.9 ms) MOOF.

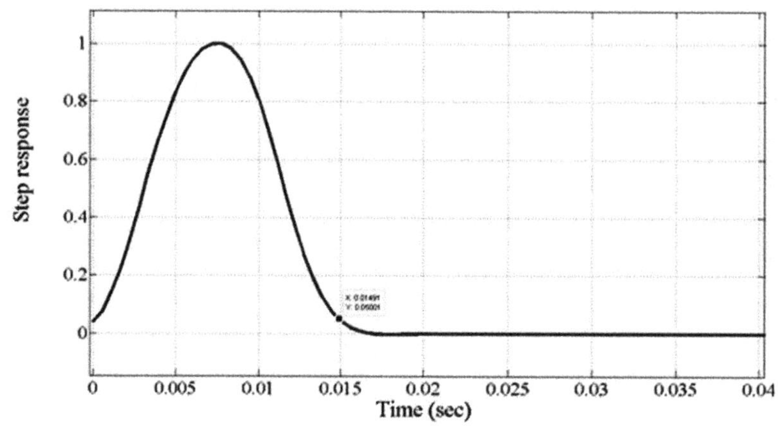

Table 2. Multi-objective optimized filter (MOOF) SOS coefficients and gains

Section	Numerator			Denominator			Gain
1	1	−1,82933833319	0,999999800000	1	−1,0914383222	0,35402500000	0,012499922808
2	1	−1,4247565639	0,99999980000	1	−1,0914383222	0,35402500000	1
3	1	−1,1469437066	0,99999980000	1	−1,1899988100	0,35402500000	1
4	1	0,227981442276607	0,999999800000010	1	−1,09143832226114	0,354025000000000	1
5	1	−0,467539619942456	−0,532460226811516	1	−1,11373644574629	0,354025000000000	1

To validate the obtained results, the MOOF optimized filter is compared with the classical filter designs widely used in literature. The filters have the same order and should meet the same requirements described earlier. The filters considered are the least p^{th} optimized filter, the elliptic and Chebyshev type II filters.

The classical filters have magnitude responses that meet well the requirements as compared to the MOOF. This can be explained by the fact that these filters are mainly designed to match a given magnitude response which can be considered to be similar to the SOOF optimized filter. However, according to Figure 10, the MOOF has lower and constant group delay over the pass-band region compared to the classical filters. In fact, the delays of the classical filters are in orders of 200 and 300 sample delay (the MOOF delay is at 14 samples) which is not acceptable for practical requirements. Also, the classical design approach relies on matching a required magnitude response followed by an all-pass filter equalizer. Hence, the classical design adds on the system order and the computational complexity. The latter approach is not suitable for real time applications.

To better assess the performance of the filter (MOOF), two tests have been performed. In the first, a step sinusoidal signal of 50 Hz is input to the filter and Figure 11 shows both filter input and output waveforms. It is seen that the filter output matches exactly the input except for a phase shift and a time delay of less than one cycle. Hence, this filter proves to be practical for high speed measurement systems where the system accuracy is of great importance. In the second test, the previous step sinusoidal signal is corrupted with a DC offset, harmonic and sub-harmonic components. The sub-harmonic component is set to 25 Hz and the harmonics to 100, 150 and 200 Hz. Both input and output signals are shown in

Figure 12. As it is clearly seen, the filter succeeded in eliminating the DC and harmonic components and mitigating the sub-harmonic component.

CONCLUSION

A filtering approach is used to completely remove or significantly mitigate the effect of fault disturbances on phasor measurement using digital band-pass filter. The digital filter has been specified to meet several objectives as constant group delay, minimum phase and sharp narrow band-width gain frequency response. A recursive digital filter is then designed using a multi-objective optimization technique based on the spiral optimization algorithm.

Figure 10. Group delay comparison between the MOOF and the classical filter designs.

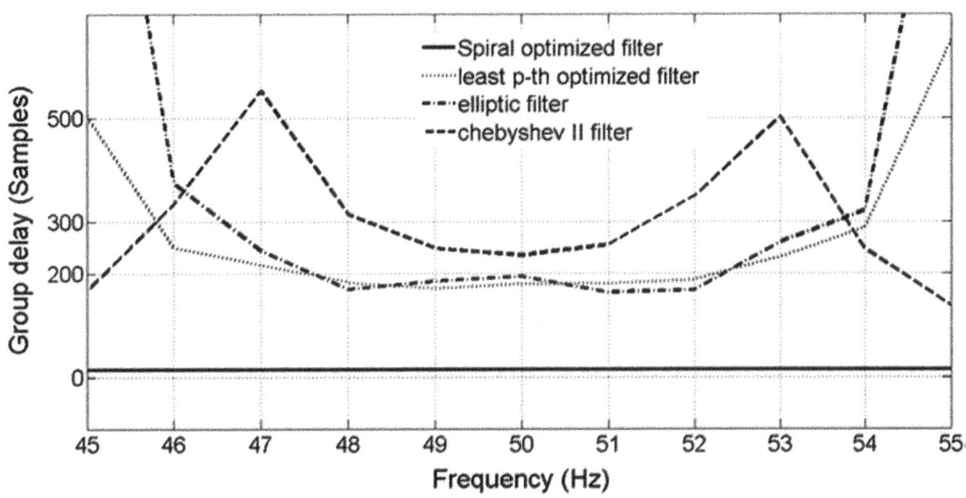

Figure 11. Test signal: step of sinusoidal signal input and filter output response.

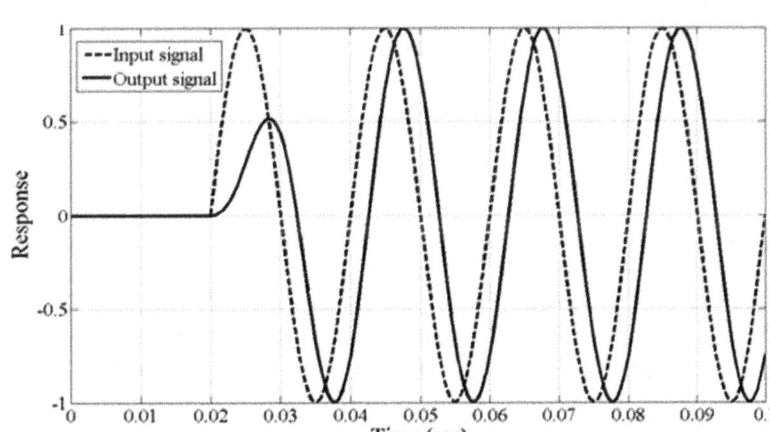

Figure 12. Second test signal by considering a step sinusoidal signal mixed with harmonic, sub-harmonic and dc offset components.

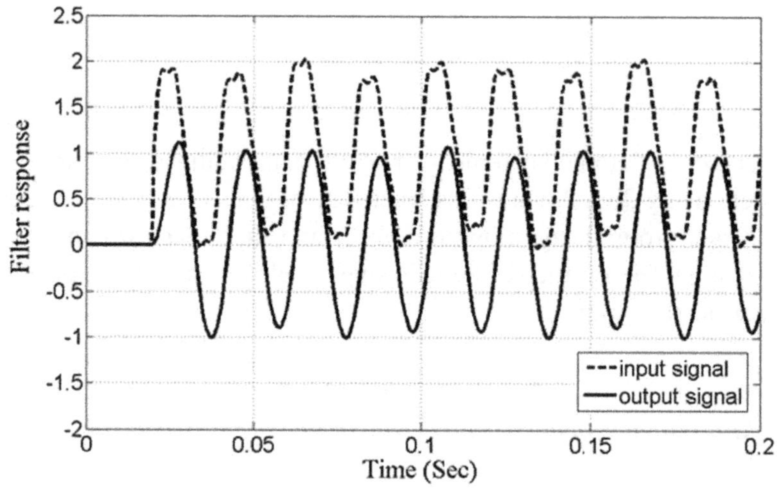

The application of the spiral optimization method to design a multiobjective digital filter has been considered in this research work. The method has been tested by considering objectives of the filter design were to match a desired magnitude response while having a minimum and linear phase. At start up, only magnitude response has been considered in the optimization task. The resulting filter was good in terms of this characteristic while it showed awful dynamic and phase performance. Next, the dynamic properties were included in the optimization algorithm to solve a multiobjective task. The spiral optimization method has succeeded in attaining the optimal design in terms of the previous requirements by achieving a compromise between them. The optimized filter has been tested and it showed good performance with required practical characteristics.

REFERENCES

Ababneh, J. I., & Bataineh, M. H. (2007). Linear phase FIR filter design using particle swarm optimization and genetic algorithms. *Digital Signal Processing*. Advance online publication. doi:10.1016/j.dsp.2007.05.011

Antoniou, A. (1993). *Digital Filters: Analysis and Design* (2nd ed.). McGraw Hill.

Antoniou, A. (2006). *Digital signal processing: signals, systems and filters*. McGraw Hill.

Antoniou, A., & Lu, W.-S. (2007). *Practical Optimization: Algorithms and Engineering Applications*. Springer.

Benmouyal, G. (1995). Removal of DC-offset in Current Waveforms Using Digital Mimic Filtering. *IEEE Transactions on Power Delivery*, *10*(2), 621–630. doi:10.1109/61.400869

Boutora, S., Bentarzi, H., & Ouadi, A. (2011). Analysis of the Disturbances in Distribution Networks using Matlab and ATP. *International Journal of Exergy*, *5*(1), 9–16.

Cortelazzo, G., & Lightner, M. R. (1984). Simultaneous design in both magnitude and group-delay of IIR and FIR filters based on multiple criterion optimization. *IEEE Transactions on Circuits and Systems*, *32*(5), 949–967.

Dumitrescu, B., & Niemisto, R. (2004). Multistage IIR filter design using convex stability domains defined by positive realness. *IEEE Transactions on Signal Processing*, *52*(4), 962–974. doi:10.1109/TSP.2004.823497

Girgis, A. A., & Brown, R. G. (1982). A new Kalman filtering-Based Digital Distance Relay. *IEEE Trans. PAS*, *101*(9), 3471–3480. doi:10.1109/TPAS.1982.317574

Gu, J., Shen, K., Yu, S., & Yu, C. (2006). Removal of dc offset and subsynchronous resonance in current signals for series compensated transmission lines using novel Fourier filter algorithm. *Electric Power Systems Research*, *76*(5), 327–335. doi:10.1016/j.epsr.2005.08.002

Gu, J.-C., & Yu, S.-Li. (2003). Removal of DC offset in current and voltage signals a novel Fourier Algorithm. *IEEE Transactions on Power Delivery*, *15*(1), 73–79.

Guo, Y., Kezunovic, M., & Chen, D. (2003). Simplified Algorithm for Removal of the Effect of Exponentially Decaying DC-offset on the Fourier Algorithm. *IEEE Transactions on Power Delivery*, *18*(3), 711–717. doi:10.1109/TPWRD.2003.813894

Ho, C. Y.-F., Ling, B. W.-K., Chi, Z.-W., Shikh-Bahaei, M., Liu, Y.-Q., & Teo, K.-L. (2008). Design of near allpass strictly stable minimal-phase real valued rational IIR filters. *IEEE Transactions on Circuits and Wystems. II, Express Briefs*, *55*(8), 781–785. doi:10.1109/TCSII.2008.922417

IEEE Standard C37.118.1-2011. (2011). *IEEE standard for synchrophasor measurements for power systems*. IEEE Publication.

Isaksson, A. (1988). Digital Protective Relaying through Recursive Least-Squares Identification. *IEE Proceedings. Part C. Generation, Transmission and Distribution*, *135*(5), 441–449. doi:10.1049/ip-c.1988.0057

Jiang, A., & Kwan, H. K. (2009). IIR digital filter design with new stability constraint based on argument principle. *IEEE Transactions on Circuits and Systems. I, Regular Papers*, *56*(3), 583–593. doi:10.1109/TCSI.2008.2002116

Jiang, A., & Kwan, H. K. (2010). Minimax design of IIR digital filters using iterative SOCP. *IEEE Transactions on Circuits and Systems. I, Regular Papers*, *57*(6), 1326–1337. doi:10.1109/TCSI.2009.2028749

Jovanovic-Dolecek, G. (2006). Design of IIR Notch filters with maximally flat or equiripple magnitude characteristics. *14th Europeon Signal Processing Conference (EUSIPCO 2006)*, Florence, Italy.

Karaboga, N. (2004). Design of minimum phase digital IIR filters by using Genetic Algorithm. *Proceedings of the 6th Nordic signal Processing Symposium-NORSIG*.

Kaur, R., Dhillon, J. S., & Singh, D. (2010). Multicriterion Optimization approach to the design of stable recursive Digital Filter. *International Journal of Engineering Science and Technology*, *2*(8), 3515–3524.

Kezunovic, M., & Kasztenny, B. (2000). Design Optimization and Performance Evaluation of the Relaying Algorithms, Relays and Protective Systems Using Advanced testing Tools. *IEEE Transactions on Power Delivery*, *15*(4), 1129–1135. doi:10.1109/61.891493

Lang, M. C. (2000). Least-Squares Design of IIR Filters with Prescribed Magnitude and Phase Response and a Pole Radius Constraint. *IEEE Transactions on Signal Processing*, *48*(11), 3109–3121. doi:10.1109/78.875468

Lutovac, M. D., & Milic, L. D. (1997). Design of computationally efficient elliptic IIR filters with a reduced number of shift-and-add operations in multipliers. *IEEE Transactions on Signal Processing*, *45*(10), 2422–2430. doi:10.1109/78.640708

Math Works Inc. (2002). *Power System Blockset User's guide*. Author.

Nagaraj, S. B., Nongpiur, R., & Antoniou, A. (2010). Design of Nearly Linear-Phase Recursive Digital Filters by Using Unconstrained Least-pth MinimaxOptimization. *Proceedings of the 4th International Symposium on Communications, Control and Signal Processing, ISCCSP 2010.*

Nagaraj, S. B. R. N., & Antonious, A. (2010). Design of Nearly Linear-Phase Recursive Digital Filters by Using Unconstrained Least-pthMinimax Optimization. *Proceedings of the 4th International Symposium on Communications, Control and Signal Processing, ISCCSP 2010*. 10.1109/ISCCSP.2010.5463429

Nongpiur, R.C., Shpak, D.J., & Antonious, A. (2013). *Improved Design Method for Nearly linear-phase filters using Constrained Optimization*. Academic Press.

Osman, A., & Malik, O. (2004). Protection of parallel transmission lines using wavelet transform. *IEEE Transactions on Power Delivery*, *19*(1), 49–55. doi:10.1109/TPWRD.2003.820419

Ouadi, A., Bentarzi, H., & Recioui, A. (2013). Optimal multiobjective design of digital filters using spiral optimization technique. *SpringerPlus*, *2*(1), 461. doi:10.1186/2193-1801-2-461 PMID:24083108

Ouadi, A., Bentarzi, H., & Recioui, A. (2014). Optimal Multi-Objective Design of Digital Filters Using Taguchi Optimization Technique. *Journal of Electrical Engineering*, *65*(1), 21–29. doi:10.2478/jee-2014-0003

Ouadi, A., Bentarzi, H., & Zitouni, A. (2018). Spiral Search Optimization Algorithm Applied to (IIR) Digital Filter Design. *International Conference On Engineering Technologies.*

Phadke, A. G., Hlibka, T., & Ibrahim, M. (1976). A digital computer system for EHV Substation: Analysis and Field Tests. *IEEE Transactions on Power Apparatus and Systems*, *PAS-95*(1), 291–301. doi:10.1109/T-PAS.1976.32105

Phadke, A. G., Hlibka, T., & Ibrahim, M. (1976). A digital computer system for EHV Substation: Analysis and Field Tests [January/February.]. *IEEE Transactions on Power Apparatus and Systems*, *PAS-95*(1), 291–301. doi:10.1109/T-PAS.1976.32105

Phadke, A. G., & Thorp, J. S. (1988). *Computer relaying for Power Systems*. Research Studies Press Ltd. John Wiley & Sons Inc.

Sullivan, J. L., & John, W. A. (1998). PCLS IIR digital filters with simultaneous frequency response magnitude and group delay specifications. *IEEE Transactions on Signal Processing*, *46*(11), 2853–2861. doi:10.1109/78.726800

Tamura, K., & Yasuda, K. (2011). Primary study of spiral dynamics inspired optimization. *IEEE Transactions on Electrical and Electronic Engineering*, *6*(S1), 98–100.

Tang & Man. (1998). Design and Optimization of IIR Filter Structure Using Hierarchical Genetic Algorithms. *IEEE Transactions on Industrial Electronics*, *45*(3).

Widraw, B. (1975). Adaptive noise cancelling; principles and applications. Proc. IEEE, 1692.

Yu, C.-S. (2006). A Discrete Fourier Transform-based Adaptive MIMIC phasor Estimator for Distance Relaying Applications. *IEEE Transactions on Power Delivery*, *21*(4), 1836–1846. doi:10.1109/TPWRD.2006.874609

Yu, S. L., & Gu, J. C. (2001). Removal of Decaying DC in Current and Voltage Using a Modified Fourier Filter Algorithm. *IEEE Transactions on Power Delivery*, *15*(4).

Zhang, X. (2008). Design of maximally flat IIR filters with flat group delay responses. *Signal Processing*, *88*(7), 1792–1800. doi:10.1016/j.sigpro.2008.01.016

KEY TERMS AND DEFINITIONS

Digital Filter: In signal processing, a digital filter is a system that performs mathematical operations on a sampled, discrete-time signal to reduce or enhance certain aspects of that signal. Digital filters can often be made very high order, and are often finite impulse response filters which allows for linear phase response.

FACT Devices: FACTS devices are static power-electronic devices installed in AC transmission networks to increase power transfer capability, stability, and controllability of the networks through series and/or shunt compensation. These devices are also employed for congestion management and loss optimization.

Harmonic: In an electric power system, a harmonic is a voltage or current at a multiple of the fundamental frequency of the system, produced by the action of non-linear loads such as rectifiers, discharge lighting, or saturated magnetic devices.

Sub-Harmonic: Sub-harmonics are harmonics having frequencies less than the fundamental frequency. Power systems can exhibit subharmonic currents under some conditions.

Section 2
Management and Control in Smart Grids

Chapter 6
Implementation of Load Control for Smart Metering in Smart Grids

Abdelmadjid Recioui
https://orcid.org/0000-0001-9028-3910
University M'hamed Bougara of Boumerdes, Algeria

Fatma Zohra Dekhandji
University M'hamed Bougara of Boumerdes, Algeria

ABSTRACT

The conventional energy meters are not suitable for long operating purposes as they spend much human and material resources. Smart meters, on the other hand, are devices that perform advanced functions including electrical energy consumption recording of residential/industrial users, billing, real-time monitoring, and load balancing. In this chapter, a smart home prototype is designed and implemented. Appliances are powered by the grid during daytime, and a photovoltaic panel stored power during the night or in case of an electricity outage. Second, consumed power from both sources is sensed and further processed for cumulative energy, cost calculations and bill establishment for different proposed scenarios using LABVIEW software. Data are communicated using a USB data acquisition card (DAQ-USB 6008). Finally, a simulation framework using LABVIEW software models four houses each equipped with various appliances. The simulator predicts different power consumption profiles to seek of peak-demand reduction through a load control process.

LIST OF ACRONYMS

AC: Alternating Current.
Ah: Ampere-Hours.
AMI: Advanced Metering Infrastructure.
As: Ampere-Seconds.

DOI: 10.4018/978-1-7998-4027-5.ch006

C: Coulomb.
DC: Direct Current.
DG: Distributed power generation.
DR: Demand Responce.
DPDT: Double Pole Double Throw.
DPST: Double Pole Single Throw.
DSM: Demand Side Management.
HAN: Home Area Monitoring.
HEMS: Home Energy Management System.
LAN: Local Area Network.
mA: Milliampere.
NC: Normally Closed.
NO: Normally Open.
PLC: Power Line Carrier.
PV: Photovoltaic.
RE: Renewable Energy.
RER: Renewable Energy Resources.
RES: Renewable Energy System.
RET: Renewable Energy technologies.
RF: Radio Frequency.
SG: Smart Grid.
SM: Smart Meter.
SPDT: Single Pole Double Throw.
SPST: Single Pole Single Throw.
ToU: Time of Use.
V: Volt.
W: Watt.
WAN: Wide Area Network.
Wh: Watt-hours.

INTRODUCTION

For a century, utility companies have used conventional power grid system to provide electricity, the traditional grid is facing numerous challenges including lack of techniques to automatically monitor the grid system and transfer useful data, besides with the ever increase in electricity demand and with the pressure from ever increasing energy demand and the reduction of fossil fuels, which pollute the environment, finding new energy resources and enhancing energy efficiency have become priority of many nations in the 21st century (Recioui & Dassa, 2017; Chellali et al., 2014).

Renewable Energies in general and solar photovoltaic (PV) in particular are a good alternative to ensure sustainability in the global energy production with environmental protection because it is a renewable and a non-polluting energy source.

The demand for power is increasing dramatically in developed and developing countries. In 2030, an increase of 50% and 40% of energy consumption is expected in the US and Europe respectively. It is expected to triple in China and India (Environnement, 2010).

In recent years, a new generation of power grid system, referred to as the Smart Grid (SG), with an aim of managing electricity demand and increasing the share of electricity generators based on renewable energies has emerged.

SG, also called intelligent grid or future grid, is a modern improvement of the existing traditional power grid. It is a network that consists of many systems and subsystems designed to provide cost-effective, reliable and efficient supply system, and create bidirectional communication between the consumer and the grid operator of electric power, built on an advanced cluster of distributed generation units, renewable energies, energy storage, equipment, and circuits for generating, transmitting, transforming, and distributing energy. It enables a real-time two-way information exchange between a customer and utility company through the exploitation of information and communication technologies (Recioui et al., 2015).

If a grid connected system is producing more power than the home consumes, the surplus is fed into the power grid. Some electricity companies meter the electricity fed into the grid by the user's system and provide a credit on his bill.

Smart meters (SM) are a key component of SG, which can be used for reducing peak demand and adapting flexible demand to fluctuating generations (Li et.al., 2011). This device enables two-way communication between suppliers and consumers in real time. Moreover, it has the ability to remotely control electrical devices to manage loads and demands (Fang et. al., 2012).

In a SG, Demand Side Management (DSM) has been developed since early 1980s to balance the time varying demand load of consumers and maximum power generation capacity in the power system. In DSM, the pricing mechanisms and direct control strategies are employed by the energy suppliers to affect consumers' consumption behaviors and reshape the total load.

1. State of the Art

Recently, the number of electricity consumers is increasing in a great extent that handling and maintaining the power as per the growing requirements became such a hard task (Preethy and Harish, 2016). It is expected that 60% of all energy consumption will be converted and used as electricity. Therefore; it is a demand that production, distribution and use of electrical energy are done as efficient as possible. Further the recent challenges with nuclear power plants are arguing to find more sustainable energy generation solutions. Of many options, two major technologies will play important roles to solve parts of those future challenges. One is to change the electrical power production from conventional, fossil based energy sources to renewable energy sources; the other is to use high efficient power electronics in power generation, transmission, distribution and end-user application (Kaźmierkowski et al., 2002).

Distributed Power Generation (DG) is emerging a new paradigm to produce on-site highly reliable and good quality electrical power. The concept is particularly interesting when different kinds of energy resources are available, such as photovoltaic (PV) panels, fuel cells (FC's) or wind turbines. The DG of different kinds of energy systems allow the integration of renewable and non-conventional energy resources. Hence, the DG is becoming a part of strategic plans for most countries to address current challenges associated with energy management (Guerrero et. al., 2010).

The main advantage of using renewable energy sources (RES) are the elimination of harmful emissions and the reduction of carbon footprints resulting from the primary energy. Denmark has a high power capacity penetration (>30%) of wind energy in major areas of the country, in 2002 (28%) of the whole electrical energy consumption is covered by this kind of energy. However the main disadvantage apart from the high cost is the uncontrollability. The availability of renewable energy sources has strong daily and seasonal patterns and the power demand by the consumers could have a very different characteristic. Therefore it is difficult to operate a power system installed with only renewable generation units due to the characteristic differences and the high uncertainty in the availability of the RES (Kaźmierkowski et al., 2002).

The authors in (Cecilia and Sudarsanan, 2016) state that the 21's century requirements cannot be achieved using the current grid and the modernization of the traditional grid is mandatory. The authors describe both the traditional and smart power grid. They define the current grid, also called the traditional grid, as a federal system where there is only one-way flow of electric power, from supplier, transmission, and distribution up to the consumers.

Smart grid studies and researches have mostly focused on conventional interconnected grid systems, works performed concerning the integration of renewable energies in smart grid are classified according to the research characteristics as follows: giving the concept of smart grid as well as pros and cons, technology adoption, optimal allocation of renewable energy in smart grid system, pricing and forecasting, and challenges resulting from the renewable energy integration (Phuangpornpitak and Tia, 2013).

(Koykul, 2011; Gaviano et. al., 2012) discussed smart grid applications and its potential study in the future by presenting the advantages and disadvantages of smart grid for renewable energy distributed generation pointing that the share of renewable energies requires coordinated efforts and standardized communication and certified products in order to ensure security, safety and reliability in a diversified future grid infrastructure.

(Reston et al., 2009; Kian and Keyhani, 2001; Iskin et al., 2012) consider the factors which can impact renewable energy investment decisions whether they are economic, technical, social political factors and obstacles.

The maintenance of the power is also important as the human operator goes to the consumer's house and produces the bill as per the meter reading, however if the consumer is not available, the billing process will be pending and the human operator needs to revisit. This process tends to be a laborious task and requires much time, human and material resources (Preethi and Harish, 2016).

In (Preethi and Harish, 2016), a wireless data communication based smart meter has been designed and implemented automatic metering and billing system where the amount of the consumed electric energy is displayed continuously on an LCD using Zigbee as a mean of data communication between the user and the controlling base. The system is developed to send a short message service (SMS) about the cost of the consumed energy along with the due date, and also to detect and control the power theft by identifying authorized and unauthorized users. As there is no human intervention, in the entire design and process, there is no chance of human error and corruption.

In (Bhimte et al., 2015), a smart meter is designed and developed in LABVIEW for domestic/industrial consumers, the prototype accounts for electricity consumption, power quality, generates monthly electrical bill along with incentives provided and penalties levied for power quality and moreover establishes communication between the consumer and the utility.

2. Current Development of Smart Metering in Different Countries

a) France

Electricity and gas markets are dominated by Electricité de France (EDF) and Electricité Réseau Distribution France (ERDF), the two companies have the ability to decide the activities on the smart meter field. The French government promulgated a decree in August 2010 that has the following policy objectives (Renner at al., 2013);

- From January 2012, every new electricity meter installed must be a smart device.
- By the end of 2014, 50% of all meters must be connected to an automated meter management (AMM) system.
- By the end of 2016, 95% of the meters must be connected to an AMM system
- EDF identified four areas in which smart meter will make great contribution to provide improvement within the energy field in 2006 [16];

The billing and customer service.

- Grid operation and monitoring.
- Remote connecting and disconnecting.
- The accounting.

ERDF planned to improve some new services by rising the potential of smart metering services and incorporating completely new information and data management systems (Zheng et al., 2013).

b) Germany

From the beginning of 2010, the national legislature requires the installation of smart meters in new and refurbished buildings. Because the Germany government follows a policy driven by customer demand smart metering service investment is very small, there are only about 15 out of 800 utilities offering smart metering products in the early 2010 and only the customers with high consumption are likely to install the smart meter and benefit from it (Zheng et al., 2013).

Other policy objectives are considered for the investment in smart metering including (Renner at al., 2013);

- By the end of 2010, the metering operator has to arrange smart meter that shows the time table of use and the real consumption of energy.
- From the beginning of 2008, the customer has the right to receive a monthly, quarterly, biannual or annual bill.
- From 2011 and on, utility companies are required to provide load-variable or time-of-use tariffs.

3. Latest Products of Smart Meters

Based on the market demand, many companies have produced meters that can be categorized into two applications; residential smart meters and commercial/industrial smart meters. The common companies producing smart meters include; General Electric (GE) company, Itron Company and Sensus Company (Zheng et al., 2013).

Smart grid implementation encourages end-users to actively take part in the grid operation and make them behave as real actors of the entire electricity production and distribution system (Miceli, 2013). Demand of electricity is increasing day by day, the demand and supply should be continuously balanced to avoid supply interruption. The overall load connected to the grid is in reality the sum of many individual working devices load profiles it is commonly variable and not stable in nature. The power demand fluctuates considerably over time (Palensky and Dietrich, 2011), resulting in a rapid increase of power consumption that stands up as the main cause of grid disruption and black-outs (Miceli, 2013). Power plants are designed to meet the peak demand resulting in an increase in the cost of generation. Hence, an effective solution for this problem is Demand Side Management (DSM) that reduces the peak demand and brings immediate benefits to the utility and the customers (Palensky and Dietrich, 2011). (Han et al., 2011) proposed a green HEMS based on energy comparison. The feedback on energy consumption to energy users was used to reduce total energy use. This HEM system figures out the efficiency of a home appliance as compared to the others by displaying energy usage information of individual appliances. So it is necessary to collect the energy usage data of home appliances in this system.

BACKGROUND

Smart Grid Renewable Energy and Storage

The electricity system; the grid, consists of electricity generation, electricity transmission, electricity distribution and electricity consumption. Traditionally, power is generated in bulk from few central generators and distributed to a large number of load centers with electricity users and customers via transmission lines (Amin and Wollenberg, 2005).

The quest for clean, secure, safe, efficient, resilient and sustainable grid with interactive services has considerable implications on the existing grid, thus; developing the network became such a necessary issue that – with huge efforts – made born of the Smart Grid. A brief definition of this intelligent solution would be the concept of transforming an electric power grid by using advanced communication, automated controls and other forms of technology. It integrates new innovative tools from generation, transmission, and all the way to consumer appliances and equipment (Jenkins et al., 2015). The concept of two-way communication is clearly defined in describing the smart grid according to the European Technology Platform "A Smart Grid is an electricity network that can intelligently integrate the actions of all users connected to it generators, consumers and those that do both in order to efficiently deliver sustainable, economic and secure electricity supplies."(Kabalci, 2016).

1. Traditional Grid vs. Smart Grid

The contribution of smart grid to conventional power grids is to provide bidirectional flow of energy and communication signals to create an advanced automatic and distributed energy delivery network (Amin and Wollenberg, 2005). (Table 1) represents a brief comparison between the existing grid and the smart grid.

Table 1. Comparison between the existing grid and the Smart grid

Existing Grid	Smart Grid
Electromechanical	Digital
One-way communication	Two-way communication
Centralized generation	Distributed generation
Few sensors	Sensors throughout
Manual monitoring	Self-monitoring
Manual restoration	Self-healing
Failure and blackouts	Adaptive and islanding
Limited control	Pervasive control
Few customer choices	Many customer choices

(Fang et al., 2012)

2. General Features and Benefits of Smart Grid

The conventional grid has been quite deteriorated from its first installation up to the recent years, and become unable to meet the actual requirements and demands. Smart grid provides quality power that fulfills customer's needs considering the changes and the challenges (Kabalci, 2016). The general benefits and features of smart grid are the following (Souran et al., 2016):

1. Improving power reliability and quality.
2. Enhancing capacity and efficiency of existing electric power networks.
3. Improving resilience to disruption.
4. Optimizing the utilization of resources and equipment.
5. Providing real time data and monitoring.
6. Enabling predictive maintenance and self-healing responses to system disturbances.
7. Facilitating expanded deployment of renewable energy sources.
8. Accommodating distributed power sources.
9. Automating maintenance and operation.
10. Reducing greenhouse gas emission by enabling electric vehicles and new power sources.
11. Reducing oil consumption by minimizing the need for inefficient generation during peak usage periods.
12. Presenting opportunities to improve grid security.
13. Increasing consumer choice and enhancing the interaction between users and markets.

14. Enabling new products, services and markets.

3. Architecture of Smart Grid / Model and Requirements

The smart grid being a complex system consists of an architecture that covers the generation, transmission and the distribution processes as well as the estimation of the residential use of energy. Different models and architectures are proposed to describe the infrastructure and hence the basic working principle of the smart grid.

a) The Conceptual Model

The conceptual model represents the planning and organization of the diverse, expanding collection of interconnected networks that compose the smart grid dividing it into seven domains as described in (Table 2) and shown graphically in (Figure 1).

Each domain comprises Smart grid actors and applications. Actors include devices, systems and programs that make decisions and exchange information and data necessary for applications performances, examples are; Smart meters, solar generators and control systems. Furthermore, applications represent tasks performed by one or more actors within a domain, home automation, solar energy generation, energy storage and management are examples of applications (Souran et al., 2016).

b) Smart Grid and the Integration of Renewable Energies

The key goal of smart grid is to promote active customer participation and decision making as well as to create the operation environment in which both utilities and electricity users influence each other. In smart grids, users can influence utilities by adding distributed generation sources based on sustainable energies at the point of use (Phuangpornpitak and Tia, 2013).

Figure 1. Interaction of actors in different Smart Grid domains
(Souran et al., 2016)

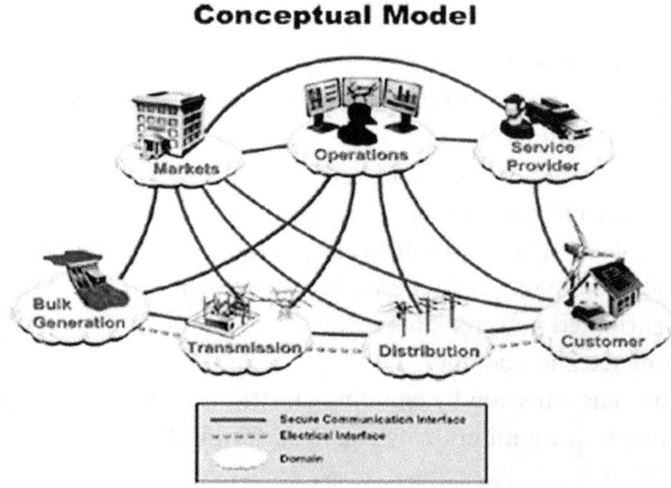

Table 2. Domains and actors in the Smart Grid conceptual model

Domain	Actors in the Domain
Customer	The end users of electricity. May also generate, store and manage the use of energy.
Markets	The operators and participants in electricity markets.
Service providers	The organizations providing services to electrical customers and utilities.
Operations	The managers of the movement of electricity.
Bulk generation	The generators of electricity in bulk quantities. May also store energy for later distribution.
Transmission	The carriers of bulk electricity over long distances. May also store and generate electricity.
Distribution	The distributors of electricity to and from customers. May also store and generate electricity.

(Souran et al., 2016).

c) Sustainable Energy Options for Smart Grid

Sustainable energy is derived from natural sources that replenish themselves over of time, sometimes called green power, because they are considered environmentally friendly and socially acceptable, they include sun, wind, hydro, biomass and geo-thermal. A power system having wind powered turbines, micro generators, fuel cell-based system and PV systems augmenting the main power lines will constitute a distributed power generation (DG) system (Momoh, 2012).

Renewable energy sources cannot support the entire grid by themselves. So, they have to be connected to the main grid acting as auxiliary power sources thus reducing the overall burden on the primary power generation units. In recent years, a significant number of programs have been implemented in various parts of the world in recent years, leading to an increase in the utilization of renewable energy sources in smart grid system. Some studies show that these technologies can provide reliable and comparatively low-cost electricity service (Phuangpornpitak and Tia, 2013).

Renewable energy options are meant to provide the smart grid with (Momoh, 2012);

1. Remote utilization and storage of renewable resources output.
2. Enhancement of functionality of grid-connected renewable energy systems (RES).
    a. Facilitating give-and-take of energy from the system.
    b. Redistribution/reallocation of unused power from the grid-connected RES.
    c. Facilitating storage of grid-generated and RER-generated energy by backup storage technologies at user end.
    d. Tracking interaction for billing.
3. Enhancement of functionality of electric vehicles and plug-in hybrids.
4. Utilization of vehicle battery packs as energy storage devices.

Renewable energy technologies (RET`s) – hydropower, biomass, wind and solar photovoltaic- have been successfully demonstrated over the years, contributing with 19% of electricity production, a vast majority (83%) of it being from hydroelectric power. Solar energy uses and applications have been justified and strongly recommended for grid electrification (Phuangpornpitak and Tia, 2013).

d) Photovoltaic Smart Grid System

The growing installations of renewable energy resources require a coordinated effort from the planning stage, all the way down to the electronic devices used for power generation, distribution, storage and consumption (Gaviano et al., 2012). The overall system is known as a PV system, (Figure 2) represents the complete PV system, whereas (Figure 3) represents different elements that constitute the PV system from the generation down to the load including the PV generator, the inverter, the charge controller and the storing element. Each component of the PV system is explained separately in the following sections.

Figure 2. A schematic of a PV system including batteries, Power conditioners, and both AC and DC loads (Vlot and Gunnink, 2011).

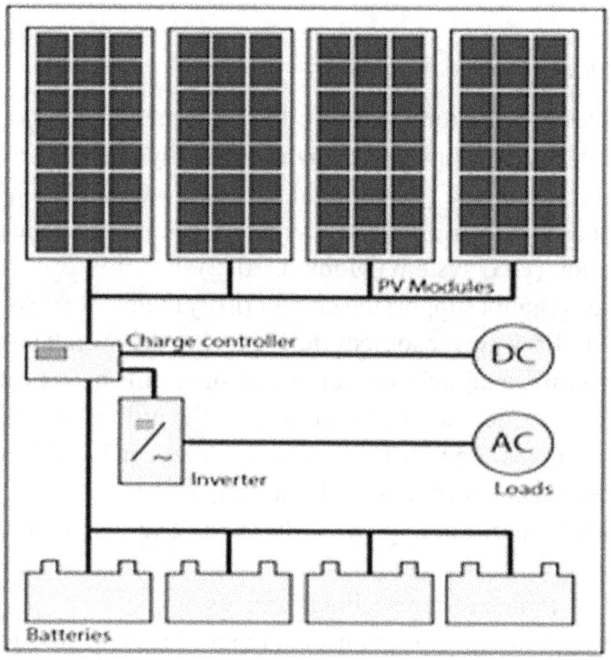

Figure 3. A schematic of the different components of a PV system (Vlot and Gunnink, 2011).

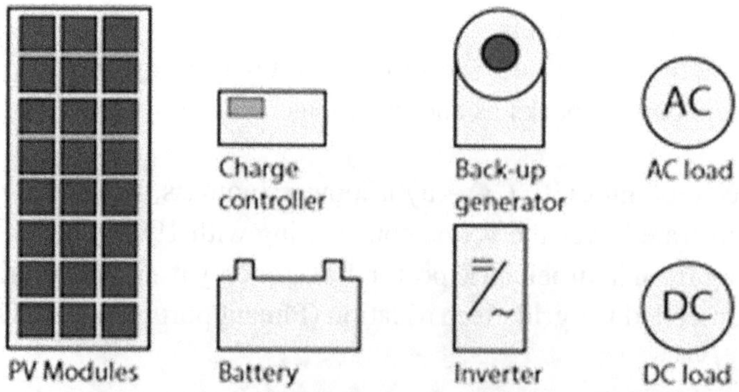

PV arrays respond dynamically to ambient conditions like irradiance, temperature and other factors like shading, furthermore; batteries have strict voltage limits which are necessary for their optimal functioning, thus; directly coupling the battery to the PV arrays and the loads is detrimental to the battery lifetime (Luque and Hegedus, 2003). The charge controller, charge regulator or battery regulator is the link and a control between the solar generator, the battery and the load, it ensures that the electrical parameters present at the battery are kept within the specifications given by the battery manufacturer, that is; preventing both overcharging and deep discharging of the battery that badly affect its performance and reduce its lifespan. The most important requirements of the charge controller are listed down below;

- Low internal consumption (<5 mA).
- High efficiency value (96% to 98%).
- Load disconnection if deep discharge occurs.
- Integrated overvoltage protection (Luque and Hegedus, 2003).

4. Smart Grid Communications and Measurement Technologies

a) Smart Grid Requirements

Modernizing the grid requires an incorporation of new innovative technologies. The five major smart grid technologies responsible for successfully implementing a smart grid are:

- Energy storage devices.
- Advanced superconducting transmission cables.
- Smart substations and smart transformers.
- Advanced Metering infrastructure (AMI).
- Home area networks (HAN`s).

Considering smart grid and smart homes, the AMI and HAN`s technologies have a direct relation with housing by influencing many new products and applications for homes; for example: smart meters, car charging stations, smart thermostats, renewable energy installations and smart appliances (Saoud and Recioui, 2017; Syal and Ofei-Amoh, 2013).

b) Advanced Metering Infrastructure AMI

AMI is an integrated two-way communication technology that makes the grid dynamic interactive system for power and real-time data exchange. AMI is the integration of several technologies, it consists of three main components that provide the infrastructure to establish the communication between the house and the utility company:

- A smart meter at the costumers` location.
- A communication network between the utility company and the smart meter.
- The HAN to connect the house with the smart meter.

The communication can enable consumer-demand response through consumer-level decisions on power supply prices. Utilities can also receive consumer usage data in real time that can enable the utilities to manage electricity demand and supply effectively (Syal and Ofei-Amoh, 2013).

c) Home Area Network HAN

The HAN forms an in-home network of smart appliances through a home gateway to link to the smart meter, and relates residential renewable energy-generating sources, a home charging battery to store excess generated energy, various equipment and appliances including lighting and security system. HAN's are commercially available and use existing communication technologies such as Bluetooth and Wi-Fi (Syal and Ofei-Amoh, 2013).

d) Smart Measurement and Metering

The modernization process of the smart grid is not only achieved by the energy infrastructures and power electronics, but also by the high level of information infrastructure, monitoring, measurement and metering operations that provide a widespread communication substructure, thus; the bidirectional flow that the smart grid provides is required for the communication as well as for energy (Ouadi et al., 2013; Ouadi et al., 2014; Chafai et al., 2013). Smart grid is constructed with its distributed information systems incorporated to power generation, transmission, distribution and consumption nodes (Kabalci, 2016).

e) Smart Metering

Utilities used to send workers to manually collect data required and this fact obviously decreases the accuracy of measurements. The smart meter is an advanced energy meter that obtains information from the end users' load devices and measures the energy consumption of the consumers and then provides added information to the utility company and/or system operator for better monitoring and billing compared to the regular energy meter (Depuru et al., 2011). Electrical data such as voltage and frequency are measured and real-time energy consumption information is recorded. Smart meter supports bidirectional communications between the meter and the central system and has a built-in ability to disconnect-reconnect certain loads remotely for better energy managing (Amin and Wollenberg, 2005). (Figure 4) represents a comparison between the conventional energy meter and the smart meter.

Figure 4. The metering architectures of a conventional energy meter and a Smart Meter *(Depuru et al., 2011).*

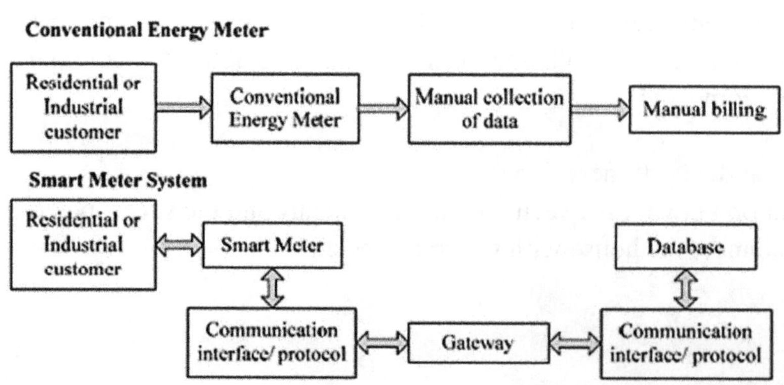

f) LABVIEW Software and Data Acquisition

LABVIEW (Laboratory Virtual Instrument Engineering Workbench) is a graphical application development environment (ADE) developed by National Instruments Corporation for the Data Acquisition (DAQ), Test and Measurement (T&M) and the Industrial Automation (IA) markets (Battacharyya et al., 1996).

For engineers, LABVIEW makes it possible to acquire information from the outside world into a computer, make decisions based on the acquired data (process, analyze) and send computed results back to control the way a piece of equipment operates, this process is known as data acquisition (Larsen, 2011). Data acquisition is what sets LABVIEW apart from numerous other software products. Data acquisition devices are designed to provide a communication bridge between a laboratory instrument or sensor and a computer system, they can be used to deal with numerous measurements including temperature, force, position, voltage, current and many other sensor output variables. Data acquisition process is configured by providing the required information prior to collecting data. LABVIEW creates a data acquisition task that contains that information. The collected data is available to the computer system and can be displayed, modified, analyzed (signal conditioning, filtering) or simply stored. Typically, DAQ devices are used to measure voltage and have predefined input ranges, [0-10V] and [-5-+5V] are common (Larsen, 2011).

SMART METERING-POWER MEASUREMENT

Circuit Implementation Description

Components: 12V PV array–12V battery–12VCharge controller–12V-220V DC/AC inverter– Light control sensor– relays–Circuit breaker – regulated voltage protection – Resistive loads (4-40-Watt light bulbs) -Fan–-Current transformer– ±12V DC power supply- resistors- NPN transistor- Data Acquisition Card (DAQ) –LABVIEW software.

Figure 5. Overall circuit

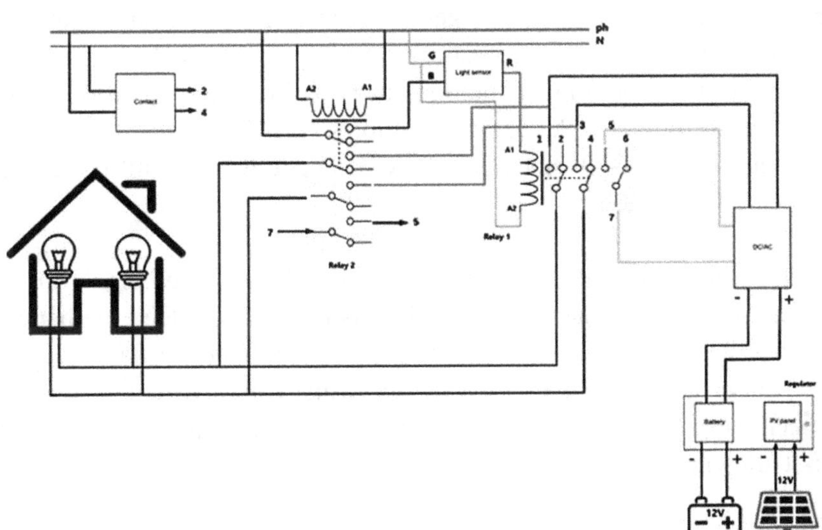

Figure 6. System Hardware implementation

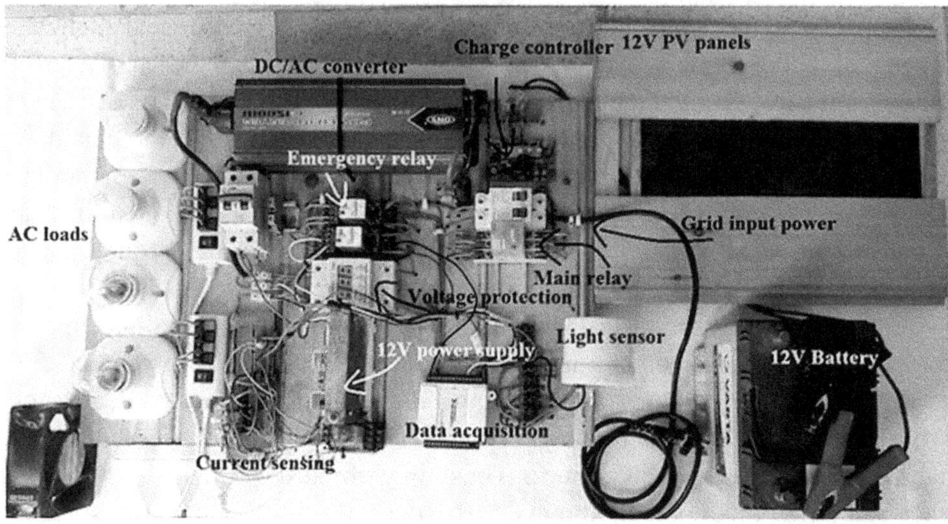

PV arrays of 12V output are used to simulate not only the integration of a RE source in a smart grid but also to clarify the influence of customers on a smart grid and their participation in energy managing and saving. The available PV panels produce an individual voltage of 6V, however the voltage required for the battery charging and the inverter is 12V, for that reason, voltage then current building configuration with four PV panels is elaborated through the steps stated here:

- Voltage is boosted by stringing two cells in a positive-to-negative-to-positive arrangement forming a group of 12V overall output voltage.
- Amperage is increased by stringing the two series cells together in parallel such that all positive leads are tied as are all negative leads.

This procedure is equivalent to forming a larger PV cell that produces an output voltage of 12V. The output is further connected to a regulator for battery charging during daytime.

The solar controller used in the implementation has two inputs, one input to connect the panel terminals **(PANEL±)**, and the other for the battery terminals **(BATT±)**. This controller is also equipped with two LED`s as state indicators (Panel ON yellow LED and Battery full Green LED). It's the link between the PV panels and the battery.

Case 1: If the battery is under the specified limits (discharged), the PANEL ON LED is switched ON, the controller allows the PV panels connected to its terminals to feed the battery, thus, avoiding the under-discharging phenomena that deteriorate the battery performances.

Case 2: When the battery is fully charged or when the voltage and current limits are exceeded (12 V and 6 A respectively), the BATT FULL LED is ON and the controller stops further charging of the battery not allowing it to overcharge.

The light control sensor can turn ON or OFF light according to ambient-light, it controls the load working only at night such as road light and garden light, it is used in this implementation to trigger the relay responsible for controlling the transition between the main (grid) and backup (battery) sources. The basic connection of a light control sensor consists of two inputs (Black-Green) and one output (Red) fed back to the Green input through a load. Its working principle is similar to a switch in a nutshell, at daylight the sensor is considered to be open, 0V appears across the load, in the dark it is considered to be closed and 220V appears across the load terminals.

The main relay used in the implementation is used as a (3 Poles - 6 throws) relay and (A1, A2 coil terminals)

- The NC throws (2-4) drive power to the load from the AC source.
- The NO throws (1-3) drive power to the load from the battery after being converted into AC power by the DC/AC inverter.
    - The coil is energized from the light control sensor, during daytime, by considering the light sensor as a switch that is open, thus, the voltage across the coil terminals is null, the relay throws remain at their normal positions connecting the 220V to the load through (2-4).
    - At night, and again by considering the light control sensor as a switch, that is closed, 220V appears across the coil, the relay throws are no longer in their normal positions, and switch to NO points and connecting (1-3) to the load, therefore the load is fed from the battery.
    - The DC/AC inverter as any other device possesses an internal switch, at daytime, the load is fed from the AC source, however the DC/AC inverter remains ON. This waste of energy that also deteriorates the device at long term, is avoided by taking the switch terminals out from the DC/AC and connecting them to the main relay in a NC point (5), so that the DC/AC is ON state only if the relay is not energized in other words; at night. The connection is represented in the schematic (Fig. 7) in a yellowish wiring.

Figure 7.

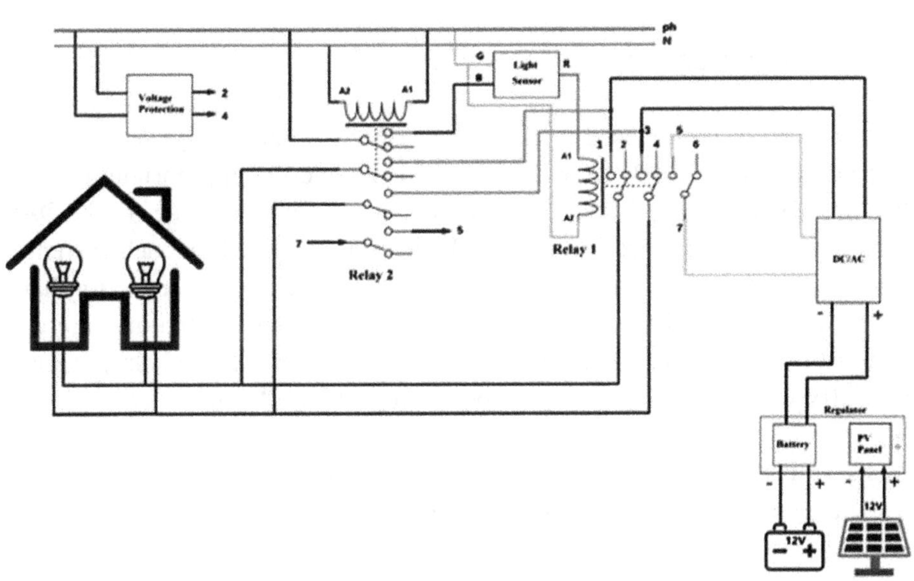

1. Emergency Circuitry

Up to this point, the load is correctly fed from either source. However, some emergency solutions should be taken into consideration;

- Electricity outage during the day.
- Electricity outage during the night.
- Battery discharged, or PV panels defect.

In the implementation only the electricity outage issues are considered, the emergency relay is used for that purpose. The relay this time is used as a (4 poles-8 throws) relay. The coil is energized from the AC source however only the NO points are considered, meaning that the emergency relay functions only in case of electricity outage.

During the day the load is fed from the grid, the light sensor is open and the main relay`s coil is at rest, points (2-4) drive power to the load, in case of an outage, the emergency relay`s coil is deenergized, and points (1-3) are forced to reconnect to the load, that is the load is fed correctly from the battery.

During the night, the load is fed from the battery, the light sensor is closed and the main relay is energized connecting points (1-3) to the load. In case of an electricity outage, the main relay comes back to rest and disconnects points (1-3), the emergency relay forces their reconnection; the load is again powered from the battery.

2. Overvoltage Control

The transition of the relay throws to connect the required source to the load is very fast. As a consequence, a defect upon the relay may join two throws to the same pole and connects two power circuitries to the same load. The latter fault is dangerous and may deteriorate some circuit components. A delay between the sources transition is required. The adjustable over/under voltage protector does not only provide the delay but voltage limits can also be adjusted to protect the load from under and over voltage issues.

3. Data Acquisition Connection Diagram

The idea behind this work was supposed to directly measure the current by the current transducer and acquire it by the DAQ-USB to LABVIEW software to calculate the appropriate power. One problem faced during this implementation and data recording was the internal resistance of the DAQ card in case of current measurements, this shunt resistance should be perfectly fixed for each current measurement, and since the load is changing, the acquired current would be somehow mistaken, for that reason; a proposed solution is to connect a resistor in the secondary circuit as represented in (Figure 8) and set it to $Rm=250\Omega$, such that the secondary current passes through it and the voltage drop is measured instead of the current.A 220V relay is used to alteranate between grid and PV input acquisition.

- The LEM sensor is powered from the ±12V grounded outputs of the switching power supply connected to the (+ and –) labelled pins of the LEM transducer.

- The Primary circuit is an AC high current (Ip) circuit, a the AC loads wires are passed inside the LEM core taking into consideration the required direction specified by an arrow on top of the transducer's core.
- A resistor *Rm*=250Ω is connected in series with the output of the current transducer (M) where a low secondary current (Is) current is flowing, one terminal of the resistor is connected to a pole of the 220V relay and the other is grounded.
- The NC throw of the relay is connected to one analog input of the DAQ (AI0), and the NO throw is connected to another analog input of the DAQ (AI1), the DAQ's common ground is the same as the power supply DC port ground.
- At normal operation the loads are fed from the grid, and the relay is at rest, the signal acquired to the DAQ at (AI0) is the voltage drop (V_m) across the resistor when a grid current, the other terminal of the DAQ (AI1) is open circuited.
- In case of PV connected, the relay is energized and the throw changes position connecting the NO terminal to the DAQ AI1 input, the signal is the voltage drop (V_m) across the resistor when PV current is flowing and this time (AI0) is open circuited. DAQ inputs connections.

Figure 8. Data acquisition circuit

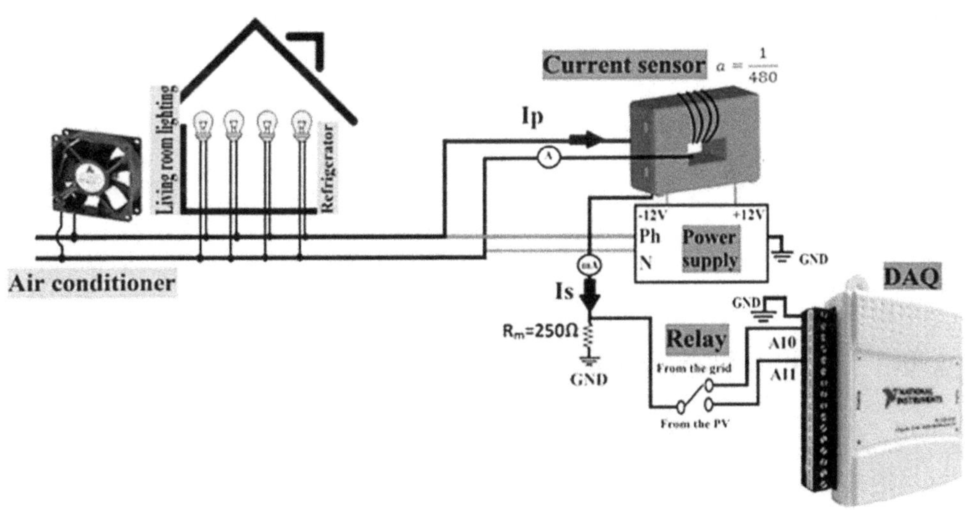

4. LABVIEW Data Acquisition and Power Calculation

The DAQ is configured to acquire an analog voltage input, the input is split into two voltage inputs; one coming from the grid (AI0) and the other coming from the PV (AI1) with respect to the ground, the RMS values are calculated for each voltage input. The current for each case is simply calculated based on Ohm's law where Rm represents the 250 Ω resistor, Vm is the voltage drop to be measured across the resistor, Ip and Is correspond to the primary and the secondary currents respectively;

$$Vm = RmIs \quad (1)$$

$$Is = \frac{Vm}{Rm} \tag{2}$$

The Primary current is calculated based on the new current ratio;

$$Is = \frac{1}{480} Ip \tag{3}$$

$$Ip = 480IS \tag{4}$$

Power is defined as:

$$P = Irms Vrms \cos\theta$$

where I and V are the instantaneous values of the current and voltage and θ is the phase angle between the voltage and the current. Because the loads are considered to be resistive loads, the voltage and the current are in phase and the appropriate power is simply calculated by multiplying the RMS value of each current by the output voltage across the loads that is constant and equal to 230V.

$$P = VIp \tag{5}$$

$$P = 480VIs \tag{6}$$

$$P = \frac{480VVm}{Rm} \tag{7}$$

Substiting for V=230V, and Rm=250Ω;

$$P = 442Vm \tag{8}$$

a) DAQ Configuration for Voltage Measurement

1. Choosing to acquire an analog voltage signal (Figure 9).
2. Selecting the input pins (Figure 10).

Pin (AI0) and (AI1) are selected to acquire input voltage in case of grid connected and PV connected respectively.

Figure 9. DAQ measurement type configuration.

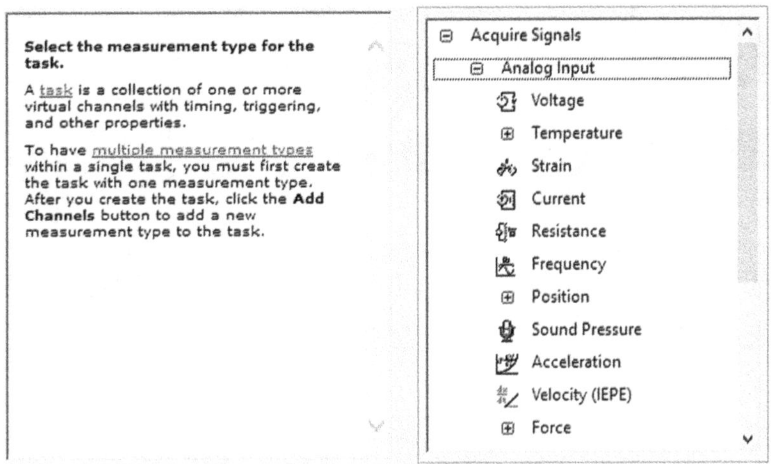

Figure 10. DAQ physical channels selection

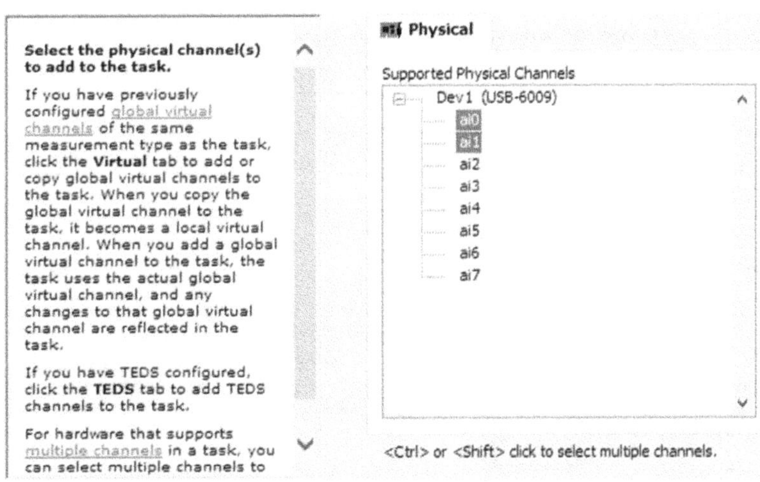

b) Data Acquisition and Recording

The designed house consists of 3 appliances; air conditioner, living room lighting and a refrigerator (Figure 11). Power ratings of each appliance and its real power consumed measured throughout this application are stated in (Table 3).

Figures down below represent a random power consumption scenario, the front panel displays the measured and the calculated parameters in case where the house is fed from the grid (Figure 12) and in case where it is fed from the PV (Figure 13), including, the secondary current Is calculated from equation (2), the primary current Ip obtained from equation (4) and finally the consumed power that could be concluded from equation (8). The front panel also displays the actual current and the power plots.

Power consumed from different and random loads combinations is recorded each one second for one hour (3600s) from the grid, and for another hour from the energy stored in the battery. The results are plotted as illustrated in (Figure 14).

Figure 11. House appliances

Table 3. Appliances specifications and power ratings

Load	Equivalent Component	Power Rating	Measured Power
Living room lighting	2 light bulbs	2×40W	Consumed power (W) 69.9
Refrigerator	2 light bulbs	2×40W	Consumed power (W) 69.93
Air conditioner	Fan	20.7W	Consumed power (W) 22.45

Figure 12. Smart meter grid parameters front panel.

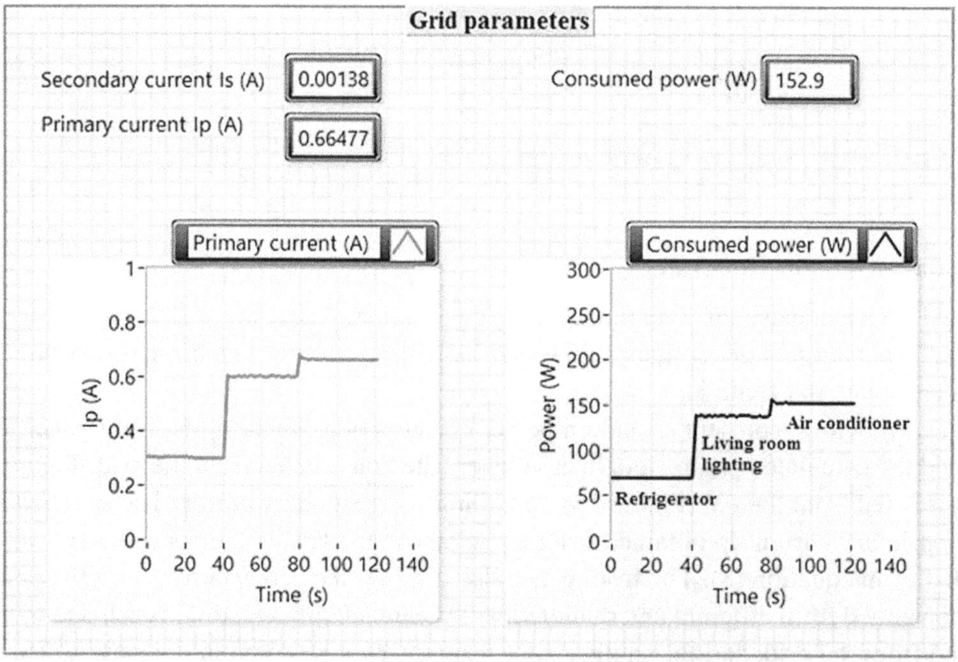

Figure 13. Smart meter PV parameters front panel

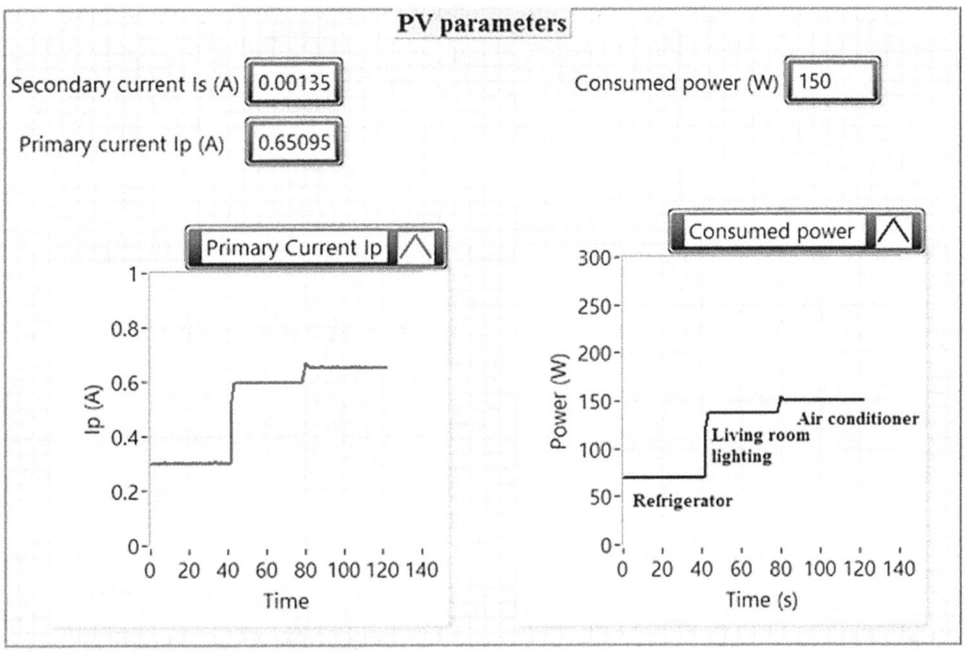

Figure 14. Grid and PV consumed power of a house recorded during 2 hours.

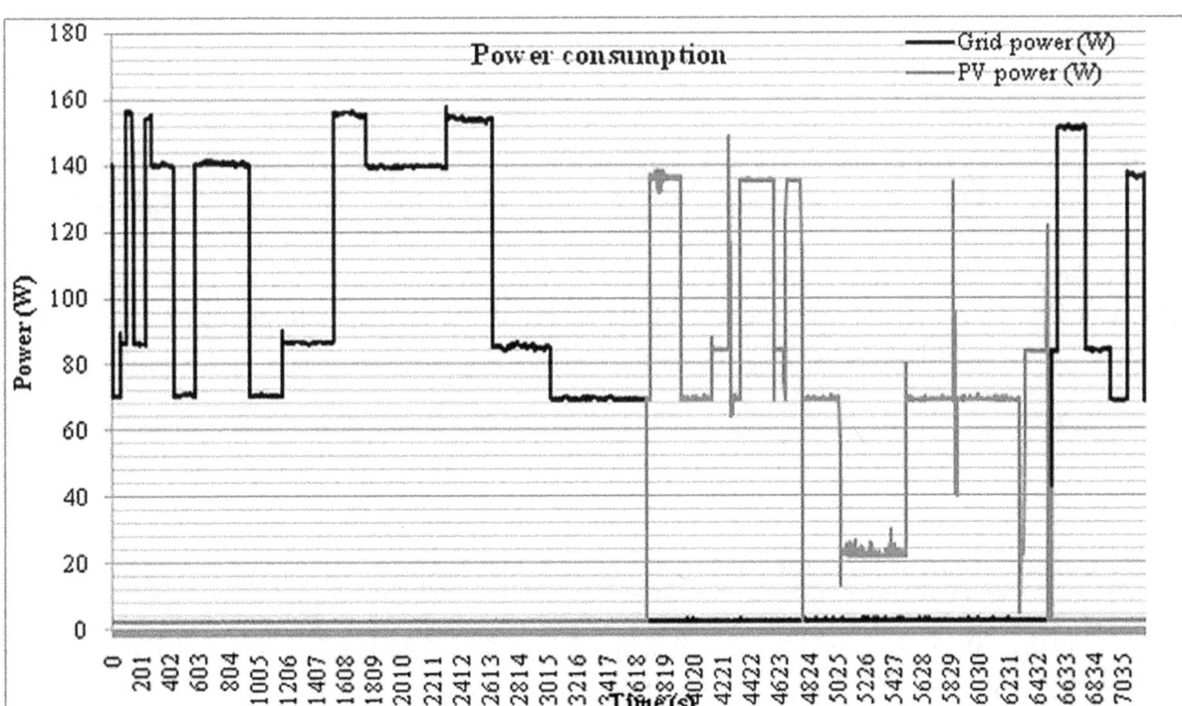

In parallel, the energy (W.s) consumed from the grid and from the PV during two hours is calculated with the power of each load combination is multiplied by its appropriate performance time duration. The total consumed energy from either source is the algebraic sum of all individual combination energies. (Figure 15) represents the different loads combinations, their individual energy consumption and the total energy consumption for two hours of recording in case of grid connected and in case of PV connected house.

Figure 15. Individual loads combination and total energy consumption from the grid and from the PV.

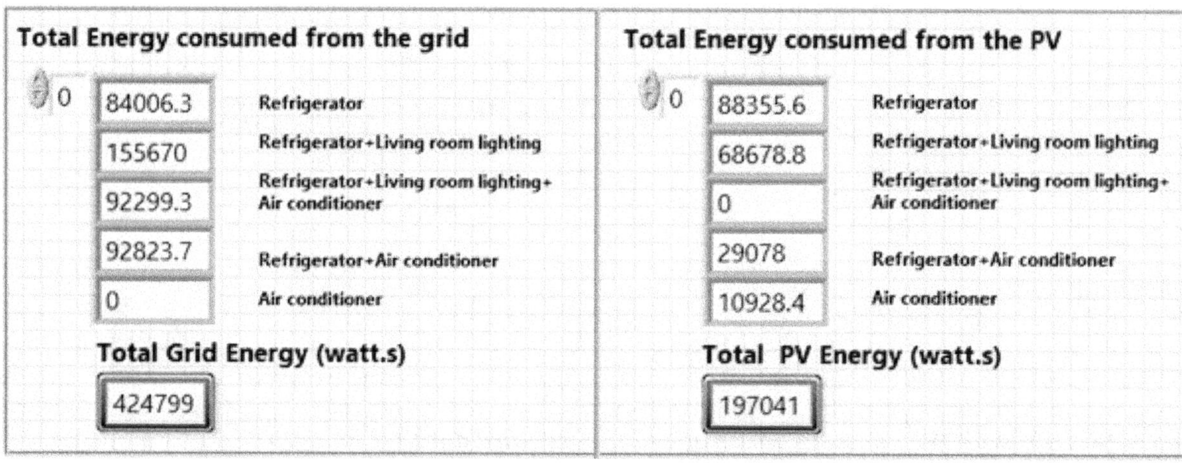

c) Smart Meter Front Panel

The smart meter designed in LabVIEW displays the instantaneous consumed energy from either sources, as well as the energy left in the battery that could be sold to the utility. The designed Smart Meter also allows the user to keep track of its energy costs, the thing that makes him more conscious and aware about his future consumptions.

The battery used in the implementation has the following characteristics: a voltage of $V=12V$ and a capacity of $C_{Bat}=45Ah$. The energy capacity of the battery is obtained by multiplying the rated battery voltage measured in Volt by the battery capacity

$$Ebat = CbatV \tag{9}$$

Substituing V and C_{Bat} yields ;

$$E = 12V \times 45Ah = 540V.Ah = 540W.h$$

Therefore the left energy is simply calculated by subtracting the total energy consumed from the battery from its energy capacity. (Figure 16) illustrates the LABVIEW designed Smart Meter front panel for the latter house power consumption.

Figure 16. Smart Meter LABVIEW front panel displayed to the user.

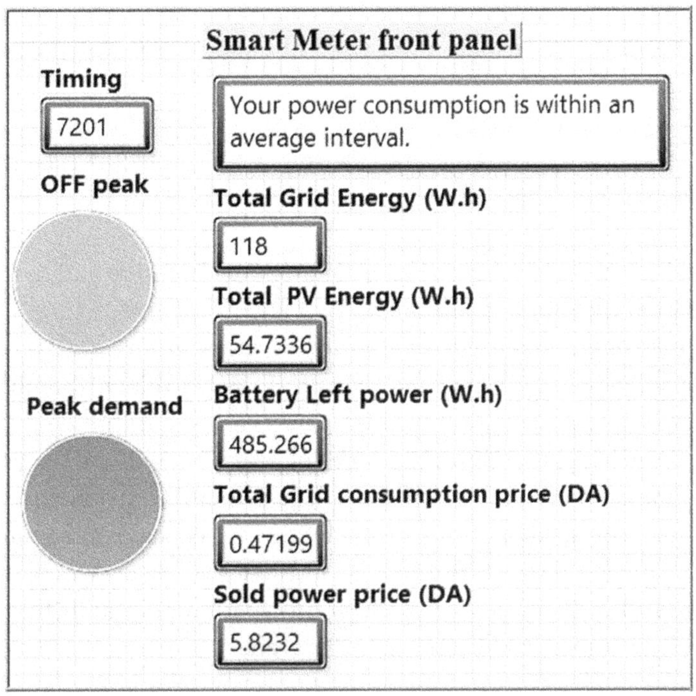

d) Bill Display

The total energy consumption cost is calculated by summing up all the energy consumptions of all connected loads, and multiplying the latter by the energy cost. The bill is constituted from the total energy consumption and cost drained from the grid and the PV.

Traditionally, utilities send workers to check the user's energy meter and note down his energy consumptions each three months, a bill is sent to the user including his consumption during the these three months. The designed smart Meter automatically displays an energy bill to the user including his energy consumption from the grid and from the PV, energy consumption cost from the grid and also the cost of the energy left in the battery in case it is injected back to the grid. (Figure 17) is the displayed energy bill generated from the previous house consumption after a time duration of two hours. Energy cost is considered to be equal to 4DA/Kwh and the grid injected energy cost is considered to be three times the real grid energy cost; that is 12DA/Kwh.

Figure 17. Consumer Electric Energy Established Bill

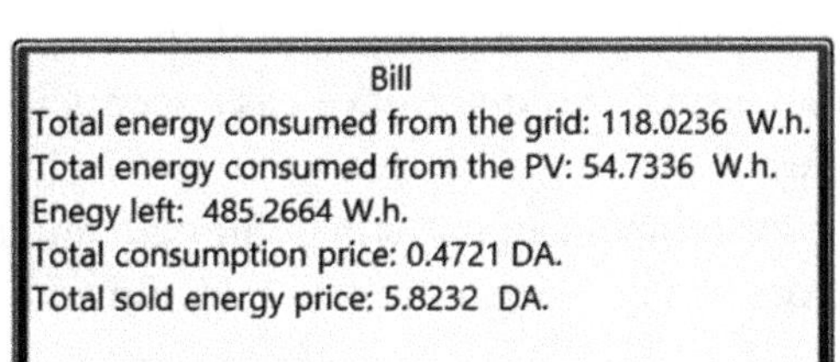
Bill
Total energy consumed from the grid: 118.0236 W.h.
Total energy consumed from the PV: 54.7336 W.h.
Enegy left: 485.2664 W.h.
Total consumption price: 0.4721 DA.
Total sold energy price: 5.8232 DA.

LOAD CONTROL EMULATION

Electricity is the most versatile and widely used form of energy, and global demand for it is continuously growing.

The insufficiency of fossil fuel resources and evidences linking carbon emissions and global warming have raised the interest in energy saving and environmental protection. To reduce our dependence on fossil fuels, two strategies are being followed:

1. Integrating renewable energy resources fostering their implantation and usage, as well as, adapting electricity demand to new supply conditions.
2. Reducing energy consumption through energy saving programs focused on energy demand reduction and energy efficiency in domestic and industrial fields (Torrent-Fontbona and Lopez, 2015).

The power shortage issue happens a lot during times of peak demand, and the power system frequency drops below its nominal value when the system loses its ability to supply power. This frequency deviation causes tragic damage events, such as blackouts. DSM is a mechanism to achieve grid stabilization via allowing consumers, incentivized by utilities, to manage their energy consumption and control of their loads by reducing electricity usage over a given time period or to shift that usage to another time period in response to utility signals and customer's preference.

1. Home Energy Management Systems

Home Energy Management Systems (HEMS) for residential customers are of recent important development. In the smart home application, interaction between users with their home appliances will be formed thereby the devices can be operated and programmed interactively either directly or remotely. Therefore, these appliances can operate automatically, multi functionally and efficiently, HEMS application is developed not only to manage energy use in households but also to conduct the management of energy supply, either from the provider of electrical energy or the own-generated one such as alternative energy sources through solar or wind power plant.

The HEMS configuration consists of household loads that are divided into scheduled and unscheduled loads, energy storage, alternative energy sources, connections to grid, electric cars and HEMS control systems which is supported by communication technology and smart meter (Fang et al., 2012).

2. Demand Side Management

Traditionally, utilities haven't focused on the customer element of DSM programs besides customers saw no incentive to participate, other than the satisfaction of helping the utility. There was no incentive for customers to shift (and meet the utility's goals) because there was little visibility into how shifting benefited the customer. Also, the customer did not have easy access to tools that informed them how they could save energy and money.

Through HEMS applications customers can participate in maintaining or improving the reliability of power systems. One of the programs on a smart grid system aimed at improving network reliability is Demand Side Management.

Through the DSM program customers can make arrangements of the use of loads that can impact on the reliability of power systems. There are several categories of DSM programs, among them: Demand Response (DR) and Time of Use (ToU) (Preethi and Harish, 2016). One of the main goals on demand side management consisting of providing the consumer with the ability to control its energy consumption and its bill and charge him based on the true price of the utility at each time is achieved.

a) Demand Response

Demand response, as defined by U. S. Department of Energy (DOE), is that consumers conduct activities to change the operating pattern of electrical energy consumption, in response to the application of tariff provisions and incentives for the use of electrical energy. Consumers change the pattern of electric energy consumption from normal usage to the usage pattern by paying attention to the different electric prices based on the operating time or incentive value provided, in order to encourage lower electricity usage when peak load time or the reliability of the power system is threatened (Kaźmierkowski et al., 2002).

b) Time of Use

In some countries tariffs for the use of electric power are differentiated for some time, especially at peak load times, where electricity rates are more expensive than out of peak load (Guerrero et al., 2010). With time-based pricing schemes, application development is to implement Time of Use programs such as Power shifting, load shifting and load shedding. Power shifting is done by moving an alternative energy source, such as from the sun, to be used when the electricity price of the network is more expensive (Cecilia and Sudarsanan, 2016). Besides that, it can be done through priority load mechanism in which peak load for less priority load will be off.

3. Smart Meter

Once meter data is available in electronic form, it becomes feasible to add communications to the meter, allowing the meter to use automatic meter reading (AMR) to access data remotely via the communication link. The result is the smart meter that are read every fifteen minutes, it helps eliminate estimated bills and the need for a meter reader to visit customer's home.

Smart metering broadens the scope of AMR beyond just meter readings with additional features enabled by two-way data communication.

SM can take real-time or near-real-time measurements, provide outage notification and power quality monitoring, and support in-home energy applications. It allows data exchange between the SM and the central system of distribution system operator, while also allowing customers to have timely and easily accessible information about their usage.

The European Commission has included ten common minimum functional requirements for electricity smart metering systems in Recommendation 2012/148/EU (Phuangpornpitak and Tia, 2013).

The two-way communication capability of the smart meter enables other functionalities:

- Automated and remote meter reading.
- Remote connection and disconnection.
- Outage detection.

- Tamper detection.
- Monitoring of power quality.
- Remote time synchronization.
- An interface for an in-home display unit.
- An interface for load control devices (Koykul, 2011).

The smart meter measures the total of household energy use and then sends these numbers back to the energy provider. If you have solar installed, it will measure the amount of energy that is being sent to the grid and will use this energy to power your home. All excess energy is then conveniently sold back to the grid.

a) Types of Customer Loads

Customer loads can be classified into two types: the first type is the shiftable load that can be scheduled at different time periods of the day, for example; the washing machine and the dishwasher. Such appliances should be able to receive a signal (either by wire or by wireless connections) and to contain automatic switches that can turn them on and off. The second type is the non-shiftable load that has to be operated during specific times of the day to meet the requirements such as the lighting system. Therefore, in the peak-hour, the customer can schedule their shiftable appliances to mitigate the extra load from the grid. According to these load types, a scheduling techniques need to shift the load as much as they can to enable customers to manage their energy bills and utility company to balance consumption levels (Bera et al., 2015).

b) Types of Load Control

In the smart grid loads may be controlled either by the supplier side or by the end user side.

Controlling of end-use loads can be carried out:

- Automatically: in response to a signal sent by a load control program operator linked to a particular event, eg high energy prices or network constraints.
- Manually: by the end user in response to an event (the initiator of the switching requires information about the event) (Crossley, 2007).

4. Simulation Setup and Analysis

a) LABVIEW Simulation

A smart metering residential load control method is designed and simulated in LABVIEW software, the framework consists of four houses with four different appliances each, power is distributed to the four houses either from the grid with a total amount of 100KW and from a backup source emitting from a PV generator with a total amount of 100KW.

Depending on the four houses total consumption, three distinct situations may occur;

- The total consumption of houses is less than the power provided from the grid, the grid power satisfies the power demand and no control procedure is considered.
- The total consumption of houses exceeds the power amount supplied from the grid but less than the power supplied from the backup, in this case, the deficiency is restored from the backup till the power demand is satisfied.
- The worst case occurs when the total consumption exceeds both power amounts supplied from the grid and from the backup, this is the case of a peak demand, the quest for power demand satisfaction implies an application of a load control procedure. Some appliances must be turned OFF, this is done by first selecting the house with largest power consumption and then swithing OFF its appliance that is consuming the largest amounts of power amongst the four appliances.

Figure 18. Overall power consumption plots

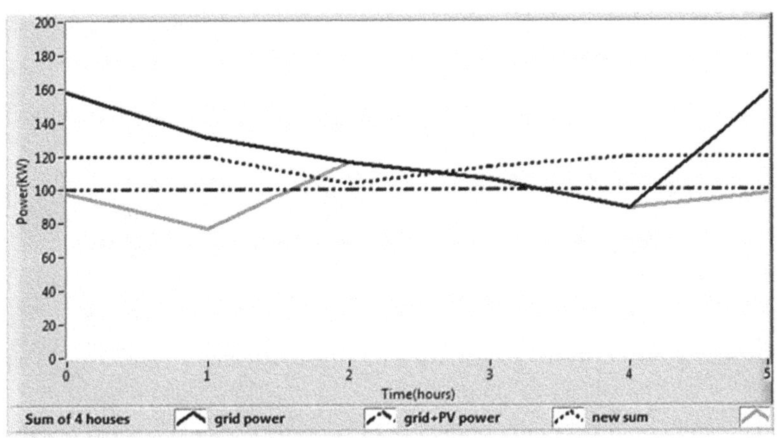

b) Interpretation of Results

The three situations discussed in the last section can be deduced from the graph simulation represented in (Figure 41), where the overall power consumption of four houses with respect to time, each case is discussed separately and illustrated by its appropriate LABVIEW front panel.

Case 1: power consumption greater than the power supplied from the grid and the PV

Referring to (Figure 18), in the interval from [0 to 1] hours, the power consumption of four houses is greater than the power supplied from both the grid and the PV, (Figure 19) and (Figure 20) are the front panels that show indicators of power consumption, the number of house with highest power consumption, the number of load with highest power consumption in this house and the new sum after optimization.

Two power consumption values are recorded, in each one, house and load number is shown, this latter is the address of the house in which load shifting to off peak technique will be taken.

This operation could be done either by utility automatically or manually.

Figure 19. First power consumption front panel for case 1.

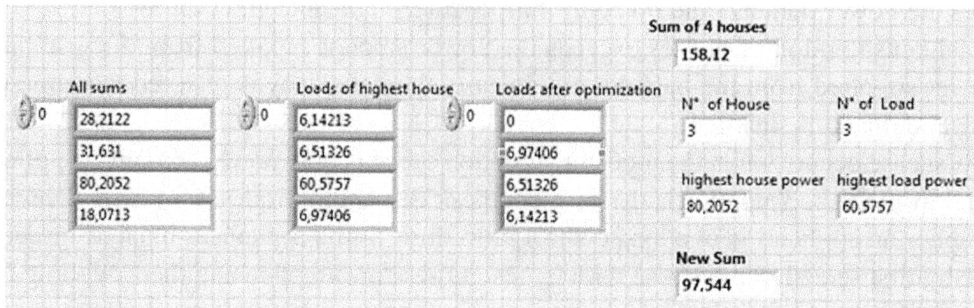

Figure 20. Second power consumption front panel for case 1.

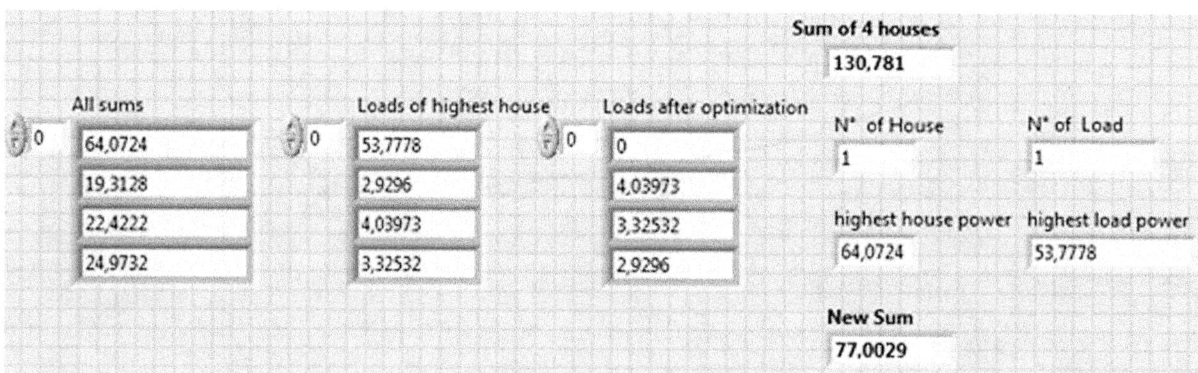

Case 2: power consumption greater than the grid power

After one hour, the power consumption starts decreasing till it reaches the values shown in (Figure 21) and in (Figure 22) between the interval 2 and 3 hours; in this case the demand exceeds the grid power (100KW) by a certain amount, the difference between 120 KW and Power of (PV+grid) indicator is the amount of power taken from PV panels.

Figure 21. First power consumption front panel for case 2

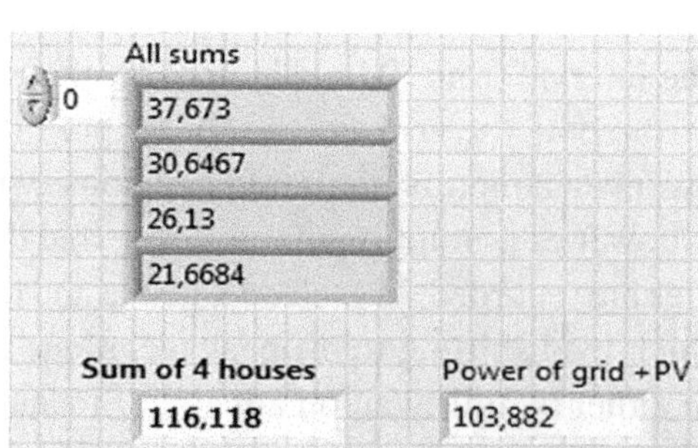

Figure 22. Second power consumption front panel for case 2

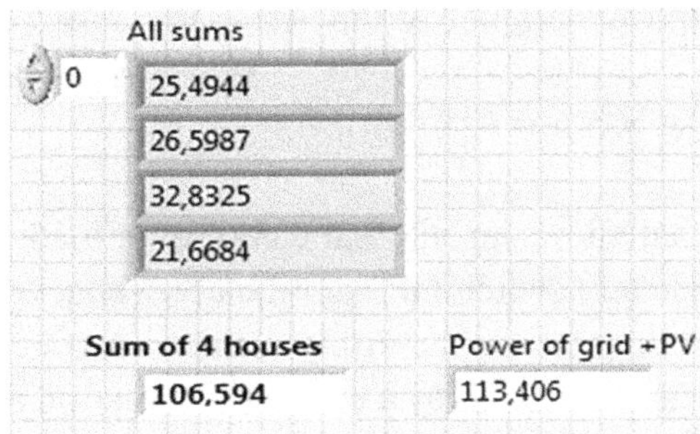

Case 3: power consumption less than grid power

As seen in the figure 18, the power consumption continues decreasing; at hour 4, it's below the grid power, it reaches the value shown in the indicator in (Figure 23), in this case the grid power meets the demand of the consumers.

Figure 23. Power consumption front panel for case 3

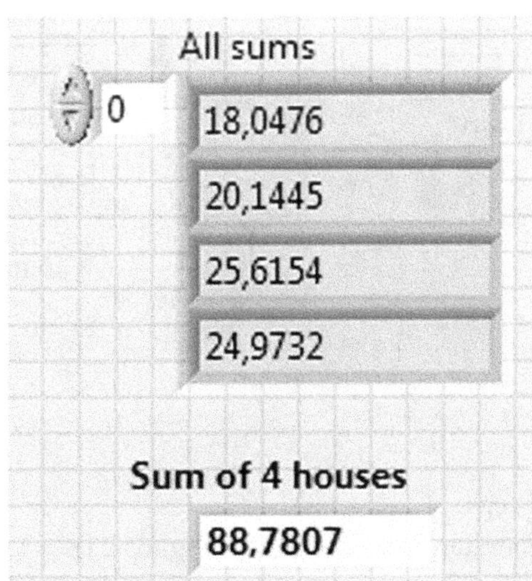

5. Load Control Application

The circuit designed previously represents a house consisting of four light bulbs, two of them simulate a living room lighting system and the two others bulbs simulate a refrigerator. The house also possesses an air conditioner. When all appliances are ON a peak demand may occur and the power consumed by the house increases in a great extent that even the back-up source will not be able to satisfy the demand. The utility interferes to reduce the peak demand and schedules the user' appliances and shifting their operation to an off-peak demand hours depending on whether they are shiftable of non shiftable loads or simply asking him to switch OFF some appliances and minimize his power demand. An application of the load control is implemented and applied to the designed house.

a) Automatic Control of One House Appliances

(Figure 24) represents a circuit that allows an automatic control of loads in case of an increase in power demand. It is designed to immediately switch OFF appliances whenever a peak demand occurs which is not the case in real smart metering where the utility cannot interfere without the user's approval to control his appliances. A 12V relay controlled by a transistor operating as a switch is used to disconnect the load in case of peak demand.

Figure 24. Automatic load control circuit design

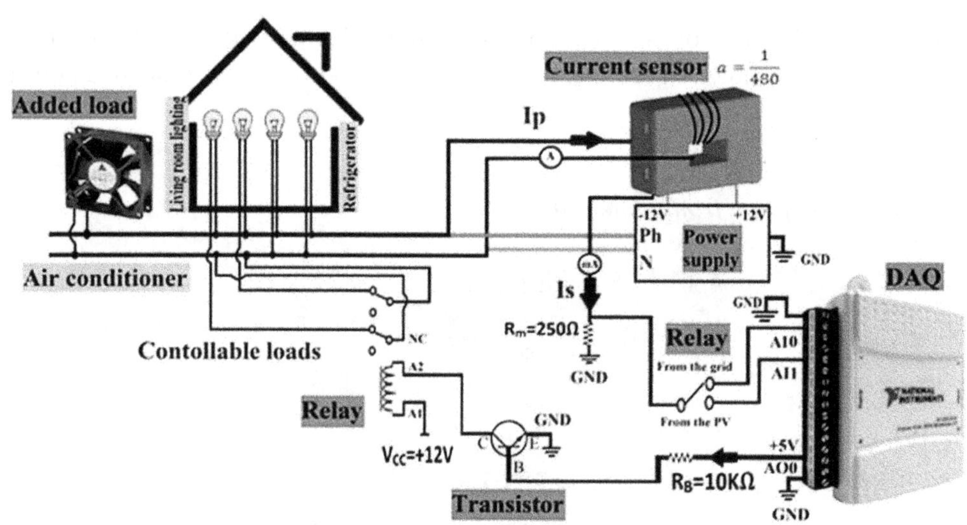

Circuit Description

- Output Signal Generation Steps:
    - A voltage signal is generated from the DAQ (Figure 25).

The signal is generated from (AO0) output of the DAQ (Figure 26) with respect to the ground (Figure 27) and connected to the transistor's base through R_B=10KΩ.

- The transistor's emitter is grounded.
- The loads for which the automatic control application is desired are connected to a 12V relay in NC points.
- V_{CC}=+12V from the switching power supply discussed in chapter two, is connected to the relay's coil A1 terminal, A2 terminal is connected to the transistor's collector.

Figure 25. DAQ measurement type configuration

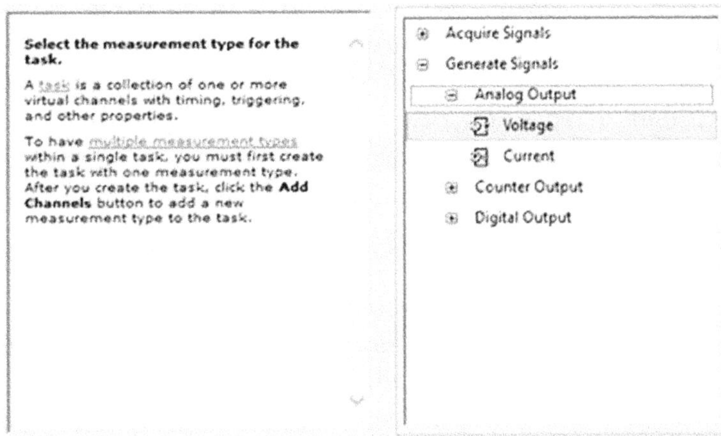

Figure 26. DAQ Pins selection

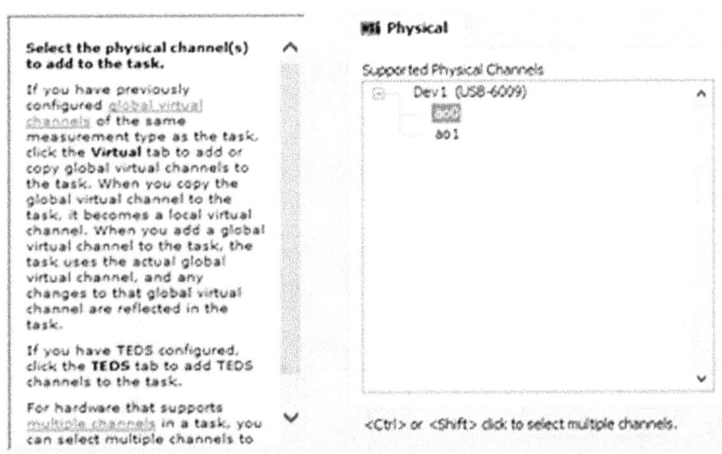

Figure 27. Signal configuration

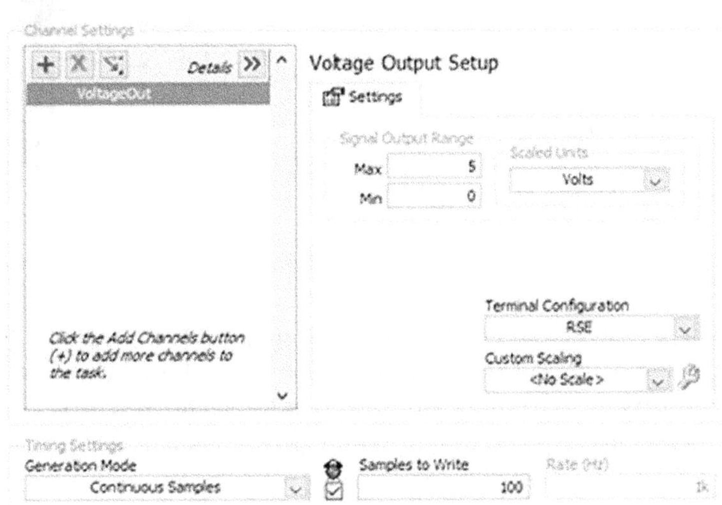

- **Operation:** When extra load is added; in this case it is a fan representing an air conditioner, the total consumed power increases to a certain level, a 5V DC signal is sent from the DAQ output pin (AO0), a current I_B flows in the transistor's base, the transistor is in saturation mode, V_{CC} is applied across the relay's coil, since the air conditioner is considered as a non shiftable load, other loads must be turned off to reduce the power, in this case two light bulbs representing the living room lighting are considered as shiftable loads and they can be controlled. Once the relay's coil is energized, the two light bulbs connected to a NC of the relay terminal are completely disconnected; the living room lighting system is switched off.

When the air conditioner is OFF, the total consumed power is below the specified level, no signal is generated, and thus no current flows in the transistor's base, the transistor operates in cutoff mode, the relay's coil remains at rest. Again at the instance when the air conditioner is switched ON, the transistor becomes saturated and completely disconnects the living room lighting. The total power swings back to its tolerated levels.

b) Manual Control

Controlling the power consumption could also be implemented manually. (Figure 28) represents the proposed circuit diagram;

Figure 28. Manual load control circuit design

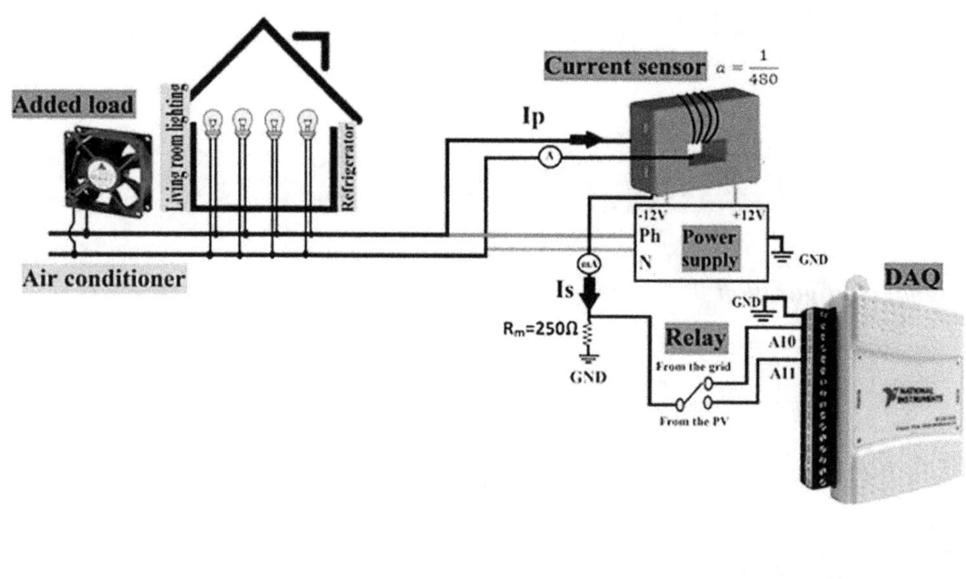

Circuit Description

When the air conditioner is switched ON, the total power jumps to a higher level and exceeds the tolerated power level, in this case, a message is displayed to the user asking him to switch OFF some loads for power balance purposes, When the power gets back to its tolerated level, a message is again displayed to the user notifying him that his consumption is within an average interval. Refer to (Figure 29).

Figure 29. Displayed message (a) in case of an OFF peak, (b) in case of a peak demand

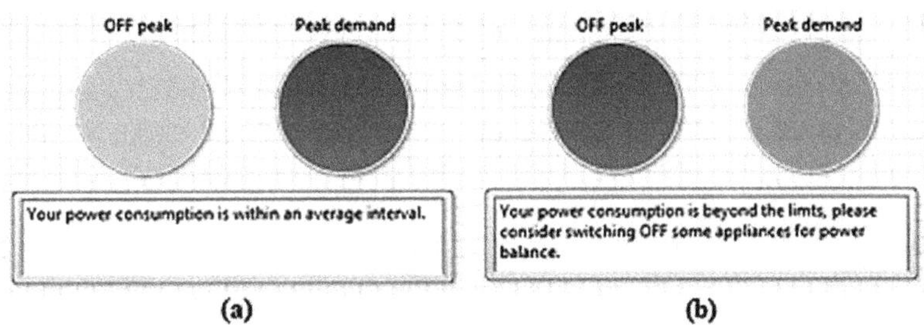

The process of controlling the loads brings immediate benefits to the utility by decreasing the peak demand and to the user by making him more aware and conscious about his power consumption and energy costs.

CONCLUSION

Smart grid merges both digital technologies and long transmission networks to manage energy consumption, as well as to open up new processes for energy production and distribution. This chapter focused on describing strategies for decreasing the load demand and power consumption costs, the active role of the customer in the smart grid is better defined and clarified. A simulation in LABVIEW software and a hardware application are performed to demonstrate the smart metering services in decreasing the pressure on the local electricity grid by taking power from RES and also managing customer energy use and save money by controlling the loads consumption.

A simulation in LABVIEW software is performed to visualize some peak demand mitigation techniques. As an application, a hardware circuit is designed to demonstrate an automatic and a manual load control technique performed on a house in case of an increase in power demand.

Simulation results based on four houses design has been presented, the work concentrated mainly on the power consumed by the four houses, different cases were analyzed and some solutions were proposed to manage the energy consumption of the system.

Since the customers are integral part of the SG, the contribution of HEMS application is conducted through demand side management (DSM) that helps to monitor and analyze the data that flows to and from customer side.

The control of loads was presented also in this chapter in order to minimize the peak demand and total energy cost, this is done either by taking the power from additional generation by RES or loads to off peak

REFERENCES

Amin, S. M., & Wollenberg, B. F. (2005). Toward a smart grid: Power delivery for the 21st century. *IEEE Power & Energy Magazine*, *3*(5), 34–41. doi:10.1109/MPAE.2005.1507024

Amos, S. W., & James, M. (2000). *Principles of transistor circuits: introduction to the design of amplifiers, receivers, and digital circuits* (9th ed.). Newnes.

Battacharyya, S. S., Murthy, P. K., & Lee, E. A. (1996). *Software synthesis from dataflow graphs*. Kluwer Academic Publishers. doi:10.1007/978-1-4613-1389-2

Bera, S., Gupta, P., & Misra, S. (2015). D2S: Dynamic demand scheduling in smart grid using optimal portfolio selection strategy. *IEEE Transactions on Smart Grid*, *6*(3), 1434–1442. doi:10.1109/TSG.2014.2386325

Bhimte, A., Mathew, R. K., & Kumaravel, S. (2015). Development of smart energy meter in LabVIEW for power distribution systems. *2015 Annual IEEE India Conference (INDICON)*, 1–6. 10.1109/INDICON.2015.7443547

Blume, S. W. (2016). *Electric power system basics for the nonelectrical professional*. John Wiley & Sons.

Cecilia, A. A., & Sudarsanan, K. (2016). A survey on smart grid. *2016 International Conference on Emerging Trends in Engineering, Technology and Science (ICETETS)*, 1–7.

Chafai, M., Bentarzi, H., Ouadi, A., Zitouni, A., & Recioui, A. (2013), PC based testing system for protective relay. *4th International Conference on Power Engineering, Energy and Electrical Drives, POWENG"2013*. 10.1109/PowerEng.2013.6635737

Chellali, F. B. M., Recioui, A., Yaiche, M. R., & Bentarzi, H. (2014). A hybrid wind/solar/diesel stand alone system optimisation for remote areas in Algeria. *International Journal of Renewable Energy Technology*, *5*(1), 12–24. doi:10.1504/IJRET.2014.059658

Crossley, D. (2007). Smart Metering, Load Control and Energy-using Behaviour. *Workshop on Smart Metering to Use Less Energy*, 10.

Depuru, S. S. S. R., Wang, L., Devabhaktuni, V., & Gudi, N. (2011), Smart meters for power grid—Challenges, issues, advantages and status. 2011 IEEE/PES Power Systems Conference and Exposition, 1–7.

Environnement, V. (2010). Smart grids. *Sci. Chron. Mag*, *18*, 1–7.

Fang, X., Misra, S., Xue, G., & Yang, D. (2012). Smart Grid — The New and Improved Power Grid: A Survey. *IEEE Communications Surveys and Tutorials*, *14*(4), 944–980. doi:10.1109/SURV.2011.101911.00087

Floyd, T. L. (2005). *Electronic devices*. Pearson.

Gaviano, A., Weber, K., & Dirmeier, C. (2012). Challenges and Integration of PV and Wind Energy Facilities from a Smart Grid Point of View. *Energy Procedia*, *25*, 118–125. doi:10.1016/j.egypro.2012.07.016

Gaviano, A., Weber, K., & Dirmeier, C. (2012). Challenges and Integration of PV and Wind Energy Facilities from a Smart Grid Point of View. *Energy Procedia*, *25*, 118–125. doi:10.1016/j.egypro.2012.07.016

Ghosh P., Maiti A. (2012). Instantaneous Power Measurement using Hall Sensor. *Int. J. Sci. Res. Publ.*, *2*(8).

Guerrero, J., Blaabjerg, F., Zhelev, T., Hemmes, K., Monmasson, E., Jemei, S., Comech, M., Granadino, R., & Frau, J. (2010). Distributed Generation: Toward a New Energy Paradigm. *IEEE Industrial Electronics Magazine*, *4*(1), 52–64. doi:10.1109/MIE.2010.935862

Han, J., Choi, C.-S., Park, W.-K., & Lee, I. (2011). Green home energy management system through comparison of energy usage between the same kinds of home appliances. *2011 IEEE 15th International Symposium on Consumer Electronics (ISCE)*, 1–4.

Hersch, P., & Zweibel, K. (2019), Basic Photovoltaic Principles and Methods. *Solar Energy Research Institute (SERI)*. Available: https://www.nrel.gov/docs/legosti/old/1448.pdf

Iskin, I., Daim, T., Kayakutlu, G., & Altuntas, M. (2012). Exploring renewable energy pricing with analytic network process— Comparing a developed and a developing economy. *Energy Econ*, *34*(4), 882–891. doi:10.1016/j.eneco.2012.04.005

Jenkins, N., Long, C., & Wu, J. (2015). An Overview of the Smart Grid in Great Britain. *Engineering*, *1*(4), 413–421. doi:10.15302/J-ENG-2015112

Kabalci, Y. (2016). A survey on smart metering and smart grid communication. *Renewable & Sustainable Energy Reviews*, *57*, 302–318. doi:10.1016/j.rser.2015.12.114

Kaźmierkowski, M. P., Krishnan, R., & Blaabjerg, F. (2002). Control in power electronics: selected problems. Academic Press.

Kian, A., & Keyhani, A. (2001). Stochastic price modeling of electricity in deregulated energy markets. *Proceedings of the 34th Annual Hawaii International Conference on System Sciences*, 7–12. 10.1109/HICSS.2001.926292

Koykul, W. (2011). Current status and action plan of utility sector on smart grid and smart community in Thailand. *Thailand-Japan Workshop on Smart Community in Thailand: Provincial Electricity Authority*.

Kumar A. (2014). *Hall-Effect Current Sensors for Power Electronic Applications: Design and Performance Validation.* Indian Institute of Science Bangalore.

Larsen, R. W. (2011). *LABVIEW for engineers*. Prentice Hall/Pearson.

Li, N., Chen, L., & Low, S. H. (2011). Optimal demand response based on utility maximization in power networks. 2011 IEEE power and energy society general meeting, 1–8.

Luque, A., & Hegedus, S. (2003). Handbook of photovoltaic science and engineering. Hoboken, NJ: Wiley.

Mahmood, A., Khan, I., Razzaq, S., Najam, Z., Khan, N. A., Rehman, M. A., & Javaid, N. (2014). Home appliances coordination scheme for energy management (HACS4EM) using wireless sensor networks in smart grids. *Procedia Computer Science*, *32*, 469–476. doi:10.1016/j.procs.2014.05.449

Miceli, R. (2013). Energy Management and Smart Grids. *Energies*, *6*(4), 2262–2290. doi:10.3390/en6042262

Momoh, J. A. (2012). *Smart grid: fundamentals of design and analysis*. Wiley. doi:10.1002/9781118156117

Nashelsky, L., & Boylestad, R. L. (1978). *Electronic devices and circuit theory*. Prentice/Hall International.

Ouadi, A., Bentarzi, H., & Recioui, A. (2013). Optimal multiobjective design of digital filters using spiral optimization technique. *SpringerPlus*, *2*(1), 461. doi:10.1186/2193-1801-2-461 PMID:24083108

Ouadi, A., Bentarzi, H., & Recioui, A. (2014). Optimal multiobjective design of digital filters using taguchi optimization technique. *Journal of Electrical Engineering*, *65*(1), 21–29. doi:10.2478/jee-2014-0003

Palensky, P., & Dietrich, D. (2011). Demand Side Management: Demand Response, Intelligent Energy Systems, and Smart Loads. *IEEE Transactions on Industrial Informatics*, *7*(3), 381–388. doi:10.1109/TII.2011.2158841

Pan, Z., Sun, H., & Guo, Q. (2013). TOU-based optimal energy management for smart home. IEEE PES ISGT Europe, 1–5.

Peretto, L. (2010). The role of measurements in the smart grid era. *IEEE Instrumentation & Measurement Magazine*, *13*(3), 22–25. doi:10.1109/MIM.2010.5475163

Phuangpornpitak, N., & Tia, S. (2013). Opportunities and Challenges of Integrating Renewable Energy in Smart Grid System. *Energy Procedia*, *34*, 282–290. doi:10.1016/j.egypro.2013.06.756

Preethi, V., & Harish, G. (2016). Design and implementation of smart energy meter. *2016 International Conference on Inventive Computation Technologies (ICICT)*, 1, 1–5. 10.1109/INVENTIVE.2016.7823225

Recioui, A., Bentarzi, H., & Ouadi, A. (2015). Application of a binary teaching learning-based algorithm to the optimal placement of phasor measurement units. *Progress in Clean Energy*, *1*, 817–830. doi:10.1007/978-3-319-16709-1_60

Recioui, A., & Dassa, K. (2017, June). Design of Standalone Micro-Grid Systems Using Teaching Learning Based Optimization. *Algerian Journal of Signals and Systems*, *2*(2), 75–85.

Renner, S., & … . (2011). European smart metering landscape report. *Imprint*, *2*, 1–168.

Reston Filho, J., Affonso, C., & Oliveira, R. (2009). Pricing analysis in the Brazilian energy market: A decision tree approach. 2009 IEEE Bucharest Power Tech, 1–6.

Saoud, A., & Recioui, A. (2017, September). A review on Data communication in smart grids. *Algerian Journal of Signals and Systems*, *2*(3), 162–179.

Segovia, R., & Sánchez, M. (2011). *Set of common functional requirements of the Smart Meter*. Jt. Contrib. DG ENER DG INFSO.

Souran, D. M., Safa, H. H., Moghadam, B. G., Ghasempour, M., & Heravi, P. T. (2016). Smart Grid Technology in Power Systems. *Soft. Computer Applications (Nottingham)*, *357*, 1367–1381.

Syal, M. M., & Ofei-Amoh, K. (2013). Smart-grid technologies in housing. *Cityscape (Washington, D.C.)*, 283–288.

Torrent-Fontbona, F., & López, B. (2015). Power re-allocation for reducing contracted electric power costs. *Energy and Building*, *89*, 112–122. doi:10.1016/j.enbuild.2014.12.030

Vlot, A., & Gunnink, J. W. (2011). *Fibre metal laminates: an introduction*. Springer Science & Business Media.

Webster, J. G. (1998). *The measurement, instrumentation and sensors handbook*. CRC press.

Zheng, J., Gao, D. W., & Lin, L. (2013). Smart Meters in Smart Grid: An Overview. Academic Press.

Zhou, B., Li, W., Chan, K. W., Cao, Y., Kuang, Y., Liu, X., & Wang, X. (2016). Smart home energy management systems: Concept, configurations, and scheduling strategies. *Renewable & Sustainable Energy Reviews*, *61*, 30–40. doi:10.1016/j.rser.2016.03.047

KEY TERMS AND DEFINITIONS

AMI: Advanced Metering Infrastructure (AMI) refers to systems that measure, collect, and analyze energy usage, and communicate with metering devices such as electricity meters, gas meters, heat meters, and water meters, either on request or on a schedule. These systems include hardware, software, communications, consumer energy displays and controllers, customer associated systems, meter data management software, and supplier business systems.

Load Profile: A load profile is a graph of the variation in the electrical load versus time. A load profile will vary according to customer type (typical examples include residential, commercial and industrial), temperature and holiday seasons. Power producers use this information to plan how much electricity they will need to make available at any given time.

Peak Demand Reduction: energy demand management, also known as demand-side management (DSM) or demand-side response (DSR) is the modification of consumer demand for energy through various methods such as financial incentives and behavioral change through education.

Smart Grids: A smart grid is an electrical grid which includes a variety of operation and energy measures including smart meters, smart appliances, renewable energy resources, and energy efficient resources. Electronic power conditioning and control of the production and distribution of electricity are important aspects of the smart grid.

Smart Metering: A smart meter is an electronic device that records consumption of electric energy and communicates the information to the electricity supplier for monitoring and billing. Smart meters typically record energy hourly or more frequently, and report at least daily. Smart meters enable two-way communication between the meter and the central system. Such an advanced metering infrastructure (AMI) differs from automatic meter reading (AMR) in that it enables two-way communication between the meter and the supplier.

Chapter 7
Decentralized Energy Management System Enhancement for Smart Grid

Hadjira Belaidi
Signals and Systems Laboratory, Institute of Electrical and Electronic Engineering, University M'hamed Bougara of Boumerdes, Algeria

Zakaria Rabiai
Signals and Systems Laboratory, Institute of Electrical and Electronic Engineering, University M'hamed Bougara of Boumerdes, Algeria

ABSTRACT

In these last decades, electrical power grids become more intelligent. Hence, sophisticated software and hardware were introduced to the power grid, which makes it a smart grid. This chapter is an introduction on smart-grid technology; thus, microgrids are explained, and the use of multiagent system in centralized/ decentralized energy management systems are discussed and compared. Smart agents are an emerging technology for decentralized computation and data storage. Hence, in this chapter, decentralized energy management system is created basing on multi-agent system technique where sources and loads are considered as separated agents each of them. After that, these sources and load create a microgrid and each microgrid can be considered as an agent. The work proposes an approach for load supplying optimization to decrease the microgrid cost and enhance its efficiency.

1. INTRODUCTION

Smart devices and processes are gradually introduced into the electric power delivery (transmission and distribution) networks. In the coming two decades, it is foreseen that the electric power grid will be electronically controlled network rather than an electromechanically controlled system. The most envisaged challenge is how to update, redesign, refit, and re-implement the existing conventional grid into a smart grid (Amin M. and Stringer J., 2008).

DOI: 10.4018/978-1-7998-4027-5.ch007

Micro-grid energy management approaches are often classified as centralized and decentralized approaches. In centralized ones, all control processing occurs in a single agent with lack of coordination between the micro-grid performers (Yimy E. et al., 2019). Their advantage is that Off-the-shelf energy management algorithms can be directly applied; however, if the dimensionality of configuration space increases, it leads to an increasing running time. Whereas, in decentralized approaches Control processing is distributed among the agents and each one computes its solution independently; then, a strategy for coordinating the micro-grid participants is implemented; where each participant evaluates its own optimal settings. The advantage of these approaches is that dimensionality of configuration space does not increase running time since each agent is independent from others; but in the other hand, coordination is not always possible, which does not complete the decoupled management.

Centralized smart-grid energy management issue has been broadly studied in the literature Farhangi H. (2010), Borlase S. (2018) and several works have tried to improve the centralized approaches. Almada J.B. et al.(2016) used the master-slave strategy for a rule-based management system applied to a microgrid composed of multiple energy resources including a PV system, a fuel cell, and a battery bank. This approach is an appropriate solution only if energy resources are close to each other. In the same context, gravitational search algorithm (GSA) method was applied in the researches of Ji B. et al. (2014), Jahan, M.S. & Amjady, N. (2013), Radosavljević, J. et al. (2014) and Jiang S. (2014) to optimize power flow in given systems and the concept of centralized microgrid management system (MMS) was used make decisions when information was collected in a central point, while cooperation and prioritization are achieved more easily when using a central controller (Planas E., 2013). Other authors such as: Almada J.B. (2016), Patrao I. (2015) and Prabhash N. (2015) had addressed the problem of multi-agent power grid operation using centralized methods; however, the best results they got even near optimal, time consuming or higher control complexity.

Several models existing in the literature such as Capitan J et al. (2012) have been judged to be highly interdependent (a system is highly interdependent if an adjustment in one of the agents' model necessitates re-calculating policies for the others). Thus, Timothy M. H. & Edwin K. P. C. (2018) and Misra S. et al., (2013) proposed advancements to decentralize multiagent and to minimize the household electricity bill in such a Real-time pricing (RTP) market by applying Partially Observable Markov Decision Process (POMDPs). Whereas, the authors Raju L. et al. (2016), Weixian L. et al. (2016) and Raju L. et al. (2016) proposed an implementation of multiagent system (MAS) for the advanced distributed energy management and demand side management of a microgrid using Java agent development environment (JADE) frame work and MATLAB Simulink (MACSimJX).

The recent relevant works have addressed the problem of energy management system for an isolated MG to minimize its operating costs (Farzin, H. et al., 2017), regulation and supervision of loads and dispatchable energy inside a MG (Zachar M. & Daoutidis P., 2018) and improving the efficiency of the hybrid renewable energy systems (HRESs) (Faccio M. et al., 2018).

Hence, this work proposes an approach based on a decentralized energy management system to enhance the performance of a smart-grid. This approach is build upon Multiagent system where the coordinator commands are merged to solve the problem of microgrid management. The approach is based on distributing the problem solving tasks between the different agents as done by Belaidi H. et al. (2018). Each agent calculates its coefficients and needs and exchanges the data with the other data to improve the performance of the overall microgrid.

2. MICROGRID VS SMART-GRID

Several researches were performed to explain the discrepancies between a microgrid and a smart grid. Principally, a microgrid is a self-reliant, small-scale power grid that can operate as a separate or as a grid-connected system with distributed energy resources (DERs) and group of controllable loads. It is supplied by conventional power sources or gas turbines via special purpose inverters, solar panels, wind sources and any kind of renewable energy. Whereas, smart grid is a large region provider with complicated and decision making potentials. Smart-grid can be composed of one or more microgrids.

To get an efficient control and management of a microgrid, the DERs coordination must be reliable, safe and cost efficient. Moreover, the update rate and the communication between the included DERs must be fast and responds to the load demand very quickly.

In this work, we consider that our microgrid is composed of power generator, a photovoltaic (PV) solar panel, a battery power bank and loads. This microgrid constitutes the smart-grid that must be controlled and coordinated to have an efficient load response. Each element in this microgrid is considered as a separated agent having enough intelligent to make decision and sends commands to the corresponding other agents.

2.1. Distributed Energy Resources (DERS)

With smart grid, the planning, investment, and operation of the distribution system change dramatically. Historically, utility investment in distribution systems ensured circuit capacity was adequate to deliver power from the bulk grid to the customer. Now, customer-owned solar PV delivers power to the distribution system, and DER from customers provides energy and capacity reduction at the bulk grid level. A host of other distributed resources, including fuel cells and energy storage, provide power that is injected at the low-voltage level and may create reverse power flows on the grid, moving power away from the customer. Platforms are being designed to host DERs at lower voltage levels to explicitly supply customers at the distribution level and to wholesale markets.

An immediate objective is to monetize the option value of DERs, which translates to more flexible DER uses in multiple markets. Multiple opportunities have emerged, and more will result as DER needs increase across the grid. We examine both the voltage context and future opportunities to provide greater understanding of these new resources.

Distributed resources can reduce peak demand, which can eliminate or defer new transmission and distribution capacity, and decrease total energy costs. Enhanced on-site peaking generation resources also improve the security and hence the reliability.

2.2. Multiagent Systems (MAS)

Multi-agent system (MAS) is referred as system built up of set of agents (a process, a robot, a human being, energy source, building etc.), located and interacting in a certain environment according to certain relationships. An agent is an entity characterized by the fact that it is, at least partially, autonomous. In smart grids, agents serve to solve several challenges such as: store the surplus energy at home, balance between the generated and the demanded energy, energy restoration and negotiate energy price between the producer and the consumer etc.

Russell S. & Norvig P. (2009) has classified the agents into five classes according to their perception ability, intelligence and capability. The classes are:

- Simple reflex agents: These kinds of agents are incented by their environments and react directly to them by executing predefined rules.
- Model-based reflex agents: These agents treat the data received from their sensors and define their actions according to their interpretations.
- Goal-based agents: They have the same functionality as the model-based reflex agents in addition to having the ability to set goals on taking decisions.
- Utility-based agents: These agents have the ability to react and coordinate with the other agents and exchange the information.
- Learning agents: These agents have the ability to work in unknown environments and to gain more knowledge.

In this work utility-based agents are used. Each source and load is considered as a smart agent able to extract the data from its environment, treat this data and take decision. After that, the agent has to exchange its own data and decisions with the other agents to coordinate with them and enhance the overall systems task execution.

2.2.1. Multiagent System (MAS) in Smart-Grid

In conventional system, Supervisory Control And Data Acquisition (SCADA) system was used for automating electrical distribution systems. Mainly, SCADA system is a large-scale remote management and central control system for processing large numbers of equipment in real time and for remote control of technical installations. SCADA systems are crucial for industrial organizations since they help to maintain efficiency, process data for smarter decisions, coordinate and communicate system issues to help mitigate downtime. Since 2010, Ignition HMI/SCADA Software has been set up in many companies in over 100 countries. SCADA system is powerful and robust permits integrators to reach the demands of their customers, Figure 1 summarizes the Ignition HMI/SCADA architecture which allows SCADA to gathers all system knowledge in central processor then makes the decision (remotely). However, in the micro-grid, uncertainty in SCADA systems arises when sensor data or inferred knowledge cannot be deemed accurate due to intermittent nature of renewable energy resources [18].

Generally, sensors information can be noisy, erroneous or uncertain; moreover, applications using multiple sources can suffer from incoherent, uncertain or divergent data; which cannot be solved even by using SCADA unless by the supervision of the human. Whereas, MAS approach for energy management does not suffer from such situations and offers better results, more energy efficiency and time reduction. MAS automates the micro-grid by associating protocols skills and communication facilities between the different agents of the micro-grid to exchange the data (Belaidi H. et al., 2019). Micro-grid based on MAS can replicate easily and seamlessly basing on plug and play adaptability to connect to external grids. MAS has the ability to add new agents (resources or loads), or dissolve them and associate them to a new environment with new resources and new abilities. Figure 2 illustrate a standard diagram of MAS for energy management.

Figure 1. Ignition SCADA standard architecture sample.

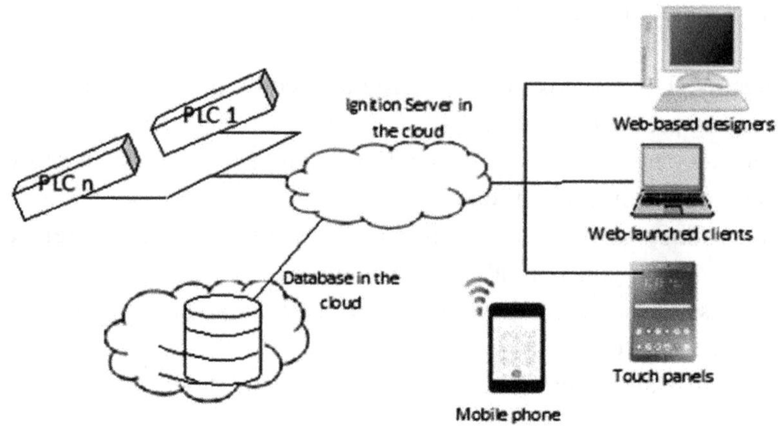

Figure 2. MAS energy management diagram

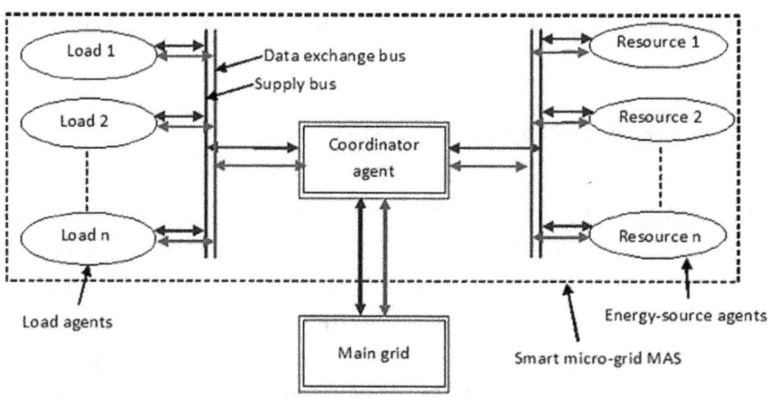

2.2.2. MAS in Centralized/Decentralized Energy Management Systems (EMS)

In the literature, both centralized and decentralized energy management systems had exploited Multi-agent techniques. When MAS are applied in centralized energy management and control systems (CEMS) of microgrids, the Central or controller Agent (CA) manages the other agent of the system. The CA is responsible for both the technical operation of the system, such as to satisfy the load demand and its economical operation (Karavas C.S.et al., 2015). Whereas, low level agents (LAs) collect necessary data and send it to CA which will process it. The processed data will be translated to commands and actions and resent to the appropriate LA which takes the final decisions.

MAS has been applied in CEMS to achieve several aims. Authors Vytelingum P. et al. (2010) proposed an energy storage management method to enhance their profit by selling energy with higher price. Authors Hongbin S. (2019) summarized how MAS are used to regulate voltage. Anvari-Moghaddam A.et al. (2017) proposed an agent-based transactive energy management framework with a comprehensive energy management system to solve the aggregated complexity induced by microgrids in distribution systems. The main benefit of the MAS centralized approach is to reduce the conflicts during operation, since all the information from the distributed energy sources are gathered in the CA level. Moreover, the action and the main decisions are taken at that level.

However, the utilization of MAS in CEMS raises several outcomes such as:

- The systems data become heavier in size and more complex, which makes the control tasks at real time very difficult. This can lead to the need for big and powerful master servers able to handle the huge amounts of data and communication, which increase the cost of the system and make it less flexible (Karavas C.S.et al., 2015).
- The installation of new power sources or loads leads to a change in the global systems and the update and configuration of the system must be done from the earlier stages.
- Any failure in the main server or in the CA results in the whole system breakdown.

On the other hand, when MAS are applied in decentralized energy management systems DEMS each component is considered as an autonomous agent having a control unit, able to control itself, react to its environment and take decision; there is no central main controller agent. The agents must have feasible tools of communication to exchange data and discuss directly between each other in order to accomplish their tasks without need to a central agent. The use of MAS as DEMS as several advantages over CEMS such:

- Reliability: the failure in one agent will not affect the overall system (the other agent still functional).
- Less complexity: each agent has acceptable amount of data to be treated which makes the system less complex than the case of CEMS.
- Advanced control: each agent in DEMS has enough intelligence and uses advanced computation techniques.
- Low cost implementation: open software and hardware platforms for electronics have been developed leading to further reductions as far as cost is concerned (Karavas C.S.et al., 2015).

According to their advantages, MAS are widely used in DEMS some works are cited hereafter. A set of agents communicating and cooperating with each other to form an intelligent distributed computing system is created by Karavas C.S. et al. (2015) for the control and management of an Autonomous Polygeneration Microgrids (APM). Anvari-Moghaddam A. et al. (2017) proposed an ontology-driven multi-agent based DEMS for monitoring and optimal control of an integrated homes/buildings and microgrid system with various renewable energy resources (RESs) and controllable loads. Authors Raju L. & Morais A.A. (2020) write a chapter where MAS is implemented with Java Agent DEvelopment (JADE) framework for advanced DEMS of a microgrid. In that chapter MAS is linked with Arduino microcontroller for practical verification of agent operations. Shirazi E. & Jadida S. (2019) designed a MAS to implement self-healing in smart distribution network to increase reliability of distribution system, minimize out-of-service area after a fault by using distributed control approaches. The proposed control structure consists of three main types of controllers: feeder, switch and zone.

3. OUR PROPOSED APPROACH

3.1. Microgrid Diagram

Several architectural styles used in the development of multi-agent systems. Four such organizations were described by Shehory O. (1999):

- ***Hierarchical multi-agent systems***, in which agents communicate according to a hierarchical structure, such as a tree.
- ***Flat multi-agent systems***, in which any agent may contact any of the others. These provide the greatest agent autonomy, but result in more communications between agents. Many smaller multi-agent systems appear to be developed using a ñat organization.
- ***A subsumption multi-agent system*** is a system in which agents are themselves made up of other agents. In this system, the subsumed agents are completely controlled by the containing agents.
- ***A modular multi-agent system*** is comprised of a number of modules. Each module normally employs a ñat structure, while inter-module communications is relatively limited. A modular multi-agent system might be useful in a situation such as power system automation, in which each substation could be categorized as a single module.

In this work, flat MAS is used to connect the PV solar panel agent, the battery agent, the conventional grid and the load agent. The diagram given in Figure 3 shows the interconnection between these agents.

Figure 3. Interconnection between the agents of our micrograd

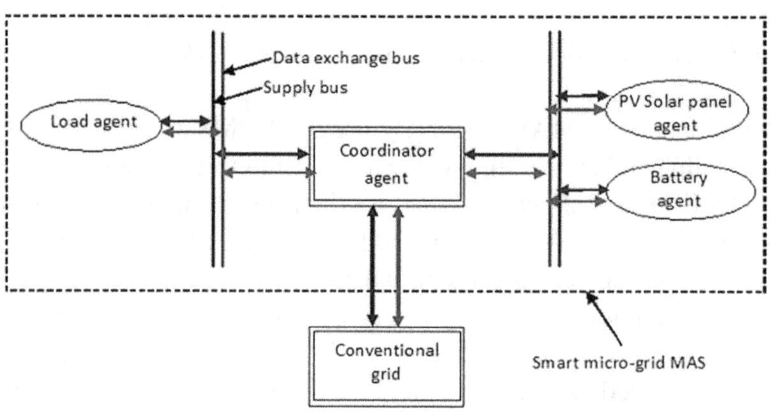

Our sample microgrid is interconnected with the conventional grid (utility grid) the "island" to remain online if weather or grid faults occur. Microgrid islanding with local DERs is chosen to insure an important increase in reliability and decrease cost. Each element in our microgrid is considered as a separated agent having enough intelligent to make decision and sends commands to the corresponding other agents. The microgrid contains two kinds of buses: supply buses to transport the energy and the data buses to exchange the data and the information between the different agents.

3.2. Case of Study (Load Supplying)

In this work, the manner the load is supplied is studied to increase the microgrid efficiency and decrease the cost. The load is first supplied by PV solar panel as it is a renewable, clean and free source; then, it will be supplied by the battery (which is charged initially from the PV solar source) when the renewable source is not available (in night for example). Finally, the load will be connected to the conventional grid (CS) in case of terrible weather conditions or if any faults occurred. The flowchart shown in Figure 4 summarizes our load supplying optimization approach.

Figure 4. Load supplying optimization approach.

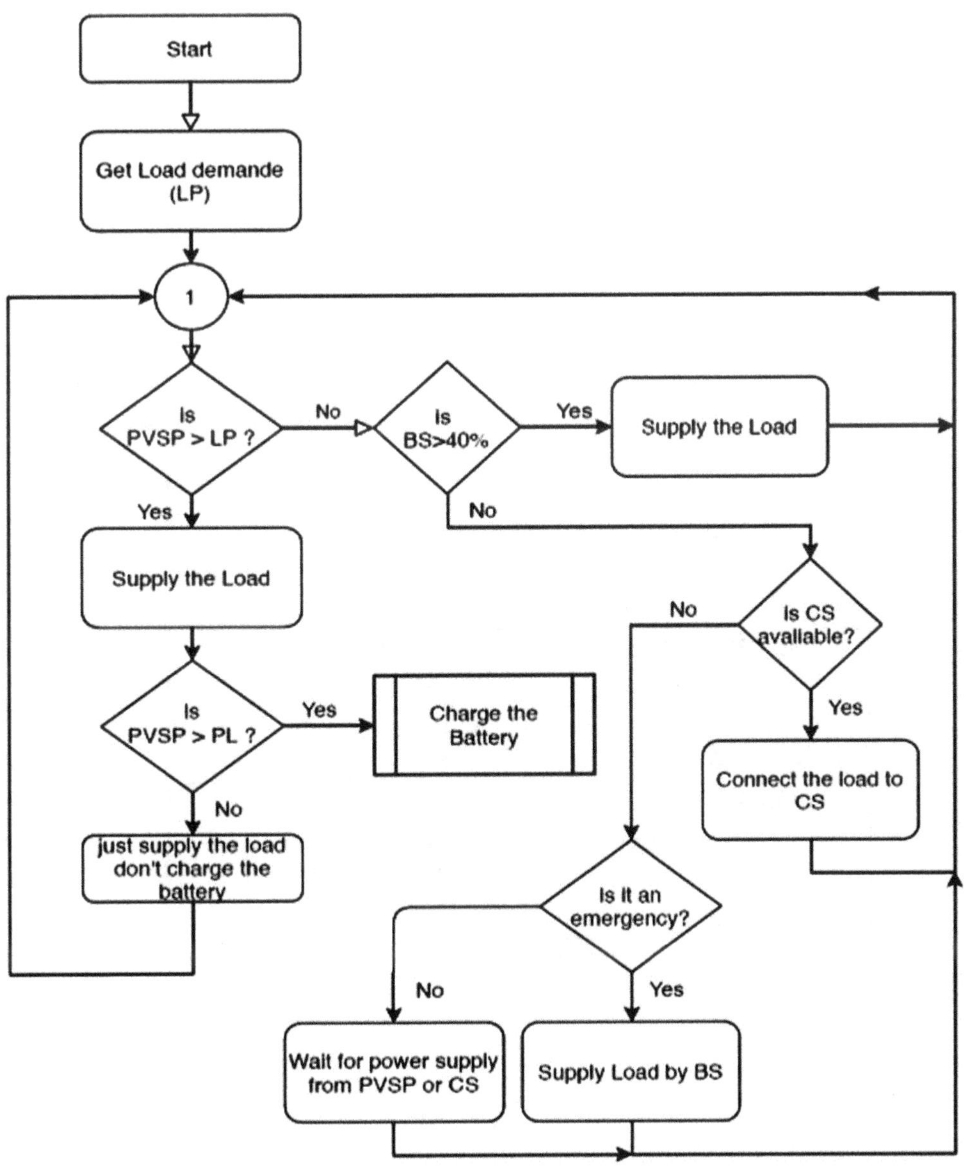

The load is primary supplied by the solar panel (PVSP). In case, the PV solar panel has a production surplus it will be stored in the battery to be used when the PVSP is not available.

The secondary load power supply is the battery (BS). The battery will not supply the load unless its charge is higher than 40% except for emergency situation and where the conventional source is also not available.

Our microgrid remains always online with the conventional grid to supply the load in case of bad weather conditions and where the battery is not charged, or in case of any faults in the microgrid.

This flowchart must be implemented in the coordinator agent (see Figure 3) in order to coordinate and exchange the information and data between the different agents. Each agent has enough intelligent to measure its needs and its surplus and to take decisions. The flowchart given in Figure 5 explains the battery charging process.

Figure 5. Battery charging flowchart.

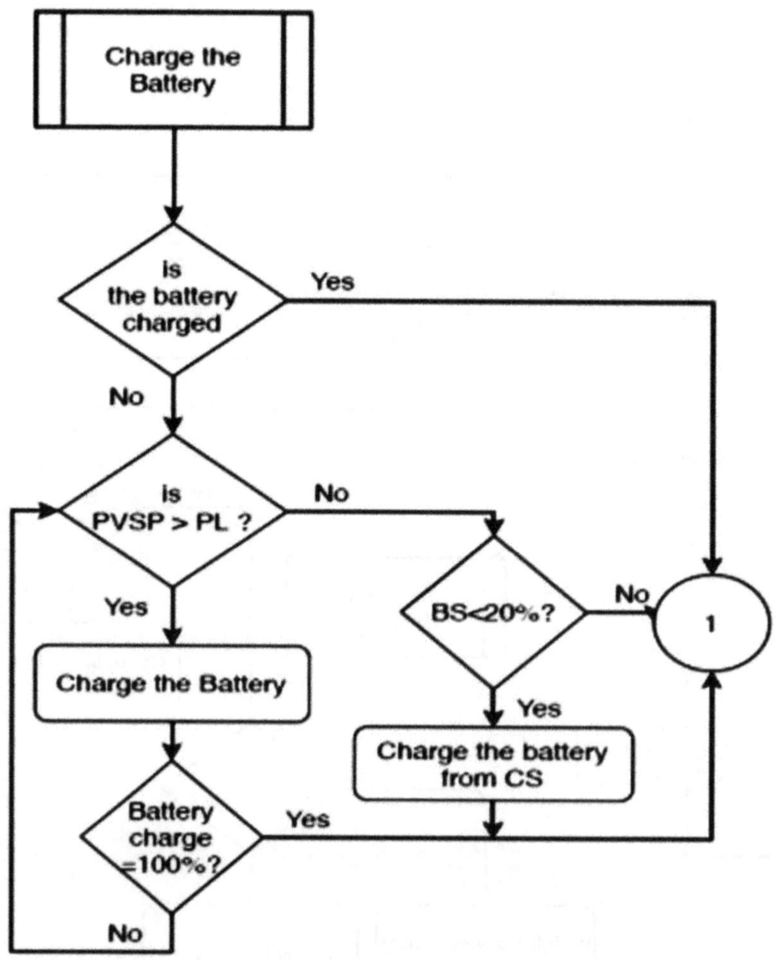

The battery is primary charged from the PV solar panel (PVSP) when this last is able to supply the load and has surplus of energy to charge the battery. However, if the battery is discharged below 20% because of its use in emergency situation, it must be charged even though from conventional supply if the PVSP has no energy to charge it.

3.3. Simulation Results

Figure 6 shows the control signals of the three sources. It is clearly shown that before 5:00 the conventional source is enabled, which means that the load will be supplied from the CS waiting for the PVSP to be activated. From 5:00 to 6:00 CS will be disabled because it is the time for the sun to shine and the load is not demanding a huga amount of energy; hence, the battery will be activated to satisfy the load demand. From 6:00 to 7:00 both the battery and the PVSP are activated because the PVSP will not be enough to supply the load alone. From 7:00 to 19:00, the PVSP will be activated to supply the load and charge the battery at the same time. From 19:00 to 20:00 both the PVSP and the battery are activated to satisfy the load demand. After 20:00 there will be no solar energy, thus, the battery will be activated to supply the load alone.

Figure 6.

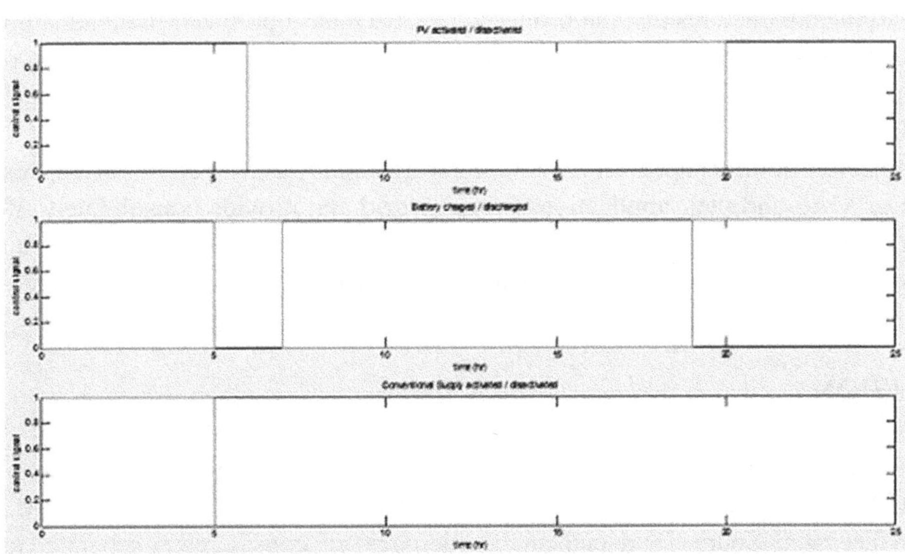

Figure 7 shows an illustration simulation of a PV solar energy production and a load consumption sample during one day. In this case, solar irradiance profile is considered typical in a period between 07:00 and 19:00 region. During the simulated period, suppose the microgrid to operate in standalone mode.

At 09:00 and 17:00, the solar panel produces exactly the power needed by the load. In side this region (from 09:00 to 17:00) the solar produces more than 300 W/hr of surplus of power. This power will be saved on the battery to be used after that to supply the load when the PVSP doesn't produce enough power as it is the case after 17:00 and before 20:00; or where the PVSP does not produce any power (in night from 20:00 pm until 06:00 am).

Figure 7. Solar panel power production and load consumption illustration.

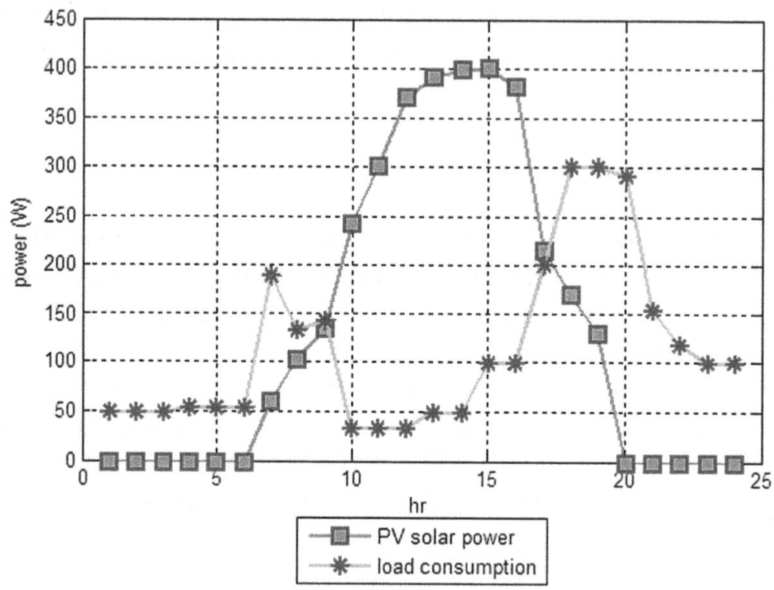

According to the obtained results, the proposed approach has optimized the load supplying so that the microgrid cost is decreased and its efficiency is enhance. Compared to the existing researches, in our proposed paradigm two sources can be used simultaneously to supply the load demand; where, the sources with low costs are primary used.

The main objective of this chapter is to review the work already attempted by various research papers in the context of MAS and their applications in smart grid and provide consolidated information for management objectives in microgrid system. This chapter can also be considered as an introduction to simple microgrid paradigm creation and load supply optimization.

4. CONCLUSION

MAS have been widely used in several fields; hence, this chapter can be considered just as an introduction on the use of MAS in smart-grid technology. Thus, microgrids are studied and the use of MAS technique in microgrid is discussed. Sources and loads in the microgrid are considered as separated agents each of them. The work proposes mainly an approach for load supplying optimization as a sample to decrease the microgrid cost and enhance its efficiency.

In order to better explain our approach, one round of operation of solar microgrid was simulated for 24 hours, considering all the possible scenarios happen in a day. Diagram of microgrid composed of Solar panel agent, battery agent, conventional agent and load agent was created. Our approach must be set up on a coordinator agent in order to link the different agent data and decision and to ensure the good function of and better control of the microgrid. However, this approach is considered as decentralized approach because each agent has enough intelligent to control itself and take decisions.

REFERENCES

Almada, J. B., Leão, R. P. S., Sampaio, R. F., & Barroso, G. C. (2016). A centralized and heuristic approach for energy management of an AC microgrid. *Renewable & Sustainable Energy Reviews*, *60*, 1396–1404. doi:10.1016/j.rser.2016.03.002

Amin, M., & Stringer, J. (2008). The Electric Power Grid: Today and Tomorrow. *MRS Bulletin*, *33*(4), 399–407. doi:10.1557/mrs2008.80

Anvari-Moghaddam, A., Rahimi-Kian, A., Mirian, M. S., & Guerrero, J. M. (2017). A multi-agent based energy management solution for integrated buildings and microgrid system. *Applied Energy*, *203*, 41–56. doi:10.1016/j.apenergy.2017.06.007

Belaidi, H., Belkalem, J., Abed, M. A., & Bentarzi, H. (2018). IoT Path Planning Approach for Mobile Robots. *2018 International Conference on Applied Smart Systems (ICASS)*, 1-8. 10.1109/ICASS.2018.8652050

Belaidi, H., Hentout, A., & Bentarzi, H. (2019). Human–Robot Shared Control for Path Generation and Execution. *International Journal of Social Robotics*, *11*(4), 609–620. doi:10.100712369-019-00520-3

Borlase, S. (2018). *Smart Grids: Advanced Technologies and solutions*. CRC Press.

Capitan, J., Spaan, M. T. J., Merino, L., & Ollero, A. (2012). Decentralized Multi-Robot Cooperation with Auctioned POMDPs. *Proceedings of the Twenty-Fourth International Conference on Automated Planning and Scheduling, 2012 IEEE International Conference on Robotics and Automation*, 14-18.

Faccio, M., Gamberi, M., Bortolini, M., & Nedaei, M. (2018). State-of-art review of the optimization methods to design the configuration of hybrid renewable energy systems (HRESs). *Frontiers in Energy*, *12*(4), 591–622. doi:10.100711708-018-0567-x

Farhangi, H. (2010). The path of the smart grid. *IEEE Power & Energy Magazine, 8*(1), 18-28.

Farzin, H., Fotuhi-Firuzabad, M., & Moeini-Aghtaie, M. (2017). Stochastic Energy Management of Microgrids during Unscheduled Islanding Period. *IEEE Transactions on Industrial Informatics, Volume*, *13*(3), 1–9. doi:10.1109/TII.2016.2646721

Hongbin, S., & Qinglai, G. (2019). Review of Challenges and Research Opportunities for Voltage Control in Smart Grids. *IEEE Transactions on Power Systems*, *34*(4), 2790–2801. doi:10.1109/TPWRS.2019.2897948

Jahan, M. S., & Amjady, N. (2013). Solution of large-scale security constrained optimal power flow by a new bi-level optimization approach based on enhanced gravitational search algorithm. *IET Generation, Transmission & Distribution*, *7*(12), 1481–1491. doi:10.1049/iet-gtd.2012.0697

Ji, B., Yuan, X., Li, X., Huang, Y., & Li, W. (2014). Application of quantum-inspired binary gravitational search algorithm for thermal unit commitment with wind power integration. *Energy Conversion and Management*, *87*, 589–598. doi:10.1016/j.enconman.2014.07.060

Jiang, S., Ji, Z., & Shen, Y. (2014). A novel hybrid particle swarm optimization and gravitational search algorithm for solving economic emission load dispatch problems with various practical constraints. *International Journal of Electrical Power & Energy Systems*, *55*, 628–644. doi:10.1016/j.ijepes.2013.10.006

Karavas, C. S., Kyriakarakos, G., Arvanitis, K. G., & Papadakis, G. (2015). A multi-agent decentralized energy management system based on distributed intelligence for the design and control of autonomous polygeneration microgrids. *Energy Conversion and Management, 103*, 166–179. doi:10.1016/j.enconman.2015.06.021

Li, W., Logenthiran, T., Phan, V., & Woo, W. L. (2016). *Intelligent Multi-Agent System for Power Grid Communication. In 2016 IEEE Region 10 Conference*. TENCON.

Misra, S., Mondal, A., Banik, S., Khatua, M., Bera, S., & Obaidat, M. S. (2013). Residential Energy Management in Smart Grid: A Markov Decision Process-Based Approach. *2013 IEEE International Conference on Green Computing and Communications and IEEE Internet of Things and IEEE Cyber, Physical and Social Computing*, 1152-1157. 10.1109/GreenCom-iThings-CPSCom.2013.200

Patrao, I., Figueres, E., Garcerá, G., & González-Medina, R. (2015). Microgrid architectures for low voltage distributed generation. *Renewable & Sustainable Energy Reviews, 43*, 415–424. doi:10.1016/j.rser.2014.11.054

Planas, E., Gil-de-Muro, A., Andreu, J., Kortabarria, I., & Martínez de Alegría, I. (2013). General aspects, hierarchical controls and droop methods in microgrids: A review. *Renewable & Sustainable Energy Reviews, 17*, 147–159. doi:10.1016/j.rser.2012.09.032

Prabhash, N., Panigrahi, C. K., & Dasgupta, A. (2015, December). Energy Management System in Smart Grid: An Overview. *International Journal of Research and Scientific Innovation, II*(XII), 31–48.

Radosavljević, J., Jevtić, M., Arsić, N., & Klimenta, D. (2014). Optimal power flow for distribution networks using gravitational search algorithm. *Electrical Engineering, 96*(4), 335–345. doi:10.100700202-014-0302-5

Raju, L., Milton, R. S., & Mahadevan, S. (2016). Multi Agent Systems based Distributed Control and Automation of Micro-grid using MACSimJX. *2016 10th International Conference on Intelligent Systems and Control (ISCO),* 7-8.

Raju, L., Milton, R. S., & Mahadevan, S. (2016). Multiagent Systems Based Modeling and Implementation of Dynamic Energy Management of Smart Microgrid Using MACSimJX. Hindawi Publishing Corporation. doi:10.1155/2016/9858101

Raju, L., & Morais, A. A. (2020). *Multi-Agent Systems Based Advanced Energy Management of Smart Micro-grid. In Multi Agent Systems - Strategies and Applications*. IntechOpen.

Russell, S., & Norvig, P. (2009). *Artificial intelligence: A modern approach. Publisher: Pearson*. Prentice Hall Pa.

Shehory, O. (1999). *Architectural properties of multi-agent systems*. Technical Report CMU-RI-TR-98-28, The Robotics Institute, Carnegie Mellon University.

Shirazi, E., & Jadida, S. (2019). A multiagent design for self-healing in electric power distribution systems. *Electric Power Systems Research, 171*, 230–239. doi:10.1016/j.epsr.2019.02.025

Timothy, M. H., & Edwin, K. P. C. (2018). A Partially Observable Markov Decision Process Approach to Residential Home Energy Management. *IEEE Transactions on Smart Grid, Volume*, *9*(2), 1271–1281. doi:10.1109/TSG.2016.2582701

Vera Yimy, E. G., Dufo-López, R., & Bernal-Agustín, J. L. (2019). Energy Management in Microgrids with Renewable Energy Sources: A Literature Review. *Applied Sciences (Basel, Switzerland)*, *9*(3854), 1–28.

Vytelingum, P., Voice, T. D., Ramchurn, S. D., Rogers, A., & Jennings, N. R. (2010). Agent-based micro-storage management for the smart grid. In *Proceedings of the 9th International Conference on Autonomous Agents and Multiagent Systems*. International Foundation for Autonomous Agents and Multiagent Systems.

Zachar, M., & Daoutidis, P. (2018). Energy management and load shaping for commercial microgrids coupled with flexible building environment control. *Journal of Energy Storage*, *16*, 61–75. doi:10.1016/j.est.2017.12.017

Chapter 8
Smart Grids and Hybrid Energy Storage Systems:
Optimization Techniques Applied in Control Strategies for Hybrid Energy Storage Systems

Muhammad Asif Rabbani
Cyprus Science University, Cyprus

ABSTRACT

It is very important that the installed renewable energy system should produce the maximum power outputs with minimum costs, and that can only be achieved with the selection of the best optimization technique applied for the best control strategies along with the introduction of the hybrid energy storage systems (HESS). This chapter presents some optimization techniques applied in control strategies for hybrid energy storage systems in distributed renewable energy systems. The integration of energy production and consumption component through the smart grid concept enables increased demand response and energy efficiency. Hybrid energy storage systems and their applications in the renewable energy systems are extensively discussed besides control strategies involved. The storages systems will play an important role in future related to smart grid.

INTRODUCTION

The increase in renewable energy systems installations and their integration with the grids is transforming the whole electric generation, transmission and distribution systems around the world. This change from classical to advance techniques is resulting in the evolution of smart grids and the introduction of new storage systems. In past the flow of electricity was one directional but now because of renewable energy power generators such as solar photovoltaic and wind, the excess energy produced by the end user is sent back to the grid using net metering making flow of electricity bi-directional. In future more

DOI: 10.4018/978-1-7998-4027-5.ch008

solar and wind generators are expected to be integrated with the grid and it is expected that the inward flow to the grids will be much more than the outward flow from the grid.

As the power generated by the solar and wind generators is not stable and is impacted by the change in metrological conditions. Therefore, energy storage is required at the site of power generation and also smart grid in order to stabilize the power system and to store and use the excess energy properly and efficiently. The microgrids (MG) compromised of distributed generators such as solar and wind are located at the prosumer end also include the storage systems and certain control units for the supply of clean energy and smooth power in the optical way(Tsikalakis and Hatziargyriou, 2011); Logenthiran et al (2010)

In case of micro grids with integration of photovoltaic and wind power systems, the power variations are basically due to the irregular power generations from renewable energy sources and improbability of load consumption Shim et al (2013). The grid connected microgrids have essentially all the support from high inertia power generators whereas off grid micro grids mainly depends on Energy Storage Systems(ESS) in order to balance the mismatch between the power generated by the renewable source and the power being consumed by the user (Tan, Li, & Wang, 2013).

The use of different energy storage systems (ESS) systems in micro grids and various benefits of using ESS such as short term power supply, peak shaving, optimizations of micro source for unit commitment, power quality improvement, ancillary services and seasonal storage are discussed in (Fathima & Palanisamy, 2015). In order to make energy storage more reliable, efficient and cost effective in renewable energy applications hybrid energy storage systems (HESS) are also used. HESS is categorized by a favorable coupling of different energy storage technologies with additional operating features such as energy and power density, self-discharge rate, efficiency, life-time etc.(Bocklisch, 2015). The control strategies play important role in renewable energy applications with HESS. These can be classified into Classical Control Strategies and Intelligent Control Strategies.

The classical control strategies include rule based control (RBC) and filtration based control (FBC) and are mostly require particular mathematical model as they can be effected by the change in parameters (Akcayol, 2004).

Whereas, for reliable and efficient control an emerging predictive modeling systems are required. The optimization plays a vital role in predicting the performance of the renewable energy system with HESS integrated with grid or standalone. The optimization helps not only to improve the performance of the system but also reduces the cost.

For optimization purpose different techniques are used including artificial neural networks (ANN) and fuzzy logic controller (FLC) to improve the dynamic behavior of the system without the need for exact mathematical modelling. These strategies are considered as intelligent although optimal performances are not guaranteed (Brka, Kothapalli, & Al-abdeli, 2015).In order to minimize the impact of parameter variations and to optimize the control strategy algorithms predictive and optimization based control strategies are also very useful. Therefore, these control strategies with higher complexity are required to be developed that will help more smooth integration of renewable energy in system Chong et al. (2016).

The aim of this paper is to review various optimization techniques that can be used to develop the adaptive controls strategies for controlling the power production form renewable energy sources mainly solar photovoltaic and the charging of HESS. The overall effect on performances of the Renewable Energy Power System (REPS) with different optimization techniques is discussed.

ENERGY STORAGE SYSTEMS: AN OVERVIEW

The energy storage systems play an important role in the performance of renewable energy systems. We can define energy storage as a change in the form of energy that can be drawn upon when needed for some useful operation. As it is difficult to store the Electrical energy therefore in order to store electrical energy it is required to be converted into another form of energy. Energy can be stored using different methods such as chemical, electrochemical, electrical, mechanical, and thermal systems (Guney & Tepe, 2017).

The energy storage technologies can be classified in terms of the form of energy stored and also on the basis of discharge duration. These can be short term, medium term and long term discharges.

The figure 1 shows only some of the possible combinations that can be used for Hybrid Energy Storage systems.

Figure 1. Types of Energy Storage Systems

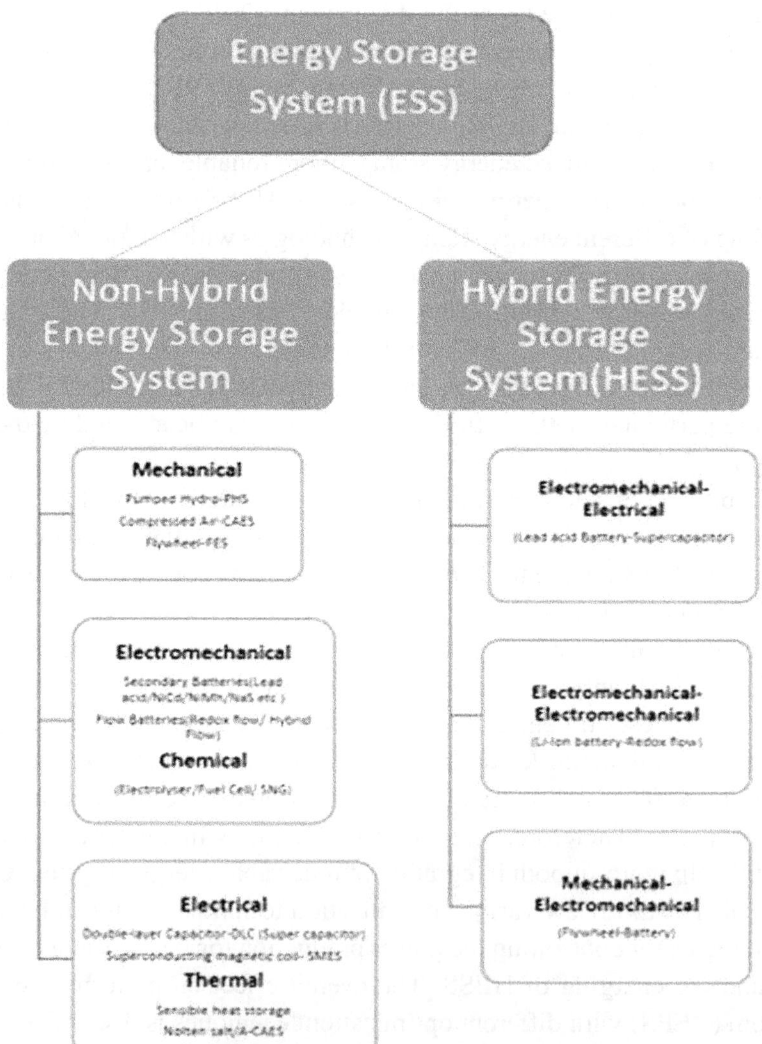

Different energy storage systems (ESS) technologies have many features and characteristics different from one another and such differences make some ESS more valuable and suitable for some specific working environment and conditions than others.

ESS technologies comparison was done for the energy density, power density, discharge time, life time along with the cost and it was found that no single energy storage technology can be considered as Ideal and each technology has an advantage over other technology depending upon the task to be achieved. Hence, any of these technologies can be utilized properly for a given condition knowing the detailed specifications and storage principles involved (Rohit, Devi, & Rangnekar, 2017)

Hybrid Energy Storage Systems

The hybrid energy storage system (HESS) can be defined as the combination of two or more energy storage systems (ESS) technologies with similar features in order to provide a joint power outputs from them suppressing the disadvantages and drawbacks of each.

The energy storage systems are used in applications where there is a requirement of supply of high power and need of high energy storage, therefore in order to provide the storage systems capable for providing both high power and high storage at the same time in one unit we use hybrid energy storage systems (HESS). Therefore keeping in view the supply of energy and power in HESS combinations according to (Hemmati & Saboori, 2016) the theoretical combinations of HESS would include the energy suppliers like battery, compressed air energy storage(CAES), fuel cell, pumped hydro and power suppliers such as super capacitors, superconducting magnetic energy storage (SMES), Flywheel and battery. HESS are based on a fact that the high power device should supply short term power needs, while the high energy device should meet the long term energy needs. According to (Aneke & Wang, 2016), real life operations and applications of each available energy storage technologies gives us the real level of the usage of these technologies in various areas of life. Also they provided the details of operational energy storage projects from around the world. The figure 2 shows the installed HESS power capacity by Technology.

Figure 2. installed HESS power capacity by Technology

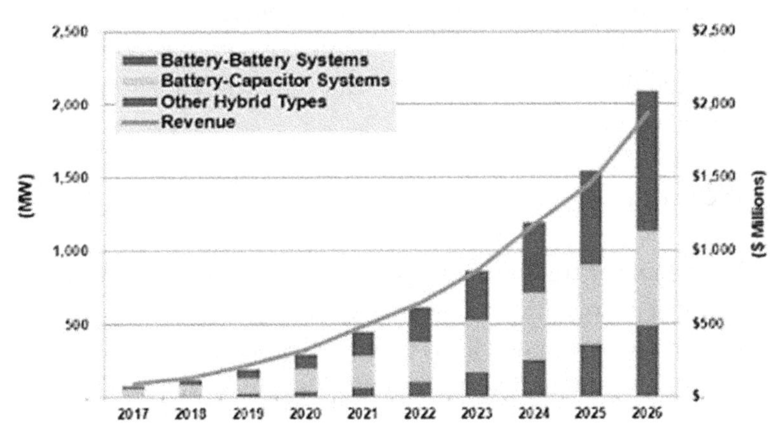

The HESS installations are on rise due to their characteristics covering the high energy and high power requirements being installed as a single unit for diversified applications and installed capacity of such systems will be about 2.1GW by the year 2026 and among all hybrid energy storage systems, Battery-battery and battery-capacitor systems are expected to be leading the HESS as 1.6GW is forecasted to be installed worldwide by 2026. (McClenny & Tokash, 2017)

Applications of Energy Storage Systems

The energy storage systems have a large number of applications and some of them are discussed below:

1. ESS in Power Sector

The conventional power systems make use of energy storage systems to overcome performance issues related to large scale generation and transmission. These storage systems can be applied in distribution networks and at the end user/customer side as well.

In power sector ESS roles is linked to proper energy management besides bridging of power along with power quality and reliability. The EES technologies provide three primary functions of energy management, bridging power and power quality and reliability (Chouhan & Ferdowsi, 2009). As discussed in (Yao et al., 2016) the energy storage technologies can be considered as a major achievement regarding fulfilling the gap during conventional energy production and consumption simultaneously.

These storages are also suitable for renewable energy integration, distributed generation, micro grids and for frequency regulations as well. The more specific applications of ESS includes peak shaving, load leveling and transmission and distribution stabilization. The ESS technologies with different combinations can meet the requirement of various power systems and network operators. (Luo, Wang, Dooner, & Clarke, 2015).

In order to have a worthy and good power quality with minimum short term requirements we can use flywheels, batteries along with capacitors and supercapacitors whereas energy management applications can make use of pumped hydro storage system (PHS),Compressed air energy storage(CAES), electromechanical batteries, flow batteries, fuel cells, solar fuels and thermal energy storage(TES) as discussed by Amrouche, et al. (2016).

It is concluded by Grantham et al. (2017),that the use of energy storage by a house hold with low energy requirements helps to utilize the maximum from the photovoltaic power systems installed at the roof top besides matching the demand requirements with the different supplies from the grid and the renewable energy source to suit the best.

2. ESS in Renewable Energy Systems

There are two types of renewable energy power systems, the grid tied power systems and off grid power systems. In off-grid renewable energy systems the Energy Storage System battery bank is charged when the power is generating from the sources such as solar or wind and can be used when there is no sun or renewable sources available. The battery bank also helps the system to remain stable during all operating hours of the day.

In case of Grid Connected systems the whole system shuts down in when the grid is turned OFF so to avoid immediate power failures the Battery Banks are used at the user's end. (Sonnenenergie, 2007).

When the grid connected renewable energy system is producing excess power during the operations this is sent back to the grid using the net metering and has many economic benefits both for the user and the utility. Eid el at. (2014). The flow of electricity in the past was a one way meaning that all produced electricity was delivered to the consumers and if not used was lost but now as the consumers can also produce their own electricity using renewable energy systems so the excess power is sent back to the grid making the flow of electricity two way. The power plants or the grid side must have energy storage banks to store the excess produced energy to be distributed in areas where there is need of energy.

The smart grids can handle bi directional communications between the utility and the customers besides improving and facilitating in the generation, transmission and distribution of energy. Also the use of ESS in order to store the excess energy at the grid side makes power network stable by supplying the stored energy during the AC power interruptions Bayindir el. at., (2016). Figure 3 below shows the integration of energy storage systems with the grids using renewable energy systems and supervisory control and data acquisition (SCADA) units.

In smart grids scenario the use of energy storage systems (ESS) plays a major role in integration of renewable energy power systems and the electric grid. ESS are mostly utilized at the grid side to absorb and accommodate high volumes of renewable energy sources (RESs) or the excess energy being sent back to the grid Katsanevakis et al.,(2017)

3. ESS in Transportation

The awareness regarding the greenhouse gas emissions and the high level of pollutions in metropolitan cities has forced the governments and people around the world to reduce their dependency on fossil fuels and to increase clean energy for their transportations.

The introduction of electric hybrid vehicles, battery electric vehicles (BEVs) and fuel cell electric vehicles (FCEVs) is considered as an alternative for the light duty transportation sector in order to lower the carbon emissions. These electric vehicles are mostly charged using renewable energy sources. Vehicle-to-Grid (V2G) is smart charging of the vehicles by the grids or renewables and thus electric vehicles are also considered as Energy Storage System (ESS) that can be utilized to store energy in grid applications as well Tarroja et al.,(2016). These vehicles can be considered as future battery banks parked in residential or commercial areas for the charging during night when excess energy is produced from the grids that will be stored for being utilized during the day.

According to(Saikong and Kulworawanichpong, 2017), for traffic in cities under frequent stop and drive conditions HESS can be considered as more viable and proper solution. Furthermore the other applications of ESS in transportation sector includes the use of supercapacitors bank in subway which leads to increases energy recovery and helps in minimizing the energy usage during the operation of the subway as discussed in Allègre et al., (2010). In railways electric trains can use ESS comprising of super capacitors, batteries and flywheels which utilized as stationary energy storage systems (SESS) and on-board energy storage systems (OESS).

The supercapacitors are used for OESS in light rail vehicles whereas flywheels can be considered as more beneficial due to their higher power and log life for heavy haul trains and the best can be achieved from ESS in trains by applying control systems designed together with optimal train operation systems Ghaviha at al.,(2017).

Figure 3. ESS applications across the grid levels including RESs and the SCADA at the dispatching center (DC)

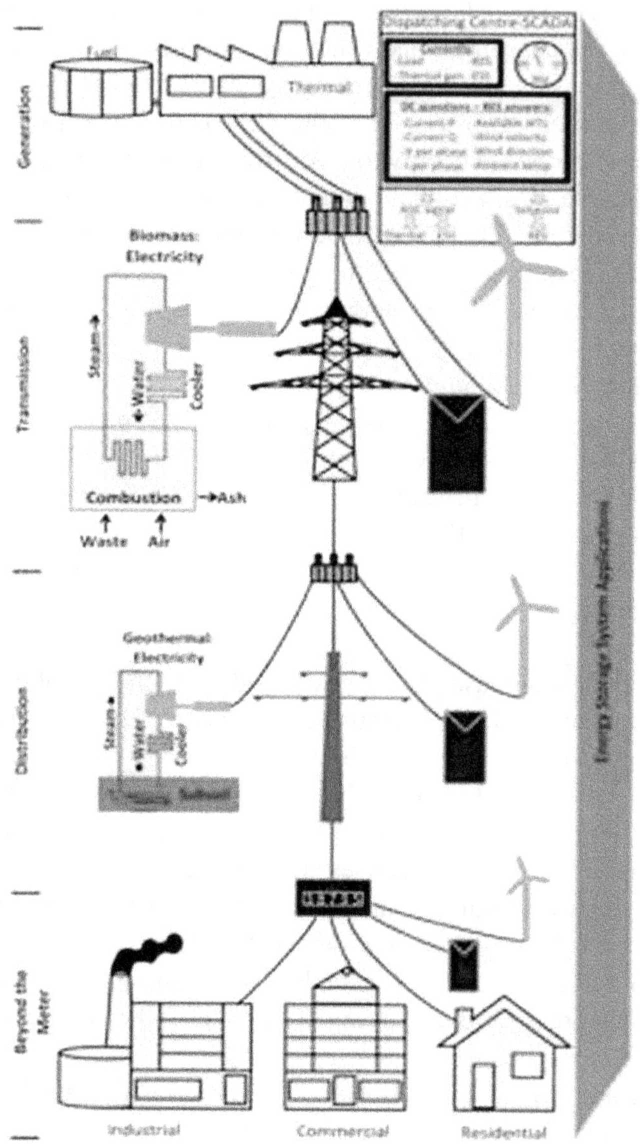

Applying Control Strategies for Smart Grids Integration

The control strategies play an important role regarding the integration of the renewable energy source with the grid and the energy storage system (ESS). Therefore, various control strategies are discussed with emphasis on the intelligent control and the optimization techniques involved in renewable energy systems with hybrid energy storage systems (HESS).

1. Classical Control Strategy

In classical control strategies buck boost converters are used in order to control the charging and discharging of the HESS integrated with renewable energy source (Hhuijuan & JianCheng, 2016; Dulout et al.,2015). The Rule based control (RBC) and Filter based control (FBC) strategies do the analysis of the power profiles differently. In RBC the rules are pre-defined before the power demands and reference powers are computed, whereas in FBC the filtration into high frequency and low frequency components is done prior to the computations. The classic control strategies are inflexible and not at all adaptive to real time system conditions.

As discussed in (M. E. E. Glavin, Chan, Armstrong, & Hurley, 2008) rule based energy control unit (ECU) has been designed for a standalone photovoltaic super capacitor-battery HESS and the rules for RBC includes the power generated from the photovoltaic (PV), State of Charging (SOC) of the battery and Super capacitor (SC). According to (Gee, Robinson, & Dunn, 2013), the low pass filter (LPF) based FBC is used to improves the ripple current along with the battery discharge depth and extends the life of battery by 19%. In (Olivares et al., 2014) different control levels and different strategies for each level are discussed for microgrids. The first level is the local or primary control which is based on local measurements and has a fast response besides do not need any communication. As they are depending on the local measurements so islanding detection, output control and power sharing controls are included in this.

The secondary control levels are basically Energy Management Systems (EMS) whereas the tertiary control is host grid based and deals with the communication and coordination between multiple microgrids according to the needs and requirements of the main grid. In (Huu, 2016), cache control for HESS and adaptive droop control for battery long term and real time operations is used and as compared to existing droop controls is more effective and reliable. The vector control of power by decoupling control of the reactive and active current components to feed the active and reactive power to the grid is proposed in Samerchur at el.(2011). The main purpose of the grid connected inverter control is to maintain the DC-link voltage and independent active and reactive power flow.

2. Intelligent Control Strategy

The most popular intelligent control approaches are comprised of fuzzy logic, artificial neural networks and genetic algorithms. In case of smoothening of the output from the PV, fuzzy logic based adaptive power management system is used more frequently Ye et al,. (2014).These intelligent controls are used basically to optimize the performance of the system. There are various optimization techniques involved in intelligent control strategies regarding the HESS and renewable energy sources.

Also from the table 1 below it can be seen that mostly the control strategies are required in order to reduce the operational cost of the system, energy management and optimization for the charging and discharging of the energy storage systems, improvement in system efficiency, reduction of stress levels of energy systems that helps to increase the life of storage and to provide overall stability for the renewable energy systems and minimization of the fluctuations caused due to the environmental and weather effects on the energy system.

Table 1. Control strategies in grid-tied renewable energy systems with hybrid energy storage systems (HESS)

Application	HESS	Control Strategy	Performance	Advantages	Reference
Wind-Grid Tied	Battery-Supercapacitors	Self-Adaptive wavelet packet decomposition technique and a two level power reference signal distribution method	Improvement in power fluctuation, smoothing and ensures that HESS works in reasonable range to guarantee the long term stable operation		(Ding & Wu, 2017)
Solar-Grid Tied	Battery-Supercapacitor	Control Logic	Extraction of maximum power form the PV panels (MPPT), Suitable task sharing between the battery and SC, Injecting power into the grid based on the direct quadrature theory		(Id & Miguel 2017)
Solar-Dc Micro Grid	Lead Acid and Li-Ion Battery	HESS distributed Control	Enhancement in system reliability, bus voltage restoration and power sharing compensation, autonomous state of charge (SoC) recovery is implemented to limit SoC variations of ES with high ramp rate.		(Xiao et al., 2016)
Solar-PV Microgrid	SC-Li-Ion	Artificial Neural Networks (ANN)	Management of Energy Storage Systems based on charging and discharging dependnig of SOC state besides satisfying the sysrem load and operation.	Reduction in dynamic stress and peak current demand of Li-Ion bateeries and Supercapacitor bank	(Dulout et al., 2015)

Table 2. Control strategies in off-grid renewable energy systems with hybrid energy storage systems (HESS)

Application	HESS	Control Strategy	Performance	Advantages	Reference
Solar-Off Grid	Battery-Supercapacitors	An energy management with the combination of the control of the bus voltage and energy management for of the storage devices	The reduction in battery stresses by the use of Supercapacitors.	The advantage of this configuration is that the battery power and the supercapacitor power can be controlled individually depending on their state of charge and supply requirement	(Cabrane,Ouassaid, Maaroufi,2016)
Solar-Off Grid	Battery-Supercapacitors	Two-stage rate limit control for battery storage	The power balance remains unaffected with the help of supercapacitor and the minimization of life time limiting effects of battery		Kollimallah et,al.(2017)
Solar-OFF Grid	Battery-Supercapacitor	Fuzzy Logic based adaptive power management system	Photovoltaic output smoothing and ramp rate control		Ye et al,.(2014)
Solar-Off Grid	Battery-Supercapacitors	Joint Control Strategy	Provided good dynamic performance during a step change in load demand and generation	Faster DC link voltage stabilization, enhancement in the life of battery by reducing the stress	(Manandhar et al., 2017)
Solar-Off Grid	Battery-Supercapacitor	Classic control using Buck-Boost DC/DC converters	The circuit controls the charging and discharging operation of the battery and supercapacitor orderly	High quality control of DC bus voltage of PV system	(Hhuijuan & JianCheng, 2016)

Table 3. Control strategies in hybrid renewable energy systems with hybrid energy storage systems (HESS)

Application	HESS	Control Strategy	Performance	Advantages	Reference
Hybrid (Wind solar)-Micro Grid	Battery-Supercapacitor	Cache control and adaptive Droop scheme	Coordination for multilevel storages in micro grids	Long term operational strategies and real time control	(Huu,2016)
Hybrid-Wind Solar	Battery-Supercapacitor	Control method based on frequency-domain analysis and an adaptive energy allocation strategy based on the isolation frequency	The isolation frequency is adjusted adaptively in order to increase energy utilization besides guaranteeing the ESS safety.		(Zhang et al., 2017)
Hybrid (Wind solar)	FuelCell-Battery-Supercapacitor	Ruled Based and Optimization based approach for Energy Management System (EMS)	EMS reduces the battery stress		(Mumtaz & Khan, 2017)
Hybrid (Wind Solar)	Battery Energy Storage System (BESS) can be used for other storage systems as well.	Rule Based Control	Rule-based controller is developed so that BESS can be integrated with renewable energy sources and inter- mittent renewable resource can be dispatched on an hourly basis like any other conventional generator.	The control method can be applied to other storage devices (HESS)as well with necessary rules. This strategy utilized 70% of the BESS and thus increases the life of the batteries	(Teleke et al., 2010)

Optimization Techniques Involved in Control Strategies

1. Fuzzy Logic

As discussed by (García, Torreglosa, Fernández, & Jurado, 2013), the use of fuzzy logic controller (FLC) controls the state of charge (SOC) of the hybrid energy storage systems (HESS) and as compared to rule based control (RBC) saved 13% by optimizing the utilization cost and the life of the HESS.

In renewable energy the sizing of the HESS, the Photovoltaic panels and the inverters is based on the optimization. According to (M. E. Glavin & Hurley, 2012) in off grid photovoltaic system with HESS there is always a need for optimization in order to prevent excess battery storage if a super capacitor with battery is used. For a given solar radiation and load profile, the optimization of different system elements has been done considering the peak power requirements. The super capacitors are used to supply the peak power requirements of the load with the average power supplied by the battery bank.

The optimization program was developed in MATLAB and the Simulink model of system was implemented for analysis purpose. (Geetha et al., 2020) proposed maximum power point tracking (MPPT) fuzzy logic based solar charge controller to track the maximum power that is produced by the solar panel and to simultaneously control the charging of the battery when power is directly being supplied to the load.

Figure 4. Fuzzy logic control components

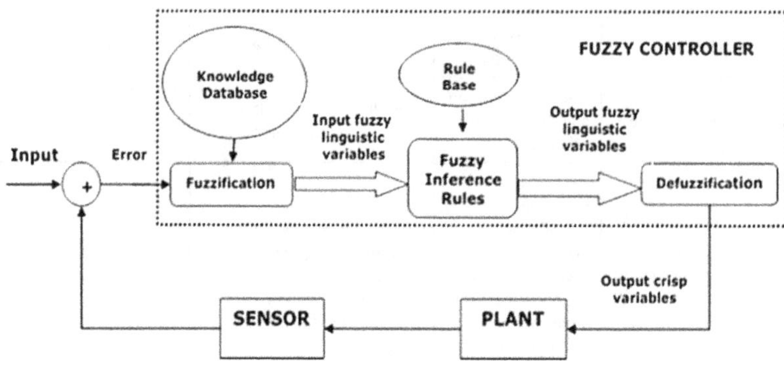

2. Artificial Neural Network (ANN)

Thiaw, at al.(2014) implementation of artificial neural networks (ANN) for maximum power point tracking (MPPT) and wind energy resource management has been provided. The MPPT principle uses optimal voltage and current values in order to extract maximum from the photovoltaic power system. The main purpose of MPPT inverter using neural network controller is to control and set the input signal value and keep the output current and voltages values as close to the optimal values of the PV power system.

In the standalone PV system comprising of 1.33KWp PV power system and a 24V /150 Ah lead acid battery as a load is used for simulation purpose. The Neural MPPT controller for a PV battery charging system is discussed and it has been concluded that the MPPT controllers using Neural Networks, the related parameters such as optimal operating point of the PV generator and the environmental conditions are known in advance and the main purpose of functioning of this inverter is to keep the system in an optimized set points which is not the case with the other techniques used in MPPT controllers.

Furthermore, even during the enormous changes in the weather and environmental conditions there is no oscillation seen around the maximum power point and this the controller operates fast and with minimum errors. The neural network based controlled for standalone photovoltaic system is shown in figure 5.

Figure 5. A standalone PV system using the neural network based controller for MPPT

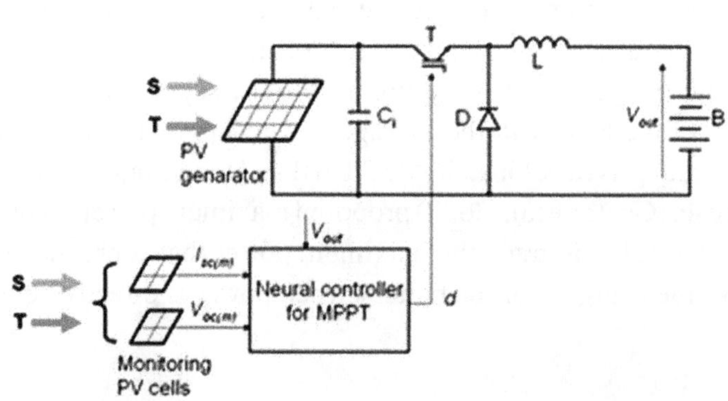

Al-Alawi et al. (2007) provided details regarding the ANN and hybrid systems, ANN architecture, operation and training. The definitions of inputs, outputs and data preparation are considered as the most important step for an ANN modeling. In ANN training the system is defined as made up of photovoltaic modules, integrated with the diesel generator and the battery bank for storage.

The training patterns were set up for the network. The back propagation (BP) algorithms are used for the supervision of the training technique. The training patterns consists of the inputs and the desired outputs and are subjected to the ANN model. For each repeated submissions of the patterns desirable adjustments for each iteration are done unless and until the required outputs of acceptable levels are obtained successfully.

The figure 6 below shows two outputs nodes representing the power from the diesel generator along with its On/Off status.

It is concluded that the developed ANN model for the optimum operation of an integrated hybrid renewable energy based water and power supply system (IRWPSS) can predict the diesel generator status (On/Off) at any given time instant, if we know the power generated from photovoltaic panels, power supplied from the Energy Storage System and the inverter respectively.

Figure 6. The architecture of ANN model

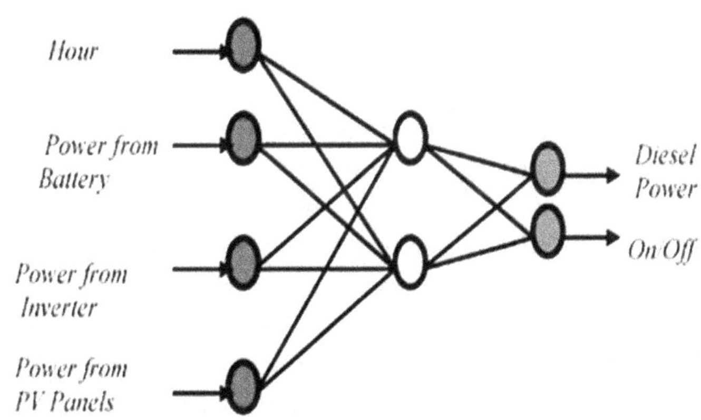

(Singh & Lather, 2020) proposed artificial neural network based control strategy, in order to increase and improve power sharing between battery and the supercapacitors connected with the micro grid as Hybrid Energy Storage System (HESS). When this ANN based controller performance is compared with the conventional controllers it is much faster, provides improved dc voltage regulation and have less settling time.

3. Genetic Algorithm

In Cano et al.,(2014), hybrid energy system with HESS, the technical optimization of system parameters is done using Simulink design optimization (SDO) of MATLAB. The optimization workflow requires series of tasks which includes the design variable, design requirements and optimization options. After testing optimization methods such as Gradient Decent, Pattern Search and Simplex search method for SDO it was concluded that both Latyn Hpercube and Genetic Algorithm in pattern search method

Figure 7. Hybrid Energy Storage Control Structure using ANN based controller

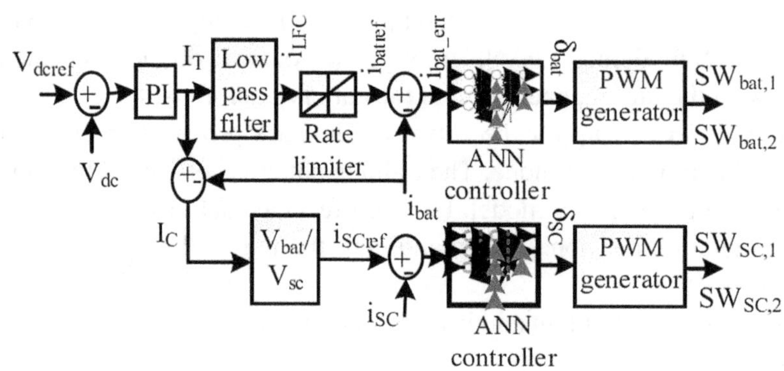

were ideal and most appropriate in finding values for the tuned parameters and satisfactory results were obtained when applied to the hybrid system. It is seen that the technical optimization using the genetic algorithm in SDO resulted in the most economic hybrid system with HESS as compared to the basic technical sizing and optimal techno-economic sizing based on HOMER and HOGA.

(Behzadi & Niasati, 2015) in their article, provided three different sizing methods regarding the power management strategies for hybrid power system, consisting of the PV panels, the fuel cell, the electrolyzer, the hydrogen tank and the battery. The sizing was done using manual calculations and the genetic algorithm optimization with HOGA and Homer software approach.

The TRNSYS software which is a simulation tool for the analysis of the energy system is also used and it was observed that HOMER and HOGA sizing approach is more suitable and cost effective as compared to the manual approach as it is not possible to meet the load demand without genetic algorithm. Also in Yang et al.,(2008) proposed method of optimization based on genetic algorithm (GA) for hybrid solar wind power generation system with the energy storage system given.

The optimal sizing method with minimum annualized cost of the system in order to achieve the loss of power supply probability (LPSP) at customer size is discussed in detail. The proposed optimization method worked well and provided good performance when it's applied on a hybrid power system with ESS which is supplying power to the telecommunication relay station. This optimization also developed relation between the system power reliability and the system configurations.

In (Jiang et al., 2014), multi-objective optimization (MOP) modelling is used for an effective power management among the SC and the battery. This MOP with linear weighted summations is then converted into single objective function. The simulation results shows that the optimization technique used is far better than the classic low pass filter (LPF) used for power management.

The flowchart of genetic algorithm for modelling and designing of hybrid renewable energy systems is shown below (Ismail, Moghavvemi, & Mahlia, 2014):

(Suresh & Meenakumari, 2019) provided an improved genetic algorithm based optimal sizing of hybrid energy systems for standalone applications. In order to find the best population strength the block diagram starts with population initialization and fitness function calculations. Crossover and Mutation rate determines the next generated population for the optimized solution.

Figure 8. The flowchart of GA for modelling and designing of hybrid renewable energy system

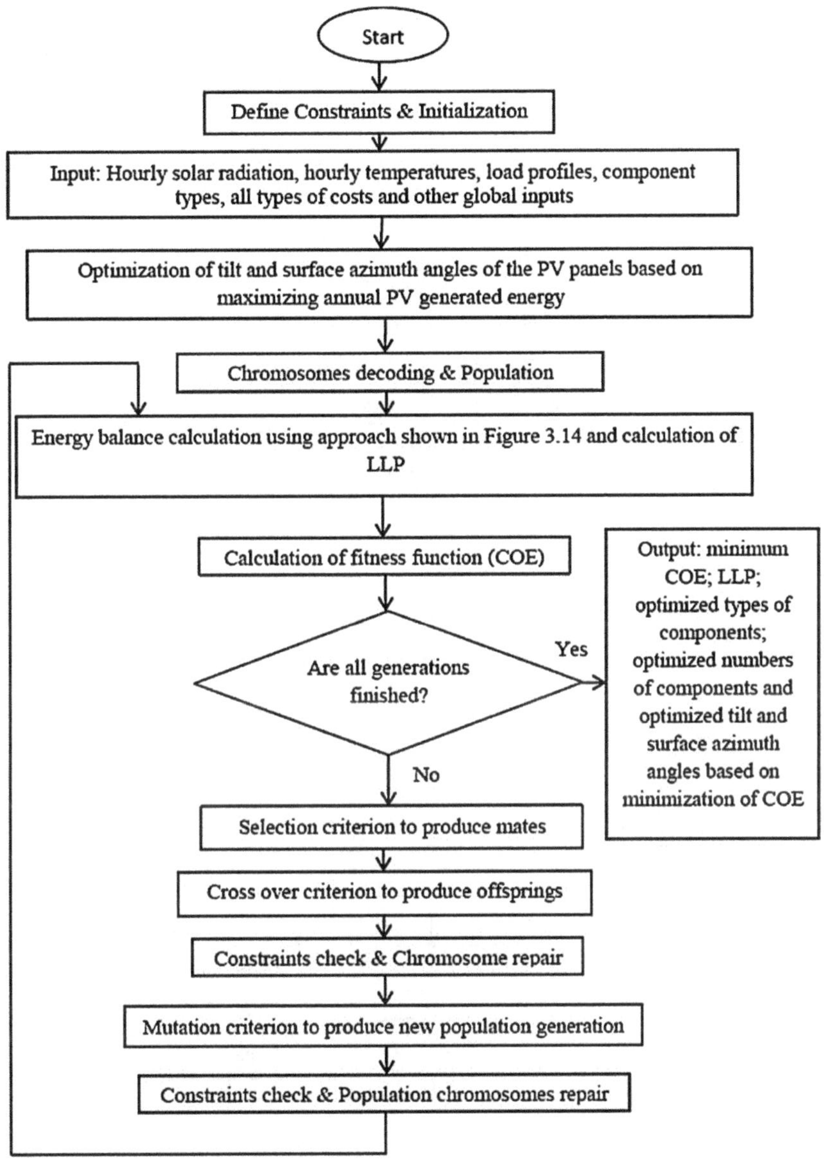

4. Particle Swarm Optimization (PSO)

Kerdphol et al.(2015) worked on particle swarm optimization algorithm for the optimal sizing of the HESS considering load shedding as a dynamic demand response for the grid connected photovoltaic power system working as a microgrids. The PSO is used to minimize the size and the cost of the battery bank besides improving the system frequency. Based on the results it is concluded that PSO algorithms provides quicker, smoother and more system stability then by optimization sizing based on analytical algorithm. The energy storage system shows good performance in load frequency control and during load shedding can be used for emergency control purpose as well. Ghazvini et al,(2015) in standalone solar

Figure 9. Flow chart of improved GA optimization

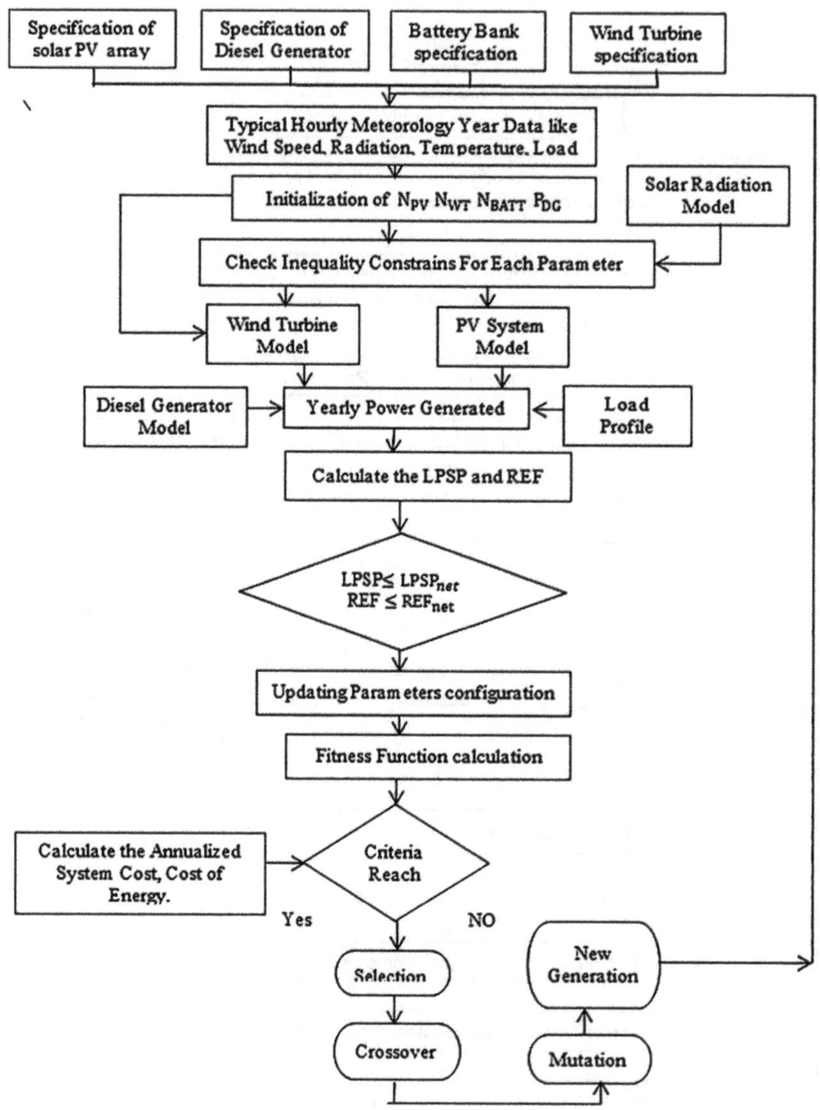

power systems the simultaneous optimization of energy storage systems, component sizing, operation strategy and slope of photovoltaic panels is done using an improved variant of particle swarm optimization (PSO), designated as passive congregation (PSOPC).

The operation strategy is based on the optimal values of the set points which are determined by the optimization algorithm in order to efficiently optimize the hybrid power system with the ESS. Different hybrid power system strategies were studied using the optimized algorithm and HOMER software package.

It is found that PSOPC gives the most cost effective and reliable results for wind/PV/diesel/battery systems. After comparison of similar method using not optimized set points and predetermined values it can be seen that proposed optimized set points are more effective. Wind speed variations and factors such as load growth should also be considered during optimization of hybrid power system.

(Delnia Sadeghi, AliHesami Naghshbandy, & Salah Bahramara, 2020) used multi-objective particle swarm optimization for optimal sizing of hybrid renewable energy systems in presence of electric vehicles. The uncertain behavior of the electric vehicle is monitored using Monte Carlo Simulation.

The electric vehicle (EV) are also used as a backup storage. That is in case when the batteries are fully charged and the excess amount of energy is stored as a charge for EV and when there is a need of charge for the battery the same is provided by the EV as a source.

Figure 10. Flowchart of the PV/wind/battery/EV system using Multi-Objective particle swarm optimization (MOPSO)

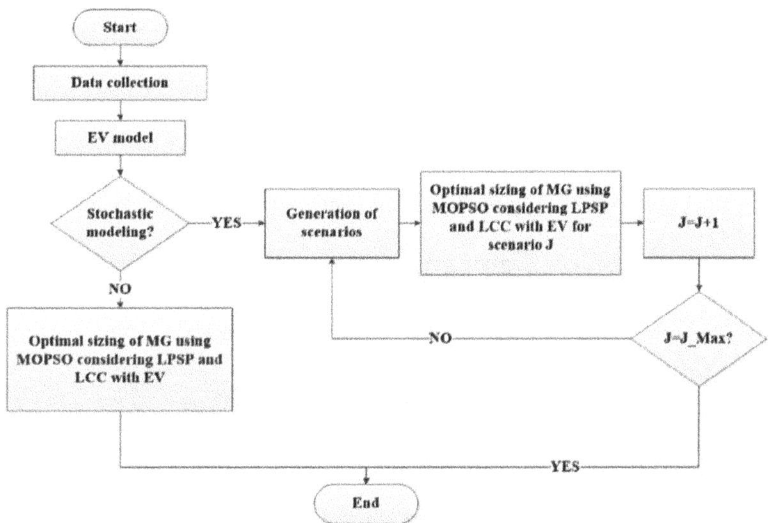

5. Cuckoo Optimization Algorithm (COA)

Also a latest Optimized Fuzzy-Cuckoo Controller for Active Power Control of Battery Energy Storage System, Photovoltaic, Fuel Cell and Wind Turbine in an isolated micro grid has been proposed in (Einan, Torkaman, & Pourgholi, 2017). The Combination of fuzzy controller and cuckoo optimization algorithm (COA) technique as compared with the genetic algorithm (GA) and particle swarm optimization algorithm (PSO) improves the performance of droop during the changes in parameters such as wind speed, sunlight. In studied microgrids the batteries are used as the energy storage system along with the fuel cell (FC) as a backup and energy supporter.

The power demand is fulfilled by managing the power flow from different power sources such as wind and solar using the Fuzzy Controller. The fuzzy controllers for the Battery and the Fuel Cell (FC) includes the membership functions (MF) for its inputs and outputs. Fuzzy controllers can be optimized using optimization of membership functions (MF) and coefficients of the controller. For optimization of membership functions (MF) different algorithms are used like PSO, water flow algorithm (WCA), Shuffled for leap (SFL) etc. COA has been used to enhance the capability of the controller which is considered as a direct search optimization method.

The optimal fuzzy structure for the proposed isolated network is shown in figure 11.

Figure 11. The optimal fuzzy structure

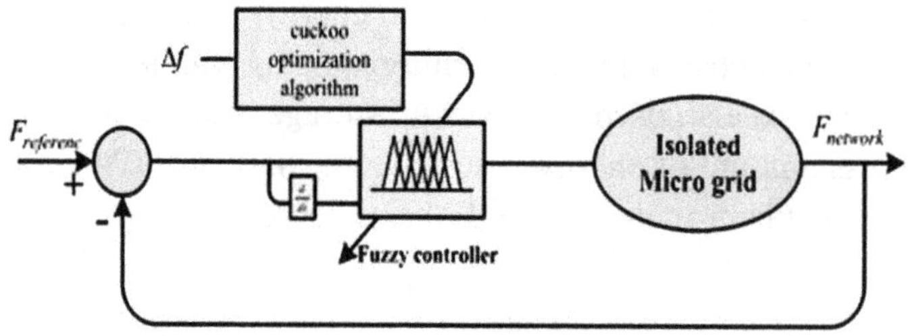

It was concluded after the simulations that 59% to 68% fewer errors were seen after setting the fuzzy controller coefficients with COA as compared to the GA and PSO.

The cuckoo optimization algorithm (COA) flowchart is summarized as flowing (Shadkam & Bijari, 2014):

Figure 12. Flow chart of Cuckoo Optimization Algorithm

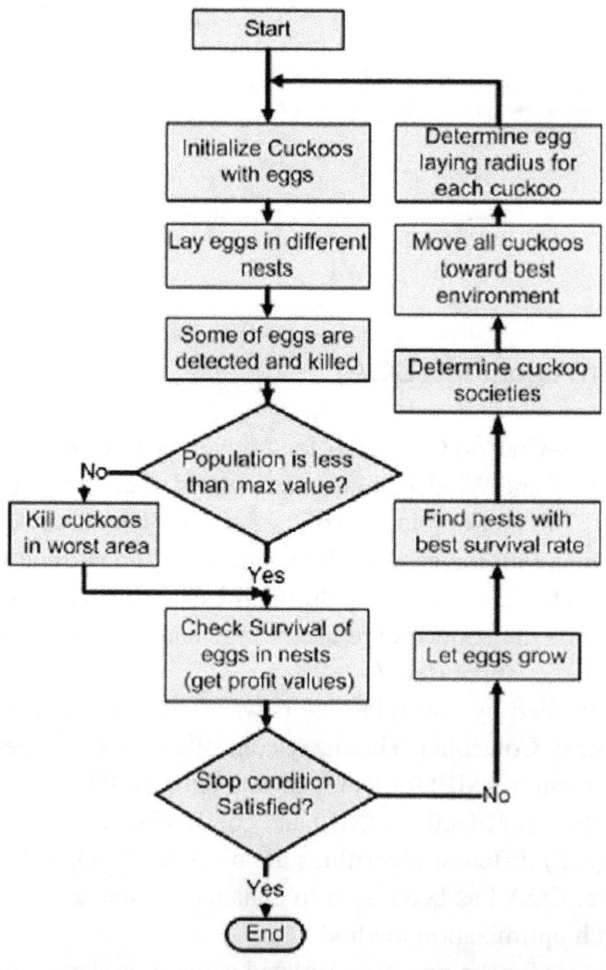

6. Simulation Annealing (SA)

(Ekren & Ekren, 2010) used simulated annealing (SA) for optimizing the size of a hybrid system with ESS. For the global optimization proposed methodology uses a stochastic gradient search. The decision variables are PV size, wind turbine, rotor swept, and the battery capacity whereas the objective function is the minimization of the hybrid energy system total cost. The SA major advantage over other methods is its flexibility by not getting trapped in local minima.

The search used by the algorithm is random and not only accepts the changes that decreases the objective function but also accepts some changes which in fact increases the objective function. A commercial software ARENA 12.0 is used in this study for simulation and optimization purpose.

The approach is very useful where there are various decision variable and large search space to optimize energy system. The SA approach for optimization is considered as less time consuming and optimum points are obtained at a reasonable time.

(Ababei, 2009) shows the block diagram of the classical simulated annealing (SA) as:

Figure 13. Block diagram of the classic simulated annealing (SA)

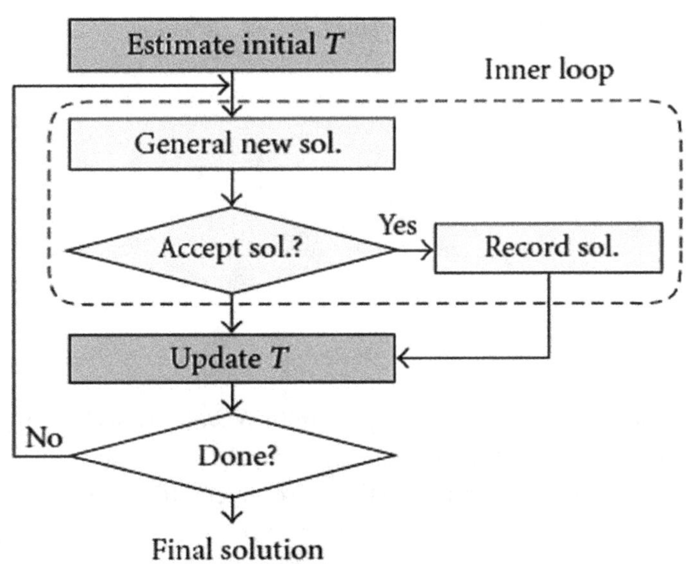

The enormous amount of EV's charging has a great impact on the power grid and to minimize this impact is to use renewable energy sources for charging of EVs. The electric vehicles (EVs) charging and its interaction with the renewable energy sources is discusses in Liu et al,(2015). Wang et al, (2017) developed adaptive mode switch strategy (AMSS) based on simulated annealing (SA) optimization of a multi-mode hybrid energy storage system (HESS) for electric vehicles. The energy management and adaptive mode selection is done using SA-AMSS. The objective function is based on the selection of most appropriate mode in order to improve the whole system efficiency of multimode HESS and the optimizations of the references SC SOC (state of charge) and battery power is done using the SA approach instead of direct power distribution optimization between the battery and the SC.

After simulation it is clear from the results that SA-AMSS not only reduces the frequency of the mode switching but also avoid the unexpected too much and excess power output from the battery. SA-AMSS is considered as flexible and effective which helps to safeguard the ESS.

(Kumar et al., 2020) proposed a novel simulated-annealing based electric bus system design, simulation and analysis for Dehrandun smart city. The simulated annealing approach is used in order to analyze the energy consumption to improve system performance and solve bound constraint optimization problems related with networks.

The power system parameters related to plug-in hybrid electric vehicle (PHEV) which is considered as an energy storage system particularly in smart grid scenario are optimized using multi-objective optimization and matching method based on generic algorithm (Song, Lei, & Fu, 2020). The multi-island genetic algorithm (MIGA) proposed is based on genetic algorithm (GA). In this modified algorithm different populations are divided into many sub-populations. Then in these sub-populations, GA selection, crossover and mutation are performed. Also targeted immigration operations among various populations is done periodically.

Figure 14. Evolution process of the multi-island genetic algorithm

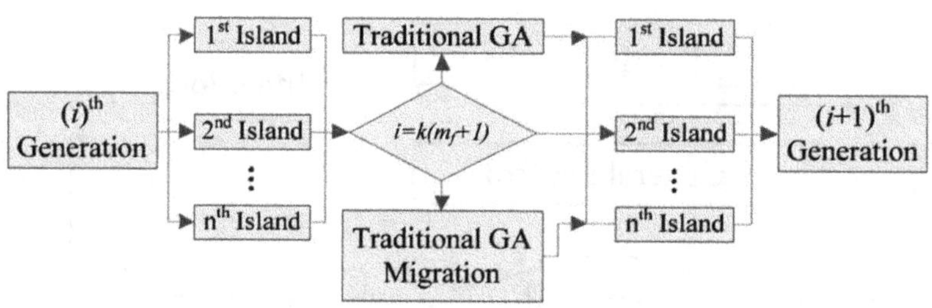

7. Wavelet Transforms

An intelligent control of photovoltaic system using Hermite Wavelet-based Adaptive Neural Fuzzy Controller (HWNF) is achieved by combination of reasoning capability of the fuzzy logic, learning capability of neural networks and the localization properties of wavelets. It's been proved using the MATLAB simulation that the performance of the controller is improved when Hermite wavelets are integrated in the proposed controller and HWANFC can track the MPP. Various simulation results and comparison indexes have shown that the HWANFC can track the maximum power point (MPP) when there is fast variations in parameters and load disturbances. (Hassan et al., 2017).

Also according to (MUMTAZ & KHAN, 2017) Hermite wavelet incorporated neurofuzzy identifier (HWNI) is used to identify the PV system. The performance of the Hermite wavelet incorporated neurofuzzy indirect adaptive controller (HWNIAC) has negligible fluctuations, overshoot, undershoot and steady state error during extensive changes in atmospheric conditions and the controller gets maximum power point (MPP) without losing control.

The maximum power is taken from the PV using HWNIAC and is delivered to DC bus to satisfy the load whereas supercapacitor (SC) and battery is charged with the access power produced by the PV system. When HESS is fully charged the surplus power is provided to electrolyzer and then to the grid. The performance of the HWNIAC is much better when compared with an IC based adaptive indirect neurofuzzy Takagi-Sugeno-Kang (TSK) control scheme, IC based adaptive direct neurofuzzy TSK control scheme, IC based adaptive proportional-integral-derivative (PID) control scheme, and IC algorithm for PV systems.(Zafar, Shafique, Nazir, & Zia, 2018)

(Guo, Liu, Zhao, Zhu, & Liu, 2020) proposed a new and effective wavelet-enabled approach for connected hybrid energy storage system (HESS) consisting of batteries and super-capacitor with the wind turbine for smoothing of generated power. In this technique, the wind power time series is decomposed within the defined limits of the physical constraints without violating them and higher efficiency is obtained in an optimal manner for power injection regulation, state of charge (SOC) of HESS and acceptable charge/discharge depth.

COMPARITIVE STUDY BETWEEN DIFFERENT OPTIMIZATION TECHNIQUES

In (Durgadevi, Sundari, Raaghavi, & Akshaya, 2019), the continuous stirred tank reactor which plays an important role in all chemical process industries for mixing uses PI & PID controllers are used for various error criteria to check the full strength of solution mixed with desired portion of water. The conventional controllers do not perform well while dealing with the problems that appear in complex non-linear process. Therefore, particle swarm optimization (PSO) techniques is used to achieve the control objective. So when the controller performance for CSTR with PSO based PI & PID controller is compared with conventional Z-N technique the result shows PSO providing a very optimal performance in terms of less settling time, rise time and %Peak overshoot.

In (Zafar et al., 2018), optimization techniques such as genetic algorithm, pattern search method and interior point algorithms were used in order to optimize fuel cost, emissions and computational time complexity in order to meet the demand from thermal power generators by integrating the renewable energy resources. In this study the interior point algorithm performance was the best as compared to genetic and pattern search algorithm

The optimization of photovoltaic system in (Dabra, Paliwal, Sharma, & Kumar, 2017) was done using P&O, Fuzzy P&O and BPSO fuzzy P&O control methods to see the power output of solar PV grid array. It was concluded that BPSO fuzzy P&O controller is more effective as compared with P&O and fuzzy P&O models.

Also in (Hosseini & Gandomi, 2010), detailed study regarding economic load dispatch comparing the performance using Heuristic Optimization Techniques with an Improved Coordinated Aggregation-Based PSO was done. The algorithm was tested on 6, 13, 15 and 40 generating units, the island power system in Crete in Greece and the Hellenic bulk power system. It was concluded that the performance was improved as compared to the state of the art heuristic optimization techniques (HOTs).

The efficiencies of GA and PSO algorithms were compared for optimization of noisy non-linear optimization problems and both showed very satisfactory results as compared to many new optimization techniques. (Sai et al., 2017)

CONCLUSION

In this chapter different optimization techniques were discussed in context with the control strategies applied in grid connected, off grid and hybrid renewable energy systems with Hybrid Energy Storages being used as a storage devices. It is concluded that each of the optimization method has its own pros and cons depending on the control strategy involved. Also using various new and latest optimization techniques such as artificial algorithms and hybrid algorithms for intelligent control can deliver more precise and correct results regarding multi objective optimization problems including the placement, sizing, design, planning and control problems in the field of renewable and sustainable energy. The provision of the basic information regarding the current control strategies and the optimization techniques being used for the development of photovoltaic systems with hybrid energy storages can help the future research regarding the optimization and control techniques for smart grid integration as well. Furthermore, we can come to conclusion that the hybrid energy storages with equal capacity is better than that of single energy storage and these storages can be improved by using the control strategies discussed above and also the hybrid energy storage systems will play a vital rule in future smart grids.

REFERENCES

Ababei, C. (2009). Speeding Up FPGA Placement via Partitioning and Multithreading. *International Journal of Reconfigurable Computing*, *2009*, 1–9. doi:10.1155/2009/514754

Akcayol, M. A. (2004). Application of adaptive neuro-fuzzy controller for SRM. *Advances in Engineering Software*, *35*(3–4), 129–137. doi:10.1016/j.advengsoft.2004.03.005

Al-Alawi, A., & Al-Alawi, M. (2007). Predictive control of an integrated PV-diesel water and power supply system using an artificial neural network. *Renewable Energy*, *32*(8), 1426–1439. doi:10.1016/j.renene.2006.05.003

Allegre, A. L., Bouscayrol, A., Delarue, P., Barrade, P., Chattot, E., & El-Fassi, S. (2010). Energy storage system with supercapacitor for an innovative subway. *IEEE Transactions on Industrial Electronics, 57*(12), 4001–4012. https://doi.org/ doi:10.1109/TIE.2010.2044124

Aneke, M., & Wang, M. (2016). Energy storage technologies and real life applications – A state of the art review. *Applied Energy*, *179*, 350–377. doi:10.1016/j.apenergy.2016.06.097

Bayindir, R., Colak, I., Fulli, G., & Demirtas, K. (2016). Smart grid technologies and applications. *Renewable & Sustainable Energy Reviews*, *66*, 499–516. doi:10.1016/j.rser.2016.08.002

Behzadi, M. S., & Niasati, M. (2015). Comparative performance analysis of a hybrid PV/FC/battery stand-alone system using different power management strategies and sizing approaches. *International Journal of Hydrogen Energy*, *40*(1), 538–548. doi:10.1016/j.ijhydene.2014.10.097

Bocklisch, T. (2015). Hybrid energy storage systems for renewable energy applications. *Energy Procedia*, *73*, 103–111. doi:10.1016/j.egypro.2015.07.582

Brka, A., Kothapalli, G., & Al-abdeli, Y. M. (2015). ScienceDirect Predictive power management strategies for stand- alone hydrogen systems : Lab-scale validation. *International Journal of Hydrogen Energy*, *40*(32), 9907–9916. doi:10.1016/j.ijhydene.2015.06.081

Cano, A., Jurado, F., Sánchez, H., Fernández, L. M., & Castañeda, M. (2014). Optimal sizing of stand-alone hybrid systems based on PV/WT/FC by using several methodologies. *Journal of the Energy Institute*, *87*(4), 330–340. doi:10.1016/j.joei.2014.03.028

Chong, L. W., Wong, Y. W., Rajkumar, R. K., Rajkumar, R. K., & Isa, D. (2016). Hybrid energy storage systems and control strategies for stand-alone renewable energy power systems. *Renewable & Sustainable Energy Reviews*, *66*, 174–189. doi:10.1016/j.rser.2016.07.059

Chouhan, N. S., & Ferdowsi, M. (2009). Review of energy storage systems. *North American Power Symposium (NAPS)*, 1–5. https://doi.org/10.1109/NAPS.2009.5484062

Dabra, V., Paliwal, K. K., Sharma, P., & Kumar, N. (2017). Optimization of photovoltaic power system: A comparative study. *Protection and Control of Modern Power Systems*, *2*(1), 3. Advance online publication. doi:10.118641601-017-0036-2

Dulout, J., Jammes, B., Seguier, L., & Alonso, C. (2015). Control and design of a hybrid energy storage system. *2015 17th European Conference on Power Electronics and Applications, EPE-ECCE Europe 2015*. https://doi.org/10.1109/EPE.2015.7309468

Durgadevi, S., Sundari, K. T., Raaghavi, D., & Akshaya, R. S. (2019). Comparative study of Controller Optimisation for CSTR using Particle Swarm Optimization Technique. *5th International Conference on Electrical Energy Systems, ICEES 2019*. https://doi.org/10.1109/ICEES.2019.8719302

Eid, C., Reneses Guillén, J., Frías Marín, P., & Hakvoort, R. (2014). The economic effect of electricity net-metering with solar PV: Consequences for network cost recovery, cross subsidies and policy objectives. *Energy Policy*, *75*, 244–254. doi:10.1016/j.enpol.2014.09.011

Einan, M., Torkaman, H., & Pourgholi, M. (2017). Optimized Fuzzy-Cuckoo Controller for Active Power Control of Battery Energy Storage System, Photovoltaic, Fuel Cell and Wind Turbine in an Isolated Micro-Grid. *Batteries*, *3*(3), 23. doi:10.3390/batteries3030023

Ekren, O., & Ekren, B. Y. (2010). Size optimization of a PV/wind hybrid energy conversion system with battery storage using simulated annealing. *Applied Energy*, *87*(2), 592–598. doi:10.1016/j.apenergy.2009.05.022

Fathima, A. H., & Palanisamy, K. (2015). Optimization in microgrids with hybrid energy systems - A review. *Renewable & Sustainable Energy Reviews*, *45*, 431–446. doi:10.1016/j.rser.2015.01.059

García, P., Torreglosa, J. P., Fernández, L. M., & Jurado, F. (2013). Optimal energy management system for stand-alone wind turbine/photovoltaic/ hydrogen/battery hybrid system with supervisory control based on fuzzy logic. *International Journal of Hydrogen Energy*, *38*(33), 14146–14158. doi:10.1016/j.ijhydene.2013.08.106

Gee, A. M., Robinson, F. V. P., & Dunn, R. W. (2013). Analysis of battery lifetime extension in a small-scale wind-energy system using supercapacitors. *IEEE Transactions on Energy Conversion*, *28*(1), 24–33. doi:10.1109/TEC.2012.2228195

Geetha, A., Usha, S., Thamizhthentral, T. M., Ganguly, U., Kumar, A., & Narnoli, P. K. (2020). *Design of a Solar Charge Controller Using Fuzzy Logic MPPT Technique For A Solar Panel*. Academic Press.

Ghaviha, N., Campillo, J., Bohlin, M., & Dahlquist, E. (2017). Review of Application of Energy Storage Devices in Railway Transportation. *Energy Procedia*, *105*, 4561–4568. doi:10.1016/j.egypro.2017.03.980

Ghazvini, M., Abbaspour-Tehrani-fard, A., & Fotuhi-Firuzabad, M. (2015). A particle swarm optimization-based approach to achieve optimal design and operation strategy of standalone hybrid energy systems. *Turkish Journal of Electrical Engineering and Computer Sciences*, *23*(2), 335–353. doi:10.3906/elk-1212-11

Glavin, M. E., & Hurley, W. G. (2012). Optimisation of a photovoltaic battery ultracapacitor hybrid energy storage system. *Solar Energy*, *86*(10), 3009–3020. doi:10.1016/j.solener.2012.07.005

Glavin, M. E. E., Chan, P. K. W., Armstrong, S., & Hurley, W. G. G. (2008). A stand-alone photovoltaic supercapacitor battery hybrid energy storage system. *2008 13th International Power Electronics and Motion Control Conference, EPE-PEMC 2008*, 1688–1695. https://doi.org/10.1109/EPEPEMC.2008.4635510

Grantham, A., Pudney, P., Ward, L. A., Whaley, D., & Boland, J. (2017). The viability of electrical energy storage for low-energy households. *Solar Energy*, *155*, 1216–1224. doi:10.1016/j.solener.2017.07.063

Guney, M. S., & Tepe, Y. (2017). Classification and assessment of energy storage systems. *Renewable and Sustainable Energy Reviews, 75*(November), 1187–1197. https://doi.org/ doi:10.1016/j.rser.2016.11.102

Guo, T., Liu, Y., Zhao, J., Zhu, Y., & Liu, J. (2020). A dynamic wavelet-based robust wind power smoothing approach using hybrid energy storage system. *International Journal of Electrical Power and Energy Systems, 116*(October), 105579. https://doi.org/ doi:10.1016/j.ijepes.2019.105579

Hassan, S. Z., Li, H., Kamal, T., Arifoğlu, U., Mumtaz, S., & Khan, L. (2017). Neuro-fuzzy wavelet based adaptive MPPT algorithm for photovoltaic systems. *Energies*, *10*(3), 1–16. doi:10.3390/en10030394

Hemmati, R., & Saboori, H. (2016). Emergence of hybrid energy storage systems in renewable energy and transport applications – A review. *Renewable & Sustainable Energy Reviews*, *65*, 11–23. doi:10.1016/j.rser.2016.06.029

Hosseini, S. S. S., & Gandomi, A. H. (2010). Discussion of "economic load dispatch - A comparative study on heuristic optimization techniques with an improved coordinated aggregation-based PSO.". *IEEE Transactions on Power Systems*, *25*(1), 590. doi:10.1109/TPWRS.2009.2037531

Huu, D. N. (2016). An adaptive control of hybrid battery-supercapacitor storage for integration of wind and solar. *2016 IEEE International Conference on Sustainable Energy Technologies (ICSET)*, 157–162. https://doi.org/10.1109/ICSET.2016.7811774

Ismail, M. S., Moghavvemi, M., & Mahlia, T. M. I. (2014). Genetic algorithm based optimization on modeling and design of hybrid renewable energy systems. *Energy Conversion and Management*, *85*, 120–130. doi:10.1016/j.enconman.2014.05.064

Jiang, W., Hu, R., Chen, W., Wang, X., Li, Z., Yang, Y., & Wang, J. (2014). Research of multipurpose optimization control in hybrid energy storage system. *2014 IEEE Energy Conversion Congress and Exposition (ECCE)*, 2550–2554. https://doi.org/ doi:10.1109/ECCE.2014.6953741

Katsanevakis, M., Stewart, R. A., & Lu, J. (2017). Aggregated applications and benefits of energy storage systems with application-specific control methods: A review. *Renewable and Sustainable Energy Reviews, 75*(August), 719–741. https://doi.org/ doi:10.1016/j.rser.2016.11.050

Kerdphol, T., Qudaih, Y., & Mitani, Y. (2015). Optimal Battery Energy Storage Size Using Particle Swarm Optimization for Microgrid System. *International Review of Electrical Engineering, 10*(2), 277–285. doi:10.15866/iree.v10i2.5350

Kumar, A., Srikanth, P., Nayyar, A., Sharma, G., Krishnamurthi, R., & Alazab, M. (2020). A Novel Simulated-Annealing Based Electric Bus System Design, Simulation, and Analysis for Dehradun Smart City. *IEEE Access: Practical Innovations, Open Solutions, 8*, 89395–89424. doi:10.1109/ACCESS.2020.2990190

Liu, L., Kong, F., Liu, X., Peng, Y., & Wang, Q. (2015). A review on electric vehicles interacting with renewable energy in smart grid. *Renewable & Sustainable Energy Reviews, 51*, 648–661. doi:10.1016/j.rser.2015.06.036

Logenthiran, T., Srinivasan, D., Khambadkone, A. M., & Aung, H. N. (2010). Multi-Agent System (MAS) for short-term generation scheduling of a microgrid. *2010 IEEE International Conference on Sustainable Energy Technologies, ICSET 2010*. https://doi.org/10.1109/ICSET.2010.5684943

Luo, X., Wang, J., Dooner, M., & Clarke, J. (2015). Overview of current development in electrical energy storage technologies and the application potential in power system operation. *Applied Energy, 137*, 511–536. doi:10.1016/j.apenergy.2014.09.081

McClenny & Tokash, W. (2017). *2Q17 Hybrid Advanced Battery Markets report.* Navigent Research. Retrieved from www.naviantresearch.com

Mumtaz, S., & Khan, L. (2017). Indirect adaptive neurofuzzy Hermite wavelet based control of PV in a grid-connected hybrid power system. *Turkish Journal of Electrical Engineering and Computer Sciences, 25*, 4341–4353. doi:10.3906/elk-1702-175

Olivares, D. E., Mehrizi-Sani, A., Etemadi, A. H., Cañizares, C. A., Iravani, R., Kazerani, M., Hajimiragha, A. H., Gomis-Bellmunt, O., Saeedifard, M., Palma-Behnke, R., Jimenez-Estevez, G. A., & Hatziargyriou, N. D. (2014). Trends in microgrid control. *IEEE Transactions on Smart Grid, 5*(4), 1905–1919. doi:10.1109/TSG.2013.2295514

Ould Amrouche, S., Rekioua, D., Rekioua, T., & Bacha, S. (2016). Overview of energy storage in renewable energy systems. *International Journal of Hydrogen Energy, 41*(45), 20914–20927. doi:10.1016/j.ijhydene.2016.06.243

Rohit, A. K., Devi, K. P., & Rangnekar, S. (2017). An overview of energy storage and its importance in Indian renewable energy sector: Part I – Technologies and Comparison. *Journal of Energy Storage, 13*, 10–23. doi:10.1016/j.est.2017.06.005

Sadeghi, Naghshbandy, & Bahramara. (2020). Optimal sizing of hybrid renewable energy systems in presence of electric vehicles using multi-objective particle swarm optimization. *Energy, 209*, 118471. https://doi.org/https://doi.org/10.1016/j.energy.2020.118471

Sai, V. O., Shieh, C. S., Lin, Y. C., Horng, M. F., Nguyen, T. T., Le, Q. D., & Jiang, J. Y. (2017). Comparative Study on Recent Development of Heuristic Optimization Methods. *Proceedings - 2016 3rd International Conference on Computing Measurement Control and Sensor Network, CMCSN 2016*, 68–71. https://doi.org/10.1109/CMCSN.2016.29

Saikong, W., & Kulworawanichpong, T. (2017). Comparative Study of Energy Consumption for Electric Vehicles with Various On-board Energy Storage Systems. *Energy Procedia*, *138*, 81–86. doi:10.1016/j.egypro.2017.10.060

Samerchur, S., Premrudeepreechacharn, S., Kumsuwun, Y., & Higuchi, K. (2011). Power control of single-phase voltage source inverter for grid-connected photovoltaic systems. *2011 IEEE/PES Power Systems Conference and Exposition*, 1–6. https://doi.org/10.1109/PSCE.2011.5772504

Shadkam, E., & Bijari, M. (2014). Evaluation the Efficiency of Cuckoo Optimization Algorithm. *International Journal on Computational Science & Applications*, *4*(2), 39–47. doi:10.5121/ijcsa.2014.4205

Shim, J. W., Cho, Y., Kim, S. J., Min, S. W., & Hur, K. (2013). Synergistic control of SMES and battery energy storage for enabling dispatchability of renewable energy sources. *IEEE Transactions on Applied Superconductivity*, *23*(3). Advance online publication. doi:10.1109/TASC.2013.2241385

Singh, P., & Lather, J. S. (2020). Accurate power-sharing, voltage regulation, and SOC regulation for LVDC microgrid with hybrid energy storage system using artificial neural network. *International Journal of Green Energy*, *17*(12), 756–769. Advance online publication. doi:10.1080/15435075.2020.1798767

Song, P., Lei, Y., & Fu, Y. (2020). Multi-objective optimization and matching of power source for PhEV based on genetic algorithm. *Energies*, *13*(5), 1127. Advance online publication. doi:10.3390/en13051127

Sonnenenergie, D. G. (2007). Planning and Installing. Infor.

Suresh, M., & Meenakumari, R. (2019). An improved genetic algorithm-based optimal sizing of solar photovoltaic/wind turbine generator/diesel generator/battery connected hybrid energy systems for standalone applications. *International Journal of Ambient Energy*, *0*(0), 1–8. doi:10.1080/01430750.2019.1587720

Tan, X., Li, Q., & Wang, H. (2013). Advances and trends of energy storage technology in Microgrid. *International Journal of Electrical Power & Energy Systems*, *44*(1), 179–191. doi:10.1016/j.ijepes.2012.07.015

Tarroja, B., Zhang, L., Wifvat, V., Shaffer, B., & Samuelsen, S. (2016). Assessing the stationary energy storage equivalency of vehicle-to-grid charging battery electric vehicles. *Energy*, *106*, 673–690. doi:10.1016/j.energy.2016.03.094

Thiaw, L., Sow, G., & Fall, S. (2014). Application of Neural Networks Technique in Renewable Energy Systems. *2014 First International Conference on Systems Informatics, Modelling and Simulation Application*, 6–12. https://doi.org/10.1109/SIMS.2014.12

Tsikalakis, A. G., Member, S., Hatziargyriou, N. D., & Member, S. (2011). Centralized Control for Optimizing Microgrids Operation. *IEEE Transfer Energy Converters*, *23*(1), 1–8. doi:10.1109/PES.2011.6039737

Wang, B., Xu, J., Cao, B., & Ning, B. (2017). Adaptive mode switch strategy based on simulated annealing optimization of a multi-mode hybrid energy storage system for electric vehicles. *Applied Energy*, *194*, 596–608. doi:10.1016/j.apenergy.2016.05.030

Yang, H., Zhou, W., Lu, L., & Fang, Z. (2008). Optimal sizing method for stand-alone hybrid solar-wind system with LPSP technology by using genetic algorithm. *Solar Energy*, *82*(4), 354–367. doi:10.1016/j.solener.2007.08.005

Yao, L., Yang, B., Cui, H., Zhuang, J., Ye, J., & Xue, J. (2016). Challenges and progresses of energy storage technology and its application in power systems. *Journal of Modern Power Systems and Clean Energy*, *4*(4), 519–528. doi:10.100740565-016-0248-x

Zafar, A., Shafique, A., Nazir, Z., & Zia, M. F. (2018). A Comparison of Optimization Techniques for Energy Scheduling of Hybrid Power Generation System. *Proceedings of the 21st International Multi Topic Conference, INMIC 2018*, (June), 1–6. https://doi.org/10.1109/INMIC.2018.8595665

Chapter 9
PMU Placement Optimization for Fault Observation Using Different Techniques

Hamid Bentarzi
Signals and Systems Laboratory, IGEE, University M'hamed Bougara of Boumerdes, Algeria

ABSTRACT

This chapter presents different techniques for obtaining the optimal number of the phasor measurement units (PMUs) that may be installed in a smart power grid to achieve full network observability under fault conditions. These optimization techniques such as binary teaching learning based optimization (BTLBO) technique, particle swarm optimization, the grey wolf optimizer (GWO), the moth-flame optimization (MFO), the cuckoo search (CS), and the wind-driven optimization (WDO) have been developed for the objective function and constraints alike. The IEEE 14-bus benchmark power system has been used for testing these optimization techniques by simulation. A comparative study of the obtained results of previous works in the literature has been conducted taking into count the simplicity of the model and the accuracy of characteristics.

INTRODUCTION

In order to make the electrical power systems more reliable, stable, and controllable; state estimation of the transmission network is necessary (Bei, X., et al, 2008; Allemong, J., et al, 1982; Abbasy, N.H., Ismail, H.M., 2009). The Phasor Measurement Unit (PMU) is a tool for measuring the voltage and current those are synchronized by using the global positioning system (GPS) satellite. Integrated with the GPS receiver (Yin, H., Fan, L., 2010; Mohammadi-Ivatloo, B., 2009), the power station including the Phasor Data Concentrator (PDC), which is used for analyzing the PMU data, can receive the synchronous data from each PMU in real time (Valenti, M. C., Feliachi, A., 2002).

State estimation of power network may allow the scheduling generation and interchange; monitoring outages and scheduling alternatives; supervising scheduled outages; scheduling frequency and time corrections; coordinating bias settings; and emergency restoration of system (Waheed Ur Rahman, et al,

DOI: 10.4018/978-1-7998-4027-5.ch009

2012). This can be achieved either by state estimation algorithms (Phadke A. G., Thorp J. S., 2008) or by means of PMUs with extreme precision, time synchronization, and excellent performance. Measuring state estimation is achieved through complex bus voltages (Baldwin, T. L., et al, 1993) that enable the estimation of the neighborhood bus voltage magnitudes and angles by using the line flow measurement (both real and reactive power). The static and dynamic behaviors of the power grid may be obtained from the information gathered by PMUs.

Faults that occur in the transmission line may be either permanent or temporary (Mosavi, M. R., Akhyani, A. A., Rahmati, A. 2012; Xu, B., 2006). Permanent fault may lead to a broken transmission line or a power generator malfunction; this causes different signals during the pre-fault and post-fault. This can be easily detected and located. However, the temporary fault may cause insulator flashover, which may lead to full insulator breakdown if it occurs frequently. This emphasizes the importance of PMU-based fault location technique (Dambhare, S., et al, 2008; Shahraeini, M., Javidi, M. H., 2011).

Nowadays, Wide Area Monitoring System (WAMS) that is considered the most advanced system to detect disturbances and avoid a bulk blackout is based on the PMUs . It aims to maintain the dynamic stability in the whole power network (Phadke A. G., 1993). This is implemented by synchronizing and recording the acquired data from systems in distributed locations through new computing and communication technologies. Upon their delivery to the central control station, these data are measured and analyzed from any point of the power network. In addition to its ability to monitor the static stability of the network (as traditional SCADA), WAMS enables the controllers to recognize unusual activities within the power network such as instability in the network voltage, to analyze the network oscillation, and to detect fault location. Thus, using the data provided by PMUs installed in some appropriate buses of a power grid, one can construct a new type of measuring system to improve the observability and the precision of the power system state estimator. The observability depends on the type, the number and the geographic distribution of measurements (Nuqui R. F. and Phadke A. G., 2005). Several methods (Milosevic B. and Begovic M. 2003; Xu B. and Abur A. 2005; Dekhandji, F. Z., et al, 2019; Chakrabarti S. and Kyriakides E. 2007; Chakrabarti S. and Kyriakides E. 2008; Hajian M., A., et al, 2007; Charu, S., and Barjeev, T. (2011) were considered when formulating the optimal placement of monitoring devices for fault location.

In this research work, a binary teaching learning based optimization (BTLBO) algorithm (Rao, R.V., et al, 2011 ; Rao, R.V., et al, 2012; Rao, V.J., Savsani, J.B., 2012; Rao, R.V., Patel, V.K., 2012; Tankasala, G. R., et al, 2012) for the optimal placement of phasor measurement units (PMUs) is proposed. The optimal PMU placement problem is formulated to minimize the number of PMUs installation subject to full network observability at the power system buses during the fault. The effectiveness of the proposed method is verified by the simulation of IEEE14-bus and compared with previous developed techniques.

FAULT ANALYSIS USING PMU

PMU-based fault location technique is achieved through monitoring the synchronized fault voltages, calculating the line currents between these buses, and forming bus injection currents at two terminals of the faulted line. Thus, calculation of fault locations can be indirectly investigated.

A PMU placed at a given bus of the network can measure both the phasor voltages and phasor currents of all lines incident to that bus. Thus, the entire parameters of a bus can be made observable by placing judiciously PMUs at specific buses of the network. As shown in Fig.1, most power transmission

Figure 1. The five possible sub-network configurations

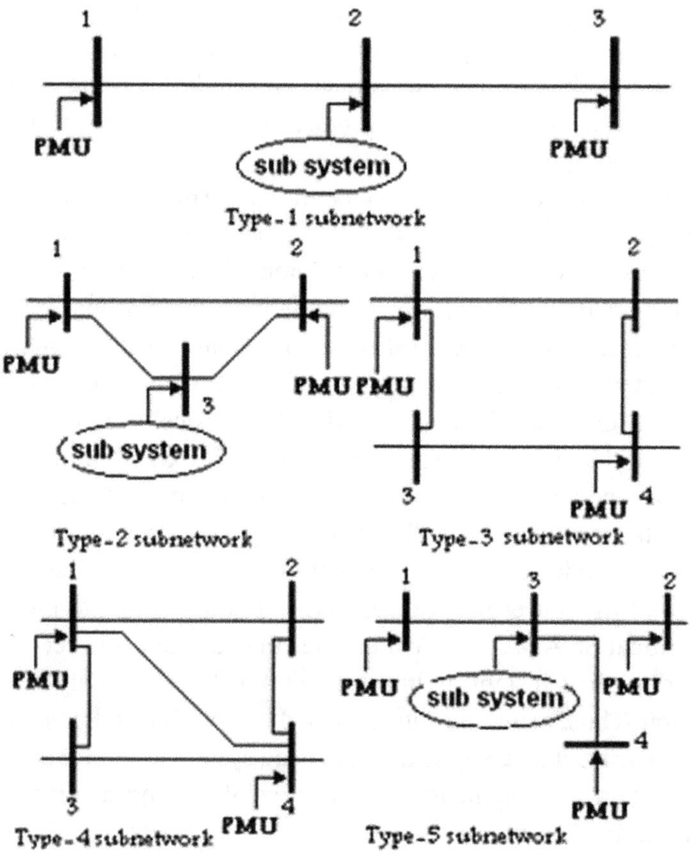

networks may consist in general of five different sub-network configurations or types (Lien, K. -P., et al, 2006). In this figure, for each of the five sub-networks, the placed PMUs obey to what is called the "one bus spaced deployment strategy".

Whatever is the considered sub-network type, it is always possible to make the fault location problem of that sub-network similar to the fault location problem of type-1 sub-network (Geramian, S. S., Abyane, H. A., Mazlumi, K., 2008). A faulty type-1 sub-network is shown on Fig. 2. In this figure, a fault occurs in line jk between bus j and bus k. The fault location problem consists thus, of finding the parameter Dk.

By using the voltage phasor of bus i and the parameters of line ij, the voltage phasor of bus j will be determined as:

$$V_j = V_i - (I_{ij} - V_i - L_{ij} Y_{ij}/2)\, L_{ij} Z_{ij} \tag{1}$$

In addition,

$$I_f = I_{kf} + I_{jf} \tag{2}$$

Where the line parameters Zij and Yij are relative to the Π line model. Current phasor Ijk however, cannot be deduced with the similar way as no PMU is connected at bus j. Therefore, the problem consists of

Figure 2. Faulty type-1 sub-network

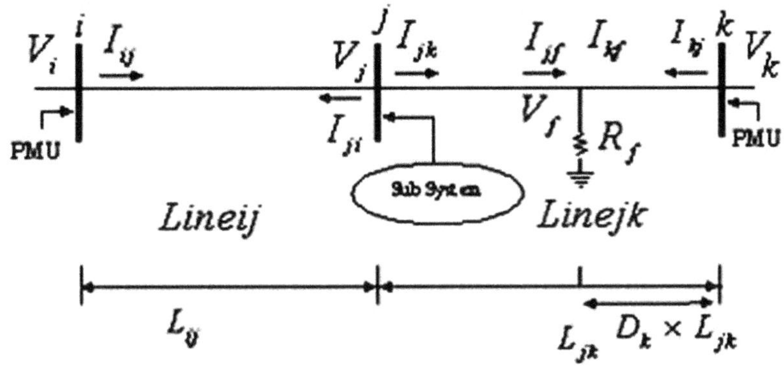

finding two unknown variables Ijk and Dk. Using the basic electrical networks theory and assumption constraints, the two sets of equations of the two unknown variables can be derived:

- Network equations, derived from equations (1) and (2)
- F1(Ijk, Dk, θ) = 0
- Constraints equations, assuming pure resistance fault impedance or Arg(If) = Arg (Vf)
- F2(Ijk, Dk, θ) = 0

Where θ represents the known variables, such as Vi, Vj, Vk, etc.

In Table 1, the two groups of equations for fault location detection of three phase faults are summarized.

Table 1. Summary of the two groups of equations for fault location detection

Group 1: Network Equation (n=0,1,2,)	
$V_{fn} = V_{kn} - (I_{kjn} - V_{kn} \times Y_{kfn}/2) \times Z_{nfn}$ $I_{jkn} = (V_{jn} - V_{fn})/Z_{jfn} + V_{jn} \times Y_{jfn}/2$ $I_{fn} = (V_{kn} - V_{fn})/Z_{kfn} - V_{fn} \times Y_{kfn}/2 +$ $(V_{jn} - V_{fn})/Z_{jfn} - V_{fn} \times Y_{jfn}/2$	
Group 2: Constraint Equation	
a-g fault	$Re(I_{f0} + I_{f1} + I_{f2}) \times Im(V_{f0} + V_{f1} + V_{f2}) =$ $Im(I_{f0} + I_{f1} + I_{f2}) \times Re(V_{f0} + V_{f1} + V_{f2})$
b-c-s fault	$Re(I_{f1} - I_{f2}) \times Im(V_{f1} - V_{f2}) =$ $Im(I_{f1} - I_{f2}) \times Re(V_{f1} - V_{f2})$
b-c-g fault	$V_{f1} = V_{f2}$
a-b-c-g fault	Re(If1) × Im(Vf1) = Im(If1) × Re(Vf1)

Solving numerically these sets of equations leads to the unknown parameter Dk, which is the location of the fault on the line between bus j and bus k.

It is not economical to install a PMU at each bus of a wide-area power network. Thus, the problem of optimal PMU placement (OPP) concerns with where and how many PMUs should be implemented in a power system to achieve full observability at minimum number of PMUs (Lin, Y.-H., et al, 2004; Lin, Y.-H., et al, 2004 ; Ibe, A., et al, 1986 ; Takagi, T., 1982 ; Müller, H.H., Castro, C.A., 2016 ; Abbasy, N.H., Ismail, H.M., 2009; Hajian M., 2011; Mahari, A., Heresh, S., 2013).

For an n-bus system, the PMU placement problem can be formulated as follows:

Objective function:

$$J = Min\sum\nolimits_{i=1}^{n} x_i \tag{3}$$

Such that:

$$f(X) \geq \hat{i} \tag{4}$$

Where x is a binary decision variable vector, whose entries are defined as:

$$x_i = \begin{cases} 1 & if\ Bus\ i\ has\ a\ PMU \\ 0 & otherwise \end{cases} \tag{5}$$

f (X) is a vector function, whose entries are non-zero if the corresponding bus voltage is solvable using the given measurement set and zero otherwise.

î is a vector whose entries are all equal to two or more.

Constraints

In order to form the constraint set, the binary connectivity matrix A, whose entries are defined as below, will be formed first.

$$A_{k,m} = \begin{cases} 1 & if\ k = m \\ 1 & if\ k\ and\ m\ are\ connected \\ 0 & if\ otherwise \end{cases} \tag{6}$$

Based on the acknowledgement that by placing a PMU on one bus, the voltage phasors of this bus and its neighboring buses can be calculated, we can get

$$f = A \times X. \tag{7}$$

Constraint functions defined by equations ensure full network fault observability while minimizing the total number of the PMUs.

The procedure for building the constraint equations (vector function f (X)) will be presented for four possible cases:

- No PMU measurements or conventional (flow and injection) measurements exist,
- Only flow measurements exist,
- Both flow and injection measurements (they may be zero injections or measured injections) exist or PMU measurements and conventional flow or injection measurements exist.

To verify the effectiveness of the proposed method, IEEE 14-bus system is taken as test system shown in Fig.3.

Figure 3. IEEE 14-bus system

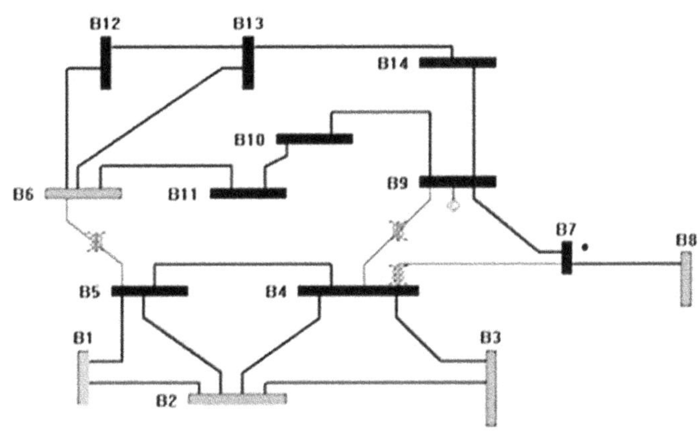

BINARY INTEGER PROGRAMMING (BIP)

The optimal PMU placement (OPP) can be obtained by two categories of methodologies: mathematical and heuristic. Integer linear programming (ILP) is the dominant mathematical optimization method used for solving the OPP problem and many studies concerning this issue have been published (Dua, D., e al, 2008; Theodorakatos, N. P., et al, 2015; Korkali, M., and Abur, A., 2009). The ILP technique was initially adopted for the OPP problem solution in (Xu, B., and Abur, A., 2004; Xu, B., and Abur, A., 2005). A voltage stability based on the contingency ranking method and a graph theoretic approach to modify the constraints under contingencies, are incorporated in (Sodhi, R., Srivastava, S. C., and Singh, S. N.; 2009). A generalized ILP formulation considering different cases of redundant PMU placement, complete and incomplete observability and the impact of zero injections, is suggested in (Gou, B., 2008).

And 1 is a vector whose elements are all one. Ax is a vector function whose entries are non-zeros if the corresponding bus voltage is observable using the given measurement set and zeros, otherwise (Xu, B., and Abur, A., 2004). The total cost to be minimized is,

$$J(x) = w^T X \tag{8}$$

subject to constraints (4)–(5), where w is the vector of PMU cost coefficients.

The steps for solving this OPP problem using the Binary Integer Linear Programming (BILP) model are:

Step 1: Read the network branch/bus data.

Step 2: Form the binary connectivity matrix A and the PMU cost coefficient vector w.
Step 3: Form the right-hand side unity vector 1 of (4).
Step 4: Solve through the use of BILP model.

The flowchart of Figure 4 depicts the BIP model.

Figure 4. Flowchart of BIP model

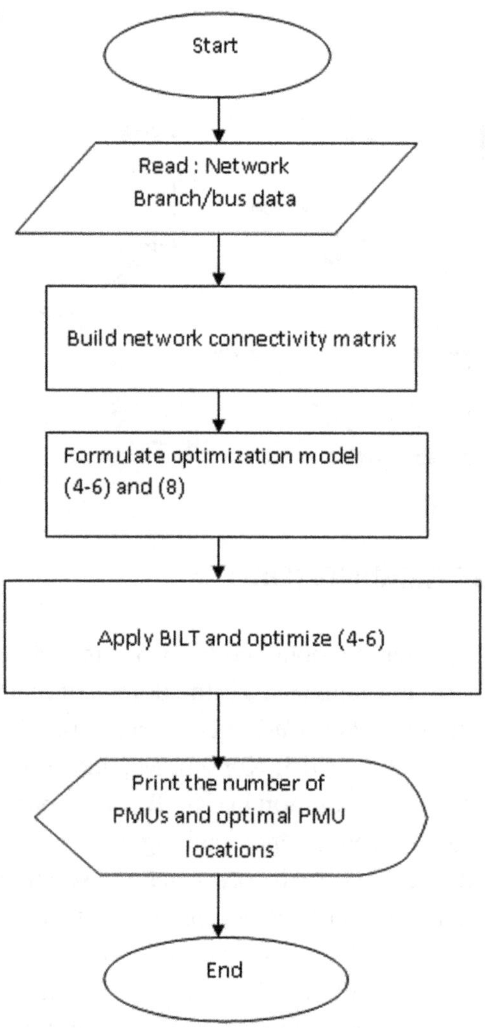

BTLBO OPTIMIZATION ALGORITHM

A binary version of TLBO that is a new method has been proposed (Recioui, A., Bentarzi, H., Ouadi, A., 2015; Zitouni, A. Bentarzi, H., Recioui, A., 2019). The algorithm in binary can be explained as follows:

Step 0: define the optimization problem, the objective function:

Minimize (objective function) y=f(x1,x2,….,xn-1,xn) such that: x1,x2,….,xn-1,xn can take only two values: either a "1" or a "0".

Decide how many students will be used or the population size, elite size and the number of generation. We consider a minimization problem; the maximization is similar.

Step 1: initialization: suggest a population (that will be developed to reach the final solution) or students randomly according to the following equation:

$$x^{1}_{(i,j)} = randi([0,1],1) \tag{9}$$

i: refer to student number, so there are ith student, i=1, 2, …,P
j:refer to the course number, $x_{(i,j)}$ is the level of the ith student at the jth course, j=1, 2,…,D

The small number 1 refers to the generation number that is the first generation.

After manipulating the above equation *P*×*D* times we obtain a *P*×*D* matrix which represents our population:

$$population^{1} = \begin{bmatrix} x^{1}_{(1,1)} & \cdots & x^{1}_{(1,D)} \\ x^{1}_{(2,1)} & \ddots & x^{1}_{(2,D)} \\ \vdots & \ddots & \vdots \\ x^{1}_{(P,1)} & \cdots\theta & x^{1}_{(P,D)} \end{bmatrix} \tag{10}$$

Choose the teacher: the best student is the one which has the minimum fitness function.

Step 2: teacher phase

As in real version each student tries to be like his teacher, so he makes his design variables as those of the teacher, of course not all of them otherwise he will be exactly as his teacher, so the student copies some components of teacher and replace his components by those copied ones. The number of copied components depends on the ability of teacher and the student; that is represented by one random number *h* that is between 0 and D. The corresponding formula is:

$$h = \text{rand } i(d,1) \tag{11}$$

This number determines how many courses that student *i*.will learn from his teacher.

The location of those courses in the teacher vector can be specified according to the equation below:

$$Course = \text{rand } i(h,[1,h]) \tag{12}$$

These randomly selected courses are now ready to be transmitted from teacher to student:

$$\begin{aligned}&X_{i,new} = X_i\\&X_{i,new}(course) = X_{teacher}(course)\\&if\ f(X_{i,new}) < f(X_i)\ then\ X_i = X_{i,new}\ else\ do\ nothing\end{aligned} \tag{13}$$

Executing may produce duplicate solutions; therefore an additional step is needed.

Step 3: remove duplicate solutions by mutation on randomly selected dimensions.
Step 4: learner phase
The interaction between students in binary interpretation will be as follows:

Choose for a student i another one j, if j is better than i then i will learn from j by trying to change components that are different from those of j, in order to be like him; otherwise he will change some components that are similar to those of j. This can be done as follows:

For $i = 1$: P do

Choose another student j and record where they are similar and where they are different in a vector,

$$q = \left[q_1\, q_2 \,........\, q_D\right] such\ that\ q_k = 1\ if\ x^g_{i,k} = x^g_{j,k}\ else\ q_k = 0; k = 1,\ldots,D$$

The vector h=rand $i(D,1)$ specifies how many courses will student i learn from j if he is better than him or change them in other case.

The vector *course*=rand $i(h,[1,h])$ specifies the location of courses which will be learnt or changed.

Now the learning process can be implemented:

$X_{i,new}$=X_i The new student is the old with some modifications.

$$\begin{aligned}&if\ f(X^g_i) < f(X^g_j)\ then\ X_{i,new}(course) = X_j(course) \times not(q(course))\\&else\ if\ f(X^g_i) \geq f(X^g_j)\ then\ X_{i,new}(course) = not(X_j(course)) \times q(course)\end{aligned} \tag{14}$$

Equation (14) indicates that some components where i and j differs will be copied from j (the highest learned one between the two) to i otherwise (j is worse than i) some of similar component in i will be changed from 0 to 1 or 1 to 0. Now, check if the new student is better than the old:

$$if\ f(X_{i,new}) < f(X_i)\ then\ X_i = X_{i,new}\ else\ do\ nothing \tag{15}$$

After completing the process for all the population, a new teacher will be determined.

Step 5: Remove duplicate solutions, replace bad solutions by elite solutions and again remove duplicate solutions.
Step 6: if $g \neq$ number of generation go to step2 else stop.

Figure 5. Flow chart of BTLBO algorithm applied to the optimal PMU placement (a) Global routine, (b) Observability subroutine

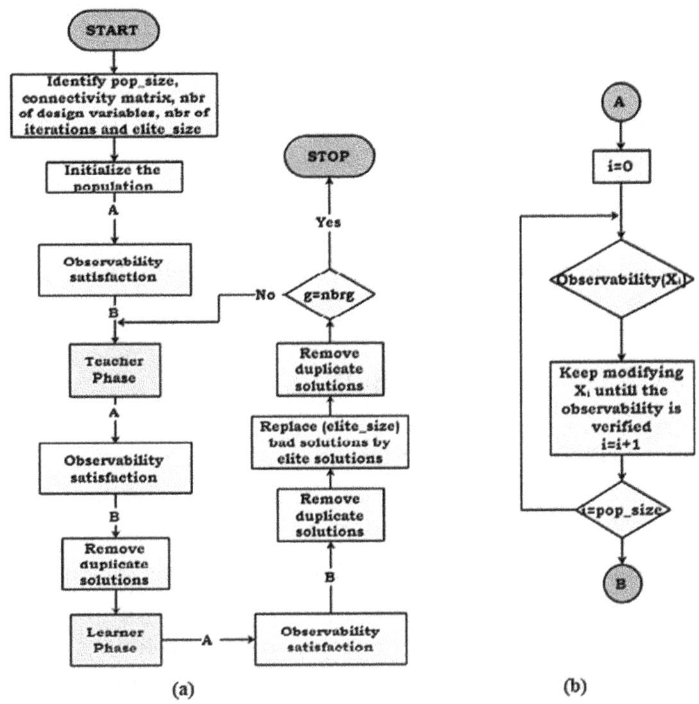

Using the binary teaching learning based optimization explained previously which is based on the problem formulation pointed out in previous sections; an algorithm for the optimal placement problem is shown below:

1. Build the binary connectivity matrix using the one line diagram and constraint modification when injection, flow measurements or already installed PMUs.
2. Define the optimization parameters; the population size (pop_size), the design variables (N_buses), and the number of generation (N_gen), the (elite size).
3. Generate random solutions within boundaries of the system.
4. Check that random solutions satisfy the inequality constraint of buses f(x)=(A*X) ≥ î
5. Modify those not satisfying the constraints.
6. Calculate the fitness of each solution based on the objective function of expression.
7. Set the best solution as teacher of the population.
8. For each student apply teacher phase, replace x by x_{new} if it gives better fitness function (less) otherwise keep the old one.
9. If the duplicate solutions exist then it is necessary to modify the duplicate solutions in order to avoid trapping in the local optima. In the present work, duplicate solutions are modified by mutation on randomly selected dimensions of the duplicate solutions before executing the next generation without losing the observability.

10. For each student apply learner phase, replace x by x_{new} if it gives better fitness function (less) otherwise keep the old one.
11. Again, remove duplicate solutions keeping the constrained satisfied (observability).
12. Replace (elite_size) bad solutions by (elite_size) elite solutions.
13. Again, remove duplicate solutions keeping the constrained satisfied (observability), then determine the teacher.
14. Repeat from step 7 for maximum number of iterations.
15. Set the best solution $x_{teacher}$ as the final solutions.

The flowchart for the optimal placement algorithm is shown in Fig.5.

PARTICLE SWARM OPTIMIZATION (PSO)

A particle swarm optimizer (PSO) is a population-based stochastic optimization algorithm modeled after the simulation of the social behavior of bird flocks and fish schooling. It is introduced firstly by (Kennedy, J., C. Eberhart, 1995), it has been successfully applied to a wide range of optimization problems of both continuous nonlinear and discrete optimization problems such as: Neural Networks, Shape topology and electric power distribution networks.

A particle Swarm Optimization is a swarm intelligence method for a global optimization. In PSO a population (called swarm) fly through the search space. This swarm is a group of individuals called particles that are looking for the best location. Each particle represents a candidate solution to the optimization problem, it adjusts its trajectory toward both of its own previous best position and the previous best position attained by any member of its topological neighborhood. So the position of a particle is influenced by the best position visited by itself and the position of the best particle in its neighborhood, when the neighborhood of a particle is the entire swarm. The best position achieved by each particle is referred to as local best position (X_i*Best*- the best position of particle *i*) while the best position in the entire swarm is referred to as the global best position (X_g*Best*). Each individual has a fitness function where the objective function is the global function its maximum or minimum should be found. The sketched diagram shown in Fig.6 clarifies the particle trajectory:

Figure 6. Particle trajectory for PSO

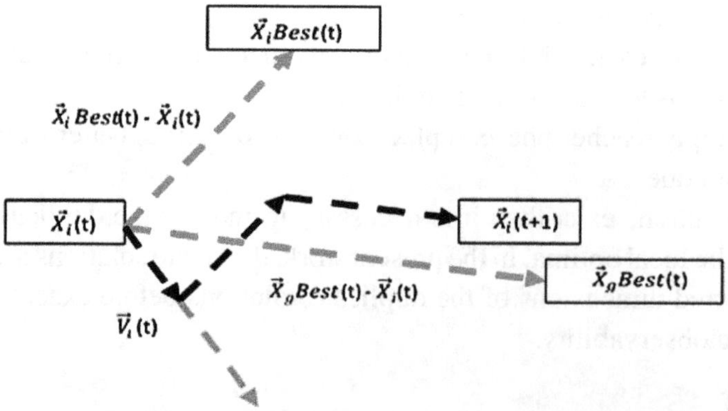

So each particle in the swarm is defined by the following characteristics:

X_i: The current position of particle *i*.
V_i: The current velocity vector of particle *i*.
X_iBest: The personal best position vector achieved by particle *i* or simply best "remembered" individual particle position.
GBest. The global best position vector for all the particles of the swarm or simply the best "remembered" swarm position.

All of these characteristics are multi-dimensional with the same dimension. It can be represented as follows:

$$X_i = \left[X_{i,1}, X_{i,2}, \ldots\ldots, X_{i,n}\right] \tag{16}$$

$$V_i = \left[V_{i,1}, V_{i,2}, \ldots\ldots, V_{i,n}\right] \tag{17}$$

Where n is the number of variables in the fitness function.
In each step of time t (iteration) each particle updates its own position as follow:

$$X_i(t + 1) = X_i(t) + V_i(t) \tag{18}$$

On the other side, the velocity is updated as follows:

$$V_i\left(t+1\right) = V_i\left(t\right) + \left(c_1 \times r_1 \times \left(X_iBest - X_i\left(t\right)\right)\right) + \left(c_2 \times r_2 \times \left(X_gBest - X_i\left(t\right)\right)\right) \tag{19}$$

Where:

$X_i(t+1)$: is the new position for the particle *i* at time step $(t+1)$.
$V_i(t+1)$: is the new velocity for the particle *i* at time step $(t+1)$.
r_2, r_1: cognitive and social parameters.
c_1, c_2: random values between 0 and 1.
X_iBest, X_gBest, $V_i(t)$, $X_i(t)$: as stated before.

Let $F(X)$ be the objective function, then the personal best position of a particle updates its self at time step t is as follows:

$$X_iBest\left(t+1\right) = \begin{cases} X_iBest\left(t\right) & F\left(X_iBest\left(t\right)\right) < F\left(X_iBest\left(t+1\right)\right) \\ X_iBest\left(t+1\right) & F\left(X_iBest\left(t+1\right)\right) < F\left(X_iBest\left(t\right)\right) \end{cases} \tag{20}$$

Simultaneously, the global best position is determined by selecting the personal best position of all particles at time step t, then the global best position is the personal best position in which the objective function has to be minimum or maximum. So $X_g Best$ is included in the fallowing set:

$$X_g Best \in \left\{X_1 Best, X_2 Best, \dots\dots\dots, X_p Best\right\} \tag{21}$$

and it satisfies:

$$F(X_g Best) = \min\left\{F\left(X_1 Best\right), F\left(X_2 Best\right), \dots\dots\dots, F(X_p Best)\right\}$$

Or,

$$F(X_g Best) = \max\left\{F\left(X_1 Best\right), F\left(X_2 Best\right), \dots\dots\dots, F(X_p Best)\right\}. \tag{22}$$

All of the previous steps are repeated until a specific number of iteration (maximum number of iteration) is reached, or velocity updates becomes too close to zero. And the optimality of particles can be evaluated through how much the objective function tends to zero (in case if we are looking to find the minimum of the objective function) (Kouzou, A.L., Mohammedi, R. D. 2017). The Pseudo code of PSO algorithm is presented in Appendix.

GREY WOLF OPTIMIZER (GWO)

In 2014, a new powerful meta-heuristic optimization algorithm known as the Grey Wolf Optimizer was introduced by (Mirjalili, S., Lewis, A., 2014; Kouzou, A. L., Bentarzi, H., et al, 2019). This algorithm is inspired from the behavior of the Grey Wolf that belongs to "Canidae family".

Grey wolf mostly prefers living in a pack of 5 to 12 maximum under strict social dominant hierarchy, the social hierarchy consists of four (04) levels as shown in Figure 7. The pack should follow the orders of the dominant wolf. The leader that can be male or female is known as Alpha (α), it is responsible for making decisions about time to walk, place where they are sleeping, hunting…etc. The Alpha wolf is the dominant wolf in all the pack and all his/her orders should be obeyed by the pack members. The Betas (β) are subordinate wolves which help the alpha in decision making. The Beta wolf is considered to be the best candidate to take the place of the Alpha wolf in case of the Alpha becomes very old or passes away. Also, the Beta disciplines the pack and work as an advisor to Alpha. The Delta (δ) wolves submit to Alpha and Beta wolves but they dominate the Omega (ω), there are different categories of Delta wolves such as Scouts, Sentinels, Elders, Hunters and Caretakers. The Omega wolves are in the lowest level of social hierarchy, they are considered to be the scapegoat in the pack and they have to submit to all the other wolves. It seems that the Omega wolves are not important individuals in the pack and they are the last allowed wolves to eat.

Figure 7. Hierarchy of grey wolf (dominance decreases from top down)
(S. Mirjalili, S. Mirjalili and A. Lewis, 2014)

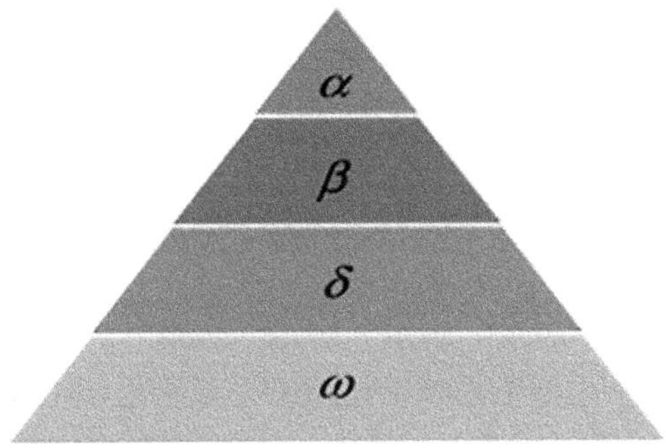

The main acts of grey wolf hunting are encircling, hunting and attacking the prey. The fittest solution is considered to be Alpha, and the second and third solutions are Beta and Delta, respectively. The other remaining solutions are considered to be Omega. The hunting is guided by α, β and δ. The other solutions (ω) follow these three wolves.

The mathematical model of the encircling behaviour is presented as follows:

$$\begin{aligned} \vec{D} &= \left|\vec{C}.\vec{X}_{prey}(t) - \vec{X}_{wolf}(t)\right| \\ \vec{X}_{wolf}(t+1) &= \vec{X}_{prey}(t) - \vec{A}.\vec{D} \end{aligned} \tag{23}$$

Where t indicates the current iteration, $\vec{X}_{prey}(t)$ is the prey position vector, $\vec{X}_{wolf}(t)$ is the grey wolf position vector, $\vec{D}$ and $\vec{C}$ are coefficient vectors calculated as follows:

$$\begin{aligned} \vec{A} &= 2\vec{a}\vec{r_1} - \vec{a} \\ \vec{C} &= 2\vec{r_2} \end{aligned} \tag{24}$$

Where the entries of $\vec{a}$ are linearly decreased from 2 to 0 over the number of iterations, and $\vec{r_1}$ and $\vec{r_2}$ are random vectors of entries varying within the interval of [0, 1]. Thus, by using equations (23) a grey wolf can update its position around the prey at any random location.

In modelling the hunting process of grey wolves, it is supposed that the Alpha wolf presents the best candidate solution. In addition to that the Beta and Delta wolves have better information about the prey potential location. Therefore, the three best solutions obtained till the actual iteration are memorized to force the other search candidates to update their positions. Figure 8 shows how a wolf updates its position in a 2D search space.

The position vector of the all the wolves can be updated with respect to Alpha, Beta and Delta as follows:

$$\overrightarrow{D_\alpha} = \left|\overrightarrow{C_1}\overrightarrow{X_\alpha}(t) - \overrightarrow{X_{wolf}}(t)\right|, \vec{X}_1 = \vec{X}_\alpha(t) - \overrightarrow{A_1}.\overrightarrow{D_\alpha}$$
$$\overrightarrow{D_\beta} = \left|\overrightarrow{C_2}\overrightarrow{X_\beta}(t) - \overrightarrow{X_{wolf}}(t)\right|, \vec{X}_2 = \vec{X}_\beta(t) - \overrightarrow{A_2}.\overrightarrow{D_\beta} \quad (25)$$
$$\overrightarrow{D_\delta} = \left|\overrightarrow{C_3}\overrightarrow{X_\delta}(t) - \overrightarrow{X_{wolf}}(t)\right|, \vec{X}_3 = \vec{X}_\delta(t) - \overrightarrow{A_3}.\overrightarrow{D_\delta}$$

The best position which can be taken by a wolf can be calculated as the average, hence it is expressed as follows:

$$\overrightarrow{X_{wolf}}(t+1) = \frac{\vec{X}_1 + \vec{X}_2 + \vec{X}_3}{3} \quad (26)$$

All of the previous steps are repeated until a maximum number of iterations are reached. The Pseudo code of PSO algorithm is presented in Appendix.

After that many applications have been developed such in PMU Optimal Placement optimization (Azzougui, Y., Recioui, A., Mansouri, A., 2019).

Figure 8. Position updating in GWO.

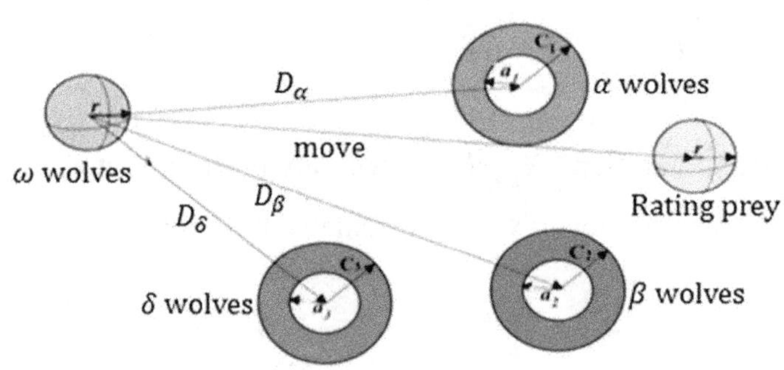

MOTH FLAME OPTIMIZATION (MFO)

Moth-flame optimization algorithm (MFO) is a new meta-heuristic optimization algorithm introduced by (Mirjalili, S., 2015), this algorithm mimics the motion of moths flight around the flames. The natural moth navigation method is known as transverse orientation. In this method, a moth flies by keeping a fixed angle with the moon, this method shows a very effective technique for travelling in a straight path because the moon is very far away from the moth. However, when the light source (for example, flame or human-made artificial light) is close,moths try to keep a fixed angle with the light to fly in straight line but moths fly spirally around it and finally converge toward it since it is extremely close compared to the moon (. Gaston, K ., 2013; Frank, K.D. et al, 2006).

Moths and flames are the key components in the MFO algorithm, the moth is the actual search body moving in the search space while the flames are the best solution achieved by the search bodies so far and they are updated in case of finding better solution. The set of moths and their corresponding fitness values are represented in matrices as follows:

$$M=\begin{bmatrix} m_{1,1} & m_{1,2} & \cdots & m_{1,d} \\ m_{2,1} & m_{2,2} & \cdots & m_{2,d} \\ \vdots & \vdots & \vdots & \vdots \\ m_{n,1} & m_{n,2} & \cdots & m_{n,d} \end{bmatrix}, OM=\begin{bmatrix} OM_1 \\ OM_2 \\ \vdots \\ OM_n \end{bmatrix} \tag{27}$$

where n and d are the number of moths and variables respectively, also the flames and their fitness values are represented initially in matrices similar to the moth matrices as follows:

$$F=\begin{bmatrix} F_{1,1} & F_{1,2} & \cdots & F_{1,d} \\ F_{2,1} & F_{2,2} & \cdots & F_{2,d} \\ \vdots & \vdots & \vdots & \vdots \\ F_{n,1} & F_{n,2} & \cdots & F_{n,d} \end{bmatrix}, OF=\begin{bmatrix} OF_1 \\ OF_2 \\ \vdots \\ OF_n \end{bmatrix} \tag{28}$$

The transverse orientation can be model mathematically through the logarithmic spiral, hence the position of each moth is updated with respect to a flame using the following equation:

$$M_i = D_i.e^{bt}\cos\left(2\pi t\right) + F_j, D_i = \left|F_j - M_i\right| \tag{29}$$

Where Mi is the ith moth, Fjis the jth flame, Di means the Euclidian distance between the ith moth and the jth flame, b is a constant that defines the logarithmic spiral shape and t is a random number in the interval of $[-1, 1]$.The t parameter specifies how close is the next moth position to the flame as shown in Figure 9. To emphasize exploitation, t is taken as a random number in the interval [r,1] where r decreases linearly from -1 to -2 during iterations, r is known as convergence constant.

In each iteration, the list of flames are updated and sorted based on the fitness values. Then, the moths update their positions toward their flames. The first moth is always updating its position to the best flame while the last moth to the worst flame of the list. Another problem here is that the position updating of moths to n different locations may degrade the exploitation of the best promising solutions, thus the number of flames to be followed is reduced during iterations as shown in the following expression:

$$number\ of\ flames = \left(N - l \times \frac{N-1}{T}\right). \tag{30}$$

Where N is the number of moths, l is the number of current iteration and T is the number of iterations. Once the maximum iteration is reached, the best moth is returned as the best approximation of the optimum. The Pseudo code of PSO algorithm is presented in Appendix.

Figure 9. Logarithmic spiral, the positions of moth around a flame with respect to t
(Frank, K.D., Rich, C., Longcore, T., 2006)

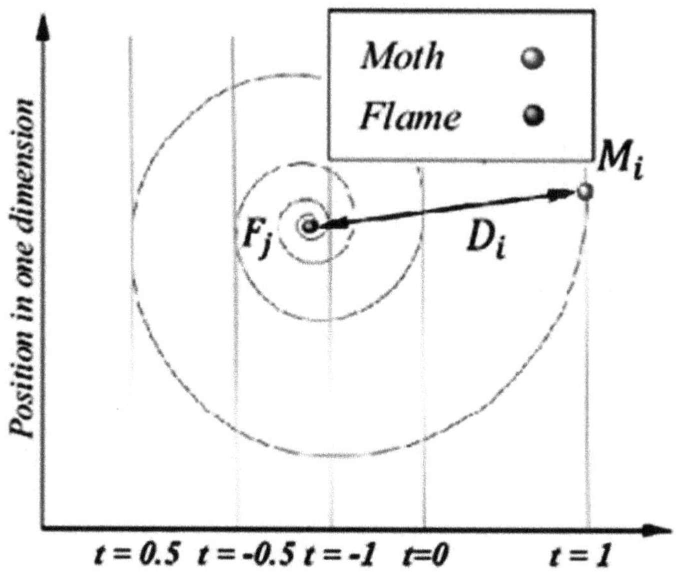

SIMULATION RESULTS AND DISCUSSION

The PMU placement optimization algorithm presented in the previous sections was implemented in MATLAB® on the IEEE 14-bus system. Network data for this system is in public domain as given in Table 5.2. The parameters used in the optimization are: The maximum iteration number in BTLBO is selected to be 1000 with the population size of 100. In the simulations presented in this research work, the best solution of the binary TLBO is concluded after 20 runs of the algorithm. The simulations are carried out on an Intel i5 (2.40 GHz) with 4 GB RAM. It should be mentioned that the corresponding configuration of the required number of PMUs is not necessarily unique. Since some heuristic algorithms, such as TLBO, Particle Swarm Optimization, the Grey Wolf Optimizer (GWO), the Moth-Flame Optimization (MFO) are based on a random search in the search space of the problem, and the result of each execution of these algorithms might be different from another one. Therefore, they must be run several times to ensure that the optimal point of the problem is found. However, from the observability point of view, there is no difference between different configurations with the same number of PMUs, so only one configuration is presented.

Two sets of simulations are carried out on the IEEE 14- bus system, which are initially assumed to have no flow measurements and no previously existed PMUs. In the first set of simulations, zero injection is simply ignored while in the second set, it is used as existing measurement.

The IEEE 14 bus test system has got only one zero-injection bus. The output of the BTLBO algorithm is 8 PMUs. The information of the system and zero injections are given in the Table.2.

The global solution is obtained with only a small number of iterations (between 5 and 10) and population size (10). Seven PMUs can be deployed to observe the IEEE14 bus system. Here, when ignoring injection, different configuration of PMUs is obtained, whereas, considering injection give only one set (1, 2, 4, 6, 9, 10 and 13). It is observed that the number of PMUs required for system observability reduces by 1, if the system observability is to be maintained after considering zero injections buses.

The number of PMUs calculated by the proposed approach is given in Table 3. The number of PMUs that guarantees 100% of observability of power system during any failure that can occur by guaranteeing that each bus is observable by at least two PMUs.

The obtained simulation results using some optimization techniques including conventional BILT (Theodorakatos, N. P., Manousakis, N. M., Korres, G. N., 2015) as well as heuristics are shown in Table4.

Table 2. System information of IEEE 14-bus system

System	Num. of Branches	Num. of Zero Injections	Zero Injection Buses
IEEE 14-bus	20	1	7

Table 3. Optimum number and location of PMU

System	Num. of PMUs	Location of PMUs
IEEE 14-bus with Zero injection buses	7	1,2,4,6,9,10,13
IEEE 14-bus with injection buses	8	1,2,4,6,9,10,13, 14

Table 4. Optimal PMUs number using various optimization techniques for the IEEE 14-bus test system.

Optimization Technique	Optimal PMU Number
BTLBO	7
BILT	8
PSO	7
GWO	8
MFO	7

CONCLUSION

The obtained simulation results are in agreement with the previous research work results (Zitouni, A., Bentarzi, H. Recioui, A., 2019; Kouzou, A. L., et al, 2019; Lin, Y.-H., et al, 2004; Lin, Y.-H., et al, 2004 ; Ibe, A., et al, 1986 ; Takagi, T., 1982 ; Müller, H.H., Castro, C.A., 2016 ; Abbasy, N.H., Ismail, H.M., 2009; Hajian M., 2011; Mahari, A., Heresh, S., 2013) and better than the conventional optimization technique BITL (Theodorakatos, N. P., Manousakis, N. M., Korres, G. N., 2015). Moreover, the results show that full network observability can be met under fault conditions using more than 50% PMUs installation; however, under normal conditions, approximately 30% PMUs installation is required.

Finally, the proposed research work represents a potential tool to monitor power systems, and will give valuable help the operators in a smart grid environment.

REFERENCES

Abbasy, N. H., & Ismail, H. M. (2009). A unified approach for the optimal PMU location for power system state estimation. *IEEE Transactions on Power Systems*, *24*(2), 806–813. doi:10.1109/TPWRS.2009.2016596

Allemong, J., Radu, L., & Sasson, A. (1982). A fast and reliable state estimation algorithm for AEP's new control center. *IEEE Transactions on Power Apparatus and Systems*, *4*(4), 933–944. doi:10.1109/TPAS.1982.317159

Aminifar, F., Khodaei, A., Fotuhi-Firuzabad, M., & Shahidehpour, M. (2010). Contingency-constrained PMU placement in power networks. *IEEE Transactions on Power Systems*, *25*(1), 516–523. doi:10.1109/TPWRS.2009.2036470

Azzougui, Y., Recioui, A., & Mansouri, A. (2019). PMU Optimal Placement in Wide Area Monitoring Systems using Grey Wolf Optimization Technique. Algerian. *Journal of Signals and Systems*, *4*(1), 1–7.

Baldwin, T. L., Mili, L., Boisen, M. B., & Adapa, R. (1993). Power System Observability with Minimal Phasor Measurement Placement. *IEEE Transactions on Power Systems*, *8*(2), 707–715. doi:10.1109/59.260810

Bei, X., Yoon, Y. J., & Abur, A. (2008). *Optimal placement and utilization of phasor measurements for state estimation*. Texas A&M University.

Chakrabarti, S., & Kyriakides, E. (2007). Optimal placement of phasor measurement units for state estimation. *Proc. 7th IASTED Int. Conf. Power and Energy Systems*, 1–6.

Chakrabarti, S., & Kyriakides, E. (2008). Optimal placement of phasor measurement units for power system observability. *IEEE Transactions on Power Systems*, *23*(3), 1433–1440. doi:10.1109/TPWRS.2008.922621

Charu, S., & Barjeev, T. (2011). An approach for optimal PMU placement using binary particle swarm optimization with conventional measurements. *International Journal of Engineering Science and Technology*, *3*(3), 56–63.

Dambhare, S., Dua, D., Gajbhiye, R. K., & Soman, S. A. (2008). *Optimal Zero: An ILP Approach*. Indian Institute of Technology.

Dekhandji, F. Z., Azzougui, Y., & Recioui, A. (2019). PMU Optimal Placement for Fault Deetction in Wide Area Monitoring Systems based on the Dron Suadron Optimization. *Algerian Journal of Signals and Systems*, *4*(1), 25–31.

Dua, D., Dambhare, S., Gajbhiye, R. K., & Soman, S. A. (2008). Optimal multistage scheduling of PMU placement: An ILP approach. *IEEE Transactions on Power Delivery*, *23*(4), 1812–1820. doi:10.1109/TPWRD.2008.919046

Frank, K. D., Rich, C., & Longcore, T. (2006). Effects of artificial night lighting on moths. *Journal of Ecological consequences of artificial night lighting. Hindawi Publishing Corporation*, *15*(1), 305–344.

Gaston, K., Bennie, J., Davies, T., & Hopkins, J. (2013). The ecological impacts of nighttime light pollution: A mechanistic appraisal. *Biological Reviews of the Cambridge Philosophical Society*, *88*(4), 912–927. doi:10.1111/brv.12036 PMID:23565807

Geramian, S. S., Abyane, H. A., & Mazlumi, K. (2008). Determination of Optimal PMU Placement for Fault Location Using Genetic Algorithm. *13th International Conference on Harmonics and Quality of Power*. 10.1109/ICHQP.2008.4668810

Gou, B. (2008). Generalized integer linear programming formulation for optimal PMU placement. *IEEE Transactions on Power Systems*, *23*(3), 1099–1104. doi:10.1109/TPWRS.2008.926475

Hajian, M., Ranjbar, A. M., Amraee, T., & Mozafari, B. (2011). Optimal placement of PMUs to maintain network observability using a modified BPSO algorithm. *Int. J. Electr. Power*, *33*(1), 28–34. doi:10.1016/j.ijepes.2010.08.007

Hajian, M. A., Ranjbar, M., Amraee, T., & Shirani, A. R. (2007). Optimal Placement of Phasor Measurement Units: Particle Swarm Optimization Approach. *Proc. Of the 14th International Conference on Intelligent System Applications to Power Systems.*

Ibe, A., & Cory, B. (1986). A travelling wave-based fault locator for two-and three-terminal networks. *IEEE Transactions on Power Delivery*, *1*(2), 283–288. doi:10.1109/TPWRD.1986.4307961

Jiangnon, P., Yuanzhang S., & Wang H. F., Optimal Placement for full Network Observability using Tabu Search Algorithm. *Int. J. Elect. Power Energy Syst.*, *23*, 1525-1526.

Kennedy, J., & Eberhart, R. (1995). Particle swarm optimization. *Neural Networks, 1995. Proceedings., IEEE International Conference*, 1942-1948.

Korkali, M., & Abur, A. (2009). Reliable measurement design against loss of PMUs with variable number of channels. *North American Power Symposium (NAPS).*

Kouzou, A. L., Bentarzi, H., Mohammedi, R. D., & Laoumer, M. (2019). Optimal Placement of Phasor Measurement Unit in Power System using Meta-Heuristic Algorithms. *Journal Electrotehnica, Electronica. Automatica*, *67*(2).

Lien, K.-P., Liu, C.-W., Yu, C.-S., & Jiang, J.-A. (2006). Transmission Network Fault Loca-tion Observability With Minimal PMU Placement. *IEEE Transactions on Power Delivery*, *21*(3), 1128–1136. doi:10.1109/TPWRD.2005.858806

Lin, Y.-H., Liu, C.-W., & Chen, C.-S. (2004). A new PMU-based fault detection/location technique for transmission lines with consideration of arcing fault discrimination-part I: Theory and algorithms. *IEEE Transactions on Power Delivery*, *19*(4), 1587–1593. doi:10.1109/TPWRD.2004.832407

Lin, Y.-H., Liu, C.-W., & Chen, C.-S. (2004). A new PMU-based fault detection/location technique for transmission lines with consideration of arcing fault discrimination-partII: Performance evaluation. *IEEE Transactions on Power Delivery*, *19*(4), 1594–1601. doi:10.1109/TPWRD.2004.835276

Mahari, A., & Heresh, S. (2013). Optimal PMU placement for power system observability using BICA considering measurement redundancy. *Electric Power Systems Research*, *103*, 78–85. doi:10.1016/j.epsr.2013.04.013

Milosevic, B., & Begovic, M. (2003). Nondominated sorting genetic algorithm for optimal phasor measurement placement. *IEEE Transactions on Power Systems*, *18*(1), 69–75. doi:10.1109/TPWRS.2002.807064

Mirjalili, S. (2015). Moth-flame optimization algorithm: A novel nature-inspired heuristic paradigm. *Knowledge-Based Systems, 89*, 228–249. doi:10.1016/j.knosys.2015.07.006

Mirjalili, S., & Lewis, A. (2014). Grey Wolf Optimizer. *Advances in Engineering Software, 69*, 46–61. doi:10.1016/j.advengsoft.2013.12.007

Mohammadi-Ivatloo, B. (2009). *Optimal placement of PMUs for power system observability using topology based formulated algorithms. Sahrif University of Technology.*

Mosavi, M. R., Akhyani, A. A., & Rahmati, A. (2012). *A PMU Placement Optimal Method in Power Systems using Modified ACO Algorithm and GPS Timing*. Iran University of Science and Technology.

Müller, H. H., & Castro, C. A. (2016). Genetic algorithm-based phasor measurement unit placement method considering observability and security criteria. *IET Generation, Transmission & Distribution, 10*(1), 270–280. doi:10.1049/iet-gtd.2015.1005

Nuqui, R. F., & Phadke, A. G. (2005). Phasor Measurement Unit placement techniques for complete and incomplete observability. *IEEE Transactions on Power Delivery, 20*(4), 2381–2388. doi:10.1109/TPWRD.2005.855457

Phadke, A. G. (1993). Synchronized phasor measurements in power systems. *IEEE Computer Applications in Power, 6*(2), 10–15. doi:10.1109/67.207465

Phadke, A. G., & Thorp, J. S. (2008). *Synchronized Phasor Measurements and Their Applications.* doi:10.1007/978-0-387-76537-2

Rao, R. V., & Patel, V. K. (2012). Multi-objective optimization of combined Brayton and inverse Brayton cycles using advanced optimization algorithms. *Engineering Optimization.* Advance online publication. doi:10.1080/0305215X.2011.624183

Rao, R. V., Savsani, V. J., & Vakharia, D. P. (2011). Teaching-learning-based optimization: A novel method for constrained mechanical design optimization problems. *Computer Aided Design, 43*(3), 303–315. doi:10.1016/j.cad.2010.12.015

Rao, R. V., Savsani, V. J., & Vakharia, D. P. (2012). Teaching-learning-based optimization: An optimization method for continuous non-linear large scale problems. *Inform Sci, 183*(1), 1–15. doi:10.1016/j.ins.2011.08.006

Rao, V. J., Savsani, J. B., & Balic, J. (2012). Teaching learning based optimization algorithm for constrained and unconstrained real parameter optimization problems. *Engineering Optimization, 44*(12), 1447–1462. Advance online publication. doi:10.1080/0305215X.2011.652103

Recioui, A., Bentarzi, H., & Ouadi, A. (2015). Application of a Binary Teaching Learning-Based Algorithm to the Optimal Placement of Phasor Measurement Units, Book chapter. *Progress in Clean Energy, 1*, 817–830. doi:10.1007/978-3-319-16709-1_60

Shahraeini, M., & Javidi, M. H. (2011). *A Survey on Topological Observability of Power Systems.* IEEE Press. doi:10.1109/PEAM.2011.6135115

Sodhi, R., Srivastava, S. C., & Singh, S. N. (2009). *Optimal PMU placement to ensure system observability under contingencies. In Power & Energy Society General Meeting, 2009 (PES '09)*. IEEE.

Takagi, T., Yamakoshi, Y., Yamaura, M., Kondow, R., & Matsushima, T. (1982). Development of a new type fault locator using the one-terminal voltage and current data. *IEEE Transactions on Power Apparatus and Systems*, *8*(8), 2892–2898. doi:10.1109/TPAS.1982.317615

Tankasala, G. R., Sanisetty, S., & Vala, V. K. (2012). Optimal Placement of Phasor Measurement Units for State Estimation using Artificial Intelligence Techniques. *International Journal of Scientific and Engineering Research*, *3*(2).

Theodorakatos, N. P., Manousakis, N. M., & Korres, G. N. (2015). Optimal Placement of Phasor Measurement Units with Linear and Non-linear Models. *Electric Power Components and Systems*, *43*(4), 357–373. doi:10.1080/15325008.2014.981319

Ur Rahman, Ali, Ullah, Ur Rahman, Iqbal, Ahmad, Zeb, Ali, Shahzad, & Beenish. (2012). Advancement in Wide Area Monitoring Protection and Control Using PMU's Model in MATLAB/SIMULINK Taj. *Smart Grid and Renewable Energy*, (3), 294–307.

Valenti, M. C., & Feliachi, A. (2002). *Communication Delays in Wide Area Measurement Systems*. IEEE.

Xu, B. (2006). *Optimal Monitoring And Visualization Of Steady State Power System Operation* (Ph.D Dissertation). Texas A&M University.

Xu, B., & Abur, A. (2004). Observability analysis and measurement placement for systems with PMUs. *IEEE PES Power Syst. Conf. Expo.*, *2*, 943–946.

Xu, B., & Abur, A., (2005). *Optimal placement of phasor measurement units for state estimation*. Final Project Report PSERC.

Yin, H., & Fan, L. (2010). PMU data-based fault location techniques. In *Int North American Power Symposium (NAPS)*. IEEE Press.

Zitouni, A., Bentarzi, H., & Recioui, A. (2019). BTLBO Based SPMUs Placement Optimization for Fault Observation in Power Grid. *Algerian Journal of Signals and Systems*, *4*(1).

KEY TERMS AND DEFINITIONS

Phasor Measurement Unit: A phasor measurement unit (PMU) is a transducer used to measure the magnitude and phase angle of an electrical phasor quantities (Voltage and Current) in the power grid using a common time source for synchronization (GPS). The time of synchronization that is less one microsecond isn important because if the grid supply and demand are not perfectly balanced, frequency imbalances can cause stress on the grid, which is a potential cause for power outages.

PMUs Placement Optimization (PPO): Optimal phasor measurement units (PMUs) placement involves the process of minimizing the number of PMUs needed while ensuring the entire power system completely observable. A power system is identified observable when the voltages of all buses in the power system are known.

Power Network Observability: Network observability is a Yes/No type of property; the network is either observable or not. This property is determined solely by the location of measurements on the network, not by measurement weights and network admittance values.

Wide Area Monitoring System: Wide Area Monitoring System (WAMS) allows the realization of new and advanced functions for support of the development and control of transmission and distribution networks, namely the evaluation of network stability, ampacity of lines and detection of critical network conditions.

APPENDIX

The Pseudo Code of PSO Algorithm

```
for each particle
     Initialize particle
end
do
for each particle
          Calculate fitness value
if the fitness value is better than the best fitness value in history
set current values as new best fitness values
end if
end for
   Choose the particle with the best fitness value of all the particles as the
X_g best
for each particle
         Calculate particle velocity according equation (18)
        Update particle position according equation (17)
end for
while maximum iterations or minimum error criteria is not attained
```

The Pseudo Code of GWO Algorithm

Initialize the grey wolf population $\overrightarrow{X_{wolf}}$ *(i = 1, 2, ..., n)*

```
Initialize a, A, and C
Calculate the fitness of each search agent
```

X_α *=the best search agent*

X_β *=the second best search agent*

X_δ *=the third best search agent*

```
while(t < Max number of iterations)
foreach search agent
Update the position of the current search agent by equation (25)
end for
Update a, A, and C
Calculate the fitness of all search agents
```

Update X_α, X_β *and* X_c *t=t+1*

```
end while
```

return X_α.

The Pseudo code of MFO Algorithm

```
 Initialize a population of n flames positions randomly in the search space.
while (maximum number of iteration not reached)
          Update the number of flames (N_flames) to be used according to equation
(29).
          Calculate the fitness of all the n moths.
if first iteration then
             Sort the moths from best to worst according to their fitness and
place the result in
flame matrix.
else
              Merge the population of past moths and flames. Sort the merged
population from
best to worst.
              Select the best N positions from the sorted merged population as
the flames.
end if
          Calculate the convergence constant r.
for each Moth i  with i≤ n do
Calculate t as t = (r - 1) *rand + 1.
ifi≤ N   then
                     Update Mothi position according to Flamei using equation
(28).
else
                      Update Mothi position according to Flame (i=N_flames) using
equation (28).
end if
end for
end while
```

Chapter 10
Evolution of Islanding Detection Methods for Microgrid Systems

Mahiraj Singh Rawat
National Institute of Technology, Uttarakhand, India

Shelly Vadhera
National Institute of Technology, Kurukshetra, India

ABSTRACT

A microgrid is a representation of controllable electrical power network, which comprises the distributed generation (DG) such as micro-hydro, wind turbines, photovoltaic arrays, bio-mass generation, battery energy storage systems (BESSs), and a group of interconnected load. A typical microgrid can be operated in either grid connected or island mode. The issue of unintentional islanding in microgrid is a major challenge for power engineers as it may harm the service personnel working during maintenance or restoration of the main grid. In literature, various passive, active, and hybrid islanding detection methods (IDMs) have been developed and implemented by researchers/engineers. In this chapter, the proposed IDMs from the recent literature have been analyzed and briefly discussed. The primary requirement of IDMs is fast and accurate detection of main grid failure and should not deteriorate the power quality of distribution networks.

INTRODUCTION

Microgrid is a group of small, interconnected electrical network consisting of distributed energy resources (DERs), energy storage devices, load and have the capability to act as a single controllable power network (Farrokhabadi et al., 2020). A Microgrid can be interconnected at low (400 V) and medium voltage (69 KV) distribution networks. Also, it can be operated in both grid connected or island mode. The term islanding is defined as the phenomena in which both generation and connected load of mirogrid becomes isolated from the main utility grid. As shown in Figure 1., the microgrid is connected to the distribution network through point of common coupling (PCC) with the main grid. To ensure the microgrid operational stability and reliability, the primary and secondary controls are used. The main function

DOI: 10.4018/978-1-7998-4027-5.ch010

of controllers includes: (i) To maintain the equilibrium among power generations and connected loads (ii) keeping the microgrid voltage, frequency within desired limits (iii) demand side management and economic dispatch (iv) smooth switching between various mode of operation (v) islanding detection. The primary control includes the active and reactive power flows, control of voltage and frequency. The microgrid central controller (MGCC) facilitates a high level management of the microgrid operation and connected to the primary controllers through communication links (Justo et al., 2013).The distributed energy sources, i.e., solar photovoltaic (PV), wind energy, microturbines, energy storage system (ESS) etc. are connected to a distribution feeder through the power electronics converters (i.e., DC/AC, AC/DC, DC/DC). When microgrid connects to grid, the DER operates in a current control mode to regulate the active and reactive power exchange with the main grid. While during island operation, the microgrid drives in voltage control mode to regulate the voltage and to maintain balance among generations and load demands by generation curtailment or load shedding. According to IEEE 1547 standard, "*An island is defined as a condition in which a portion of an area electric power system (EPS) is energized solely by one or more local EPSs through the associated PCCs while that portion of the area EPS is electrically separated from the rest of the area EPS*" (IEEE Application Guide, 2009).

Figure 1. Typical configuration of Microgrid

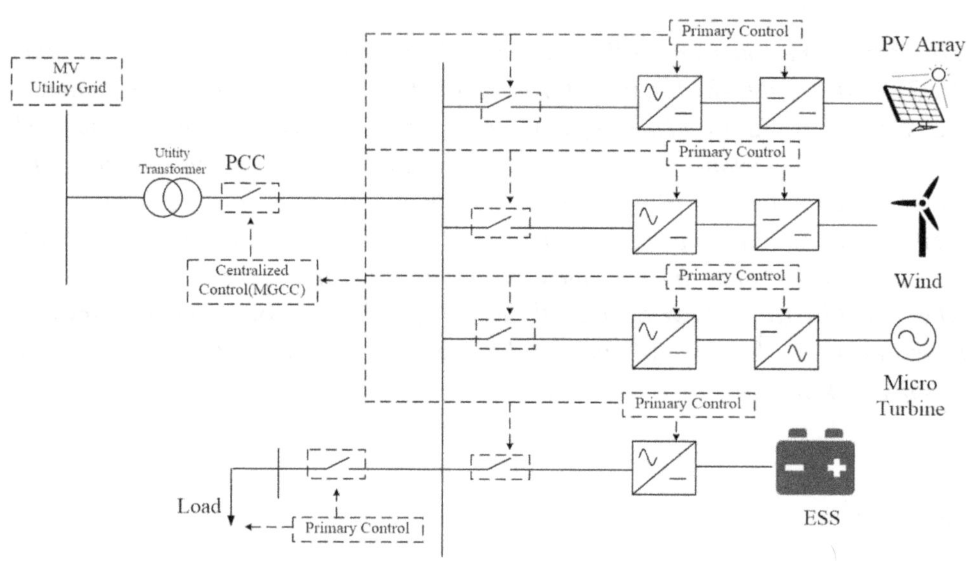

The unintentional islanding is an undesirable phenomena, it may cause of safety hazard to service personnel working on the main grid and can also lead to damage of utility equipment due to asynchronous reclosing of circuit breakers. The islanding operation includes two scenarios: (i) intentional/planned islanding (ii) unintentional/unplanned islanding. The intentional/planned island operation is done for maintenance purposes, long term voltage dips or general faults in the main grid. The unintentional/ unplanned islanded operation can lead to threat to the emergency service personnel, utility workers, and public because the DER represent an unknown source(Bevrani et al., 2014). According to IEEE 1547 standard, "*For an unintentional island in which the DR energizes a portion of the Area EPS through the*

PCC, the DR interconnection system shall detect the island and cease to energize the Area EPS within two seconds of the formation of an island". The unintentional islanding operation will cause the variations in frequency/voltage at the output of DGs e.g.,if the microgrid is importing power from utility grid, then at the event of islanding the microgrid frequency will decrease whereas if micrgrid exports power to utility grid, the frequency will increase.Therefore, the various international standards such as IEEE 929–2000, IEC 62116 and IEEE 1547 must be adopted for detection of islanding phenomena in typical microgrid (Table 1).

Table 1. A various standard of islanding detection

Parameter	IEEE Std. 929-2000		IEEE 1547		IEC 62116	
Frequency	Frequency	Detection time (Cycles)	Frequency	Detection time# (sec)	Frequency	Detection time (sec)
	f <60.5	6	59.3 Hz <*f*<60.5 Hz (DR£30kW)	0.16	*f* <58.5	1
	59.3<*f*	6	*f*<59.8 to 57 (DR>30kW) Adjustable set point	0.16 to 300	*f*> 61.5	1
	-	-	*f*<57 (DR>30kW)	0.16	-	-
Voltage	Voltage (% of the nominal voltage)	Detection time (Cycles)	Voltage (% of the nominal voltage)*	Detection time (sec)	Voltage (% of the nominal voltage)	Detection time (sec)
	V<60	10	V<50	0.16	V > 115	2
	60£V<106	120	50£V<88	2.0	V < 85	2
	132£V³165	120	110<V<120	1.0	-	-
	V³ 165	2	V³120	0.16	-	-
Quality factor	2.5		1		1	

DR£ 30 KW, Maximum detection time DR³ 30 KW, Default detection time

Nowadays, the power systems are reshaping into small microgrids integrated with renewable and non-renewable DGs to increase the system reliability. The islanding detection is a major issue in microgrid which causes protection and safety problems. In recent literature, the review of different islanding detection methods (IDMs) has been presented by various researchers (Li et al., 2014; Ahmad et al., 2013). The chapter briefly evaluates the reliability and accuracy of different IDMs proposed by researchers, IDMs performance parameters, various international standards on anti-islanding and future scope. Also, the chapter will help to researchers/scholars to gain the fundamental knowledge of islanding detection methods from bird's eye view. The issue of islanding has been observed and analyzed by various researchers and developed various islanding detection methods (IDMs). Primarily, the IDMs is classified as follows (i) Active methods (ii) Passive methods (iii) Remote methods (iv) hybrid methods. The brief classification has shown in Figure 2. The fundamental concept of passive IDM is based on the monitoring of DG's output parameters such as voltage, current, frequency, harmonics etc., at PCC under islanding. The passive methods exhibit large non detection zone (NDZ) characteristics. The NDZ represents the imbalance between the generated power from DG units and load demand. The passive methods fail to detect island situation when the load demand is balanced by the DGs power output. The active methods use the concept of perturb and observe where, a disturbance is inserted into the inverter's output of the DG in order to detect islanding. In grid connected mode, the injected disturbance does not affect the distortion in system parameters at PCC whereas a significant disturbance is observed during island mode. Usually, the active IDMs affect the network power quality. The remote method utilizes the existing communication channels such as supervisory control and data acquisition system(SCADA), microwave, telephone line, power line carrier communication (PLCC) and advanced signal processing equipment for islanding detection.The remote methods possess a high reliability and small NDZ compared to active and passive method. But due to high cost, difficult to implement and high complexity, these methods are not preferred. In recent year, several authors have proposed hybrid methods having advantages of both active and passive methods.

LITERATURE REVIEW

The islanding detection methods are classified into local and remote methods. The local methods are based on measurement of electrical parameters, i.e., voltage, current, frequency and phase at the point of common coupling (PCC) and are further divided into passive and active methods. The passive methods are simple and do not affect the power quality at PCC, whereas the active methods inject the disturbance at the inverter's output and the islanding event is detected by taking measurement of electrical parameters i.e., voltage, frequency and impedance at PCC. The remote methods are based on the communication infrastructure between main grid and microgrid. The remote methods require large amount of investment and are suitable for multi-inverter systems. The recently developed hybrid methods are based on passive and active methods and utilizes the benefits of both methods. The traditional passive methods proposed in literature are over/under voltage or over/under frequency (Ye et al., 2004), rate of change of frequency (Freitas et al., 2005), rate of change of frequency over power (Pai & Hung, 2001), voltage/current harmonic monitoring (Kim et al., 2019), rate of change of frequency over power (Li et al., 2014), voltage unbalance (Iong & Kim, 2004), phase jump detection (Singam & Hui, 2006). Recently, the improvement in traditional passive methods have been achieved by utilizing signal processing based methods (Kim et al., 2019). The researchers have utilized some of the signal processing-based methods such as Fourier transform (Hashemi et al., 2017), wavelet transform (Ray et al., 2012, Barros et al., 2006), time-frequency transform (Samantaray et al., 2011), and Kalman filter (Haider et al., 2018). The intelligent islanding detection methods associated with signal processing techniques include the artificial neural network (Hartmann et al., 2018), decision tree (El-Arroudi et al., 2007; Heidari et al., 2013), probabilistic neural network (Khamis et al., 2015; Samantaray et al., 2011), support vector machine (Alshareef et al., 2014),, fuzzy logic (Dash et al., 2012), and adaptive neuro-fuzzy inference system (Hashemi et al., 2013). The active methods use the concept of perturb and observe where, a disturbance is inserted into the inverter's output of the DG in order to detect islanding. A significant variation in the electrical parameters is observed during islanding condition, which triggers the relay to operate. The active methods may arise the power quality problems and hence degraded the performance of system. The active methods proposed in literature are active frequency drift (Lopes & Sun, 2006), frequency jump (Justo et al., 2013), active frequency drift with positive feedback (Ropp et al., 1999), Sandia frequency shift (Stevens et al., 2000), Sandia voltage shift (Liu et al., 2010), sliding mode frequency shift (Steven et al., 2000), active/reactive power variation (Trujillo et al., 2010), impedance measurement (Kane &

Figure 2. Classification of Islanding detection methods (IDMs)

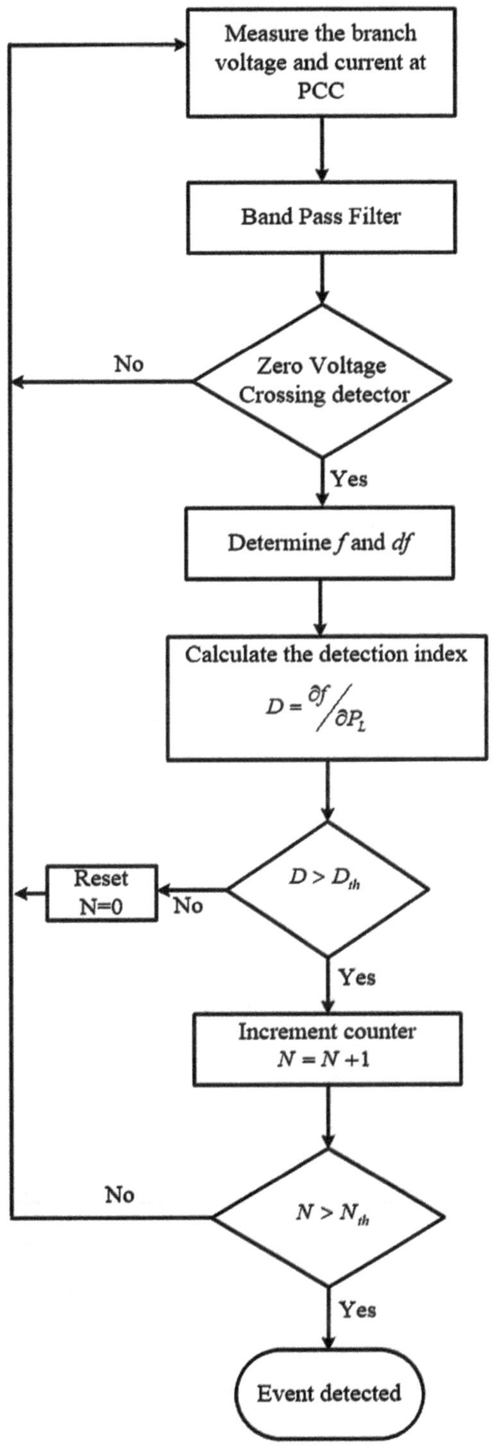

Fox, 1997), impedance detection at specific frequency (Li et al., 2014), reactive power error detection (Karimi et al., 2008). The remote methods possess a high reliability and small NDZ compared to active and passive method. The remote methods include power line carrier communication (Xu et al., 2007), signal produced by disconnect (Steven et al., 2000), and SCADA (Kim et al., 2019). Due to high cost and complexity, these methods are not preferred.

In literature, the hybrid methods are emerged as the most promising method for islanding detection in terms of negligible NDZ, high efficiency and improved reliability. The hybrid method merges the benefits of both active and passive methods. Recently, the hybrid methods i.e., SFS & ROCOF (Khodaparastan et al., 2017), VU/THD and BRPV (Wang et al., 2020), VU and frequency set point (Menon & Nehrir, 2007), Average rate of voltage change & real power shift of DGs (Mahat et al., 2009) have been proposed by researchers.

IDMs PERFORMANCE PARAMETERS

The accuracy and efficiency of any IDM depend upon the performance parameter, i.e., non-detection zone, error detection ratio, system power quality and detection time.

Non-Detection Zone

The non-detection zone (NDZ) is the region defined in power mismatch space (ΔP and ΔQ) under which IDMs fails to detect islanding. The NDZ region has shown in Figure 3 in which x and y axes are represented in terms ofchange in active power (ΔP) and reactive power (ΔQ)(Roop et al, 2000).

Figure 3. NDZ region for typical IDM

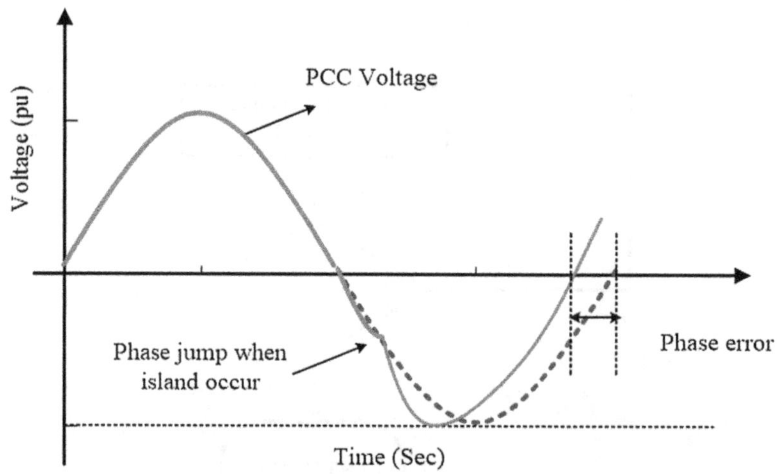

The range of power mismatch ΔP and ΔQ can be defined as follows:

$$\left(\frac{V}{V_{\max}}\right)^2 - 1 \le \frac{\Delta P}{P} \le \left(\frac{V}{V_{\min}}\right)^2 - 1 \tag{1}$$

$$Q_f\left[1 - \left(\frac{f}{f_{\min}}\right)^2\right] \le \frac{\Delta Q}{P} \le Q_f\left[1 + \left(\frac{f}{f_{\max}}\right)^2\right] \tag{2}$$

Where, f_{min} and f_{max} are minimum and maximum frequency respectively; V is the rated voltage; V_{min} and V_{max} are the minimum and maximum allowable voltage at PCC; P & Q are active and reactive power respectively; Q_f is the quality factor of RLC load. When the load demand becomes equal to the DGs output ($\Delta P = \Delta Q = 0$), the NDZ region becomes zero. Since, certain IDMs or combinations of methods have no NDZ, in such cases the NDZ is defined in load parameter space.

$$f_1\left(cf, G, Q_f\right) < \Delta C_{norm} < f_2\left(cf, G, Q_f\right) \tag{3}$$

Where, *cf* is the chopping fraction, *G* and Q_f represents the accelerating gain and quality factor respectively, C_{norm} is the normalized capacitance which is represented as the load capacitance as a fraction of capacitance that resonant with respect to load inductance at the resonant frequency.

$$C_{norm} = \frac{C}{C_{res}} = C\omega_0^{\,2}L \tag{4}$$

Detection Time

It is defined as the time elapsed between the instant the microgrid disconnecting fromthe main grid and the instant IDM have detected islanding(Kim et al. 2019). The detection time can be defined as follows:

$$\Delta t = t_{IDM} - t_{CB_{trip}} \tag{5}$$

Where, Δt defines the detection time, t_{IDM} is the instant IDM have detected islanding, t_{CBtrip} is the instant microgrid is disconnected from the main grid.

Error Detection Ratio (*EDR*)

Due to load switching or other disturbances, the monitored parameters at PCC exceeds beyond their set limit,which further causes false detection of islanding. The EDR is defined as the ration of number of time IDM detected false islanding to the total number to times islanding detected.

$$EDR = \frac{N_{false}}{N_{false} + N_{accurate}} \tag{6}$$

Power Quality

The active IDMs injects disturbances during islanding detection, which deteriorate the power quality of output power. Therefore, power quality is an important parameter while selecting IDMs. A power quality comparison of different active IDMs has been presented by Hamzeh & Mokhtari (2009).

PASSIVE METHODS

Over/Under Voltage (OUV) or Over/Under Frequency (OUF)

The standard (IEEE 1547 std.) microgrid system for performance checking of IDMs is shown in Figure 4.The load is represented as a parallel RLC circuit connected at PCC. The power flow from grid, DG, and to connected load are shown in Figure 5. In OUV/OUF technique, the inverters will be disconnected at PCC, if the observed voltage/frequency exceeds a predefined thresholds(Ye et. Al. 2004). After being islanded, the deviation in voltage/frequency occurs due to power mismatch between DGs output and load demand in microgrid. The power flow to the load can be represented as follows:

$$P_{load} = P_{DG} + \Delta P_{grid} \tag{7}$$

$$Q_{load} = Q_{DG} + \Delta Q_{grid} \tag{8}$$

The amplitude of load impedance and power factor angle can be given as follows:

$$\left|Z\right| = \frac{1}{\sqrt{\frac{1}{R^2} + \left(\frac{1}{\omega L} - \omega C\right)^2}} \tag{9}$$

$$\theta_{load} = \tan^{-1}\left[Q_f\left(\frac{f_0}{f} - \frac{f}{f_0}\right)\right] \tag{10}$$

Where, f_0 is the resonant frequency and Q_f represents the quality factor. The parameterQ_f and f_0 can be solved using Equation (11) and (12) respectively.

$$Q_f = R\sqrt{\frac{C}{L}} \tag{11}$$

$$f_0 = \frac{1}{2\pi\sqrt{LC}} \tag{12}$$

Figure 4. A standard microgrid system for testing IDMs

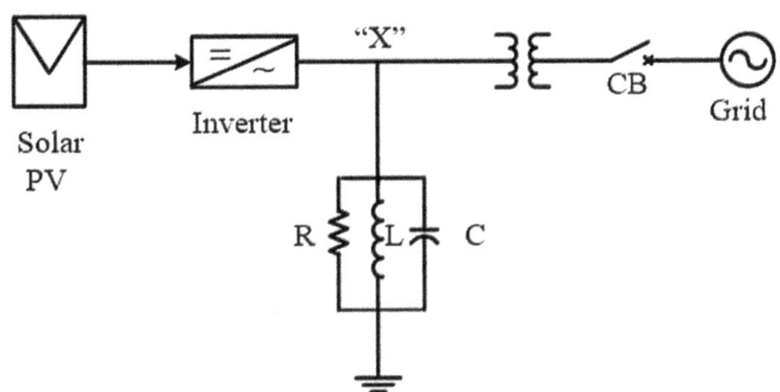

Figure 5. Equivalent power flow representation of microgrid

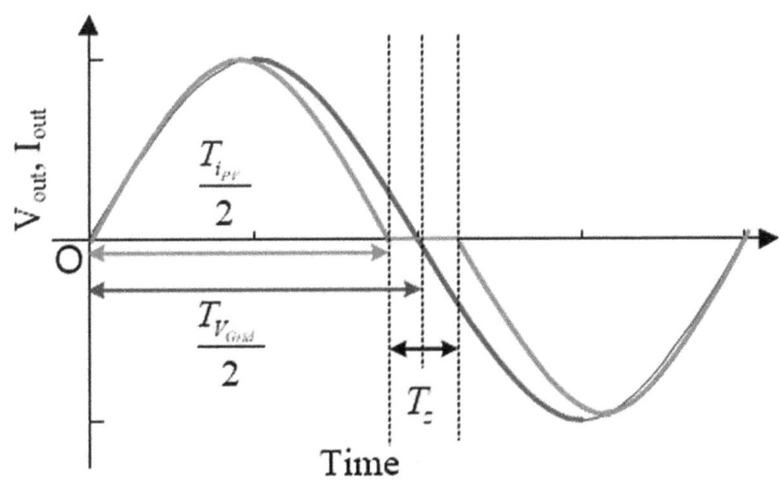

The NDZ for reactive power can be defined by Equation (13).

$$Q_f\left[1-\left(\frac{f}{f_{\min}}\right)^2\right] \le \frac{\Delta Q}{P} \le Q_f\left[1+\left(\frac{f}{f_{\max}}\right)^2\right] \tag{13}$$

For IEEE 1547 standard f_{min}=59.3 Hz and f_{max} = 60.5 Hz and Q_f = 1

$$-2.37\% \le \frac{\Delta Q}{P} \le 1.65\% \tag{14}$$

The NDZ foractive power can be represented by Equation (15).

$$\left(\frac{V}{V_{\max}}\right)^2 - 1 \le \frac{\Delta P}{P} \le \left(\frac{V}{V_{\min}}\right)^2 - 1 \quad (15)$$

For IEEE 1547 V_{min} 88% and V_{max} 110%

$$-17.36\% \le \frac{\Delta P}{P} \le 29.13\% \quad (16)$$

This method have the advantage of low cost and do not influence the system power quality. The large NDZ and difficulty in prediction of detection time are the demerits with this method. The detection time varies from 4 ms to 2 sec even some times more than 2 sec. The detection time depends upon the power imbalance between DGs output and load demand. The NDZ region for OUV/OUF is shown in Figure 6.

Figure 6. NDZ of OUV and OUF method

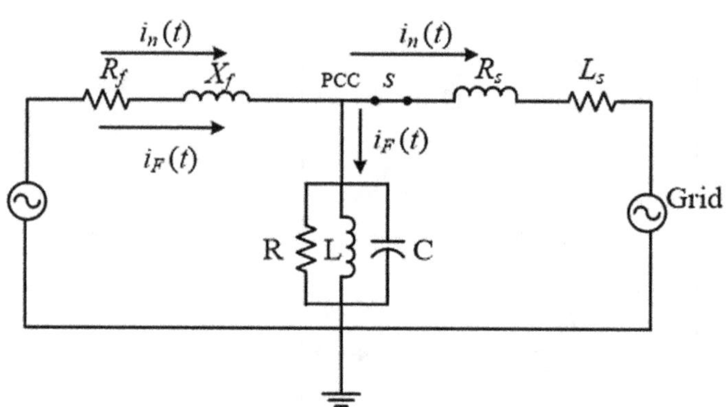

Rate of Change of Frequency (ROCOF)

In this method, the rate of change in frequency (i.e., *df/dt*) over a few (2-50) cycles at PCC is monitored to track the state of island in microgrid. The inverter of distributed generator should be shut down, if *df/dt* exceeds the pre-set or threshold value.The ROCOF technique is much faster and reliable compared to over/under voltage or over/under frequency based passive technique. The detection time for ROCOF technique is up to 24 ms. For small and medium scale DG units, the *df/dt* settings with 0.3 Hz/sec has found to be the optimum value with 0.3 - 0.7 sec operating time. The ROCOF based technique is sensitive to load switching and fluctuations which may lead to erroneous results. Therefore, it is difficult to predict whether islanding occurs due to grid disconnect or load switching. Moreover, the setting of threshold values is also difficult. The implementation of ROCOF relay is shown in Figure 7(Freitas et al. 2005).

Figure 7. Implementation of ROCOF Relay

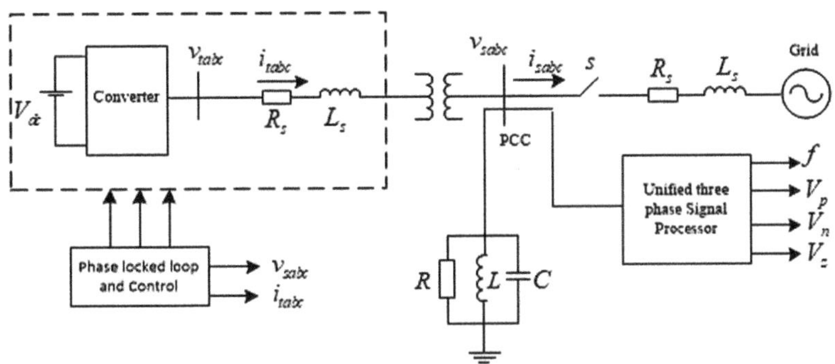

The system frequency is determined from measuring the speed ω_r (rad/s) of electrical generator. The obtained frequency is processed through derivative block and filters. The filter block is represented by a first order transfer function ($1/1+T_a s$) with Ta time constant. Usually, Ta can be set to few (5-6) cycles. If the *df/dt* (Hz/s) is greater than the relay setting and the voltage is also greater than the minimumvalue, the relay will send a trip signal to circuit breaker at PCC. The detection time of ROCOF relay depends upon the active power imbalance (Figure 8). The term active power imbalance represents the difference between the generation and load demand at pre-island condition. The detection time of relay increases as the active power imbalance decreases and it change abruptly as the power imbalance approaches towards zero. The variation in frequency with time becomes negligible at this point. In ROCOF relay, the maximum operating detection time is defined for a minimum active power imbalance also called critical active power imbalance. The detection time of ROCOF relay is independent of reactive power imbalance in the system. It was observed that the operation of ROCOF relay also depends upon the following system characteristics.

- Generator inertia constant
- Voltage dependent loads
- System loading
- Electrical distance between the generator and fault location
- Fault clearing time
- Substation short circuit level
- Feeder X/R ratio

Rate of Change of Frequency Over Power (ROCOFOP)

In this method, the index $\partial f / \partial P_L$ is measured at PCC to identify the state of islanding in microgrid. Let us consider a grid connected microgrid system linked through a tie line. (Figure 9). If utility circuit breaker is open state, the system X and Y becomes two isolatedsystems. Let, the power imbalance (∂P) due to change in generation (∂P_G) and sudden load increase (∂P_L) is represented by Equation (17).

$$\partial P = \partial P_G - \partial P_L \qquad (17)$$

Figure 8. Islanding detection time vs Active power imbalance

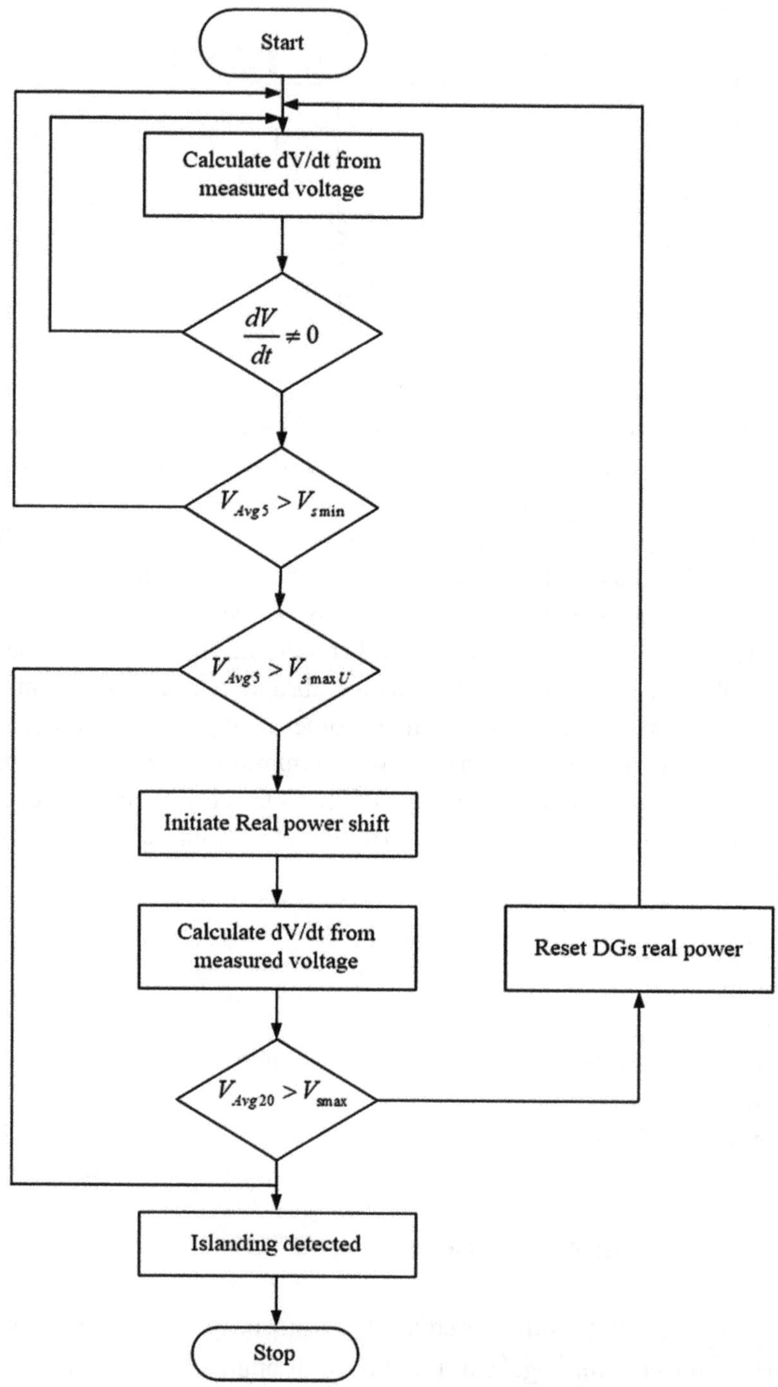

Figure 9.Grid connected to microgrid system

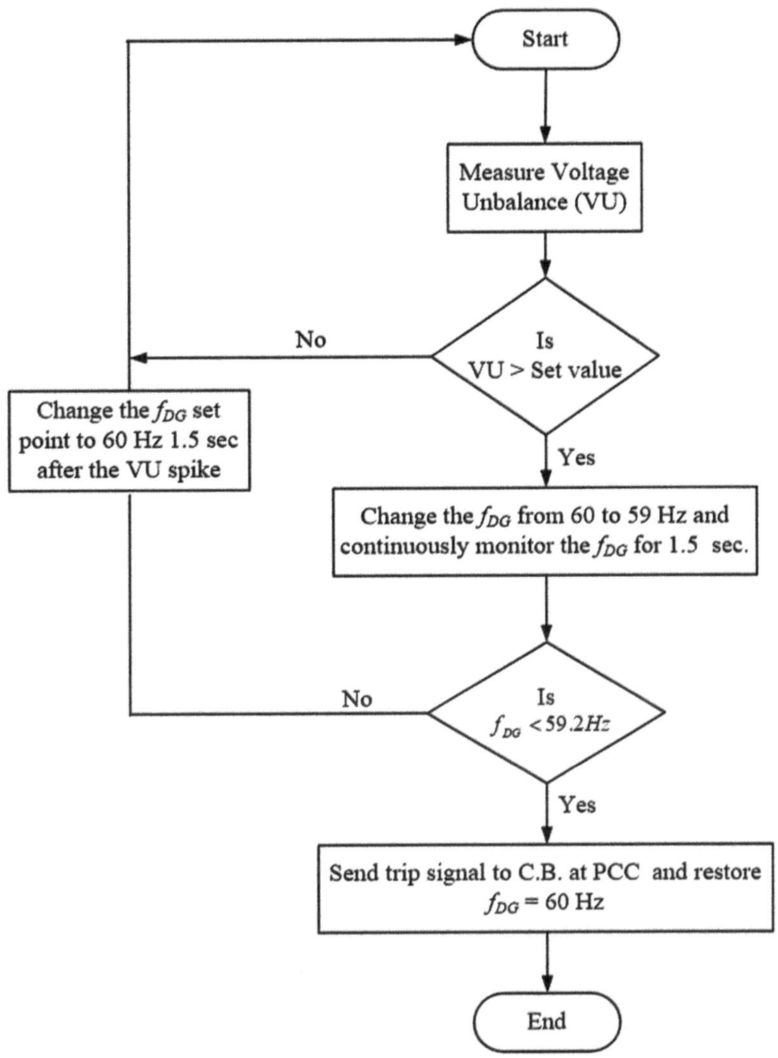

For system X and Y, the power frequency characteristics are defined as follows:

$$K_X = \frac{\partial f_X}{\partial P_X},\ K_Y = \frac{\partial f_Y}{\partial P_Y} \tag{18}$$

Where, ∂P_X and ∂P_Y represents unbalance active power of system$X$ and system$Y$; f_X and f_Y are frequency of system X and system Y. The K_X and K_Y are inversely proportional to system generation capacity(Pai & Hung, 2001). Let, the power transferred from system X to Y is ∂P_t for an additional load demand variation of ∂P_L in system Y, which is represented by Equation (19).

$$\frac{\partial P_Y}{\partial f_Y} = \frac{\partial P_t}{\partial f_Y} - \frac{\partial P_L}{\partial f_Y} \tag{19}$$

$$\frac{\partial P_X}{\partial f_X} = -\frac{\partial P_t}{\partial f_X} \tag{20}$$

Where, the load variation of ∂P_L in system Y does not affect the output power of dispersed generation. Now, assume the frequency change in each subsystem must be equal and can be represented by Equation (21).

$$\partial f_X = \partial f_Y \tag{21}$$

Substituting Equations (19) and (20) in Equation (21) results into following:

$$\frac{\partial P_Y}{\partial f_Y} = -\left(\frac{\partial P_X}{\partial f_X} + \frac{\partial P_L}{\partial f_Y}\right) \tag{22}$$

From Equations (18) and (22)

$$\frac{\partial f_Y}{\partial P_L} = \frac{-K_X K_Y}{K_X + K_Y} \tag{23}$$

Equation (22) represents the frequency variation with respect to load change in system Y when DGs are connected to the grid. Now, let us consider the system X and Y are isolated and a load change ∂P_L in system Y occurs which is represented as follows:

$$\partial P_Y = \partial P_L \tag{24}$$

Substituting Equation (24) into Eq. (18) results into following:

$$\frac{\partial f_Y}{\partial P_L} = K_Y \tag{25}$$

From Equations (23) and (25), it is observed that change in ∂P_L in system Y have different values of $\partial f_Y / \partial P_L$ during grid and island mode of system Y. The implementation of ROCOFOP method is shown in Figure 10.

Figure 10. Flow chart for ROCOFP method

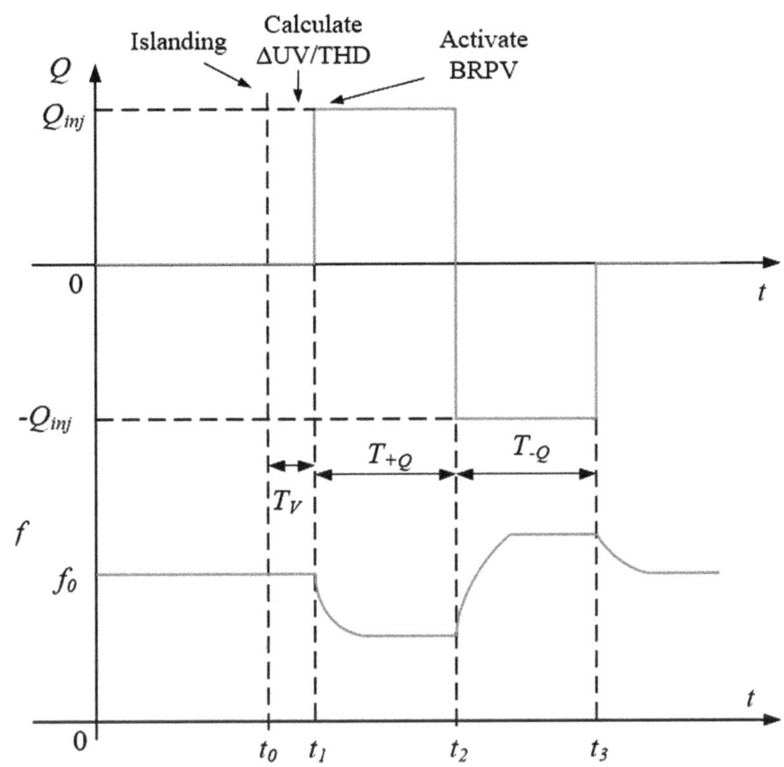

The voltage and current at PCC is measured through the current transformer (CT) and potential transformer (PT). A band pass filter is used to separate the disturbance in the measured data. The output dataare followed by a zero crossing detector which measure the frequency *f*. The computation of df/dP_L can be represented as detection index D. Further, this index is compared from the threshold detection index, D_{th}. If detection index D is greater than D_{th}, then the counter is triggered and incremented by one. The islanded mode will be detected if detection index D is greater than D_{th} and the counter is greater than number threshold N_{th}. The reasonable value of threshold N_{th} was found 4. After detecting the island mode, a trip signal is given to circuit breaker to separate the connected DG from the utility grid. The islanding detection time for the method is around 100 ms.

Voltage/Current Harmonic Monitoring

The passivemethod monitors the total harmonic distortion (THD) in the output current of the DG's inverter at PCC. The island state is detected when the THD exceeds a threshold value. During grid connected mode, the low impedance of grid allows the current harmonics generated from DGs inverters to transmit to the grid without voltage defromation at PCC.However, during island mode, the current harmonics generated from DGs inverter flows to the load and produces harmonic distortion at PCC. Let at time *t*, the THD of inverter's output current is given by Equation (26).

$$THD_t = \frac{\sqrt{\sum_{h=2}^{H} I_h^2}}{I_1} \times 100 \tag{26}$$

Where, I_h represents the RMS value of harmonic component h in current and I_1 is fundamental component (RMS value).

$$THD_{avg,t} = \frac{1}{N}\sum_{i=0}^{N-1} THD_{t-i} \tag{27}$$

Where, the term N represents the samples over one cycle. The ΔTHD from steady state value under normal loadingat any time t can be defined by Equation (28).

$$\Delta THD_t = \frac{THD_{avg,ss} - THD_{avg,t}}{THD_{avg,ss}} \times 100 \tag{28}$$

$$THD_{DE,t} = THD_{avg,t} - THD_{avg,t-p} \tag{29}$$

Where, $THD_{avg,ss}$ represents the reference value of THD under steady-state and nominal loading. After setting the $THD_{avg,ss}$, if ΔTHD_t lies between -100% to 75% over cycle, the $THD_{avg,ss}$ is updated by $THD_{avg,t}$ in order to adopt normal loading condition. To avoid maloperation during short transient period, if over ¼ cycle (4.17 ms) as defined in Equation (27), the value of $THD_{avg,t}$ goes above 0.1%, the passive method rejects the value and will move to next time step.

The detection time for this passive method is around 45 ms. The advantage is that the effectiveness of this approach does not affected by multiple DGs connected to same PCC. The disadvantage of this method is that the difficulty in calculating threshold, sensitive to network disturbances.

Rate of Change of Power Output (ROCOP)

At the event of loss of main grid, a sudden load change leads to the change in output power of DG in microgrid. In ROCOP, the rate of change in output power (*dP/dt*) is measured over few cycle(Pai & Huang 2001). If the measured quantity (*dP/dt*) exceeds a set threshold, the output power from DGwill cease to the load. The method possesses NDZ in power balance condition andbecomes more effective under increasing of power imbalance. Compared to OUV/OUF based passive method, this method have fast response and unaffected by the power imbalance between DG's output powerand load demand. The approximate islanding detection time for ROCOP method is about 24 ms to 26 ms(Redfern et al. 1993).

Voltage Unbalance

During the loss of the main grid, the DGs installed in distribution network should supply the load demands, therefore the loading of DG will changed abruptly after islanding. Normally, a large percentage of the loads connected in distribution network are single phase loads, therefore there will be the possibility of unbalance in supply voltage of DGs. It is possible to effectively detect the islanding condition of DGs, if the unbalance voltage (UV) of DG output voltage exceed the threshold setting(Iong & Kim 2004).The VU at instant *t* can be calculated by Equation (30).

$$UV_t = \frac{V_{NS_t}}{V_{PS_t}} \times 100 \tag{30}$$

Where V_{PSt} and V_{NSt} represents the magnitudes of positive and negative sequence voltage at instant t. The voltage unbalance over one cycle is defined by Equation (31).

$$UV_{avg,t} = \frac{1}{N}\sum_{i=0}^{N-1} UV_{t-i} \tag{31}$$

Where, N represents the sample numbers over one cycle and *t* indicate themonitoring time. The deviation of voltage unbalance under the steady-state and nominal loads is defined as follows:

$$\Delta UV_t = \frac{UV_{avg,s} - UV_{avg,t}}{UV_{avg}} \times 100 \tag{32}$$

Where, $UV_{avg,s}$ represents the unbalance voltage reference value at steady-state under nominal loads. After setting the $UV_{avg,s}$, if ΔUV_t lies between -100% to 50% over cycle, the $UV_{avg,s}$ is updated by $UV_{avg,t}$ in order to adopt normal loading condition. To avoid maloperation during transient states, if over ¼ cycle (4.17 ms) as defined in Equation (32), the value of $UV_{avg,t}$ goes above 0.05, VU method remove the value and moves to next time step.

$$UV_{DE,t} = UV_{avg,t} - UV_{avg,t-p} \tag{33}$$

The islanding detection time with VU method is about 53 ms. The disadvantage of this method is that the difficulty in calculating threshold, insensitive to network disturbances and extraction of negative sequence component is affected by harmonics.

Phase Jump Detection (PJD)

To detect islanding, the method monitors the phase difference between the voltage and current at the terminal of inverter. Under normal operation of microgrid, the phase locked loop (PLL) detects the zero crossing of voltage signal and synchronized the inverter current with the voltage (Singam & Hui 2006). Ifislanding occurs, the inverter current remains unaffected, but due to the load phase angle, the phase jump will be observed at inverter output voltage. Figure 11 shows the new phase jump of voltage at PCC as phase angle of load remains unaltered. Moreover, a phase error will be monitored between PCC voltage and inverter current at the next zero crossing of the voltage. If the phase error exceeds the set threshold, the inverter ceases its operate.

The advantages of this method are its simplicity, fast detection speed and have no effect on power quality. The islanding detection time for phase jump method is about 10 to 20 ms. The setting of threshold value is difficult in this method.

Figure 11. Operation of phase jump detection method

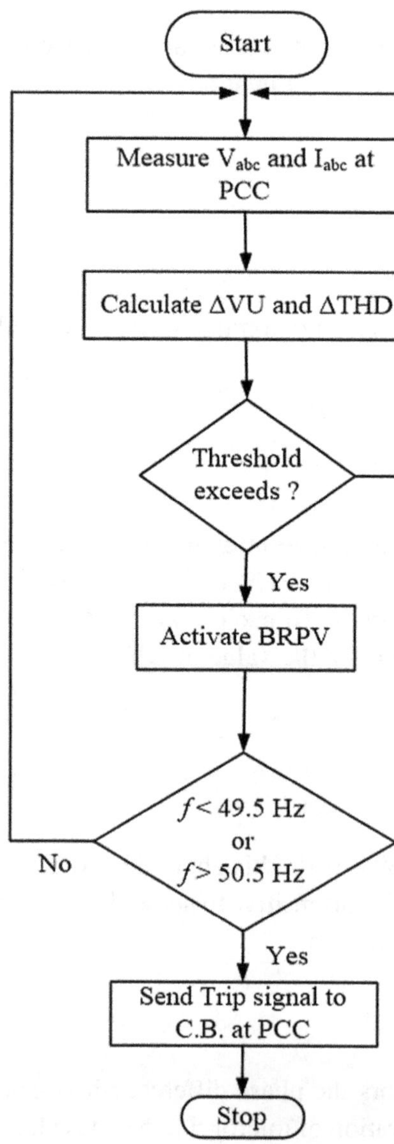

ACTIVE METHODS

These methods are based on the concept of perturb and observe techniques in which a disturbance is intentionally injected at inverter's output and a variation in electrical parameters e.g, voltage, frequency and current at PCC shall be observed.

Active Frequency Drift (AFD)

In theAFD method, the inverter injects a slightly distorted current at PCC. This method is also known as frequency bias method because it triggers the islanding detection by forcing the frequency of voltage at PCCto shift up or down(Lopes & Sun 2006). The Figure 12 shows a PV system with parallel RLC load in grid connected mode. The active and reactive power consumed by the load is represented by the following:

$$P_{load} = \frac{V_x^2}{R} \tag{34}$$

$$Q_{load} = V_x^2\left(\frac{1}{\omega L} - \omega C\right) \tag{35}$$

At the time of circuit breaker (C.B.) opening, if the active power output of the PV unit is not matched with the active power load demand, then the voltage at node "X" must be increased or decrease until P_{load} becomes equal to P_{PV}. Similarly, if reactive power mismatch is found between generation from PV system and load demand, then the frequency of the voltage at node "X" must change until $Q_{load}=Q_{PV}$. The PV system search for a frequency where the power factor of the connected load becomes equal to the PV system. Under unity power factor (u.p.f) operation of PV unit, the condition at which $Q_{load}=Q_{PV}$ is given by Equation (36).

$$\arg\left\{\frac{1}{R} + \frac{1}{j\omega L} + j\omega C\right\}^{-1} = 0 \tag{36}$$

In grid connected mode, the frequency ofthe injected current and voltage are remainconstant irrespective of disturbance. Whereas under island mode, due to distortion in injected current waveform at node "X", the zero crossing is observed before the actual time in voltage waveform. Therefore, a phase error will be observed between voltageand inverter output current.The distortion in inverter output current is described by the parameter chopping fraction, which can be defined by Equation (37).

$$Cf = \frac{2T_z}{T_{Vgrid}} \tag{37}$$

Where, C_f is the chopping frequency and defined as the ratio of the dead time T_z to half of the period of the voltage waveform. T_{Vgrid} is the time period of utility grid voltage and T_{ipv} is the period of output injected current form inverter of PV system. From Figure 13, it is observed that during the first half cycle, the inverter output current frequency is slightly higher than the utility grid voltage and the current remains at zero for a time period of Tz before the start of the second half cycle. The effectiveness of the AFD method depends largelyto the load parameter.For loadother than resistive load, the NDZ and detection time increases for large value of quality factor (Q). This method will not work properly if multiple inverters are connected in parallel due to deviation of frequency bias.

Figure 12. Photovoltaic (PV) unit integrated with grid

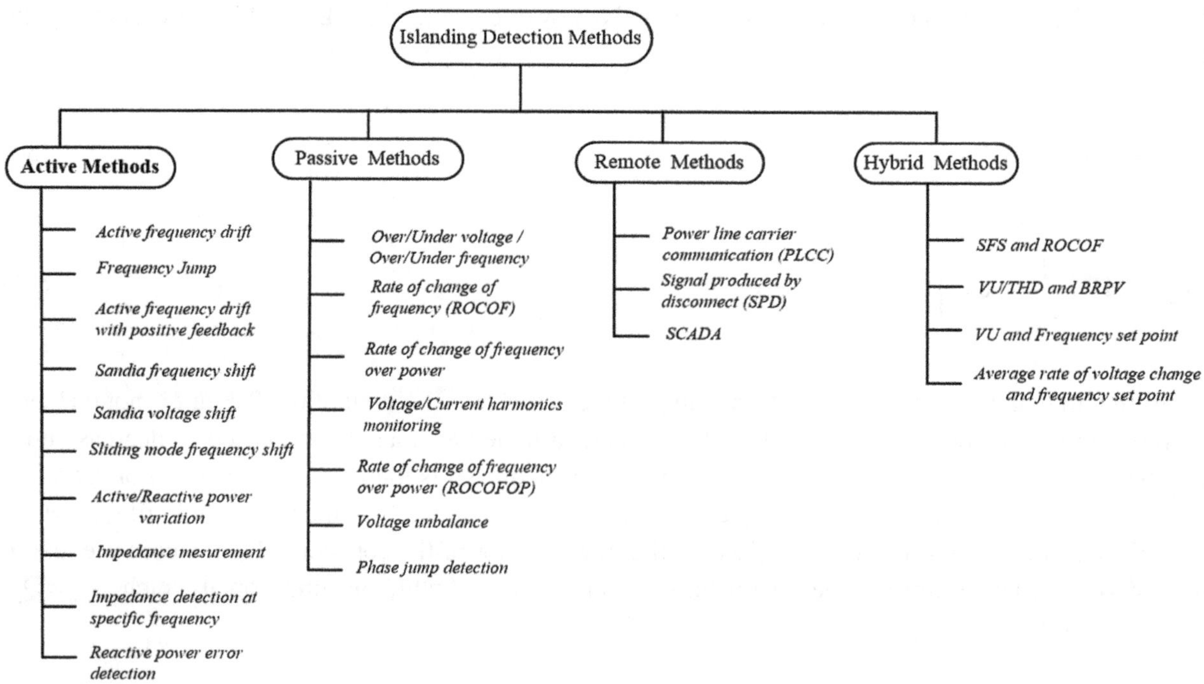

Figure 13. Implementation of AFD method

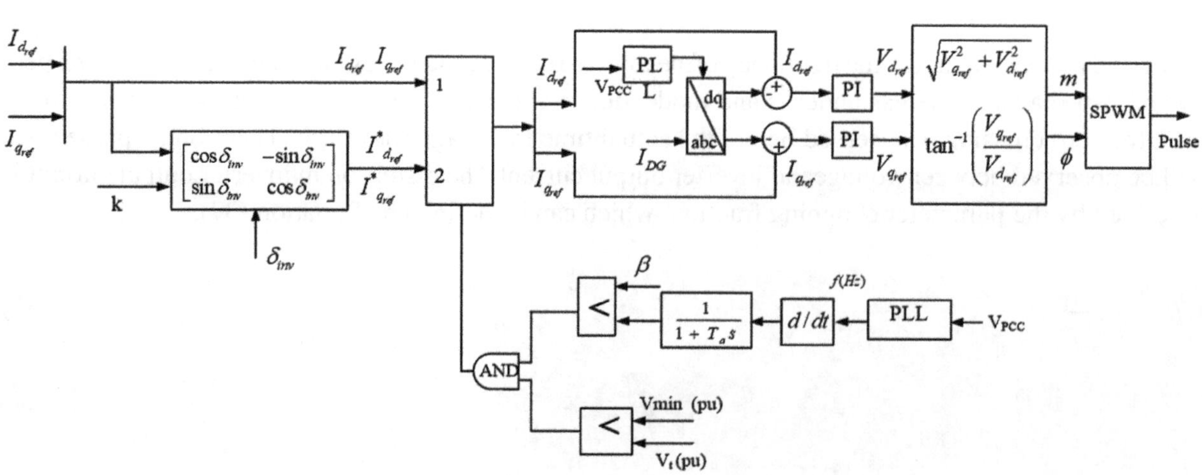

Active Frequency Drift With Positive Feedback (AFDPF)

The AFDPF method uses a positive feedback which increases the chopping fraction (*Cf*) with increasing frequency deviation from the nominal value(Ropp et al. 1999). The modified value of chopping fraction is represented as follows:

$$Cf_i = Cf_{i-1} + F\left(\omega_{i-1} - \omega_0\right) \tag{38}$$

Where, Cf_i and Cf_{i-1} represents the chopping frequency at i^{th} and $i\text{-}1^{th}$ cycle; ω_{i-1} is the angular frequency at $i\text{-}1^{th}$ cycle; F represents a linear function which maps sampled frequency error ($\Delta\omega_i = \omega_{i-1} - \omega_0$) on the real number.The AFDPF method prevents levelling off of the frequency until much larger frequency errors are detected which further reduces the size of NDZ compared to AFD method. The value of *Cf*in AFDPF can also be negative if the frequency error is negative.The AFDPF method shows improved performance over AFD method if multiple inverter is connected into the microgrid.The disadvantage of this method is that it introduces subharmonic in inverters output due to noise or insufficient sample rate.

Frequency Jump

The frequency jump (FJ) method is the modification of AFD method.The method injects a dead zone in every 3 cycles ofthe inverter's output current (Justo et al. 2013).During grid connected, the voltage at PCC will not be distorted inspite of distortion of inverter current. During disconnection from the main grid the change in frequency of the voltage at PCC will be used to detect islanding. The effectiveness of FJ method decreases with multiple inverters connected in parallel at PCC.

Sandia Frequency Shift (SFS)

The SFS method for islanding detection was developed by Sandia National Laboratories, USA. The SFS method is a modification of frequency bias/AFD method, which also utilizes the positive feedback in order to detect islanding. In SFS method, a positive feedback is applied to the frequency of the voltage at node "X" (Figure. 11).Equation (39) represents the implementation ofpositive feedbackfor SFS method (Reis et al. 2015).

$$Cf = Cf_0 + k\left(f_x - f_{grid}\right) \tag{39}$$

Where, Cf_0 is the chopping frequency at no error in line frequency; f_x and f_{grid} is the frequency at node "X" and grid respectively. The term "*k*" represents the accelerating gain. In a grid connected mode, when the minor frequency error is observed the method increases the chopping frequency but grid stability will prevent it. Whereas, during islanded operation, the frequency error increases, which led to increases in chopping frequency and the inverter, thus acts to strengthen the frequency deviation (Stevens et al. 2000). The recursive procedure continues till the frequency becomes equal to the set threshold of over frequency relay. The SFS method is easy to implement and have smallest NDZ among other active islanding methods. This method slightly affects the power quality and can cause undesirable transient behavior in the system when a weak utility grid is connected.

Sandia Voltage Shift

This active islanding method is similar to SFS method. In SVS method, a positive feedback is applied to the magnitude of voltage at node "X". If the voltage magnitude decreases at node "X", the PV inverter reduces its output current which further reduces the power output. Under grid connection, there will be no change whenever power is reduced.

During grid disconnection, the decrease in output current of inverter causes further reduction in voltage amplitude (V_X).This further reduction in voltage magnitude leads to further reduction in inverter output currentand this process is again repeated. When the voltage magnitude becomes less than a threshold value, the under voltage relay sends a trip signal to CB at utility side. In this method, the inverter output current magnitude can be increased or decreased, leading to the corresponding use of over or under voltage relay (Liu et al. 2010).

The SVS method can be implemented easily with microcontroller based inverter. The SVS method is normally implemented with SFS which increases the effectiveness of the method during islanding detection. The demerit of SVS method is that the power quality of inverter's output degraded due to positive feedback. The small reduction in operating efficiency of inverters is also observed. The method is also insensitive to grid transient response and power quality. Moreover, the load quality factor has negligible effect on operation of SVS method.

Sliding Mode Frequency Shift

In sliding mode frequency shift method, the phase angle of inverter's output current is controlled as a function frequency of the voltage at PCC (Stevens et al. 2000). The inverter output current (i_{inv}) is represented as follows:

$$i_{inv} = I \sin\left(\omega t + \theta_{inv}\right) \tag{40}$$

Where,$\omega = 2\pi f$ and f represents the frequency of voltage at PCC; θ_{inv} is the phase angle of inverter's output current. The phase angle of inverter's output current is represented by Equation (41).

$$\theta_{inv} = \frac{2\pi}{360}\theta_m \sin\left(\frac{\pi}{2}\frac{f - f_g}{f_m - f_g}\right) \tag{41}$$

Where, f_g is the grid frequency,θ_m is the maximum value of phase angle at frequency f_m; f represents the frequency of the previous cycle. Under grid connected mode, the phase angle of inverter's output current and voltage at PCC is almost zero. Under island mode of operation, thefrequency of the voltage at PCC will vary and if thefrequency variation exceeds a set threshold, the islanding will be detected. The islanding detection time is about 0.4 sec. The methodperformance for multiple inverter connected power system is also high. However,the method reduces the power quality and also affect the grid transient stability.

Active/Reactive Power Variations

In this method, the inverter injects the power variations in the output and monitors the voltage magnitude and frequency at PCC. Under grid connected mode, the injected power flows into the grid due to the low impedance of grid. However, in island mode of operation, the injected power flows into the load and produce voltage and current variations at the PCC (Trujillo et al. 2010). The variations in voltage with respect to change in active power at linear load RLC can be represented as follows:

$$P_{DG} = P_{load} = \frac{V^2}{R} \tag{42}$$

Where, P_{DG} represents distributed generated power; P_{load} represents the load demand and V is the voltage at PCC. Differentiating P_{DG} with respect to V can be represented by Equation (43).

$$\frac{dP_{DG}}{dV} = 2.\frac{V}{R} = 2.\sqrt{\frac{P_{DG}}{R}} = K_V \tag{43}$$

Thus the relationship between voltage and power variation can be represented as follows:

$$\Delta V = \frac{\Delta P_{DG}}{2}\sqrt{\frac{R}{P_{DG}}} \tag{44}$$

From equation (), the ΔV is directly proportional to ΔP_{DG} as R and P_{DG} are constant. Therefore the voltage at PCC can be varied directly by varying injected power from the inverter. Hence, if the voltage reaches the set threshold, the relay will send the trip signal to CB at PCC. The detection time of the method can be adjusted with factor Kv. The initial value of factor Kv is represented as follows:

$$K_{v0} = 2\sqrt{\frac{P_{DG}}{R}} \tag{45}$$

The reference value of current in current loop of inverter can be represented as follows:

$$I_{ref} = \frac{K_v\left(V - V_{grid}\right) + P_{DG}}{V} \tag{46}$$

Where, V and V_{grid} represents the amplitude of measured actual value and grid nominal voltage. Similarly, the relationship between reactive power and frequency exists. As per standard 929-2000, when DG is operated at unity p. f., the generated reactive power must set equal to zero at nominal condition.

$$Q_{DG} = Q_{Load} = 0 = V^2.\left(\frac{1}{\omega L} - \omega C\right) \tag{47}$$

$$\frac{Q-0}{\omega - \omega_0} = \left.\frac{dQ}{d\omega}\right|_{\omega=\omega_0} \tag{48}$$

Where, ω_0 is the resonant frequency. Solving the Equaions (48), the variation of frequency at resonant frequency as a function of ΔQ can be represented by Equation (49).

$$\Delta\omega = -V^2\left(\frac{\omega_0^2 L}{1+\omega_0^2 L}\right)\Delta Q \tag{49}$$

In terms of quality factor Q_f=$RC\omega_0$ and DGs active power generation (P_{DG}), the Equation (49) can be represented by Equation (50).

$$\Delta f = \frac{f_0}{2P_{DG}Q_f}\Delta Q \tag{50}$$

The reference frequency for the inverter current control loop can be calculated as follows:

$$f_{ref} = f_{grid} + K_f\left(f - f_{grid}\right) \tag{51}$$

Where,f_{grid} and f refers to nominal grid frequency and measured frequency at PCC respectively, the factor K_f represents the acceleration time constant. The demerits of the method is that it will maloperate when multiple inverters are integrated to PCC.

Impedance Measurement

In this method, the perturbation in current amplitude at inverter output is used to detect islanding event. In grid connected mode, the perturbation in voltage magnitude that results from perturbation in amplitude of output inverter current, which leads to perturbation in power can be represented as follows:

$$\Delta V = \frac{\Delta P}{2}\sqrt{\frac{R}{P}} \tag{52}$$

The equivalent impedance(dV_x/di_{pv}) can be detemine from the monitored voltage and current perturbations and can be further used to detect islandingevent (Kane & Fox 1997).The islanding detection time for the impedance measurement method is around 0.77 - 0.95 sec. This method shows small NDZ for single PV inverter and local loads having a larger impedance than the grid impedance. The demerits of this method is that the effectiveness of method decreases in multiple inverter cases.

Impedance Detection at Specific Frequency

The methods injects current harmonics of specified frequency at the point of common coupling via PV inverter and is a specific case of voltage/current harmonics monitoring, a passive method. Since, the utility impedance has lower impedance thanthe local load impedance, the injected current harmonics will flow into the grid without developing the abnormal voltage at PCC under grid connected mode(Li et al. 2014).

However, under island mode of operation, the injected current harmonics at specific frequency will flow into the linear RLC load and generates the harmonics in voltage proportional to current harmonic frequency flows into the load impedance. This generated harmonics in voltage can be detected by a relay. The advantages and disadvantages of this method are similar to voltage/current harmonic monitoring method. The method fails, if multiple inverters are connected at PCC.

Negative Sequence Current Injection

The RLC load draws a fundamental component of a balanced positive sequence current $i_F(t)$ supplied by inverter at unity p.f.. If a negative sequence current component$i_n(t)$ is injected by the controller of inverter, it will absorb by the grid due to its low impedance. However, when the grid is disconnected, the injected current component will flow into load resistor and results into positive and negative sequencevoltages across the resistive load at PCC. Therefore, under island mode, the net voltage is an unbalance voltage at PCC. If the unbalance voltage reaches to a set threshold than islanding will be detected(Karimi et al. 2008). The voltage unbalance can be calculate by Equation (53).

$$VI = \frac{V_n}{V_P} \times 100\% \tag{53}$$

Where, V_n and V_p are the amplitude of negative and positive sequence voltages respectively. Since,V_n and V_p are proportional to corresponding negative and positive sequence currents. Hence, the Equation (53) can be represented by Equation (54).

$$VI = \frac{V_n}{V_p} \times 100\% = \frac{I_n}{I_F} \times 100\% \tag{54}$$

Figure 14. Schematic diagram illustrating positive and negative sequence current injection.

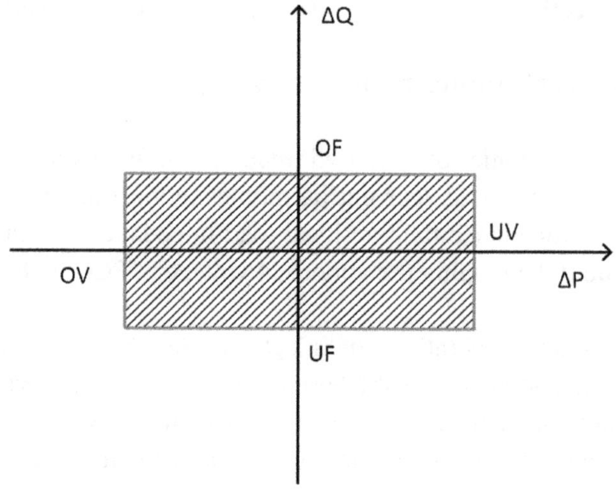

Figure 15. Schemetic diagram of anti islanding test

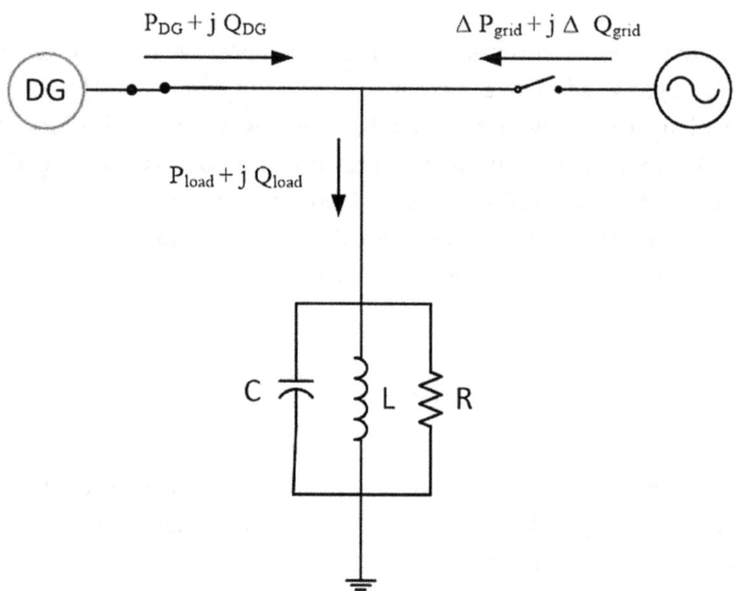

REMOTE METHOD

Power Line Carrier Communication (PLCC)

In this remote method, the power line carrier communication(PLCC) is used as the communication channel for detecting islanding event. A signal transmitter and its receiver are positioned at grid and PCC respectively. When connectedwith the grid, the receiver gets a continuous signal from the transmitter. However, if PLCC signals gets discontinued for three consecutive cycles, it indicates that the microgrid is disconnected from the utility grid(Kan & Fox, 1997). The PLCC receiver can be of different types such that it can be included within the inverter or it can be separated from the inverter but must be connected with some communication medium or else it separate circuit interrupter and remains detached with inverter. In order to make this method more effective, the PLCC method should have the following characteristics (Xu et al., 2007):

- The PLCC signal must be sent from the utility grid to the receiver at PCC. The existing utility PLCC signal that is to be utilized for the purpose of load shedding and harmonic detection needs to be sent from customers to the utility grid.
- The signal should be carried by a continuous carrier signal otherwise it will be difficult to distinguish that the signal loss is dueto grid disconnection or termination of transmitting signal.
- The signal must be the capability to propagate throughout the distribution network.Thehigher frequency signals are blocked by the transformer inductance, therefore low frequency signal should be used for communications.

Advantages

- The method does not have NDZ within the normal operation of load and therefore considered as the effective method for islanding detection.
- The method does not deteriorate the power qualityand transient stability of grid.
- Under small disturbances, the PLCC based islanding prevention provide support to system voltage through fault ride through (FRT) capability of distributed generation.
- The method is unaffected by the number of parallel inverters, DGs penetration level, size and types of DGs.
- The existing PLCC line can be utilized for islanding detection.

Disadvantages

The requirement of the PLCC transmitter at the utility grid with the capability to send signals to all inverters at distribution network is expensive.This method can be applied successfully to the multiple inverter case system.

Signal Produced by Disconnect (SPD)

Similar to PLCC method, this method also utilizes the communication medium between grid and inverters to detect islanding event. In this method, the transmitter installed at utility CB sends the status of CB through microwave, telephone lines etc. to the PV inverters. This method may fail due to malfunctioning of the transmitter or receiver.Moreover, the method allows the additional control to utility grid for coordination between DGs inverter and grid.

Supervisory Control and Data Aquition (SCADA)

In this method, SCADA system is used to send the signals to DGs inverter. The SCADA system is extensively used by the utility grid to monitor the status of various electrical devices within the system and respond to system disturbances. In this method, a voltage sensor is installed at PCC and used to measure the voltage during grid failures. Similar to SPD method, the grid has additional control of DGs. The SCADA method have no NDZ.

HYBRID METHODS

The passive method do not disturb the system parameters but have large NDZ, whereas the active method have small or no NDZ but perturb the system parameters by injecting disturbance at PCC. The remote methods are effective but expensive for implementation. Therefore, recently different hybrid methods have been developed by various researchers. Some of the hybrid methods from recent literature have been discussed in this section.

Average Rate of Voltage Change and Real Power Shift of DGs

In this hybrid method,a formulation of passive and active method is used to detect island event(Mahat et al., 2009). The method uses the passive method(average rate of voltage change, dV/dt) over a few cyclesto initiate the active method (change in real power output of DGs) when the passive method is not able to discriminate between the island event with other events. The PCC voltage (DG output voltage) is measured continuously for every period. IfdV_{DG}/dt^{1} 0, then the magnitude of dV_{DG}/dtover 5 cycles is used to identify the islanding event. The process is represented by Equation (55).

$$V_{Avg5} = \left| \frac{1}{5} \sum_{m=1}^{5} \left(\frac{dV_{DG}}{dt} \right)_m \right| \tag{55}$$

If V_{Avg5} is between the minimum (V_{smin}) and maximum (V_{smax}) threshold values, then the rate of change in voltage may be due to islanding or any other events such as switching or sudden load change. In such case, the change in real power output of DG will be used to detect island mode. However, if $V_{Avg5} > V_{smax}$, then it is certain that islanding event has occurred on distribution network.While increasing and decreasing the real power output of DGs, it is assumed that DGs are operating at unity power factor. After initiating the change in real power output, the average rate of change in output voltage over 20 cycle is used to determine the islanding event. If V_{Avg20} is larger than V_{smaxU} (threshold value with real power shift), then system is islanded.

$$V_{Avg20} = \left| \frac{1}{20} \sum_{m=1}^{20} \left(\frac{dV_{DG}}{dt} \right)_m \right| \tag{56}$$

Therefore, this method can detect islanding events in 25 cycles, when load and generation are closely matched. Moreover, the amount in real power variation, V_{smax}, V_{smin} and V_{smaxU} depends upon system parameters and can be specified accordingly. The flow chart for this hybrid algorithm has shown in Figure 16.

Figure 16. Flowchart for hybrid method (average rate of voltage change and real power shift)

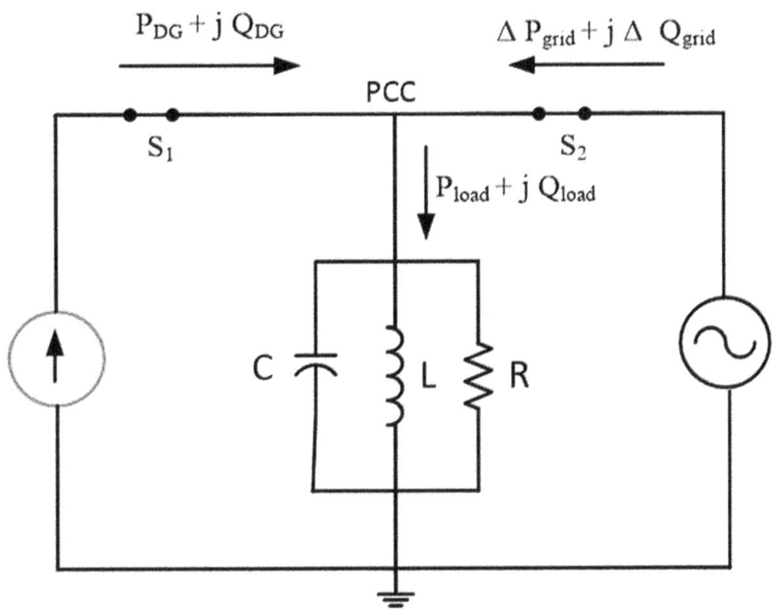

Voltage Unbalance and Frequency Set Point

This hybrid methodtakes the advantages of voltage unbalance (passive method) and positive feedback (active method) to detect island event in the microgrid. The method works correctly with microgrid having synchronous DGs. The procedure of developed hybrid method (Menon & Nehrir, 2007) is given as following: The 3ϕ output voltage of each DG is continuously monitored and voltage unbalance, which is a ratio of negative and positive sequence component of output voltage is calculated. The spikes in VU can be observed, if there is any disturbance applied to DGs such as switching of load or islanding occurs. To discriminate between islanding and other events in system, a feature has been incorporated into the algorithm and discussed as follows: If a VU spike is above a threshold value, then DG's frequency set point will be decreased gradually from its nominal value (e.g., 60 Hz to 59 Hz) in 1 second.However, if the time span is too short (1 sec), then the average value of VU spike closely follows the instantaneous values and therefore undetected. The VU spikes can only be detected, if the instantaneous value is much larger than the average value. After the DG frequency set at 59 Hz, the frequency of the output voltage is continuously measured. If frequency decreases to 59.2 Hz in next 1.5 s, then it shows the disconnection from the main grid or occurrence of islanding event. Further, a tripsignal will be sent to circuit breaker installed at PCC in order to isolate the microgrid from the main grid. Moreover, the frequencyof DG will again set to 60 Hz for continuing the microgrid operation in the island mode. If after the frequency set point at 59 Hz, the frequency of the DG terminal voltage remains close to 60 Hz, it means that themicrogrid is connected to the main grid. The flow chart for this hybrid algorithm has shown in Figure 17.

Figure 17. Flowchart for hybrid method (voltage unbalance and positive feedback)

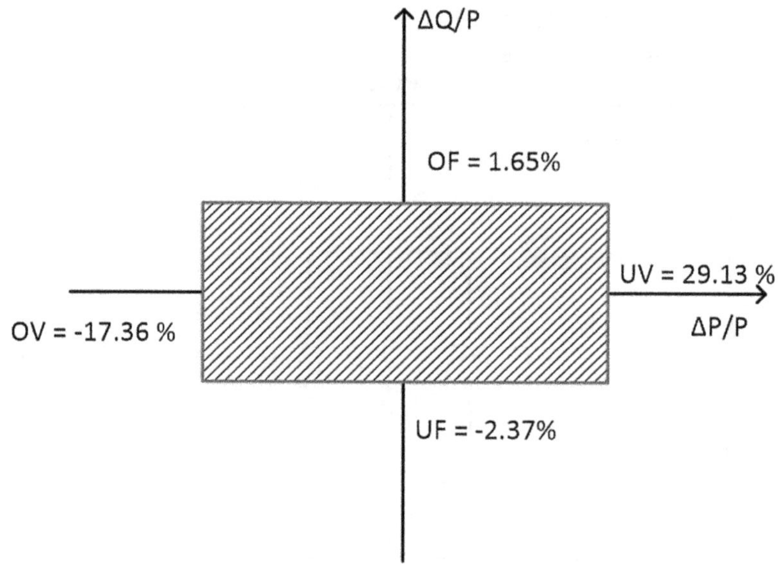

Voltage Unbalance/THD and Bilateral Reactive Power Variations

This hybrid method utilized the passive method (VU/THD) and bilateral reactive power variations (BRPV), active method for accurate detections of islanding event in inverter based microgrid (Wang et. al., 2020). The VU and THD based method are already discussed under passive method for islanding detection and are not discussed here. The VU method is very sensitive passive method and may maloperate in some situations. Therefore, a modified VU/THD method has been used in this hybrid method to achieve better performance. In modified method, the negative and positive sequence of voltage at PCC are obtained by decoupled double synchronous reference frame PLL, which has a good dynamic response. To avoid inaccurate detection caused by normal voltage fluctuation, a low pass filter is used. Unlike other active methods, the BRPV method does not introduce any harmonic component.In this method, theinverter's reactive power output is controlled to drift the PCC frequency outside the permissible limits. This method eliminates the NDZ without affecting the power quality. As shown in Figure 18, the intermittent BRPV method changes the amplitude of the reactive power between positive ($+Q_{inj}$), negative ($-Q_{inj}$) and 0 values periodically.

Figure 18. Procedure for VU, THD and BRPV based hybrid IDM

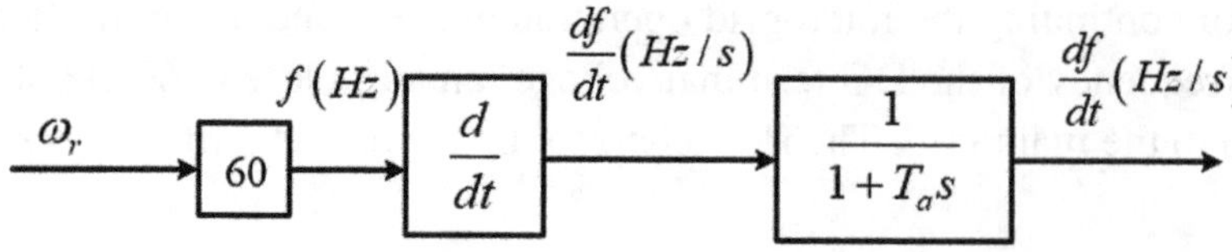

During grid connected mode, the active and reactive power consumed by RLC load can be represented as by Equation (57).

$$P_L = P + P_g = 3\frac{V_{PCC}^2}{R} \tag{57}$$

$$Q_L = Q + Q_g = 3V_{PCC}^2\left(\frac{1}{2\pi fL} - 2\pi fC\right) \tag{58}$$

Where, V_{PCC} and f represents the voltage & frequency at PCC; Q & P represents the inverter'sreactive and active power output and RLC are parameters of local load. The resonant frequency (f_0)and quality factor (Q_f) are defined by Equation (59) and (60) respectively.

$$f_0 = \frac{1}{2\pi\sqrt{LC}} \tag{59}$$

$$Q_f = R\sqrt{\frac{C}{L}} \tag{60}$$

The system frequency in grid connected mode is defined by Equation (61).

$$f = \frac{f_0}{2}\left[\sqrt{\left(\frac{Q_L}{Q_f \times P_L}\right)^2 + 4} - \frac{Q_L}{Q_f \times P_L}\right] \tag{61}$$

Similarly, the system frequency in islanded mode is defined as follows:

$$f_{is} = \frac{f_0}{2}\left[\sqrt{\left(\frac{Q}{Q_f \times P}\right)^2 + 4} - \frac{Q}{Q_f \times P}\right] \tag{62}$$

After islanding, the system frequency has been shifted outside the allowable limit by controlling the output reactive power of inverter and can be derived as follows:

$$f_{is} = \frac{f}{4}\left[\sqrt{\left(\frac{Q_L}{Q_f \times P_L}\right)^2 + 4} - \frac{Q_L}{Q_f \times P_L}\right]\left[\sqrt{\left(\frac{Q}{Q_L \times P}\right)^2 + 4} - \frac{Q}{Q_f \times P}\right] \tag{63}$$

Let the frequency has allowable limit $[f_{min}, f_{max}]$, the NDZ can be defined by Equation (64).

$$\left(\frac{f_{\min}\sigma}{f}-\frac{f}{f_{\min}\sigma}\right)\leq\frac{\frac{Q_L}{P}}{\frac{P_L}{P}}\leq Q_f\left(\frac{f_{\max}\sigma}{f}-\frac{f}{f_{\max}\sigma}\right) \tag{64}$$

Where,

$$\sigma=\frac{1}{2}\left[\sqrt{\left(\frac{Q}{Q_f\times P}\right)^2+4}+\frac{Q}{Q_f\times P}\right] \tag{65}$$

In BRPV method, the NDZ can be expressed as Z_{+Q} and Z_{-Q} When the output reactive power of inverter switches between $+Q_{inj}$ and $-Q_{inj}$ respectively. The overall NDZ for the method would be the overlapping area between Z_{+Q} and Z_{-Q}. In order to eliminate NDZ, the Z_{+Q} and Z_{-Q} should not overlap. The variation in output reactive power should fulfill the following criterion.Therefore, the NDZ can be eliminated in BRPV method without the introduction of the harmonic component into the system.

$$Q_{inj}>\frac{f_{\max}-f_{\min}}{\sqrt{f_{\min}f_{\max}}}Q_f P \tag{66}$$

In this hybrid method, the VU/THD and BRPV methods are combined together to get a reliable IDM for inverter based DG microgrid system. The flow chart for the hybrid method has shown in Figure 19. The three phase voltage at PCC is continuously monitored over one cycle and calculated the ΔVU and ΔTHD values. If the calculated values are greater than the threshold values, the BRPV method will be activated.

The BRPV method changes the output reactive power of inverter between Q_{inj} and Q_{-inj}. During islanding mode of operation, if the frequency will drift outside the allowable limit then the inverter will be shut down. Otherwise in grid connected mode, the inverter will work under normal operation. From Figure18, the islanding occurs at t_0 and the islanding condition is suspected by VU/THD method and at $t_1$if ΔVU/ΔTHD values exceed the threshold after time period T_v, the BRPV is activated to change the output reactive power and frequency will drift outside the allowable limit. Once the frequency will reaches the threshold, the inverter will immediately shut down. The threshold values for ΔVU/ΔTHD passive method is set as ΔVU>50% or ΔTHD>100%.

SFS and ROCOF

This hybrid method combines the features of ROCOF, a passive method and SFS along with active method to improve the efficiency of IDM in microgrid (Khodaparastan et al., 2017). In this method, initially the islandingevent is detected by ROCOF relay, which further activate the SFS method. The phase locked

Figure 19. Flow chart of VU, THD and BRPV based hybrid IDM

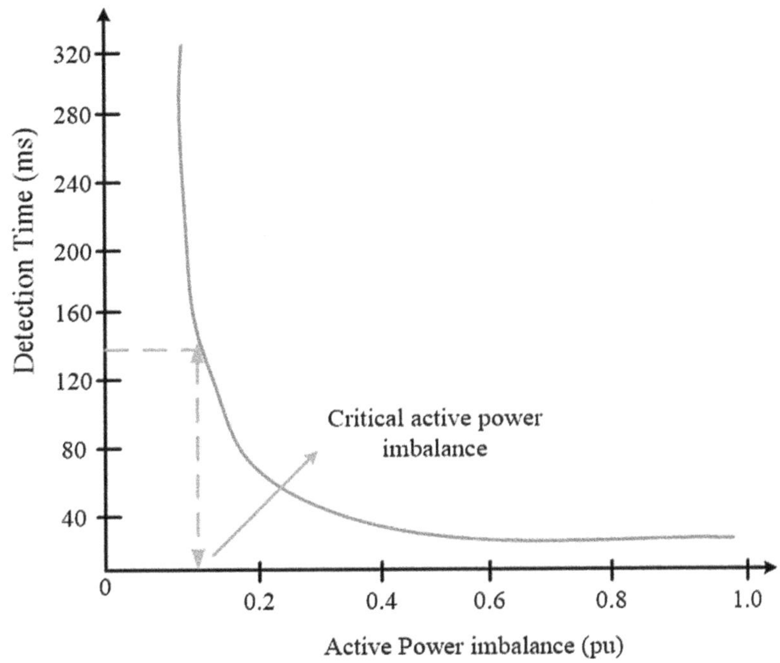

loop (PLL) is used for measuring the system frequency.The*df/dt*is detected by ROCOF relay anda trip signal is sent to multiple switches which further activate the SFS signal. The major problem with ROCOF relay is that the islanding detection becomes difficult when the generation and load are almost equal. Therefore, special care should be taken while selecting the setting of the threshold value. The relay will maloperate for low value of threshold, whereas a high value of threshold results inthe non - detection of the island.The effectiveness of SFS method depends on the gain (k) of positive feedback. The high value of gain increases the islanding detection speed and lowers the NDZ region. Figure 20 shows the control scheme for hybrid method based on SFS and ROCOF.

Figure 20. Block diagram representation of SFS and ROCOF based hybrid method

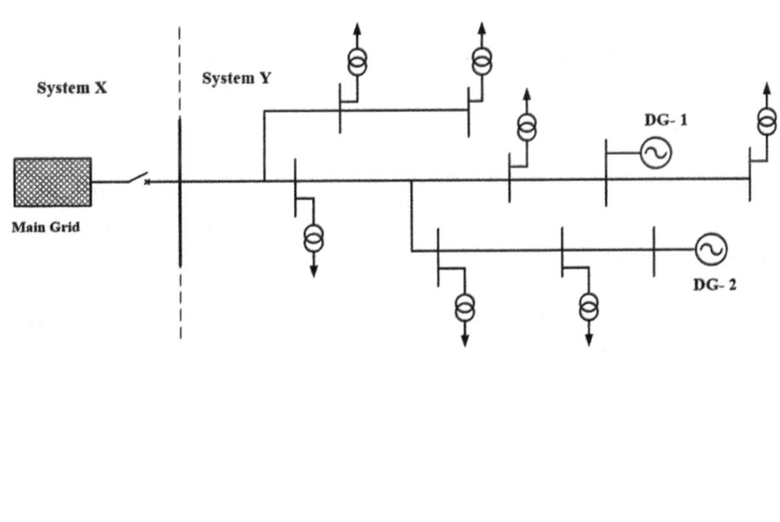

FUTURE RESEARCH DIRECTION

The unintentional islanding is a major issue for reliable operation of microgrid. The islanding detection methods (IDMs) have been categorized into active, passive, remote and hybrid methods.The performance of any IDM depends upon NDZ, detection time, error detection ratio and effect on power quality of concerned system. The active, passive and remote methods have certain advantages and disadvantages. The hybrid method combines advantages of passive and active methodsin order to improve the performance of IDM.Recently, some authors have used fuzzy, wavelet transform, S-transform method as passive IDM having very fast detection time and small/No NDZ. In Futurem, the artificial intelligence techniques such as artificial neural network (ANN), decision tree (DT), support vector machine (SVM) etc. can also be used for detecting island mode in microgrid.

CONCLUSION

The chapter enlightens on broad categories of various IDMs proposed in the literature. The definitions, standards, issues and methods of unintentional islanding detections are discussed. Based on comprehensive analysis it was observed that each islanding detection method (IDM) has certain advantages and disadvantages. The passive methods are simple and do not affect the power quality of the system. However, the method suffers from large NDZ and may fail to detect islanding when output power of DGs and load demand are balanced. The active methods provide fast detection speed and small NDZ, but affect the power quality of power system. The remote methods use existing communication link and have fast detection speed, high accuracy and reliability. However, the implementation cost is very high. Recently, various hybrid methods are proposed and analyzed by various researchers. In hybrid methods, the passive method is used to trigger the active method for accurate detection of islanding. It is observed that majority of proposed hybrid methodshave used VU/THD method as passive method.

REFERENCES

Ahmad, K. N. E. K., Selvaraj, J., & Rahim, A. (2013). A review of the islanding detection methods in grid-connected PV inverters. *Renewable & Sustainable Energy Reviews*, *21*, 756–766. doi:10.1016/j.rser.2013.01.018

Alshareef, S., Talwar, S., & Morsi, W. G. (2014). A new approach based on wavelet design and machine learning for islanding detection of distributed generation. *IEEE Transactions on Smart Grid*, *5*(4), 1575–1583. doi:10.1109/TSG.2013.2296598

Barros, J., & Diego, R. I. (2006). Application of the wavelet-packet transform to the estimation of harmonic groups in current and voltage waveforms. *IEEE Transactions on Power Delivery*, *21*(1), 533–535. doi:10.1109/TPWRD.2005.848437

Bevrani, H., Watanabe, M., & Mitani, Y. (2014). *Power System Monitoring and Control*. John Wiley & Sons, Inc. doi:10.1002/9781118852422

Dash, P. K., Padhee, M., & Panigrahi, T. K. (2012). A hybrid time–frequency approach based fuzzy logic system for power island detection in grid connected distributed generation. *International Journal of Electrical Power & Energy Systems*, *42*(1), 453–464. doi:10.1016/j.ijepes.2012.04.003

El-Arroudi, K., Joos, G., Kamwa, I., & McGillis, D. T. (2007). Intelligent based approach to islanding detection in distributed generation. *IEEE Transactions on Power Delivery*, *22*(2), 828–835. doi:10.1109/TPWRD.2007.893592

Farrokhabadi, M., Canizares, C. A., Simpson-Porco, J. W., Nasr, E., Fan, L., Mendoza-Araya, P. A., Tonkoski, R., Tamrakar, U., Hatziargyriou, N., Lagos, D., Wles, R. W., Paolone, M., Liserre, M., Meegahapola, L., Kabalan, M., Hajimiragha, A. H., Peralta, D., Elizondo, M. A., Schneider, K. P., ... Reilly, J. (2020). Microgrid Stability definitions, Analysis &examples. *IEEE Transactions on Power Systems*, *35*(1), 13–29. doi:10.1109/TPWRS.2019.2925703

Freitas, W., Xu, W., Affonso, C. M., & Huang, Z. (2005). Comparative analysis between ROCOF and vector surge relays for distributed generation applications. *IEEE Transactions on Power Delivery*, *20*(2), 1315–1324. doi:10.1109/TPWRD.2004.834869

Haider, R., Kim, C. H., Ghanbari, T., & Bukhari, S. B. A. (2018). Harmonic-signature-based islanding detection in grid-connected distributed generation systems using Kalman filter. *IET Renewable Power Generation*, *12*(15), 1813–1822. doi:10.1049/iet-rpg.2018.5381

Hamzeh, M., & Mokhtari, H. (2009). Power quality comparison of active islanding detection methods in a single phase PV grid connected inverter. *Proceedings of IEEE International Symposium on Industrial Electronics*. 10.1109/ISIE.2009.5219766

Hashemi, F., Ghadimi, N., & Sobhani, B. (2013). Islanding detection for inverter-based DG coupled with using an adaptive neuro-fuzzy inference system. *International Journal of Electrical Power & Energy Systems*, *45*(1), 443–455. doi:10.1016/j.ijepes.2012.09.008

Hashemi, F., Mohammadi, M., & Kargarian, A. (2017). Islanding detection method for microgrid based on extracted features from differential transient rate of change of frequency. *IET Generation, Transmission & Distribution*, *11*(4), 891–904. doi:10.1049/iet-gtd.2016.0795

Heidari, M., Seifossadat, G., & Razaz, M. (2013). Application of decision tree and discrete wavelet transform for an optimized intelligent-based islanding detection method in distributed systems with distributed generations. *Renewable & Sustainable Energy Reviews*, *27*, 525–532. doi:10.1016/j.rser.2013.06.047

IEEE Application Guide for IEEE Std 1547(TM) (2009). IEEE Standard for Interconnecting Distributed Resources with Electric Power Systems. *IEEE Std 1547.2-2008*, 15, 1-217. doi:10.1109/IEEESTD.2008.4816078

Iong, S., & Kim, K. H. (2004). An islanding detection method for distributed generations using voltage unbalance and total harmonic distortion of current. *IEEE Transactions on Power Delivery*, *19*(2), 745–752. doi:10.1109/TPWRD.2003.822964

Justo, J. J., Mwasilu, F., Lee, J., & Jung, J. W. (2013). AC-microgrid versus DC-microgrid with distributed energy resources: A review. *Renewable & Sustainable Energy Reviews*, *24*, 387–405. doi:10.1016/j.rser.2013.03.067

Kane, P. O., & Fox, B. (1997). Loss of mains detection for embedded generation by system impedance monitoring. *Proceedings of 6th International Conference on Development of Power System Protection*.10.1049/cp:19970037

Karimi, H., Yazdami, A., & Iravni, R. (2008). Negative-sequence current injection for fast islanding detection of a distributed resource unit. *IEEE Transactions on Power Electronics*, *23*(1), 298–307. doi:10.1109/TPEL.2007.911774

Khamis, A., Shareef, H., Mohamed, A., & Bizkevelci, E. (2015). Islanding detection in a distributed generation integrated power system using phase space technique and probabilistic neural network. *Neurocomputing*, *148*, 587–599. doi:10.1016/j.neucom.2014.07.004

Khodaparastan, M., Vahedi, H., Khazaeli, F., & Oraee, H. (2017). A hybrid islanding detection method for inverter based DGs using SFS and ROCOF. *IEEE Transactions on Power Delivery*, *32*(5), 2162–2170. doi:10.1109/TPWRD.2015.2406577

Kim, M. S., Haider, R., Cho, G. J., Kim, C. H., Won, C. Y., & Chai, J. S. (2019). Comprehensive review of islanding detection merhod for distributed generation systems. *Energies*, *12*(837), 1–21. doi:10.3390/en12050837

Li, C., Cao, C., Cao, Y., Kuang, Y., Zeng, L., & Fang, B. (2014). A review of islanding detection methods for microgrid. *Renewable & Sustainable Energy Reviews*, *35*, 211–220. doi:10.1016/j.rser.2014.04.026

Liu, F., Kang, Y., Zhang, Y., Duan, S., & Lin, X. (2010). Improved SMS islanding detction method for grid connected converters. *IET Renewable Power Generation*, *4*(1), 36–42. doi:10.1049/iet-rpg.2009.0019

Lopes, L. A. C., & Sun, H. (2006). Performance assessment of active frequency drifting islanding detection methods. *IEEE Transactions on Energy Conversion*, *21*(1), 171–180. doi:10.1109/TEC.2005.859981

Mahat, P., Chen, Z., & Bak-Jensen, B. (2009). A hybrid islanding detection technique using average rate of voltage change and real power shift. *IEEE Transactions on Power Delivery*, *24*(2), 764–771. doi:10.1109/TPWRD.2009.2013376

Menon, V., & Nehrir, M. H. (2007). A hybrid islanding detection technique using voltage unbalance and frequency set point. *IEEE Transactions on Power Systems*, *22*(1), 442–448. doi:10.1109/TPWRS.2006.887892

Pai, F. S., & Huang, S. J. (2001). A detection algorithm for islanding prevention of dispersed consumer owned storage and generation units. *IEEE Transactions on Energy Conversion*, *16*(4), 346–351. doi:10.1109/60.969474

Ray, P. K., Kishor, N., & Mohanty, S. R. (2012). Islanding and power quality disturbance detection in grid-connected hybrid power system using wavelet and S-transform. *IEEE Transactions on Smart Grid*, *3*(3), 1082–1094. doi:10.1109/TSG.2012.2197642

Redfern, M. A., Usta, O., & Fielding, G. (1993). Protection against loss of utility grid supply for a dispersed storage and generation unit. *IEEE Transactions on Power Delivery*, *8*(3), 948–954. doi:10.1109/61.252622

Reis, M. V. G., Barros, T. A. S., Moreira, A. B., Nascimento, F. E. R., & Villalava, M. G. (2015). Analysis of the Sandia Frequency Shift (SFS) islanding detection method with a single phase photovoltaic distributed generation system. *Proceedings of IEEE PES Innovative Smart Grid Technologies Latin America (ISGT LATAM)*, 10.1109/ISGT-LA.2015.7381140

Roop, M. E., Bagovic, M., Rohtgi, A., Kern, G. A., Bonn, R. H., & Gonzalez, S. (2000). Determining the relative effectiveness of islanding detection methods using phase criteria and nondection zones. *IEEE Transactions on Energy Conversion*, *15*(3), 290–296. doi:10.1109/60.875495

Ropp, M. E., Begovic, M., & Rohatgi, A. (1999). Analysis and performance assessment of the active frequency drift method of islanding prevention. *IEEE Transactions on Energy Conversion*, *14*(3), 810–816. doi:10.1109/60.790956

Samantaray, S. R., Babu, B. C., & Dash, P. K. (2011). Probabilistic neural network based islanding detection in distributed generation. *Electric Power Components and Systems*, *39*(3), 191–203. doi:10.1080/15325008.2010.526986

Samantaray, S. R., Samui, A., & Chitti Babu, B. (2011). Time-frequency transform-based islanding detection in distributed generation. *IET Renewable Power Generation*, *5*(6), 431–438. doi:10.1049/iet-rpg.2010.0166

Singam, B., & Hui, L. Y. (2006). Assessing SMS and PJD schemes of anti islanding with varying quality factor. *Proceedings of the IEEE International Power and Energy Conference*. 10.1109/PECON.2006.346645

Stevens, J., Bonn, R., Ginn, J., Gonzalez, S., & Kern, G. (2000). *Development and testing of an approach to anti-islanding in utility interconnected photovoltaic systems*. Sandia National Laboratories. doi:10.2172/759506

Trujillo, C. L., Velasco, D., Figueres, E., & Garcera, G. (2010). Analysis of active islanding detection methods for grid connected microconverter for renewable energy processing. *Applied Energy*, *87*(11), 3591–3605. doi:10.1016/j.apenergy.2010.05.014

Wang, G., Gao, F., Liu, J., Li, Q., & Zhao, Y. (2020). Design consideration and performance analysis of hybrid islanding detection method combining voltage unbalance/ toal harmonic distortion and bilateral reactive power variation. *CPSS Transactions on Power Electronics and Applications*, *5*(1), 86–100. doi:10.24295/CPSSTPEA.2020.00008

Xu, W., Zhang, G., Li, C., Wang, W., Wang, G., & Kliber, J. (2007). A power line signaling based technique for anti-islanding protection of distributed generators part I scheme & analysis. *IEEE Transactions on Power Delivery*, *22*(3), 1758–1766. doi:10.1109/TPWRD.2007.899618

Ye, Z., Kolwalkar, A., Zhang, Y., Du, P., & Walling, R. (2004). Evaluation of anti-islanding schemes based on nondetection zone concept. *IEEE Transactions on Power Electronics*, *19*(5), 1171–1176. doi:10.1109/TPEL.2004.833436

Section 3
Reliability Studies and Applications

Chapter 11
A Comparative Study of Power Quality Monitoring Using Various Techniques

Fatma Zohra Dekhandji
University M'hamed Bougara of Boumerdes, Algeria

Mohamed Cherif Rais
University M'hamed Bougara of Boumerdes, Algeria

ABSTRACT

In recent years, power quality (PQ) has become an increasingly major concern for both electric utilities and the end users. Accordingly, the electrical engineering community has to deal with the analysis, diagnosis, and solution of PQ issues using system approach rather than handling these issues as individual problems. This chapter describes the analysis of PQ using advanced signal processing tools represented in Hilbert and wavelet transforms (HT-WT) and artificial intelligence tools represented in artificial neural network and support vector machine (ANN-SVM) for detection and classification of power quality disturbances, respectively. These techniques were successfully simulated using LABVIEW software capabilities. The results of simulation indicate that the signal processing techniques are effective mechanisms to detect and classify power quality disturbances. At the end, the combination of WT as a tool of detection and features extraction with SVM as a classifier tool resulted as the best combination for PQ monitoring system.

LIST OF ABBREVIATIONS

PQ: Power Quality.
PQD: Power Quality Disturbances
PQM: Power Quality Monitoring
IEEE: Institute of Electrical & Electronic Engineers
ANSI: American National Standards Institute

DOI: 10.4018/978-1-7998-4027-5.ch011

WT: Wavelet Transform
CWT: Continuous Wavelet Transform
DWT: Discrete Wavelet Transform
DWC: Discrete Wavelet Coefficients
MRA: Multi Resolution Analysis
HT: Hilbert Transform.
NN: Neural Network
ANN: Artificial Neural Network
FFNN: Feed-Forward Neural Network
RNN: Recurrent Neural Network
SV: Support Vector
SVM: Support Vector Machine
ST: Stock well Transform
CZT: Chirp Z-Transform
FT: Fourier Transform
DT: Decision Tree
FL: Fuzzy Logic
HP & LP: High Pass & Low Pass Filters
RMS: Root Mean Square
THD: Total Harmonic Distortion

INTRODUCTION

Power quality is known as one of the very serious issues in electric power transmission and distribution, manufacturers and end users because of its bad effects on electricity suppliers.

Power quality is a term of many meanings; each meaning relies on the reference to which it is attributed. For example, a utility may define power quality as reliability of its system. A manufacturer of load equipment may define power quality as those features of the power supply that enable the equipment to work properly. The end user's point of view of power quality is that any power problem manifested in voltage, current, or frequency deviations that result in failure or malfunction of customer equipment (McGranaghan et. al., 1996; Dekhandji et al., 2017; Dekhandji, 2018; Recioui, 2018).

Power quality is generally meant to express the quality of voltage and/or the quality of current which can be defined as: the measure, analysis, and improvement of the bus voltage to maintain a sinusoidal waveform at rated voltage and frequency; this definition includes all momentary and steady-state phenomena (Masoum and Fuchs, 2011).

All electrical devices are likely to fail or not work properly when exposed to one or more power quality problems; these devices react negatively to power quality issues, depending on the severity of problems. A simpler and may be more concise definition might state: "Power quality is a set of electrical boundaries that allows a piece of equipment to function in its intended manner without significant loss of performance or life expectancy." This definition contains things needed from electrical devices which are performance and life expectancy (Sankaran, 2001,).

It is common experience that electric power of poor quality has harmful effects on different equipment and systems. More than that, power system stability, continuity and reliability fall with the degradation of quality of electric power (Kennedy, 2000).

There are four main reasons for engineers to increase concern about the need for energy quality (McGranaghan et. al., 1996):

- Newer-generation load equipment, with microprocessor-based controls and power electronic devices, are more sensitive to power quality variations than was equipment used in the past.
- The increasing focus on overall power system efficiency has resulted in continued growth in the use of devices such as high-frequency, adjustable-speed motor drives and shunt capacitors for power factor correction to reduce losses. This is resulting in increasing harmonic levels on power systems and has many people worried about the future effect on system abilities.
- End users have an increased knowledge of power quality issues. Utility customers are becoming better informed about such issues as interruptions, sags, and switching transients and are challenging the utilities to improve the quality of the delivered power.
- Many things are now interconnected in a network. Integrated processes mean that the failure of any part has much more important results (Dekhandji et al., 2019).

The common thread running through all these reasons for increased concern about the quality of electric power is the continued push for increasing productivity for all utility customers. Manufacturers want faster, more productive, more efficient machinery. Utilities encourage this effort because it helps their customers become more beneficial and also helps put off large investments in substations and generation by using more efficient load equipment. Interestingly, the equipment installed to increase the productivity is also often the equipment that suffers the most from common power disturbances. And the equipment is sometimes the source of adding more power quality problems. When whole processes are automated, the efficient operation of machines and their controls becomes more and more dependent on quality power (McGranaghan et. al., 1996).

The power system faces many issues; some of them shut it down causing a blackout of the network, the main problems that can cause harm to the power system are:

- Voltage sag or dip.
- Very short interruptions.
- Long interruptions.
- Voltage spike.
- Transient.
- Voltages swell.
- Harmonic distortion.
- Voltage fluctuation.
- Noise.
- Voltage unbalance.

Power quality disturbances have become a serious issue in power world, and this is because of the great damage that they caused to the system from the generation to the end users. For this reasons a power quality monitoring system must be included in the electrical network in order to: detect, classify and mitigate these problems (Dekhandji et al., 2019, Azzougui and Recioui, 2018).

Many researchers have focused their attention on this goal, on how to create a monitoring system capable of detecting the disturbances the moment they appear and then classify them according to their parameters. Detection techniques are numerous, each one distinct from the others; but the fastness time response to detect the disturbance is the major concern of all the studies. Classification is also an important step in the monitoring, after the system detects a disturbance it must be recognized. The objective of the classifier is to identify the events and must have a perfect percentage of accuracy and precision.

THEORETICAL BACKGROUND

Power quality involves voltage and frequency waveform, electricity need to travel from the generator to the transmission lines till it reaches the end users. In real life no electrical signal is pure; it must contain at least one of these disturbances that are frequently appearing in the voltage signal (Aggoun et al., 2017, Recioui et al., 2017).

The Power Quality problems to be examined are transients, short term and Long- Duration Voltage Variations, harmonic distortion, Voltage unbalance, and Voltage Fluctuations. In this chapter each disturbance will be treated alone, definitions, causes and effects on the power system.

Types of Power Quality Disturbances

1. Transients

The term transient in the point of view of an engineer is the short-duration oscillatory or non-oscillatory event which is undesirable and usually highly damped. Based on waveform shape of a current or voltage; Power system transients, can be classified into (McGranaghan et. al., 1996):

a. Impulsive Transients

An impulsive transient is a sudden non-power frequency change in the steady-state condition of voltage, current, or both that is unidirectional in polarity (primarily either positive or negative). Impulsive transients are usually characterized by their rise and decay times, which can also be showed by their spectral content (McGranaghan et. al., 1996).

b. Oscillatory Transients

An oscillatory transient is a sudden, non–power frequency change in the steady-state condition of voltage, current, or both, that includes both positive and negative polarity values. An oscillatory transient consists of a voltage or current whose instantaneous value changes polarity rapidly. It is described by its spectral content (predominate frequency), duration, and magnitude (McGranaghan et. al., 1996, Recioui et al., 2016).

There are different causes for power system transients. For example, lightning strokes to the wires in the power system or to ground and component switching either of network components or end user equipment. These transients can affect the power system either in operation or performance of different part of it, during a transient a high current passes in the circuit, this high current can burn out devices and instruments; Transients can cause mal-operation of relays and mal-tripping of circuit breakers. Frequent number of direct or indirectly induced oscillatory transients may change the magnetic properties of core materials used in electric machines (Mitra et. al., 2011).

2. Short-Duration Voltage Variations

There are three types of short-duration voltage variations, namely, instantaneous, momentary and temporary, depending on its duration. Each category is divided into interruption, sag, and swell. Principal cases of short-duration voltage variations are large load energization, fault conditions, and loose connections (Masoum and Fuchs, 2011).

a. Voltage Sag (Dip)

IEEE defines voltage sags as a reduction in voltage for a short time; this reduction is between 10 percent and 90 percent of the normal root mean square (rms) voltage. The duration of voltage sag is less than 1 minute and more than 8 milliseconds (0.5 cycles) (Kennedy, 2000).

b. Voltage Swell

Voltage swells, or momentary overvoltages, are RMS voltage variations that exceed 110 percent of the nominal voltage and last for less than 1minute (Kennedy, 2000).

There are many causes of voltage sags (Mitra et. al., 2011):

- Energization of heavy loads.
- Starting of large induction motors.
- Single line-to-ground faults.
- Line-line and symmetrical fault.
- Load transferring from one power source to another.

As for voltage swells, the reasons are:

- Switching off of a large load.
- Energizing a capacitor bank.
- Voltage increase of the unfaulted phases during a single line-to-ground fault.

These causes affect the power system and their results are (Mitra et. al., 2011):

- Voltage stability because of reduction of bus voltage for short duration.
- Electrical low-voltage devices will not work correctly.
- Malfunctions of uninterruptible power supply.
- Malfunction of measuring and control equipment.
- Interfacing with communication signals.

c. Interruptions

Interruptions are a complete loss of voltage (a drop to less than 10 percent of nominal voltage) in one or more phases. IEEE defines three types of interruptions according to the time of their occurrence: momentary, temporary, and long-duration interruptions. Momentary interruptions are the complete loss of voltage on one or more phase conductors for a time period between 8 milliseconds and 3 seconds. A temporary, or short-duration, interruption is a drop of voltage below 10 percent of the nominal voltage for a time period more than 3 seconds and less than 1 minute. Long-duration, or sustained, interruptions last longer than 1 minute (Kennedy, 2000).

There are many reasons for power interruption. Some of the general causes of interruption are (Mitra et. al., 2011):

- Equipment failures.
- Control mal-function.
- Blown fuse.
- Breaker opening.

Interruptions can cause a loss of production, for the factories loss of production is loss of money and loss of time will lead to bad results (Kennedy, 2000).

3. Long-Duration Voltage Variations

Long-duration variations can be either overvoltages or undervoltages. They contain RMS deviations at power frequencies for a period of time longer than 1 min. They are usually not caused by system faults but system switching operations and load variations on the system (McGranaghan et. al., 1996).

a. Undervoltage

An undervoltage is a decrease in the rms ac voltage to less than 90 percent at the power frequency for a duration longer than 1 min. Undervoltages are the result of switching events that are the opposite of the events that cause overvoltages. A load switching on or a capacitor bank switching off can cause an undervoltage until voltage regulation equipment on the system can bring the voltage back to within tolerances. Overloaded circuits can result in undervoltages also (Mitra et. al., 2011).

b. Overvoltage

An overvoltage is an increase in the rms ac voltage greater than 110 percent at the power frequency for duration longer than 1 min. Overvoltages are usually the result of load switching. The overvoltages result because either the system is too weak for the desired voltage regulation or voltage controls are inadequate. Incorrect tap settings on transformers can also result in system overvoltages (McGranaghan et. al., 1996).

The main reasons of undervoltages and overvoltages (Kennedy, 2000):

- Capacitor switching.
- Load dropping.
- Mis-setting of voltage tap in transformers.

- Loss of transmission lines.
- Overloading.

Both disturbances (undervoltages and overvoltages) can have the following effects (Mitra et. al., 2011):

- Over stress on insulation.
- Problems of voltage instability.
- High demand for reactive power.
- Drawl of high current by motors.

4. Harmonic Distortion

Waveforms in a non-sinusoidal wave having frequencies other than fundamental frequency are called harmonics. The frequency of each harmonic component is known as harmonic frequency. In most of the cases of periodic and well defined waves, where waveform can be expressed by Fourier series, harmonic frequencies are integer multiple of fundamental frequency (however it may be fractional). The integer factor is known as the order of harmonic component (Mitra et. al., 2011, Dekhandji et al.,2020).

Harmonic distortion is normally attributed to the application of nonlinear loads (i.e loads that when supplied by a sinusoidal voltage; do not draw a sinusoidal current). The current in these loads can be heavily distorted compared to a sine wave, and usually containing the odd harmonics, 3, 5, 7, 9, ...etc. These nonlinear loads not only have the potential to create problems within the facility that contains the nonlinear loads but also can adversely affect the neighboring facilities (Skvarenina, 2018, Azzougui and Recioui, 2019).

Harmonics have many causes, some important causes are (Sankaran, 2001):

- The non-linear loads like computers, rectifier banks, fluorescent lighting....etc
- The non-uniformly distribution of coils around the generator stator slots.
- Current generated by end-user equipment.

There are many effects of harmonics on the power system (Chakravarthi and Ortmeyer, 1985):

- Overheating neutral conductors and transformers.
- Nuisance tripping of circuit breakers.
- Malfunction of different equipment and generator systems.
- Computer malfunctions.
- Overvoltage problems.

5. Voltage Unbalance

This type of power quality disturbance is caused by unequal distribution of loads among the three phases. At three-phase distribution level, unsymmetrical loads at industrial units can result in voltage unbalance. Voltage unbalance is very significant for three-phase equipment such as transformer, motors and rectifiers, for which it results in overheating due to a high negative sequence current flowing into the equipment. The asymmetry can also have a bad effect on the performance of converters, as it results in the production of harmonic (Muneta and Rey, 2011).

The major causes of unbalance of the voltage are (McGranaghan, 1996, Dekhandji and Bentarzi, 2019):

- Single-phase loads on a three-phase circuit.
- Blown fuses in one phase of a three-phase capacitor bank.
- Single-phasing conditions.

The equipment use and efficiency is negatively influenced by voltage unbalance on the load side which is explained as follows (Fritz and Gupta, 2016):

- Temperature rises in motor.
- The rotating machine in presence of voltage unbalance results in a reverse magnetic field, which results in decrease of useful torque due to machine windings heating.
- It also damages power supply wiring, transformers and generators.
- The unbalanced voltages across motor terminals causes phase current unbalance, 6-10 times the percent voltage unbalance for a fully loaded motor.
- It also reduces torque capability.

6. Voltage Fluctuations (Flicker)

Voltage fluctuations are defined as the cyclic variation of voltage with amplitude that does not exceed 10%. This variation in magnitude is usually much lower than the sensitivity threshold of most equipment and, consequently, operating problems are experienced only in rare cases. Except for these very particular cases, the main disturbing effect of voltage fluctuations is producing changes on the illumination intensity of light sources (Baggini, 2008).

Voltage fluctuations can be attributed to various causes (Blume et. al., 2001):

- Pulsed power output where there is burst-firing control.
- Resistance welders.
- Start-up of drives.
- Pulsed power output with thermostat controls.
- Drives with steeply-changing loading.
- Arcfurnaces.

Voltage Fluctuation can affect the power system (Blume et. al., 2001):

- Control actions for control system acting on the voltage angle.
- braking or acceleration moments from motors connected directly to the system
- impairment of electronic equipment where the fluctuation of the supply voltage passes through the power supply assembly to the electronic equipment.

Classification of Power Quality Problems

The classification of power quality problems can be performed according to (McGranaghan, 1996):

- The source of the problem (e.g. converters, magnetic circuit nonlinearities…)
- Wave-shape of problematic signal (e.g. harmonic, flicker…)
- Frequency spectrum (e.g. radio frequency interference)

ANS1 C84.1- 1982 and other standards more specifically depict duration and other characteristics of different categories listed above. These phenomena listed above can be described further by listing appropriate attributes.

For steady-state phenomena, the following attributes can be used:

- Amplitude, Frequency.
- Spectrum, Modulation, Source impedance.
- Notch depth, Notch area.

For non-steady-state phenomena, other attributes may be required:

- Amplitude, Duration.
- Spectrum, Frequency.
- Rate of occurrence, Energy potential.
- Source impedance.

Table 1 summarizes all the disturbances with their parameters:

DETECTION AND CLASSIFICATION METHODS OF POWER QUALITY PROBLEMS

The important issues in power quality (PQ) problems are to detect and classify disturbance waveforms automatically in an efficient manner (Sarma and Venkatesh, 2010). In order to improve electric power quality, the sources and causes of disturbances must be known before appropriate mitigating action can be taken, and continuous recording of disturbance waveforms is necessary (Ram et. al., 2010). Therefore, the monitoring equipment needs to firstly and accurately detect and identify the disturbance types.

Thus, the uses of new and powerful tools of signal analysis have enabled the development of additional methods to accurately characterize and identify several kinds of power quality disturbances (Bansal et.al., 2011). Therefore, two methods of both detection and classification are introduced and used for power quality monitoring.

Detection Methods

Detection methods are designed to identify the occurrences of disturbances. An effective method of detecting power quality events such as (sag, swell, harmonics, interruption, flicker and transient) is dependent on their accurate measurements in time which is still a challenge for researchers and engineers (Bollen et. al., 2000).

Signal processing techniques are widely used in analyzing PQ events to extract the most sensitive and interesting features concerning disturbances, these techniques are classified as using time domain, frequency domain, and time-frequency domain methods (Bicker and Jabar, 2014; Abdoos et. al., 2009). Some examples are the fast Fourier Transform method, Fractal-Based method, S-Transform method (ST), Time- Frequency Ambiguity Plane method, Short Time Power and Correlation Transform method, Chirp-Z Transform (CZT), Wavelet Transform (WT) method, and Hilbert Transform (HT) (Abdoos et. al., 2009). These two later have been proven to be effective signal processing tools for the detection and analysis of power signals.

Table 1. Different types of PQD and their parameters

Categories	Spectral Content	Typical Duration	Voltage Magnitude
1. Transients • Impulsive • Nanosecond • Microsecond • Millisecond • Oscillatory • Low frequency • Medium frequency • High frequency	5ns rise 1µs rise 0.1ms rise <5kHz 5-500kHz 0.5-5MHz	<50ms 50ns-1ms >1ms 0.3-50ms 20µs 5µs	0-4pu 0-8pu 0-4pu
2. Short duration variations • Instantaneous • Sag • Swell • Momentary • Interruptions • Sag • Swell • Temporary • Interruptions • Sag • Swell		0.5-30cyc 0.5-30cyc 0.5-3s 0.5-3s 0.5-3s 3s-1min 3s-1min3s-1min	0.1-0.9pu 1.1-1.8pu <0.1pu 0.1-0.9pu 1.1-1.2pu <0.1pu 0.1-0.9pu 1.1-1.2pu
3. Long duration variations • Interruption, sustained • Undervoltage • Overvoltage		>1min >1min >1min	0.0pu 0.8-0.9pu 1.1-1.2pu
4. Voltage Imbalance		Steady state	
5. Waveform distortion			
• DC offset • Harmonics	0-100th H	Steady state Steady state	0-0.1% 0-20%
• Inter-harmonics	0-kHz	Steady state	0-2%
• Notching		Steady state	
• Noise	Broad-band	Steady state	0-1%
6. Voltage fluctuations	<25Hz	Intermittent	0.1-7%
7. Power frequency variations		<10s	

1. Wavelet Transform

The Wavelet Transform (WT) has been found to be particularly useful for analyzing signal which is mostly non-stationary and can be best described as aperiodic, noisy, intermittent, transient and so on (Addison, 2017).Therefore, it represents one of the powerful signal processing algorithms that is used in the field (Abbod et. al., 2015). It has the ability to analyze different Power quality problems simultaneously in both time and frequency domains in a distinctly different way from the traditional tools such as Fourier Transform (Sarma and Venkatesh, 2010; Addison, 2017), and this is why it overcomes the resolution limitation of FT. The main advantage of wavelets is that they have a varying window size, which is wide for slow frequencies and narrow for fast, thus leading to an optimal time frequency resolution in all frequency ranges (Daubechies, 1992).WT has two categories primarily called Continuous Wavelet Transform (CWT) and Discrete Wavelet Transform (DWT) (Daponte, 1998).

a. Basic Concepts of Wavelet

The Wavelet Transform procedure is to adopt a wavelet prototype function, called an analyzing wavelet or mother wavelet W*(t)*(Ram et. al., 2010). This function has a mean of zero and sharply decays in an oscillatory fashion, i.e. it rapidly falls to zero either side of its central path (Aggarwal and Kim, 2000). Each wavelet is created by scaling (stretching and squeezing) and translation (moving) operations in this mother wavelet (Radhika et. al., 2006; Addison, 2017). Some of the popular wavelet families such as: Haar, Daubechies, biorthogonal, Coiflet, Symlet, and Gaussian.

The selection of the mother Wavelet function is one of the main factors in the successful application of Wavelets. Shorter Wavelet functions of the Daubechies family being the most widely used in the detection and analysis of voltage events (Diego et al., 2012).

The signal can be represented in terms of wavelet function and scaling function. Basically, it is represented by one set of scaling coefficients, and one or several sets of wavelet coefficients:

$$\left(t\right) = \sum_{n} Aj\left(n\right)\varnothing(t-n) + \Sigma \sum_{n,j} Dj\left(n\right)2^{\frac{j}{2}}\Psi\left(2^{j}t-n\right) \tag{1}$$

Where:

A j: Represents level scaling coefficient,
D j: Represents j level wavelet coefficient,
ϕ: Represents scaling function,
ψ(t): Represents wavelet function,
j: Can be any higher level wavelet transform and *t* is time.

b. Continuous Wavelet Transform (CWT)

CWT algorithm uses the window function or the Wavelet function to compute individually each part of the time-domain signal according to the length of the window. The advantage of the CWT is that the window length is changed as the window function translating the entire signal (Abidin et. al., 2017), and it divides a continuous time signal into an equal time intervals and equal frequency intervals (Gawre et al., 2014).

c. Discrete Wavelet Transforms (DWT)

The equation of CWT has great theoretical interest for the development and comprehension of its mathematical properties. However, its discretization is necessary for practical applications (Radhika et. al., 2006). This is because the possibility of reconstruction the original signal using infinite summation of discrete wavelet coefficients rather than continuous integral, and this leads to a fast wavelet transform for the rapid computation of DWC and its inverse (Addison, 2017). The expression of CWT of a continuous signal is given as:

$$CWT\left(a,b\right) = \int_{-\infty}^{+\infty} x\left(t\right)\Psi^{*}_{a,b}\left(t\right)dt \text{ with } a,b \in \mathbb{R} \tag{2}$$

Where $\Psi^{*}_{a,b}$ is the dilated and translated mother wavelet expressed as:

$$\Psi^{*}_{a,b} = \frac{1}{\sqrt{a}}\Psi\left(\frac{t-b}{a}\right) \tag{3}$$

$x(t)$: is signal to be analyzed,
a is scaling or dilation parameter and b is translation parameter.

d. Multi Resolution Analysis (MRA)

The wavelet Multi-Resolution Analysis is a tool that utilizes DWT for the analysis of waveforms and images. In power quality, MRA can be used for detection as well as for distortion features extraction. The goal of MRA is to develop representations of a signal at various levels of resolution by decomposing the time domain signal (Bicker and Jabar, 2014; Radhika et. al., 2006). Wavelet functions and scaling functions are used as building blocks to decompose and reconstruct the signal at different resolutions in MRA (Gawre et al., 2014).

2. The Hilbert Transform

The Hilbert Transform (HT) has been used widely in the telecommunication research for signal modulation and demodulation, and in various medical image processing applications. In power quality signal analysis, the HT has not been investigated, although it shows an accurate tracking of the changes in the power quality signals.

The Hilbert Transform of a signal is equivalent to a ±90° phase shift in all frequency components of the signal (Hossen, 2001).

Hilbert Transform is used to generate a complex signal from a real signal. And it is a multiplier operator. Unlike other transform techniques, the HT of time-domain function is still in time domain [30].

According to mathematical reasoning, it is confirmed: H[x(t)] has the effect of shifting the phase of the negative frequency components by +90° (π/2 radians) and the phase of the positive frequency components by −90°. And H[x(t)] has the effect of restoring the positive frequency components while shifting the negative frequency ones an additional +90°, resulting in their negation (Huihui et. al., 2012).

Classification Methods

The main task of power quality (PQ) disturbance recognition is to accurately identify the disturbance type, and provide the references for parameter estimation and control strategies of power system (Cui et. al., 2015). Therefore continuous monitoring and classification are required for these disturbances due to increasing demand of pure power (Zin et. al., 2013). To perform the classification of these disturbances different algorithms have been defined in order to relate the signal characteristics with the group they belong to: Decision Trees (DTs), Fuzzy Logic (FL), Neural Networks (NNs), and Support Vector Machine (SVM), among others. These methodologies play an important role on the disturbance classification because their performance depends on the extracted features and the classifier utilized; if the disturbance characteristics are not accurately captured, the performance is also limited (Romero-Troncoso et. al., 2011).

1. Artificial Neural Network

Artificial Neural Networks have been developed as generalizations of mathematical models of human cognition or neural biology. A neural network is characterized by (Fausett, 1994):

- Its pattern of connections between the neurons (called architecture).
- Its method of determining the weights on the connections (called training or learning algorithm).
- Its activation function.

All definitions of Artificial Neural Networks emphasize the idea of highly interconnected units comprising simple nonlinear elements (Ekwue et. al., 1997).

Neural Networks have been extensively used for the classification because of their large data handling capability. They are used to recognize and classify complex fault patterns without much knowledge about the system they deal with. The neural networks are described by the transfer function of their neurons, by training algorithm and by the connecting formula (Rao et. al., 2016).

Artificial Neural Networks (ANNs) have attracted a great deal of attention because of their pattern recognition capabilities, and their ability to handle noisy data; however, its ability to perform well is greatly influenced by the weight adaptation algorithm and the amount of noise in the data (Mishra et al., 2000).

2. Support Vector Machine

The Support Vector Machine (SVM) is one of the most prominent techniques in supervised machine learning that have attracted a significant amount of attention in the field of machine learning over the past decade by proving themselves to be very effective in a variety of real-world pattern classification and regression estimation tasks (Lam, 2012).It has been applied to the problems of dependency estimation, forecasting, and constructing intelligent machines (Vapnic, 1998).

SVM is a statistical learning technique based on structural risk minimization method. Its main objective is building a model with the use of training set, where each sample belongs into one of the two possible classes, class 1 or class 2. With the trained SVM, a prediction model can be achieved (Taskovski and Markovska, 2018), which can deal with linear and nonlinearly separable models using a hyperplane based on theoretical results from the statistical learning theory (Wu et. al., 2008).

SVM is an algorithm with great advantages for classification developed from Statistical Learning Theory by Vapnik; Firstly, it provides accurate classifier with support vectors, which is robust to noise (Chilukuri and Li, 2014; Abdoos et. al., 2009). Secondly, SVMs have the potential to handle very large feature spaces, because the training of SVM is carried out so that the dimension of classified vectors does not have as a distinct influence on the performance of SVM as it has on the performance of conventional classifiers (Ekici, 2008). Also, SVM-based classifiers are claimed to have good generalization properties compared to conventional classifiers, because in training the SVM classifier, the so-called structural misclassification risk is to be minimized, whereas traditional classifiers are usually trained so that the empirical risk is minimized (Lam, 2012; Ekici, 2008). Finally, SVM is capable to solve nonlinear problems without constructing a feature space with explicit high dimension and to be applied for both binary and multiclass classification (Chilukuri and Li, 2014).

PRACTICAL SIMULATION RESULTS AND DISCUSSIONS

Simulation studies have been performed to investigate the performance of the power quality detection and classification methods. In this section, the description of the methods shown in the block diagram, simulated circuits and their results are discussed including speed (time response) of the detection methods and the precision of the classification ones. Two methods for detection and four combinations for classification were simulated.

Figure 1. Block diagram of PQDs detection & classification

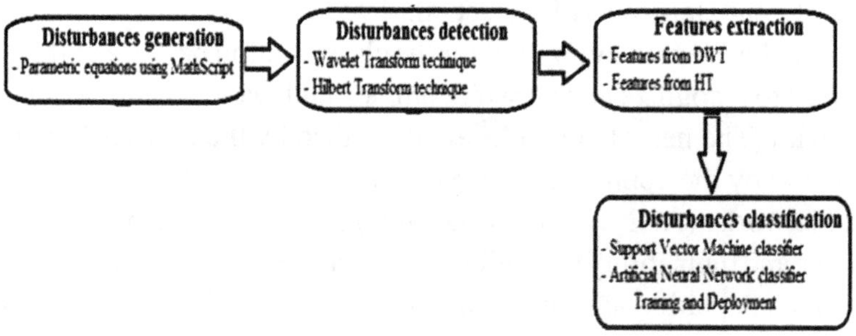

The software used for the simulation is the Laboratory Virtual Instrument Engineering Workbench (LABVIEW). It is a development environment based on the graphical programming language G. LABVIEW programs are called Virtual Instruments, or VIs. LABVIEW contains a comprehensive set of tools for acquiring, analyzing, displaying, and storing data. A VI contains the three components: the front panel, the block diagram, and the icon and connector pane; The front panel serves as the user interface whereas the block diagram contains the graphical source code that defines the functionality of the VI, The icon and connector pane identifies the VI so that a VI can be used in another VI.

1. Disturbance Generation

The generation of power quality events was performed through a program written in the MATLAB language called by a MATLAB Script node which offered in LABVIEW. In addition to a pure sine wave with amplitude of 230V and frequency of 50Hz, eight types of PQDs that are commonly occur in power system were generated including sag, swell, interruption, harmonics, transient, flicker, sag with harmonics, and swell with harmonics. The different parameters of signals represented in the starting time, duration and distortion magnitude were generated randomly. Signals simulated that way are very close to reality because none of these are fixed for real power system events. On the other hand, different signals belonging to the same class give the opportunity to estimate the generalization ability of classifiers. During the detecting and classifying of PQ disturbances, 100 different cases for every type of PQ disturbance waveforms are generated.

2. Detection Methods

After generating the disturbed signals, the event detection is performed via Discrete Wavelet Transform VI and Fast Hilbert Transform VI for the DWT and HT's detection methods respectively. For DWT, the mother wavelet chosen is fourth-order Daubechies (db4) and the MRA is of 5 levels. The detection is made using the third level detailed coefficient signal (D3). The simulated circuit is shown in Figure 2.

The results of detection are shown in the Figure (3)

For the HT, the phase shifting algorithm has been applied to the different disturbances sing the envelope of the transform as shown in Figure (4).

The disturbances have been successfully detected the results are illustrated in Figure (5).

Table 2 compares the measured time to the real starting time used in the disturbance generation program for the eight disturbances.

Figure 2. Detection circuit of the DWT

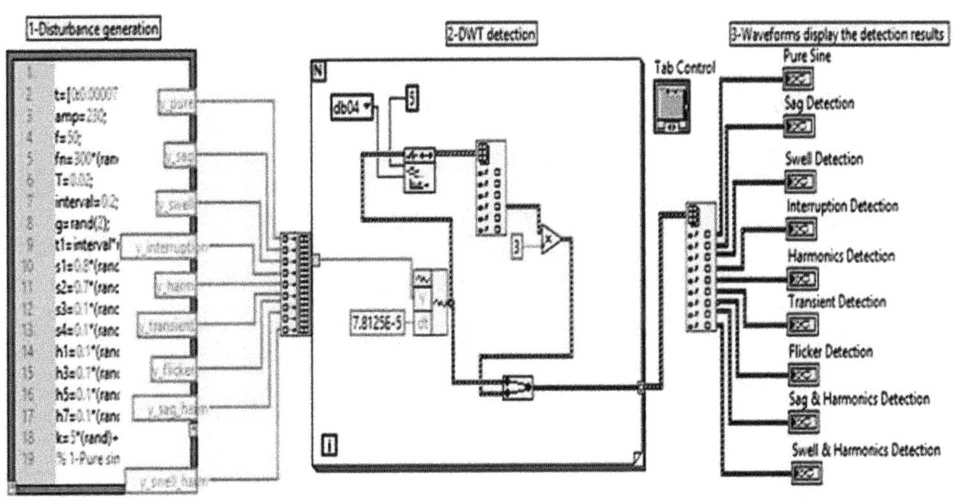

- As it is mentioned in Figure (4) and Figure (5), the two proposed methods detected the eight disturbances successfully and for the pure sine signal the detector signals stayed a straight line along the x-axis to indicate the absence of any disturbance.
- For DWT method, the detection is made by spikes at the starting and ending time of disturbance. Therefore in addition to detect the disturbance, it measures the period of disturbance occurrence and that is an advantage for it.
- For HT method, the detection is made by reshaping the envelope according to the shape of disturbance depending on its magnitude. Therefore the advantage of this method is giving information about the amplitude of disturbance.

Figure 3. Detection results of DWT

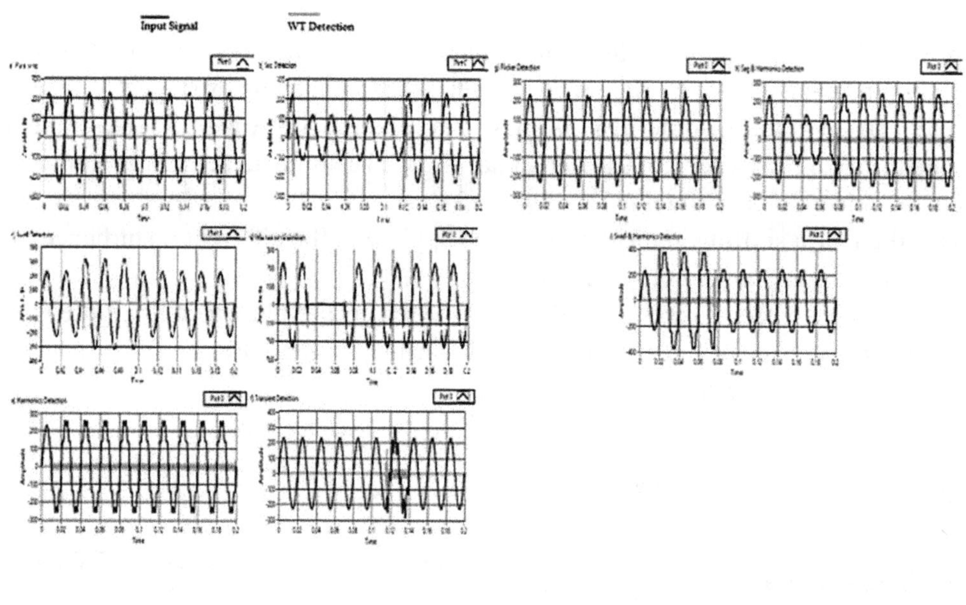

Figure 4. Detection circuit of HT

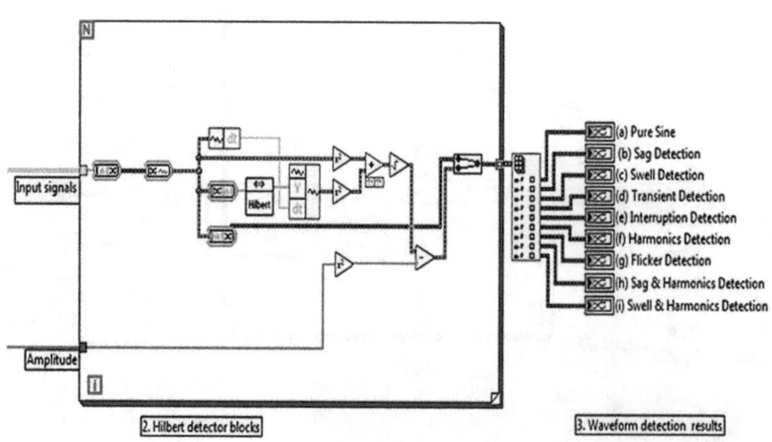

Figure 5. Detection results of HT

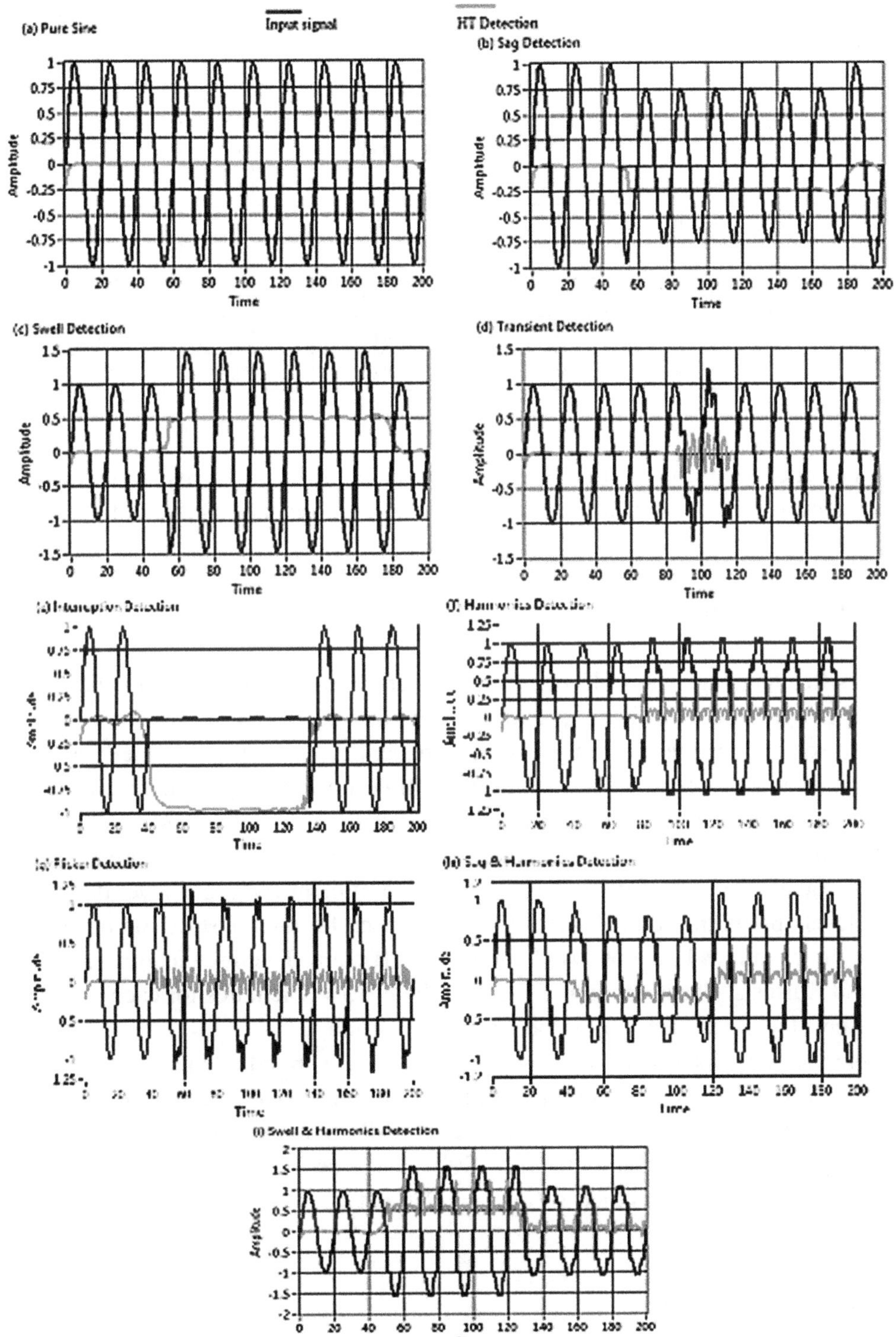

Table 2. Detection time for DWT and HT

Disturbance	Detection of Starting Time for DWT			Detection of Starting Time for HT		
	Actual Time (ms)	Measured Time (ms)	Time Delay (ms)	Actual Time (ms)	Measured Time (ms)	Time Delay (ms)
Sag	22.671	22.758	**0.087**	54.42	55.2	**0.78**
Swell	116.967	117.043	**0.076**	31.18	32.1	**0.92**
Interruption	59.141	59.601	**0.46**	14.56	15.5	**0.94**
Harmonics	46.040	46.612	**0.572**	60.24	61.14	**0.90**
Transient	95.187	95.247	**0.06**	106.66	107.23	**0.57**
Flicker	42.263	42.682	**0.419**	45.6	46.33	**0.73**
Sag & harmonics	77.945	78.358	**0.413**	87.14	88.02	**0.88**
Swell & harmonics	35.859	35.888	**0.029**	137.88	138.5	**0.62**

- For the detection of starting time, the Wavelet detection has a higher time response than the Hilbert detection which has also an acceptable time response. This is due to the spikes generated by DWT.

3. Features Extraction

Before the classification stage takes place, a stage of feature extraction stage must be done. The input of the classifier is a preprocessed signal generated from this stage. Feature extraction is the key for pattern recognition so that it is the most important component of designing the intelligent system based on pattern recognition since even the best classifier will perform poorly if the features are not chosen well. In this work, the preprocessed signal was generated using features of DWT and HT once a time.

For the WT, the wavelet coefficients obtained from discrete wavelet and multiresolution analysis of the disturbance signal represented in 1st level detailed coefficient (D1), 5th level detailed coefficient (D5), and 5th level approximated coefficient (A5) were used for the construction of feature vector by calculating the three features based on DWT for each coefficient. The three features are Energy, Shannon Entropy and Kurtosis. The number of features is than 3*3 = 9 features for each disturbance signal; Figure 6 represents the block diagram for the feature extraction.

1000 samples of each disturbance class were randomly generated. Hence the size of the training data is 1000*9 = 9000 data samples per class. For HT, nine extracted features were calculated for the envelope of the transform. The features are Mean value, RMS, Standard deviation, Variance, Maximum and Minimum Values, THD, and Duration. Also 1000 disturbances for each class were randomly generated and the size of training data is then 1000*9 = 9000 data samples per class. The block diagram is shown in Figure 7.

4. Classification Methods

After preparing two different training data, they were utilized as the inputs of the untrained models SVM and ANN. These two late are available blocks in LABVIEW. Four proposed methods resulted from the possible combination between training data and classifiers. The methods are:

- Training data obtained from DWT as trained data for SVM classifier denoted as DWT-SVM.
- Training data obtained from HT as trained data for SVM classifier denoted as HT-SVM.
- Training data obtained from HT as trained data for ANN classifier denoted as HT-ANN.
- Training data obtained from DWT as trained data for ANN classifier denoted as DWT-ANN.

Figure 6. Feature extraction circuit based on DWT

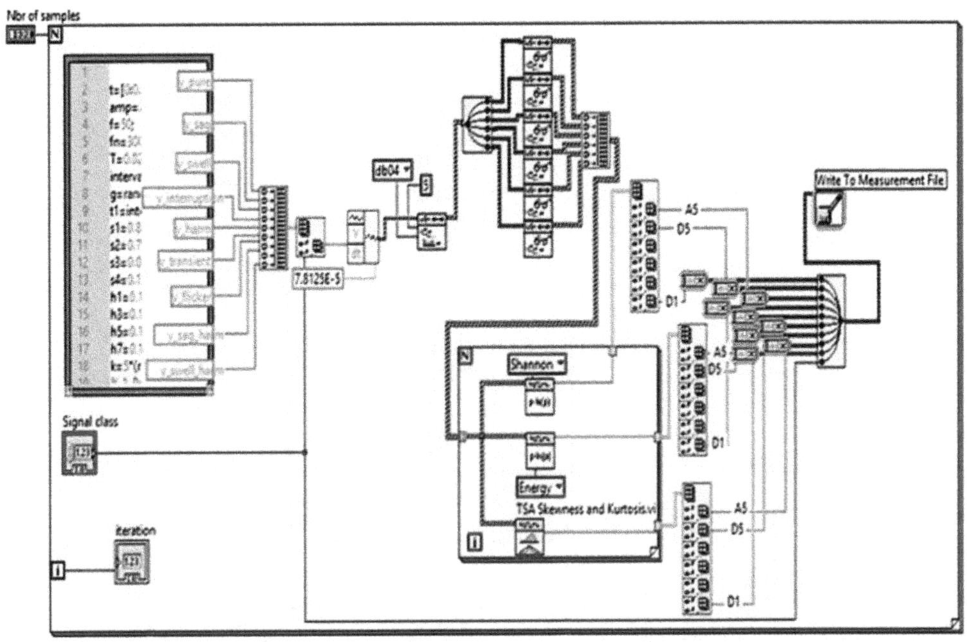

Figure 7. Features extraction from the Envelope of HT

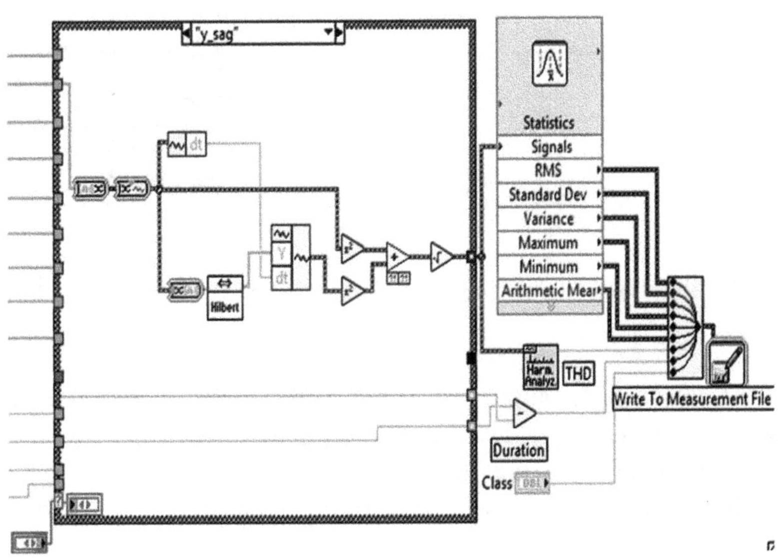

The training process of the two classifier are shown in Figures (8) and (9), where the test data file in the figures is the CSV file contains the training data obtained from the features extraction.

SVM Classifier is shown in Figure 8.

ANN classifier is shown in Figure 9.

The classification model parameters were set automatically by the model itself using cross validation techniques. 100 testing data sample for each class were used to test the performance of each method. The classification results are shown in Figure (10).

Figure 8. Testing circuit of SVM classifier

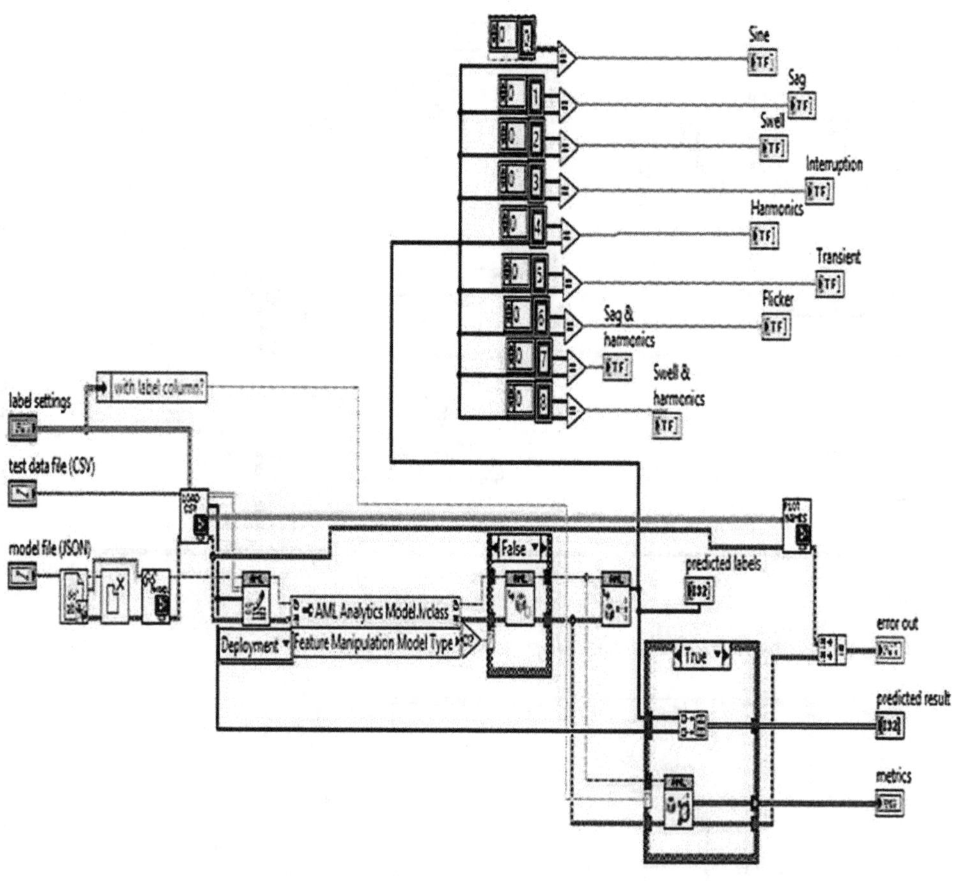

The following tables illustrate the results including the number of correct classified and misclassified events and the accuracy.

- First, DWT- SVM which has the higher value of precision over all the used methods
- Second, HT-SVM also has acceptable results.
- Third, HT-ANN almost has as the previous method.
- Finally, DWT-ANN which has the lowest value of precision among the used methods.

Figure 9. Testing circuit of ANN classifier

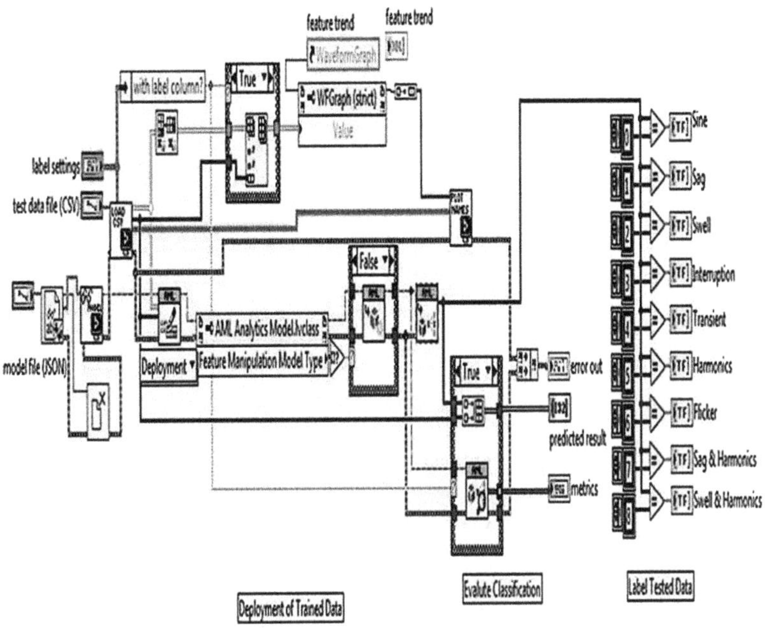

Table (7) indicates the accuracy of the four proposed methods.
The results reveal the following remarks:

- All the proposed classification methods give good accuracy results that are between 96.7% and 99.4%.
- These results show the effectiveness of the features chosen for both DWT and HT.
- The features extracted from DWT are more suitable for SVM classifier whereas the features extracted from HT are more suitable for ANN.
- The lowest accuracy was resulted in DWT-ANN, and this indicates that the DWT features utilized are not much suitable for the ANN classifier.
- ANN classifier had low accuracy in classifying Sag & Harmonic events with DWT (91%).
- SVM classifier had low accuracy in classifying Sag & Harmonic events with HT (88%).
- SVM classifier proves its robustness and efficiency over ANN.
- The method DWT-SVM expresses a very high accuracy that makes it to be the best one over the other methods, which proves the effectiveness of DWT features for SVM classifier.

5. Overall PQ Monitoring System

From the previous results of detection and classification, a proposed method for a power quality monitoring system was deduced. The method utilizes the combination between DWT and SVM; DWT is used for detection and features extraction of disturbances, and SVM classifier is used for identifying the type of disturbances. The system had been successfully simulated and the resulted front panels for the nine events are shown in Figure (11).

Figure 10. Testing results of the four proposed methods (a) WT-SVM,(b) HT-SVM,(c) HT-ANN,(d) WT-ANN

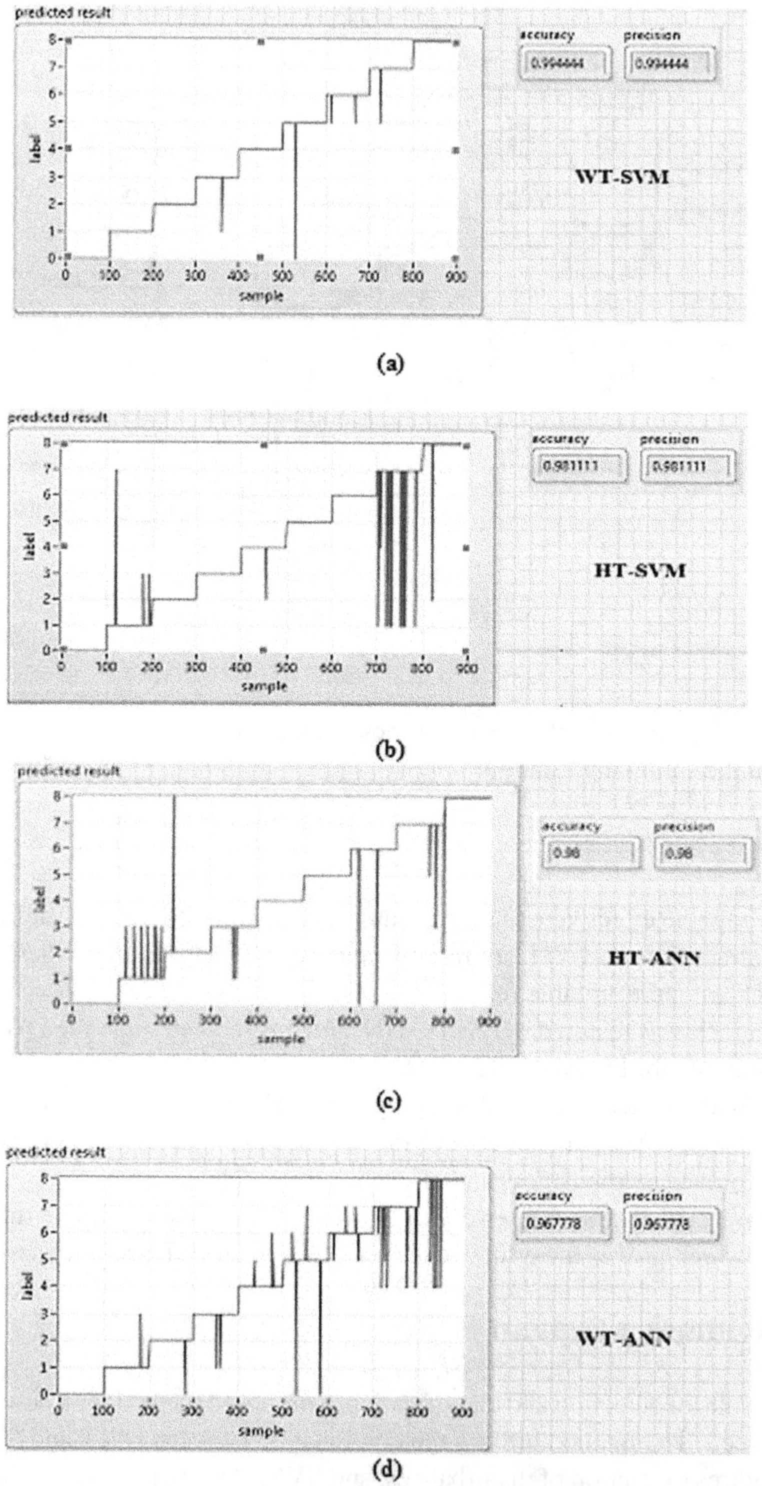

Table 3. Classification results using the proposed WT-SVM classifier

Signals	Tested Samples	Correctly Identified	Fault Identified	Precision (%)
Pure Sine	100	100	0	100
Sag	100	100	0	100
Swell	100	100	0	100
Transients	100	99	1	99
Interruptions	100	99	1	99
Harmonics	100	100	0	100
Flicker	100	98	2	98
Sag & Harmonics	100	99	1	99
Swell & Harmonics	100	100	0	100
Total	***9000***	***895***	***5***	***99.44***

Table 4. Classification results using the proposed HT-SVM classifier

Signals	Tested Samples	Correctly Identified	Fault Identified	Precision (%)
Pure Sine	100	100	0	100
Sag	100	97	3	97
Swell	100	100	0	100
Transients	100	100	0	100
Interruptions	100	100	0	100
Harmonics	100	99	1	99
Flicker	100	100	0	100
Sag & Harmonics	100	88	12	88
Swell & Harmonics	100	99	1	99
Total	***9000***	***883***	***17***	***98.11***

Table 5. Classification results using the proposed HT-ANN classifier

Signals	Tested Samples	Correctly Identified	Fault Identified	Precision (%)
Pure Sine	100	100	0	100
Sag	100	90	10	90
Swell	100	99	1	99
Transients	100	100	0	100
Interruptions	100	98	2	98
Harmonics	100	100	0	100
Flicker	100	98	2	98
Sag & Harmonics	100	97	3	97
Swell & Harmonics	100	100	100	100
Total	***900***	***882***	***18***	***98.00***

Table 6. Classification results using the proposed WT-ANN classifier

Signals	Tested Samples	Correctly Identified	Fault Identified	Precision (%)
Pure Sine	100	100	100	100
Sag	100	99	1	99
Swell	100	99	1	99
Transients	100	95	5	95
Interruptions	100	98	2	98
Harmonics	100	97	3	97
Flicker	100	96	4	96
Sag & Harmonics	100	91	9	91
Swell & Harmonics	100	96	4	96
Total	***9000***	***871***	***29***	***96.78***

Table 7. Accuracy results of all simulated methods

Method	DWT-SVM	HT-SVM	HT-ANN	DWT-ANN
Accuracy (%)	99.44	98.11	98.00	96.78

CONCLUSION

The objectives of this chapter were to develop a PQM system that had the ability to detect power quality disturbances at the minimum possible of time and to correctly identify them with high precision, which is the major area of research in power system field. To do that, WT and HT were introduced as signal processing tools used for detecting and characterizing PQDs and they have proved their efficiency and flexibility in detecting different disturbances, either the amplitude variation disturbances or the frequency variation ones. Also the machine learning tools ANN and SVM were proposed in the literature to classify the different PQDs, as a result, the accuracy of the classifiers was encouraging by more than 96% that shows their effectiveness. From the analysis done in the literature, it may be concluded that:

- One of the main purposes in the power quality field is the development of high speed-accuracy automatic system for the identification, detection and classification of disturbance.
- DWT with MRA technique is an effective tool for analyzing non-stationary signals as the ones of PQDs and detect any type of them at a faster rate (0.029ms ~ 0.572ms) by choosing suitable mother wavelet and multi-resolution level.
- Feature extraction is the key for pattern recognition to design the best performance classifier.
- Features extracted from HT and DWT are more efficient for a classifier than the ones extracted directly from the disturbed signal.

Figure 11. PQM Front Panel (a) Sine, (b) Sag, (c) Swell, (d) Interruption, (e) Transients, (f) Harmonics, (g) Flicker, (h) Sag & Harmonics, (i) Swell & Harmonics

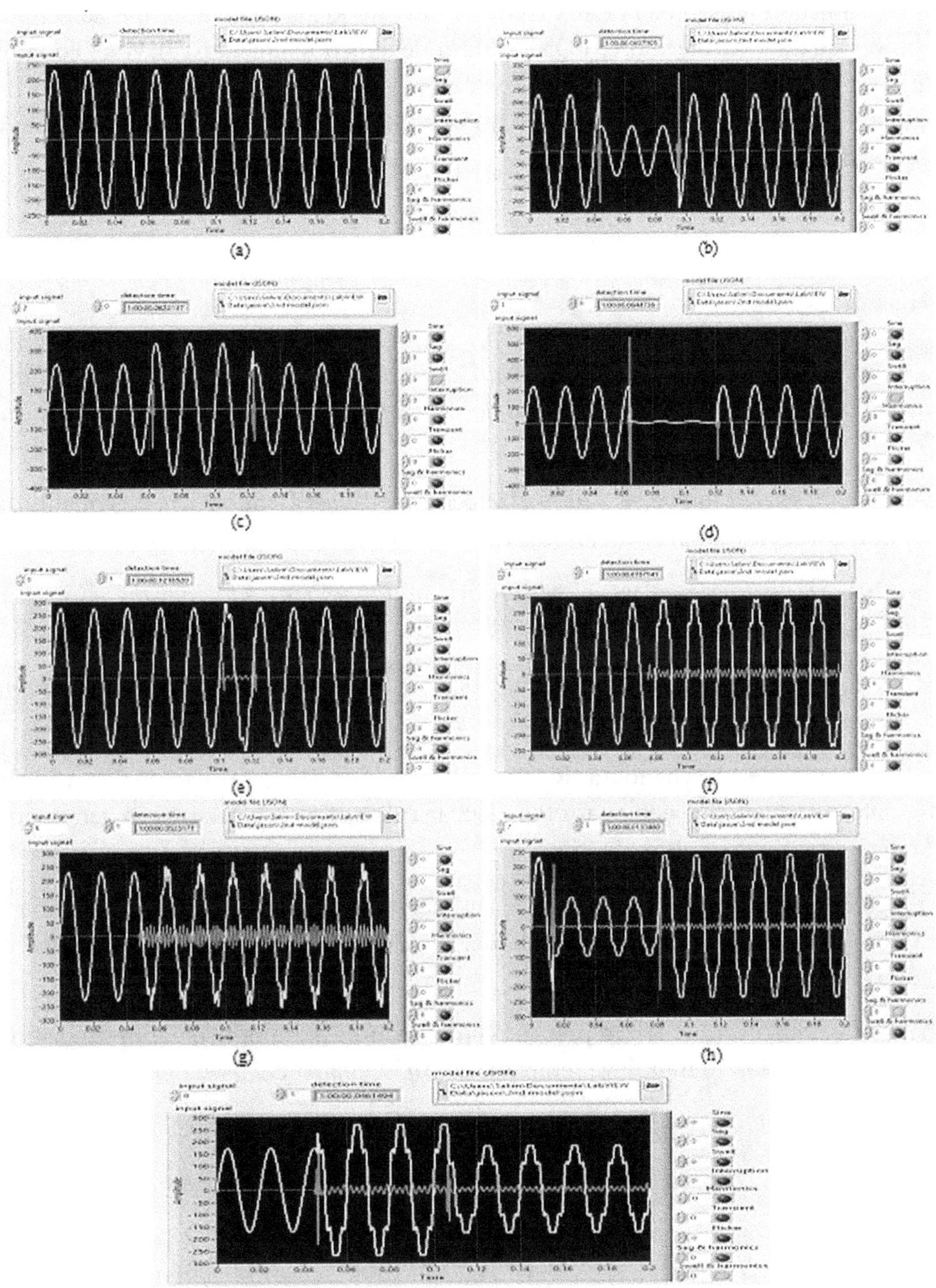

- ANN and SVM classifier are efficient tools in automatic classification of disturbances (96.78% ~ 99.44%), for best accuracy results; features that are the training data for the classifier must be well chosen by trying different sets of features combination.

- The combination DWT-SVM proposed in the literature shows high robustness and accuracy (99.44%) for detection and classification of the nine PQDs.
- Virtual instrument represented in LabVIEW is a very good instrument for the detection and analysis of power quality, furthermore, PQM system designed using LabVIEW is economic, highly flexible, user-friendly and easily upgradable comes up as a revolutionary tool with its improved performance and more reliability than the traditional PQ analyzers which themselves may be tedious to handle and time consuming.

REFERENCES

Abbod, M., Alamri, B., & Alshahrani, S. (2015). *Detection and classification of power qualityevents based on wavelet transform and artificial neural networks for smart grids, In 2015 Saudi Arabia Smart Grid.* SASG.

Abdoos, A. A., Pazoki, M., & Moravej, Z. (2009). Detection and classification of power quality disturbances using wavelet transform and support vector machines. *Electric Power Components and Systems, 38*(2).

Addison, P. S. (2017). *The illustrated wavelet transform handbook: introductory theory and applications in science, engineering medicine and finance*. CRC Press.

Aggarwal, R., & Kim, C. H. (2000). Wavelet transforms in power systems, Part 1: General introduction to the wavelet transforms. *Power Engineering Journal, 14*(2).

Aggoun, G., Boubouzel, N., Kachenoura, R., & Fellag, S. A. (2017, April). Quality of energy resulting from association converter- machine for the electric drive of a system of wind generation. *Algerian Journal of Signals and Systems, 2*(1), 1–11.

Azzougui, Y., & Recioui, A. (2018). Hardware and software load power control in smart home applications based on Taguchi optimisation technique. *Int. J. Intelligent Systems Design and Computing, 2*(3/4), 203–223.

Azzougui, Y., & Recioui, A. (2019). (in press). Application of the moth flame optimisation to the selective harmonic elimination in multilevel converters. *Int. J. Smart Grid and Green Communications*.

Baggini, A. (2008). *Handbook of power quality*. John Wiley & Sons. doi:10.1002/9780470754245

Bansal, R., Manana, M., & Zobaa, A.F. (2011). *Power quality: Monitoring, analysis and enhancement*. Academic Press.

Berkowitz, J., Elkan, C., & Lipton, Z. C. A. (2015). *Critical review of recurrent neural networks forsequence learning*. arXiv:1506.00019

Bicker, R., & Jabar, A. A. A. (2014). Simulation of Non-stationary Signal Analysis Using Wavelet Transform Based on LabVIEW and Matlab. In *2014 European Modelling Symposium* (pp. 138-144). IEEE.

Blume, D., Stephanblome, T., & Schlabbach, J. (2001). *Voltage quality in electrical power systems*. IET.

Bollen, M. H., Watson, N. R., & Arrillaga, J. O. (2000). Power quality following deregulation. *Proceedings of the IEEE*, *88*(2).

Chakravarthi, K. R., & Ortmeyer, T. M. (1985). *The effects of power system harmonics on power system equipment and loads. IEEE Trans. Power Appar. Syst.*

Chandra, P., & Pahuja, P. (2015). Power Quality Monitoring Using LABVIEW. *International Journal Of Electrical Electronics & Computer Engineering*, *4*(2), 59.

Chilukuri, M. V., & Li, J. (2014). Power supply quality nalysis using S-transform and SVM classifier. *Journal of Power and Energy Engineering*, *2*(04), 438–447. doi:10.4236/jpee.2014.24059

Cui, Y., Li, A., & Liu, Z. (2015). A classification method for complex power quality disturbances using EEMD and rank wavelet SVM. *IEEE Transactions on Smart Grid*, *6*(4), 1678–1685. doi:10.1109/TSG.2015.2397431

Daponte, P., D'apuzzo, M., Testa, A., & Angrisani, L. (1998). A measurement method based on the wavelet transform for power quality analysis. *IEEE Transactions on Power Delivery*, *13*(4).

Daubechies, I. (1992). *Ten lectures on wavelets*. Siam. doi:10.1137/1.9781611970104

Davim, P., & Ozel, T. (2009). *Intelligent maching*. Wiley-IEEE Press.

Dekhandji, F. Z. (2017). Signal processing deployment in power quality disturbance detection and classification. Acta Physica Polonica A, 132(3), 415-419.

Dekhandji, F. Z. (2018), Optimal placement of power factor correction capacitors using a discrete teaching learning based optimization. WSEAS Journal of Power Systems, 13, 226-234.

Dekhandji, F. Z., & Bentarzi, H. (2019, December). Theoretical and Experimental Investigation of the Supply Voltage Unbalance Effects on Squirrel Cage Induction Motors. *Algerian Journal of Signals and Systems*, *4*(2), 51–60.

Dekhandji, F. Z., Bouaraki, M., Zaidi, N., & Recioui, A. (2019). Demand Response and Optimal Sizing of Hybrid Renewable Energy Systems. *International Journal of Control Energy and Electrical Engineering*, *7*, 35–40.

Dekhandji, F. Z., Douche, M., & Zebidi, N. (2017). DVR and D-STATCOM Mitigation Techniques of Power Quality Effects on Induction Motors. *Algerian Journal of Signals and Systems*, *2*(2), 110–129.

Dekhandji, F. Z., Kassah, M. T., & Bensalah, A. (2020). Harmonic Reduction and Elimination in Three Phase PWM Inverters using a Spiral-Inspired Optimization Technique, *Algerian J. Env. Sc. Technology*, *6*(3), 1436–1447.

Dekhandji, F. Z., Talhaoui, S., & Arkab, Y. (2019, December). Power Quality Detection, Classification and Monitoring Using LABVIEW. *Algerian Journal of Signals and Systems*, *4*(2), 101–111.

Diego, R. I., De Apràiz, M., & Barros, J. (2012), Applications of wavelets in electric powerquality: Voltage events. Electric Power Systems Research, 130-136.

Ekici, S. (2008), A Pattern Recognition System For Power disturbance classification. *The 5th International Conference on Electrical and Computer Systems (EECS'08).*

Ekwue A, Aggarwal R Warwick K (1997), Artificial Intelligence Techniques in Power Systems (Energy Engineering). *The Institution of Engineering and Technology.*

Fais, M., Ghani, A., Abidin, A. F., & Hannoon, N. S. (2017). Real time detection and classification of single and multiple power quality disturbance based on embedded S-transform algorithm in labview. *Proceedings of the 2017 international symposium on Industrial Engineering and Operations Management (IEOM)*, 280–289.

Fausett, L. V. (1994). *Fundamentals of neural networks: architectures, algorithms, and applications.* Prentice-Hall.

Fritz, W., & Gupta, G. (2016). *Voltage unbalance for power systems and mitigation techniques a survey.* In *2016 IEEE 1st International Conference on Power Electronics, Intelligent Control and Energy Systems (ICPEICES).* IEEE.

Gawre, S. K., Kumar, T., & Kumar, V. (2014). Power quality analysis using wavelet transform: a review. *International Journal of Innovative Research in Science, Engineering and Technology, 3*(3).

Hahn, S. L. (1996). *Hilbert transforms in signal processing.* Artech House.

Haykin, S. (1994). *Neural Networks: a comprehensive foundation* (Vol. 2). Prentice Hall PTR.

Hossen, A. (2001). A new fast approximate hilbert transform with different applications. IIUM Engineering Journal, 2(2).

Huihui, Q., Wei, Z., Pengfei, W., & Chunling, C. (2012). Transient power quality signal generatorand detector platform. *Energy Procedia, 16*, 1380–1385. doi:10.1016/j.egypro.2012.01.219

Kennedy, B. W. (2000). *Power quality primer.* McGraw Hill Professional.

Lam, H. K. (2012). *Computational intelligence and its applications: evolutionary comutaion, fuzzy logic, neural network and support vector machine techniques.* World Scientific.

Lobos, T., & Janic, P. (2006). Automated classification of power-quality disturbances using SVM and RBF networks. *IEEE Transactions on Power Delivery, 21*(3), 1663–1669. doi:10.1109/TPWRD.2006.874114

Mao, J., Mohiuddin, K. M., & Jain, A. K. (1996). Artificial neural networks: A tutorial. *Computer, 1*(3).

Martens, J., Hinton, G. E., & Sutskever, I. (2011). Generating text with recurrent neural networks. *Proceedings of the 28th International Conference on Machine Learning (ICML-11)*, 1017-1024.

Masoum, M. A., & Fuchs, E. (2011). *Power quality in power systems and electrical machines.* Academic Press.

McGranaghan, M. F., Beaty, Hw., Santos, S., & Dugan, R. C. (1996). *Electrical power systems quality* (2nd ed.). McGraw-Hill.

Mishra, S., Salama, M. A., Liew, A. C., & Dash, P. K. (2000). Classification of power system disturbances using a fuzzy expert system and a Fourier linear combiner. *IEEE Transactions on Power Delivery, 15*(2).

Mitra, M., Sengupta, S., & Chattopadhyay, S. (2011). *Electric power quality*. Springer.

Morcos, M. M., & Ibrahim, W. A. (2002). Artificial intelligence and advanced mathematical tools for power quality applications: A survey. *IEEE Transactions on Power Delivery*, *17*(2).

Muneta L. M. & Rey G. R. (2011). *Electrical generation and distribution systems and power quality disturbances*. Academic Press.

Poularikas, A. D. (2010). *Trasforms and applications handbook*. CRC press. doi:10.1201/9781420066531

Radhika, P., Subasri, V., Kanagavalli, R., & Devaraj, D. (2006). Power Quality monitoring usingwavelet transform and artificial Neural Networks. *2006 India International Conference on Power Electronics, IEEE*, 425-430.

Ram, G. T., Laxmi, A. J., & Sushama, M. (2010). Detection of power quality disturbances using wavelet transforms. *International Journal of the Computer, the Internet and Management, 18*(1).

Rao, K.U., Jade, S., & Sridhar, S. (2016). Detection and classification of power quality disturbances inthe supply to induction motor using wavelet transform and neural networks. *Balkan Journal of Electrical and Computer Engineering*, *4*(1).

Recioui, A. (2018). *Optimal Placement of Power Factor Correction Capacitors Using Taguchi Optimization Method. In Handbook of Research on Emergent Applications of Optimization Algorithms*. IGI Global.

Recioui, A., Bentarzi, H., & Tsebia, M. (2016). PMU deployment in power system oscillation monitoring. *Sustainable Aviation*, 313-322.

Recioui, A., Tazibt, M., & Benhenni, H. (2017, September). Fault Detection in Wide Area Monitoring Systems via PMU Optimal Placement. *Algerian Journal of Signals and Systems*, *2*(3), 149–161.

Romero-Troncoso,, R.J., Osornio-Rios,, R.A., Garcia-Perez,, A., & Cabal-Yepez,, E., & Granados-Lieberman, D. (2011). Techniques and methodologies for power quality analysis and disturbances classification in power systems: A review. *IET Generation, Transmission & Distribution*, *5*(4).

Samanta, C. K., & Mishra, D. P. (2013). Power Quality Monitoring by Disturbance Detection using Hilbert Phase Shifting. *Int. Conf. on Recent Trends in Power, Control and Instrumentation Engineering*.

Sankaran, C. (2001). *Power quality*. CRC Press.

Sarma, D. S., & Venkatesh, C. (2010). Techniques used for Detection of Power Quality Events-a Comparative Study. *16th National Power Systems Conference*.

Sazli, M. H. (2006). A brief review of feed-forward neural networks. University of Ankara. doi:10.1501/0003168

Sharmeela, C., Rajeshwari, V., Keerthana, B., & Murugesan, P. (2017). Localizing Power Quality Disturbances Using Hilbert Transform In labview. *National Conference On Innovations And Research In Electrical & Electronics Engineering*.

Skvarenina, T. L. (2018). *The power electronics Handbook*. CRC Press. doi:10.1201/9781420037067

Taskovski, D., & Markovska, M. (2018). The effectiveness of wavelet based features on power quality disturbances classification in noisy environment. *18th International Conference on Harmonics and Quality of Power (ICHQP)*, 1-6.

Ucar, A., Demir, Y., & Eristi, H. (2010). Wavelet-based feature extraction and selection for classification of power system disturbances using support vector machines. Electric Power Systems Research, 80(7), 743-752.

Vapnic, V. (1998). The support vector method of function estimation. In *Nonlinear Modeling* (pp. 55–85). Springer. doi:10.1007/978-1-4615-5703-6_3

Wu, C. H., Lin, C. H., Cheng, F. S., & Lin, W. M. (2008). Detection and classification of multiple power- quality disturbances with wavelet multiclass SVM. *IEEE Transactions on Power Delivery*, *23*(4), 2575–2582. doi:10.1109/TPWRD.2008.923463

Yang, Y., Xu, T., Cao, Y., Wang, X., & Chen, C. L. (2009). Power quality disturbances detection based on hilbert phase-shifting. 2009 Asia-Pacific Power and Energy Engineering Conference, 1-4.

Zin, A. M., Mokhtar, A. S., Ismail, N. M., Zaren, N., & Khokhar, S. (2013). *Automatic classification of power quality disturbances: A review. In 2013 IEEE Student Conference Research and Developemen.* IEEE.

KEY TERMS AND DEFINITIONS

Disturbances (Power Quality Problems): The reasons by which the power quality is affected. The occurrence of such problems in the power system network is almost indispensable. To maintain the quality of power care must be taken that suitable devices are kept in operation to prevent the consequences of these problems.

Power Quality: Power quality is referred as maintaining near sinusoidal power distribution at rated voltage magnitude and frequency.

Chapter 12
Fault Tree-Based Root Cause Analysis Used to Study Mal-Operation of a Protective Relay in a Smart Grid

Hamid Bentarzi
Signals and Systems Laboratory (LSS), University M'hamed Bougara of Boumerdes, Algeria

ABSTRACT

This chapter presents how root cause analysis based on fault tree technique can be used to study mal-operation of protective relay in a smart power grid. This approach is used to identify disturbances first and root cause of mal operation of protective relays which may lead to unnecessary tripping and hence to blackout. Once the critical root causes are identified, mitigation measures first such as blocking protection functions and digital filters may be used in view of increasing the reliability of the considered protection system. The proposed approach has many advantages as it allows obtaining an important quantitative figure (security). These permits to strongly strengthen the elements of the protecting system which are most likely appropriate to failure and hence the impact on the overall system's cost is significant. Another main advantage is that it takes into account the reliability of the software part of the system which is considered to have a significant contribution in the overall reliability of the protecting system.

INTRODUCTION

A protection system is a vital part of any electric power system and plays an incredible role in maintaining high availability required in present day. The mal-operation (false trip) of one relay of the protective scheme to operate as intended may place at risk the stability of the whole power grid and hence may lead to blackout.

In this chapter, Root Cause Analysis (RCA) based on Fault Tree Analysis (FTA) is used to identify disturbances first and root cause of false trips of protective relays which may lead to unnecessary tripping in cascading and then to the bulk blackout. Once the critical root causes are identified for this mal-

DOI: 10.4018/978-1-7998-4027-5.ch012

operation of these relays, mitigation measures such as blocking functions and digital filters may be used for increasing the reliability of the considered protection system. Some research works (Chena, J. et al, 2005; Jonsson, M., & Daalder, J., 2001; Lin, X. et al, 2008; Su, B., et al, 2005; Jiao S., et al 2001; Rao, J. G., & Pradhan A. K.; 2012) proposed some efficient solutions (based mainly on digital filters) to obtain accurate and disturbance-free phasor measurements. As a result, the reliability of the measuring block of the protection system can be significantly improved. However, this improvement concerns one term of the reliability (the dependability) of the protection system. In the present chapter, how RCA based FTA is used to improve the reliability of the global protection system. This proposed approach has many advantages compared to those published in previous works as it allows obtaining an important quantitative figure (security). This permits to strongly strengthen the elements of the protective system which are most likely appropriate to failure and hence the impact on the overall system's cost. Another main advantage, it takes also into account the reliability of the software part of the system which is considered to have a significant contribution in the overall reliability of the digital /numerical protective system.

A Cigré study found that 27% of bulk power system failure (blackout) resulted from false trips of protection system. Besides, an analysis of 17 years data provided in NERC reports revealed that 63% of major disturbances are protection related protective relays which suffers from two types of mal-operation: unnecessary tripping (false trip) and fail to operate (Rao, J. G., & Pradhan A. K., 2012; Sachdev, M.S., & Baribeau M. A., 1979; Tziouvaras, D., 2007).

Major power system disturbances are more likely to be caused by unnecessary tripping rather than by the failure of a relay to take action (Jonsson, M. & Daalder, J.; 2001). In fact, a major cause of power system instability is due to the relative long time delay for fault clearance of zone three backup relays in distance relay that may used to protection the power transmission lines. In order to overcome the drawbacks of these relaying systems, many mitigation techniques have been developed such as a blinder that blocks the relay to operate during the power swing (Bentarzi H., Ouadi A., and Maun J.C., 2014). A more advanced new blocking and unblocking function based on power swing detector using Phasor Measurement Units (PMUs) has been proposed. In addition to the local phasors information, the proposed scheme uses remote phasors information from different locations for power swing detection, which can be provided at high speed by PMUs. These measurements are used for calculating the apparent power absorbed by power line and the difference in phase angles of voltages that may be used for detecting power swings and faults. Hence, the proposed scheme blocks tripping signal during the power swing and unblocks it during fault condition (Ouadi, A. Bentarzi, H., Chafai, M.; 2016).

The second order of major causes of instability of the power system is due to the magnetization current in differential relay used to protect the power transformers. In order to overcome the drawbacks of these relaying systems, many techniques have been developed such as an introduction of delay that blocks the relay to operate during the inrush phenomena (Sonnemann, W.K., Wagner C.L., & Rockefeller G.D., 1958). The mal-operation of this differential relay is generally due to not only unnecessary tripping during inrush current that reduces the security of protection system and hence its reliability but also unnecessary tripping when external fault occurs accompanying an overexcitation (Kennedy, L. F., & Hayward, C. D., 1938). Also this latter action may affect on a security of the relay and hence its reliability. Many techniques have been developed to block the operation of the relay during the saturation associated with an external fault condition such as a harmonic restrains (Hayward C. D., 1941). A more advanced new blocking function based on fourth harmonic detector using digital signal processing has been proposed (Guzman, A., & Zocholl S., 2001; Bouderbal, R., Bentarzi H. Ouadi, A., 2011). In addition to the second harmonic magnitude information, the proposed system uses other harmonic orders magnitudes information for different disturbances detection, which can be provided at high speed by digital differential relay. Hence, the proposed system blocks tripping signal during the inrush current as well as when an overexcitation occurs during external fault condition.

Moreover, protective relays can easily discriminate between fault and non-fault conditions when power quality is good. When power quality is poor, the threshold between normal and fault conditions becomes imprecise. In the relaying application, engineers can no longer rely on the performance of the relay under conventional normal system conditions. However, under conditions of poor power quality especially in the smart power grid including TCSC, they have improved the relay performance by designing digital filters which in turn make the relay more secure and dependable. In several cases, transmission line protective relays have undesirably operated in response to harmonics or dc offset in the power grid (Tziouvaras, D., 2007).

In this study, Root Cause Analysis based on fault tree analysis is used to identify disturbances first and root cause of false trips of a protective relay which may lead in cascading to blackout. Once the critical root causes are identified, conventional mitigation measures have been used first and then blocking and unblocking functions and digital filters in view of increasing the reliability of the considered protection system are proposed. As a result, the reliability of the measuring block of the protection system has been significantly improved. However, this improvement concerns the reliability of one part (the phasor measurement block) of the protection system. In the present work the reliability of the global protection system is considered. This new approach has many advantages compared to those published in previous works as it allows obtaining an important quantitative figure (security). This permits to selectively reinforce the elements of the protecting system which are most likely apt to failure and hence the impact on the overall system's cost. Another main advantage is that unlike former works, our approach takes also into account the reliability of the software part of the system which is considered to have a significant contribution in the overall reliability of the numerical protective system.

PROTECTIVE SYSTEM DESIGN CONSIDERATIONS

A protective system should be designed to recognize certain system abnormalities which, if undetected, could lead to damage equipments or extended loss of service. The design and specification of the system components are important parts of the protective strategies and power system that are designed to withstand the usual operating contingencies that accompany load changes and line switching operation. There are several design considerations that must be weighed against cost in devising a protection plan; the following fundamental requirements that are considered in designing the protective systems with a good performance of a relay are (Phadke, A. G., 1993; Lin, X., et al, 2008):

- **Reliability:** it is the ability of the relay system to operate under the predetermine condition. Without reliability, the protection will be rendered largely ineffective and could even become a liability. The reliability of a relay is directly in correspondence with the concepts of dependability and security. A relay is said to be dependable when it operates in the occurrence of a fault relevant to its protection zone. In other words, dependability is a measure of the relay ability to operate when it is supposed to operate. Security is defined as "the degree of certainties that a relay or relay system will not operate incorrectly". Security is reached either when the relay does not operate for a fault outside its operating zone, or when the system is in a healthy state.

- **Selectivity:** it is the ability of the protective system to disconnect correctly a part of the system in trouble (a faulty part) and the healthy parts of the power system remain connected. Selectivity can be achieved by time grading or by unit protection. Selectivity by time grading means that different operation zones of protection are graded by time and that in the occurrence of a fault, although a number of protective equipment responds, only those relevant to the faulty zone complete the tripping function. Selectivity by unit protection means that the relay will only operate under certain faults occurring within a well defined zone.
- **Speed:** the relay system must disconnect the faulty section rapidly due to the following reasons. During a fault, the longer the time in which the fault is affecting the power system, the risk is greater that the power system falls into an unstable operation point. Relays are therefore required to clear the fault as quickly as possible otherwise the following may happen:
    - Electrical apparatus may be damaged if the fault current persists for a long time.
    - A failure on the system leads to a great drop in the system voltage. If the faulty section isn't disconnected rapidly, then the low voltage created by the fault may shut down the customers' motors and the generation stations on the system may be unstable.
    - The high speed relay system decreases the possibility of development of more sever type of fault.
- **Sensitivity:** it is the ability of the relay system to operate with low value of actuating quantity. In other words, the relay is said to be sensitive if it operates with minimum value of faulted input signals.
- **Simplicity:** the relaying system should be simple so that it can be easily maintained. Simplicity has great impact on the reliability. The simpler the protection scheme will have the greater reliability.
- **Economy:** the most important factor in the choice of a particular protection scheme is the economic aspect. Sometimes it is economically unjustified to use an ideal protection scheme and compromise methods have to be adopted. As a rule, the protective gear should not cost more than 5% of total cost. However, when the apparatus to be protected is of utmost importance, economics considerations are often subordinate to reliability.

PROTECTIVE SYSTEM CLASIFICATION

The relay application to protection of power system date back nearly 100 years ago. Since then, the technology has been developed to construct relays that have improved dramatically relay size, weight, cost and functionality. Based on the technology employed for their construction, relays can be classified as follows (IEEE Guides and Standards for Protective Relaying Systems. (1991); Kasztenny, B., & Kezunovic, M., 1998; Abdelmoumene, A., & Bentarzi H., 2014):

1. **Electromechanical relays:** The first relays have been employed in the electric industry were electromechanical devices. These relays operation is based on creating a mechanical force to operate the relay contacts in response to a fault. The mechanical force is established by the flow of a current that reflects the fault current through windings mounted in magnetic cores. Electromechanical relays are relatively heavier and bulkier than relays constructed with other developed technologies. Besides, the burden of these relays can be extremely high, affecting protection purposes. However, electromechanical relays have been so largely employed, tested and known that even modern re-

lays employ their principle of operation, and still represent a good choice for certain application conditions.

2. **Solid-state relays (static relays):** With the advances on electronics, the electromechanical relays as the first generation have been replaced by static relays in the early 60's. The operating characteristic of static relays is based in analog circuitry rather than in the action of windings. The advantages of this generation compared to electromechanical relays are a reduced size, weight and electrical burden.
3. However, static relays have some disadvantages due to analog circuitry that is extremely affected by electromagnetic interference and the ranges of current and voltages values are strongly restricted in analog circuits, affecting the sensitivity of the relay.
4. **Digital relays:** implementation of this type relay incorporates a microprocessor and logic functions, which was started happening in the 80's. Digital relays integrate analog-to-digital converter (ADC) to sample the analog signals incoming from instrument transformers, and used in microprocessor to define the logic of the relay. Digital relays presented an improvement in accuracy and control over incoming signals, and the use of more complexes relay algorithms, extra relay functions and complementary task.
5. **Numerical relays:** The difference between numerical relays and digital relays lies in the kind of used microprocessor. Numerical relays use digital signal processors (DSP) cards, which contain dedicated microprocessors especially designed to perform digital signal processing.

DISTANCE PROTECTION TECHNIQUE

Distance relay may be used in the power grid protection. The basic principle governing the operation of a distance relay is the impedance Z at the relaying point as shown in Fig.1. Then, the measured impedance is compared to the set impedance, and if this impedance is within the reach of the relay then it operates.

Figure 1. A typical single line AC connection of a protective distance relays

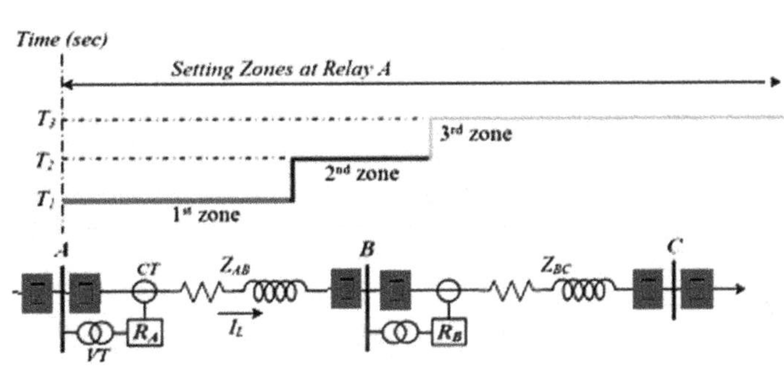

Distance relay can be implemented through the use of phase comparators of two input quantities for obtaining the operating characteristics which may be circles when they are plotted on an R-X diagram. The impedance elements of the distance relay are based on the two quantities being compared that are the voltage and current measured by the relay.

The signals S_1 and S_2 that may be introduced to the phase comparator are in the following form:

$$S_1 = -KI + K_1 \left|V\right| \angle \left(\theta - \varphi\right) \tag{1}$$

$$S_2 = KI + K_2 \left|V\right| \angle \left(\theta - \varphi\right) \tag{2}$$

The general equation of phase comparator at threshold condition (S_1-S_2=0) can be obtained:

$$-K^2 I^2 + -KK_2 \cos\left(\theta - \varphi\right) + K_1 K \cos\left(\theta - \varphi\right) VI + K_1 K_2 V^2 = 0 \tag{3}$$

Letting $Z=V/I$, we get:

$$-K^2 + K_1 K_2 Z^2 + KZ\left(K_1 - K_2\right)\cos\left(\theta - \varphi\right) = 0 \tag{4}$$

Replacing Z^2 by R^2+X^2 and after some Algebraic manipulations, we finally obtain:

$$\left(R - \frac{K'\left(K_1 - K_2\right)}{2K_1 K_2}\cos\theta\right)^2 + \left(X - \frac{K\left(K_1 - K_2\right)}{2K_1 K_2}\sin\theta\right)^2 = \left(\frac{K\left(K1 + K_2\right)}{2K_1 K_2}\right)^2 \tag{5}$$

Figure 2. The characteristic of the offset mho relay

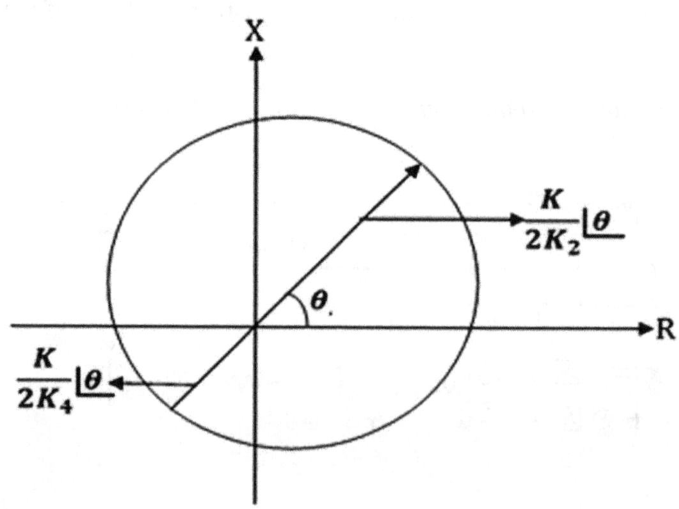

This equation represents a circle with a radius of $\left(\frac{K}{2K_2}+\frac{K}{2K_1}\right)$ and centered at $\frac{K\left(K_1-K_2\right)}{2K_1K_2}\angle\theta$ on the R-X plan. Figure 2 shows the characteristic of the offset MHO relay.

The distance relay can be set to protect different zones of a transmission line either in forward direction (MHO characteristic) or in both forward and backward direction (Offset MHO characteristic).

The zone 1 protection for the MHO relay should never fall outside of the line to be protected as shown in Fig.1. Normally it is protecting up to 80-90% of the line. Therefore, the remote transmission line end is not protected by zone 1 relays. Zone 2 relays protect the entire transmission line section and a half of the next section but for coordination purpose a time delay may be introduced. Zone 3 relay is normally a backup protection for an adjacent transmission line sections associated with longest time delay.

Traditional impedance-based techniques for detecting power swings in the transmission network as shown in Fig.3, involve measuring apparent impedance and introducing a time delay between two measuring elements. The impedance measurement is typically a phase-phase impedance, where all three phase-phase loops are required for operation, or it is positive-sequence impedance. A timer is started when the apparent impedance enters the characteristic as illustrated in Fig.3. If the apparent impedance remains between the inner and outer characteristics for the set time delay, the Power Swing Block (PSB) element operates and selected distance element zones are blocked from operation for a period of time. A timer determines if the change in impedance is due to a fault or a power swing. If it is an unstable power swing, then one can select tripping on the way into the characteristic or on the way out of the characteristic. Selection of tripping on the way in or out is determined from numerous stability studies and the ability of the circuit breakers to isolate the circuit with a significant voltage angle across the breaker (Ouadi, A. Bentarzi, H., Chafai, M., 2016).

Lines currents disturbances in a transmission line which contain a dc offset which decays exponentially, and/or a large number of unwanted significant harmonics due to switching actions may cause false trip of the relay. In numerical relay, Discrete Fourier Transform (DFT) is the most widely used filtering algorithm (Giray, M.M., & Sachdev, M.S., 1989; Girgis A.A., & Brown R.G., 1981; Phadke, A. G., & Thorp, J. S., 1988; Gurusinghe, et al, 2012; Johns.A.T., & salman S.K., 1995; Yalla, V. S., 1992) for computing the fundamental components and illuminating these disturbances.

Figure 3. Different power swing protection schemes: (a) Double blinder, power swing characteristic, (b) Offset mho, power swing characteristic

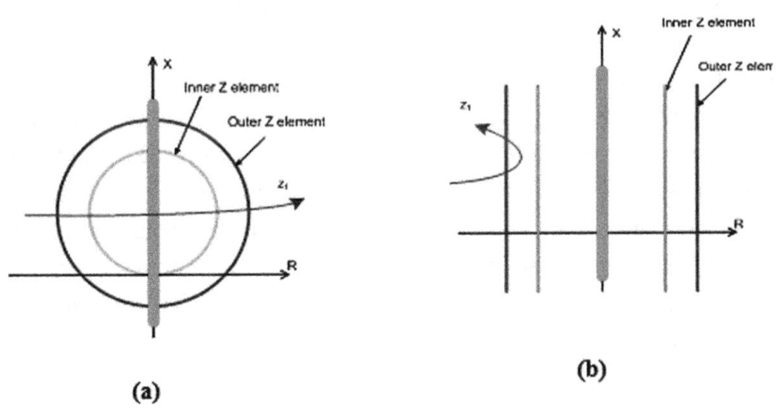

DIFFERENTIAL PROTECTION TECHNIQUE

Differential protection replaces overcurrent protection and Buchholz relay as the main protection for large power transformers.

A typical differential protection system is shown in Fig.4. Multiple circuits may exist, but the example is sufficient to explain the basic principle of differential protection (Guzman, A., & Zocholl S., 2001). It can be observed from Fig.2 that the protection zone is delimited by current transformers. Due to its very nature, differential protection does not provide backup protection to other system components. For this reason, differential protection is categorized as a unit protective scheme. The conductors bringing the current from the current transformers to the differential relay are in some situations called pilot wires.

Figure 4. A typical connection of a protective differential relays with power transformer.

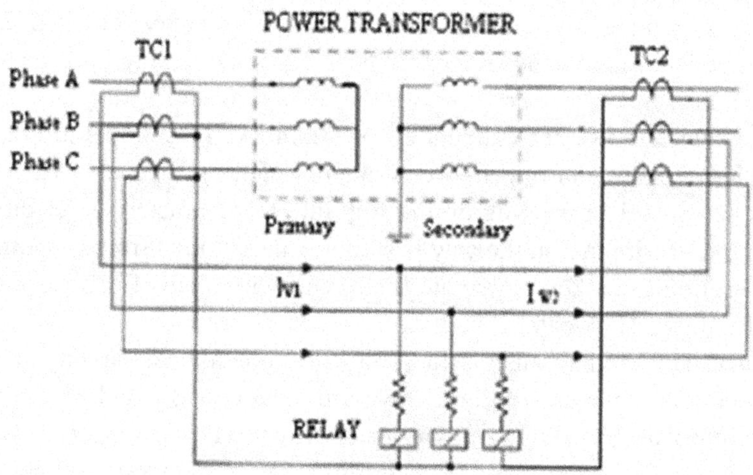

Figure 4 shows a diagram illustrating the principle of the differential relay protection. Current transformers with similar characteristics and ratio are connected on the both sides of the transformer and a relay is connected between the two current transformers by using pilot wires. Under healthy or external fault conditions, the current distribution as shown in Fig.5 (a), no current flowing in the relay. When the internal fault occurs as shown in Fig.5 (b), the conditions of balance are upset and current flows in the relay to cause operation. It can be noted also that the protected zone of this differential relay is between the two current transformers. If the fault had occurred beyond, as shown in Fig.2 (a), than the operation will not occur as the fault current would then flow through both current transformers thus maintaining the balance. Differential relays perform well for external faults as long as the current transformers reproduce the primary currents correctly (Bouderbala, R., Bentarzi, H., & Ouadi A.,2011).

When one of the current transformers saturates, or if both current transformers saturate at different levels, false operating current appears in the differential relay and causes relay mal-operation. Some relays use the harmonics caused by the current transformer saturation for added restraint (Bouderbala, R., Bentarzi, H., & Ouadi A.,2011).

Figure 5. Principle of Differential Relay Protection.

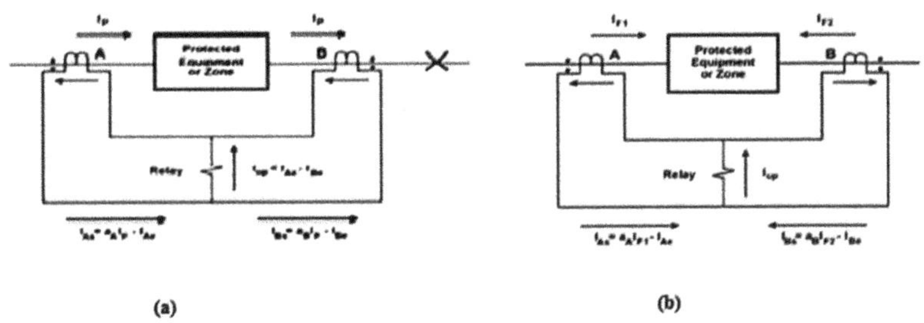

Under normal conditions, the current I_p entering the protected unit would be equal to the current leaving it at every instant as shown in Fig.5. Consider current transformer A. The secondary current of current transformer A is equal to,

$$I_{As} = \alpha_A I_p - I_{Ae} \tag{6}$$

where, α_A is the transformation ratio of current transformer A, and I_{Ae} is the excitation current of current transformer A on the secondary side.

For current transformer B, the equation is similar and is as follows.

$$I_{Bs} = \alpha_B I_p - I_{Be} \tag{7}$$

where, α_B is the transformation ratio of current transformer B, and I_{Be} is the excitation current of current transformer B on the secondary side.

Assuming equal transformation ratios, $\alpha_A = \alpha_B = \alpha$, the relay operation current I_{op} is given by:

$$I_{op} = I_{Ae} - I_{be} \tag{8}$$

During normal system operation and during external faults, the relay operating current I_{op} is small, but it is never zero ($I_{op} \neq 0$).

In the event of a fault in the protection zone, the input current is no longer equal to the output current. The operating current of the differential relay is now the sum of the input currents feeding the fault as shown in Fig.5.

$$I_{op} = \alpha(I_{F1} + I_{F2}) - I_{Ae} - I_{be} \tag{9}$$

Transformer differential relay are subject to several factors that can cause maloperation such as: magnetizing inrush currents, which differential relay sees as internal faults, overexcitation due to the external faults and current transformer saturation.

Traditional techniques for detecting inrush current in the power transformer when the first time is energized as shown in Fig.6, introduce the second harmonic pass filter (Kennedy, L. F., & Hayward, C. D.; 1938).

Figure 6. Blocking during the inrush current in the Differential relay

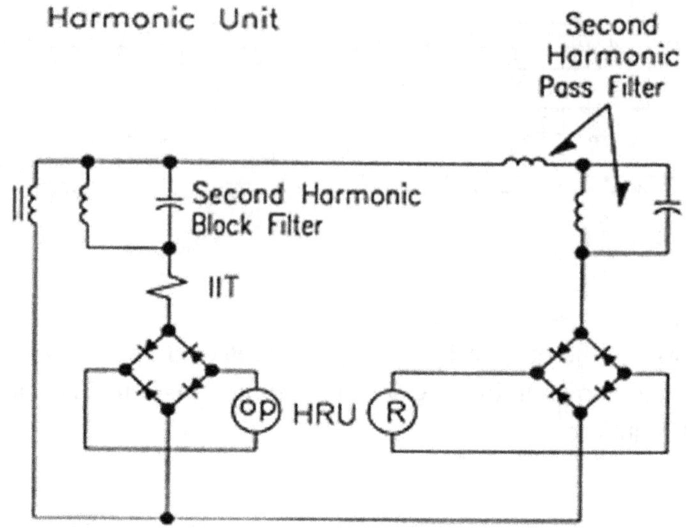

Currents of disturbances entering the power transformer which contain a dc offset which decays exponentially, and/or a large number of unwanted significant harmonics due to switching actions may cause false trip of the relay. In digital relay, Discrete Fourier Transform (DFT) is the most widely used filtering algorithm (Giray, M.M., & Sachdev, M.S., 1989; Girgis A.A., & Brown R.G., 1981; Phadke, A. G., & Thorp, J. S., 1988) for computing the fundamental components and illuminating these disturbances.

FAULT TREE ANALYSIS

A fault tree (FT) is a top-down logical diagram that displays the interrelationshos between a critical system event and its causes. The main elements of a fault tree are (Ebling, C. E., 2000):

- TOP event, which is the description of the critical system event.
- Basic events, which are the lowest level of identified causes,
- Logic gates, such as OR or AND gates, which give the logical relationship between the TOP event and the basic events.

Fault tree analysis (FTA) is the qualitative and quantitative analyses that can be carried out on the basis of a fault tree.

Fault tree analysis is carried out in five steps (Ebling, C. E., 2000):

1. Definition of the problem, system, and boundary conditions of the analysis,
2. Construction of the fault tree,
3. Identification of minimal cut sets,
4. Qualitative analysis of the fault tree,
5. Quantitative analysis of the fault tree.

FTA can be used as a valuable design tool, can identify potential accidents, and can eliminate costly design changes. It can also be used as a diagnostic tool, predicting the most likely system failure in a system breakdown. FTA is used in safety engineering and in all major fields of engineering (Waghmode, L. Y., & Patil, R. B., 2013).

ROOT CAUSE ANALYSIS

Root cause analysis is a step by step approach that leads to identify a disturbance's first or root cause. There are specific successions of events (disturbances) that lead to a failure such as false trip. A root cause analysis investigation follows the cause and the effect path from the final outcome back to the root cause (Abdelmoumene A. and Bentarzi H., 2012). Steps that may be followed for investigating this analysis are:

Step 1: Identify Possible Causal Factors. During the situation analysis, the researchers set the vision, identified the problem and collected data needed to better understand the current situation.
Step 2: Identify the Root Cause.
Step 3: Identify Communication Challenges.
Step 4: Prioritize Communication Challenges.

RELAY RELIABILITY

A reliable relaying system can be achieved by redundancy i.e. duplicating the relaying system. Obviously redundancy can be a costly proposal. However, it is important to realize that back-up protection for safe operation of relaying system. Redundancy in protection also depends upon the criticality of the power apparatus to be protected.

A quantitative measure for reliability is defined as follows:

$$R = \frac{N_c}{N_d + N_i} \tag{10}$$

Where, N_c: Number of corrected trips, N_d: Number of desired trips, N_i: Number of incorrect operations (false trip).

Protection system reliability is characterized by following two important terms: 1-Dependability, and 2-Security.

Dependability

A relay is said to be dependable if it trips only when it is expected to trip. This happens either when the fault is in its primary control (primary protection) or when it is called upon to ensure the back-up protection. However, false tripping of relay due to faults that are either not within its jurisdiction or within its purview may lead to power system instability. Power system may get unnecessarily stressed or else there can lose its service. Dependability (D) is the degree of certainty that the relay will operate correctly:

$$D = \frac{N_c}{N_d} \tag{11}$$

Security

On the other hand, security is a property used to characterize false tripping of the relays. A relay is said to be secure if it does not trip when it is not expected to trip. It is the degree of certainty that the relay will not operate incorrectly during a given time interval according to the IEEE/PSRC Working Group (ANSI/IEEE C37.91-1985).

$$S = 1 - \frac{N_i}{N_t} \tag{12}$$

Where S: security, and $N_t = N_c + N_i$: total number of trips.

Software Reliability Model

In software engineering, most of failures are due to errors, 'Bugs' or defect built into software from the beginning at design stage or exploitation and this may cause malfunction or halt in digital relaying system.

Assuming that an error is eliminated without causing any new errors, the simplest model that takes this into account is the following; let the number of errors N initially equal to No and let each have the same probability of detection p (and hence elimination) per unit time, then the failure rate is expressed as follows (Abdelmoumene A., Bentarzi H., Chafai. M., Ouadi A., 2014):

$$\lambda(t) = pN(t) = pNo\exp(-pt) \tag{13}$$

Under this model the failure rate is decreasing with time in exponential fashion with the essential tasks of debugging in order to eliminate the errors. The expression for reliability growth is given as follows:

$$Rs = \exp(+N) = \exp[\exp(-pt)] \tag{14}$$

CASE STUDY (I)

In this case, a Fault Tree is constructed to determine all possible causes that can lead to the undesirable top event that is false trip of distance relay.

A small power system integrating one source and three busses with three zones relays at each bus can be considered in this research paper as shown in Fig.1.

From field data, false trips may generally be caused by stable power swings, post-faults, the relay system failures or other unknown phenomena as given in Table 1 (IEEE Power System Relay Committee Working Group I17 Report, 2005; Chena, J., Thorpa, J. S., Dobsonb, I., 2005). These phenomena may include harmonics, DC offset and sub-harmonics generated by power electronic devices in the modern power grid. The fault tree diagram of the false trips (incorrect operations) has been illustrated in Fig.7.

Figure 7. Root cause Analysis based on FTA of false trip.

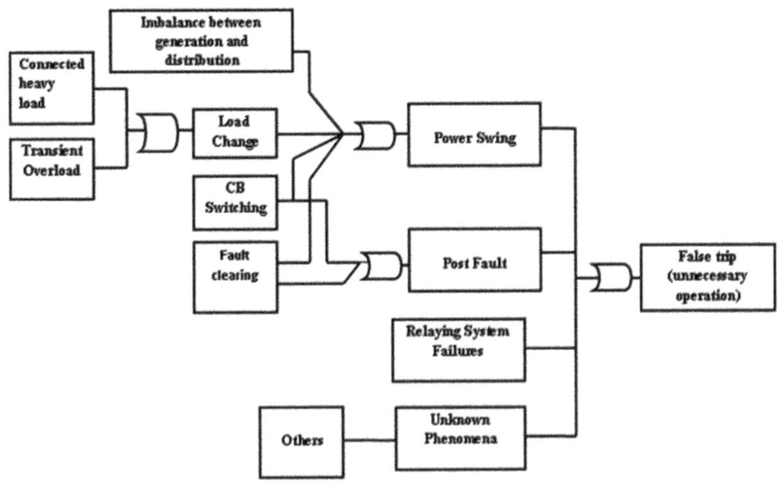

Table 1. Disturbances causes with their weights (proportions).

Disturbances	Abbreviation	Causes	W (%)
Power swing	PS	Power generation unbalance and load change.	30
Post-fault	PF	Switching actions	30
Relaying system	RS	Self technical failures in hardware or software parts and or incorrect setting, current transformer saturation	20
Others	UP	Unknown phenomena	20

Distance Relay Reliability (Security)

Subsequent to the identification of potential root cause of false trips (incorrect trips), the preventive mitigation (barriers) for enhancing the security and hence reliability of the relay has been proposed.

The weighted important factor of the disturbance cause, obtained previously, will provide the priority on mitigations and barriers which can be provided first against fast power swing, post fault and relaying system failures as shown in Fig.8.

These false trips may be produced from power swing, post-fault or switching actions. They do not just create trouble. They can even compromise system security. One way to categorize the data from the chosen utilities' relay operations is as follows: Over (Ne=1425) total events or disturbances per year we have the following data indicated in table 2 (Moxley, R., 2010; Phadke, A.G. et al 1976).

The probability of occurrence of power system disturbances leading to false tripping relative to the total number of disturbances or events can be evaluated as follows:

$$PDT = N_i / N_e \quad (15)$$

The security for the given protective system can be obtained using Eq. (7) associated with statistics of table 2 which gives Se=0.9632 (Zitouni A., Ouadi A., Bentarzi H., and Chafai M., 2018).

Figure 8. Root cause (FTA) of false trip after providing barriers

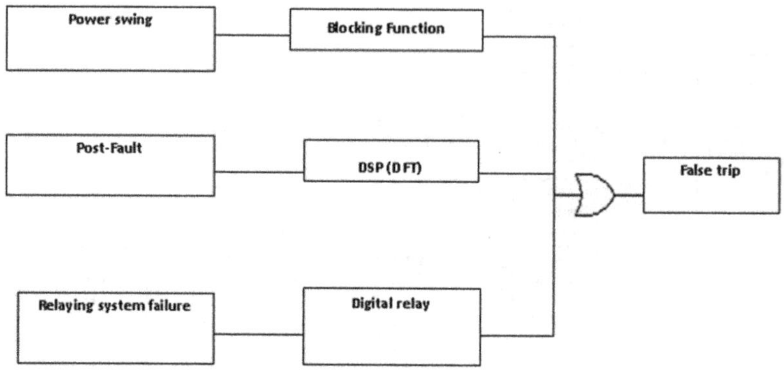

Table 2. Statistics for different operations

Operation	Symbol	Value
Corrected trips	N_c	1346
incorrect operations (False Trips)	N_i	66
Desired trips	N_d	1359
Failure to operate	N_f	13

CASE STUDY (II)

In this study, Root Cause Analysis based on fault tree analysis is used to identify disturbances first and root cause of false trips of a differential relay which may lead to power transformer damage. Once the critical root causes are identified, mitigation measures have been used in view of increasing the reliability of the considered protection system. As a result, the reliability of the measuring block of the protection system has been quantified.

Root cause analysis is used to develop new model of successions of events (disturbances) that lead to a false trip. A root cause analysis investigation follows the cause and effect path from the final outcome back to the root cause (Abdelmoumene A. and Bentarzi H., 2012). In this case, a Fault Tree is constructed to determine all possible causes that can lead to the undesirable top event that is false trip as shown in Fig.9.

A differential relay of power transformer can be considered in this research. The relay can monitor the current line electrical variables using appropriate number of current Transformers (CT). From field data, false trips may generally be caused by inrush current, overexcitation due to heavy external faults, the relay system failures or other unknown phenomena as given in Table 1 (Sonalgaz Company internal report, 2013; Anderson, P. A., et al, 1997). These phenomena may include harmonics and sub-harmonics generated by power electronic devices in the modern power grid. The fault tree diagram of the false trips (incorrect operations) has been illustrated in Fig.9.

Figure 9. Root cause Analysis based on FTA of false trip.

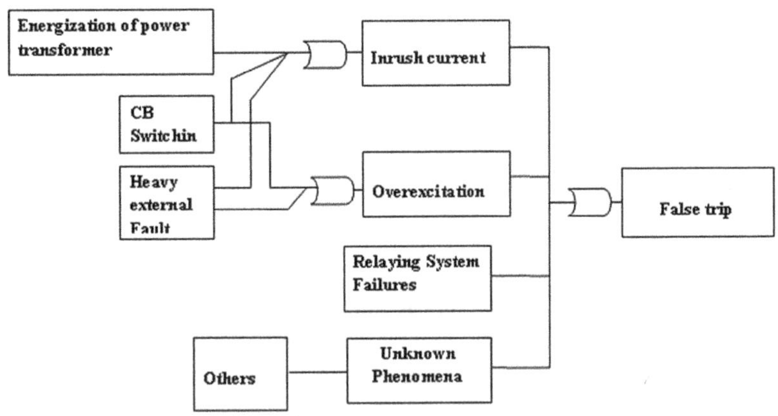

Table 3. Disturbances causes with their weights (proportions).

Disturbances	Abbreviation	Causes	W (%)
Inrush current	IC	Energization of Power Transformer.	45
Over-Excitation	OE	Switching actions	20
Relaying system	RS	Self technical failures in hardware or software parts and or incorrect setting, current transformer saturation	15
Others	UP	Unknown phenomena	20

Figure 10. Root cause (FTA) of false trip after providing barriers

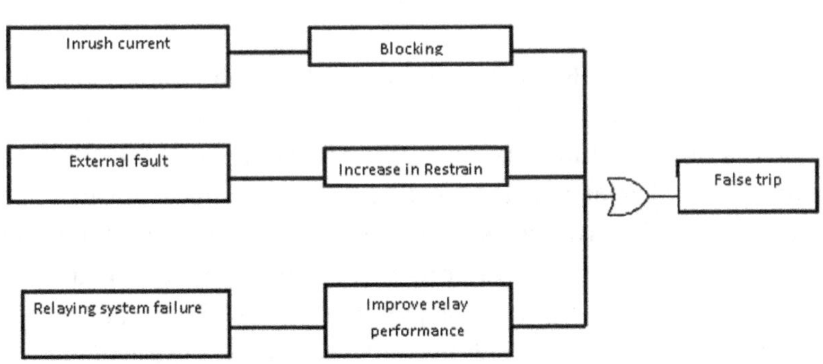

Differential Relay Reliability (Security)

Subsequent to the identification of potential root cause of false trips (incorrect trips), the preventive mitigation (barriers) for enhancing the security and hence reliability of the relay has been proposed.

The weighted important factor of the disturbance cause, obtained previously, will provide the priority on mitigations and barriers which can be provided first against inrush current, overexcitation and relaying system failures as shown in Fig.10.

Table 4. Statistics for different operations

Operations	Symbol	Value
Corrected trips	N_c	9
incorrect operations (False Trips)	N_i	124
Desired trips	N_d	10
Failure to operate	N_f	1

These false trips may be produced from inrush current and overexcitation due to external fault or switching actions. They do not just create trouble. They can even compromise system security. One way to categorize the data from the chosen utilities' relay operations is as follows: Over (Ne=134) total events or disturbances during a period of 5 years: from 2006 till 2010, we have the following data indicated in table 4 (Anderson, P. A., et al, 1997).

The probability of occurrence of power system disturbances leading to false tripping relative to the total number of disturbances or events can be evaluated using Eq.(15) that is 0.925

The security for the given protective system can be obtained using Eq. (7) which is S=0.26. This new approach has many advantages it allows obtaining an important quantitative figure (security). This permits to selectively reinforce the elements of the protecting system which are most likely apt to failure and hence the impact on the overall system's cost is significant. Another main advantage takes also into account the reliability of the software part of the system which is considered to have a significant contribution in the overall reliability of the numerical/digital protecting system.

CONCLUSION

In this chapter, Root Cause Analysis (RCA) based on fault tree analysis has been used to identify root cause of false trips of distance relay. Once the critical root causes have been identified and their weights quantified, barriers such as blocking functions and DSP Algorithms have been introduced. By applying recent developed techniques to the barriers, a considerable increase in the security and hence reliability has been noticed. One of the main advantages of the RCA is to make possible to perform quantitative analysis on the protection system before and after using improved barriers. Thus, the overall cost of the system may be significantly optimized, since it is possible to selectively reinforce the elements of the protecting system which are most likely apt to failure and hence avoid the use of unnecessary redundancy.

In this chapter, different blocking techniques for differential protection were also discussed in context with the blocking function applied in large power transformer and smart transformer. It is concluded that each of the blocking technique has its own pros and cons depending on the application involved. Besides, the application of various latest hardware implementation techniques can deliver more reliable protective differential relay regarding the dependability and security including the rapidity.

A relaying system has to be reliable. Reliability can be achieved by redundancy i.e. duplicating the relaying system. Obviously redundancy can be a costly proposition. Another way to improve reliability is to make an existing protection scheme, an apparatus A to backup protection of apparatus B. Both the approaches are used (simultaneously) in practice. However, it is important to realize that back-up protection must be provided for safe operation of relaying system. Redundancy in protection also depends upon the criticality of the power apparatus.

NOMENCLATURE

CT: current transformer,
D: Dependability,
PDT: Probability of occurrence of disturbances leading to tripping,
P_{ft}: probability of causes leading to 'false trips',
P_{ut}: Probability of occurrence of disturbances leading to 'unnecessary tripping',
OE: Overexcitation,
IC: Inrush Current,
R: Reliability,
RS: Relay System
S: security,
S_n: new security,
S_e: existing security,
N_c: Number of corrected trips,
N_d: Number of desired trips,
N_e: total number of events of disturbances,
N_f:Number of Failures to operate,
N_i: Number of incorrect trips (false trip),
N_t: total number of trips,
UP: Unknown Phenomenon.

REFERENCES

Abdelmoumene, A., & Bentarzi, H. (2012). Reliability Enhancement of Power Transformer Protection System. *Journal of Basic and Applied Scientific Research*, *2*(10), 10534–10539.

Abdelmoumene, A., & Bentarzi, H. (2014). A Review on Protective Relays' Developments and Trends. *Journal of Energy in Southern Africa*, *25*(2), 91–95. doi:10.17159/2413-3051/2014/v25i2a2674

Abdelmoumene, A., Bentarzi, H., Chafai, M., & Ouadi, A. (2014). *Reliability assessment and improvement of digital protective relays. Int J Syst Assur Eng Manag*. doi:10.100713198-014-0300-z

Anderson, P. A., Flemming, B., Lee, T. J., & Schweitzer, E. O. (1997). Reliability Analysis of Transmission Protection Using Fault Tree Methods. *Proceedings of the 24th Annual Western Protective Relay Conference*.

ANSI/IEEE C37.91-1985 (1985). *IEEE Guide for Protective Relay Applications to Power Transformers*. IEEE Publication, New York.

Bentarzi, H., Ouadi, A., & Maun, J. C. (2014). A New PMU Based Power Swing Detector to EnhanceReliability of Distance Relay. *Proceedings of CIGRE Session 45*.

Bouderbala, R., Bentarzi, H., & Ouadi A. (2011). Digital Differential Relay Reliability Enhancement of Power Transformer. *International Journal of Circuits, Systems and Signal Processing*, *5*, 263-270.

Chena, J., Thorpa, J. S., & Dobsonb, I. (2005). Cascading dynamics and mitigation assessment in power system disturbances via a hidden failure model. *Electrical Power and Energy Systems*, *27*(4), 318–326. doi:10.1016/j.ijepes.2004.12.003

Ebling, C. E. (2000). *Introduction to Reliability and Maintainability Engineering*. McGraw Hill Publishing House.

Giray, M. M., & Sachdev, M. S. (1989). Off-normal frequency measurements in electric power systems. *IEEE Transactions on Power Delivery*, *4*(3), 1573–1578. doi:10.1109/61.32645

Girgis, A. A., & Brown, R. G. (1981). Application of Kalman filter in computer relaying. *IEEE Trans. on PAS*, *PAS-100*(7), 3387–3395. doi:10.1109/TPAS.1981.316681

IEEE Guides and Standards. (1991). *IEEE Guides and Standards for Protective Relaying Systems*. IEEE Publication.

Gurusinghe, D. R., Rajapakse, A. D., & Narendra, K. (2012). Valuation of Steady-State and Dynamic Performance of a Synchronized Phasor Measurement Unit. In *Proceedings IEEE elect. Pow. And Ene. Conf. 2012*. IEEE. 10.1109/EPEC.2012.6474980

Guzman, A., Zocholl, S., Benmouyal, G., & Altuve, H. J. (2001). A current based solution for transformer differential protection – part 1: Problem Statement. *IEEE Transactions on Power Delivery*, *16*(4), 485–491. doi:10.1109/61.956726

Hayward, C. D. (1941). Harmonic-Current-Restrained Relays for Transformer Differential Protection. *Transactions of the American Institute of Electrical Engineers*, *60*(6), 377–382. doi:10.1109/T-AIEE.1941.5058349

IEEE Power System Relay Committee Working Group I17 Report. (2005). *Transmission Relay System Performance Comparison*. IEEE Publication.

Jiao, S., Bo, Z., Liu, W., & Yang, Q. (2001). New Principles to Detect Faults During Power Swing. *Seventh IEE Int. Conf. on Dev. in Power Syst. Protect.*, 515 – 518.

Johns.A.T., & salman S.K. (1995). Digital protection for power system. *IEE Power Series,* 15.

Jonsson, M., & Daalder, J. (2001). A New Protection Scheme to Prevent Mal-Trips Due to Power Swings. *IEEE/PES Transmission and Distrib. Conf.*, *2*, 724 – 729. 10.1109/TDC.2001.971327

Kasztenny, B., & Kezunovic, M. (1998). Improved Power Transformer Protection UsingNumerical Relays. *IEEE Computer Applications in Power*, *11*(4), 39–45. doi:10.1109/67.721702

Kennedy, L. F., & Hayward, C. D. (1938). Harmonic-Current-Restrained Relays for Differential Protection. *Transactions of the American Institute of Electrical Engineers*, *57*(5), 262–266. doi:10.1109/T-AIEE.1938.5057808

Lin, X., Gao, Y., & Liu, P. (2008). A Fast Unblocking Scheme for Distance Protection to Identify Symmetrical Fault Occurring During Power Swings. *IEEE Transactions on Power Delivery*, *1*, 73–77. doi:10.1109/TPWRD.2007.911154

Moxley, R. (2010). *Analyse relay fault data to improve service reliability*. Schweitzer Engineering Laboratories, Inc.

Ouadi, A., Bentarzi, H., & Chafai, M. (2016). A new PMU based power swing detector to prevent mal-operation of distance relay. *Russian Electrical Engineering*, *87*(10), 572–578. doi:10.3103/S1068371216100059

Phadke, A. G. (1993). Synchronized phasor measurements in power systems. *IEEE Computer Applications in Power*, *6*(2), 10–15. doi:10.1109/67.207465

Phadke, A. G., Hlibka, T., & Ibrahim, M. (1976). A digital computer system for EHV Substation: Analysis and Field Tests. *IEEE Transactions on Power Apparatus and Systems*, *PAS-95*(1), 291–301. doi:10.1109/T-PAS.1976.32105

Phadke, A. G., & Thorp, J. S. (1988). *Computer Relaying for Power Systems*. John Wiley and Sons.

Rao, J. G., & Pradhan, A. K. (2012). Differential Power-Based Symmetrical Fault Detection During Power Swing. *IEEE Transactions on Power Delivery*, *27*(3), 1557–1564. doi:10.1109/TPWRD.2012.2196527

Sachdev, M. S., & Baribeau, M. A. (1979). A New Algorithm for Digital Impedance Relay. *IEEE Transactions on Power Apparatus and Systems*, *98*(6), 2232–2240. doi:10.1109/TPAS.1979.319422

Sonalgaz Company internal report (2013), Algiers, Algeria

Sonnemann, W. K., Wagner, C. L., & Rockefeller, G. D. (1958). Magnetizing Inrush Phenomena in Transformer Banks. *Transactions of the American Institute of Electrical Engineers*, *77*, 884–892.

Su, B., Dong, X. Z., Bo, Z. Q., Sun, Y. Z., Caunce, B. R., Tholomier, D. J., & Apostolov, A. (2005). Fast Detector of Symmetrical Fault During Power Swing for Distance Relay. *IEEE Power Eng. Soc. Gen. Meet.*, *2*, 1836 – 1841.

Tziouvaras, D. (2007). *Relay performance during major system disturbances*. Schweitzer Engineering Laboratories, Inc. doi:10.1109/CPRE.2007.359905

Waghmode, L. Y., & Patil, R. B. (2013). An Overview of Fault Tree Analysis (FTA) Method for Reliability Analysis. *Journal of Engineering Research and Studies*, *4*(1), 6–8.

Yalla, V. S. (1992). A digital multifunction protective relays. *IEEE Transactions on Power Delivery*, *7*(1), 193–201. doi:10.1109/61.108907

Zitouni A., Ouadi A., Bentarzi H., & Chafai M. (2018). Distance relay reliability enhancement using false trip root cause analysis. *International Journal of System Assurance Engineering and Management*, 1-9.

KEY TERMS AND DEFINITIONS

Dependability: Dependability is a measure of the relay ability to operate when it is supposed to operate.

Fault Tree Analysis: Fault tree analysis is a top-down, deductive failure analysis in which an undesired state of a system is analyzed using Boolean logic to combine a series of lower-level events.

Root Cause Analysis: Root cause analysis is a step by step approach that leads to identify a disturbance's first or root cause.

Security: Security is defined as the degree of certainties that a relay or relay system will not operate incorrectly.

Chapter 13
Advanced Real-Time Tester for a Smart Power Grid

Abderrahmane Ouadi
Laboratory Signals and Systems, University M'hamed Bougara of Boumerdes, Algeria

Abdelkader Zitouni
University M'hamed Bougara of Boumerdes, Algeria

Ahmed Maache
University M'hamed Bougara of Boumerdes, Algeria

ABSTRACT

The testing approach is facing many difficulties regarding the actual implementations in the modern smart power grids. One of these challenges is the testing of hardware devices such as protective relays, PMUs, and smart meters before its final deployment to the power grid. One way to overcome this is the real-time simulation of power grid. The hardware-under-test (HuT) is plugged to a real-time simulator via signal conditioning circuit (SCC). SCC is an interface circuit involving power amplifier and measurement sub-circuit between the real-time power grid simulator and the HuT. In this chapter, some advanced developed techniques and approaches will be presented.

INTRODUCTION

The reliability of the power system depends upon the performance of the thousands of smart devices that may be used in monitoring, protective and control system. The failure of an apparatus to operate correctly as intended may jeopardize the stability of the entire power system and its bulk elements.

The testing approach is facing many difficulties regarding the actual implementations in the smart power systems. One of these challenges is the testing of hardware devices such as protective relays, PMUs and smart meters before its final deployment to the power grid. One way to overcome this is the real-time simulation of power grid. The Hardware-under-Test (HuT) is plugged to a real-time simulator via Signal Conditioning Circuit (SCC). SCC is an interface circuit involving power amplifier and

DOI: 10.4018/978-1-7998-4027-5.ch013

measurement sub-circuit between the real-time power grid simulator and the HuT (Faruque, M.O. et al., 2015; Hernandez, M.E. et al., 2018).

In fact, major system failures during a disturbance are more likely to be caused by unintended protective relay operation rather than by the failure of a relay to take an action. Reliability in this case has two aspects: dependability and security. Dependability is known as the degree of certainty that a relay system will operate correctly as intended. Security is the degree of certainty that a relay will operate unnecessary during any transient disturbance (Phadke, A.G., et al., 1976; Zitouni A., et al., 2018). Appropriate relay testing provides first guard against relay false-tripping. Relay testing can aid to confirm the design of relay, compare the performance of different relay typess, verify relay settings, classify system conditions that might cause unplanned relay operation, and carry out post-event analysis to understand the causes of unintentional or incorrect relay actions. Relay tester enhancements need to continue because of the use of relays in smart power grids where the conditions that are not the same as in the conventional one. This leads to new relay technologies (Dierks, K.C.A., 2006).

Disturbances include transient distortion in the voltage due to post fault, potential transformer saturation or compensator switching may affect on transmission line relays and relaying systems in various ways. The mal-operation of this relay is generally unnecessary tripping during post fault or compensator connection which produces DC offset and harmonics. This may reduce the security of protection system and hence its reliability (Abdelmoumene A., Bentarzi H., Chafai. M., Ouadi A., 2014).

This work focuses mainly on the design and the implementation of the Class D amplifier which is the amplification part of the testing system. The Class D amplifier has been used in this work to amplify the simulated disturbances that are generated by Simulink power system model to be then injected to the protective relay and monitor its response. Labview has been used for developing the graphical user interface and controlling NI 7851Board

In this research work, we have used new technology that allow designing an enhanced relay testing system which in turn can be used for improving the performance of protective relay. In order to test both security and dependability and hence the reliability, this work propose a new frame work of tester based on FPGA associated with acquisition card NI.

TESTABILITY

Protection System Reliability and Testability require in part that the protection system be designed for high functional reliability and in-service testability commensurate with the safety function to be performed. They also require a design that permits on-line periodic testing of the functioning of the protective system.

The testing of protection schemes faces many problems. Since the main function of protection equipment is taken into consideration under operation during system fault conditions, and it cannot readily be tested under normal conditions (Relay testing and commissioning, Chapter Online). This situation is aggravated by increasing the complexity of protection schemes and use of relays containing software.

Type tests are required to prove that a relay meets the published specification and complies with all relevant standards. Since the principle function of the protection relay is to operate correctly under abnormal power conditions, it is essential that the performance be assessed under such conditions. Comprehensive type tests simulating the operational conditions are therefore conducted at the manufacturer's works during the development and certification of the equipment.

Figure 1. FPGA based tester

Different testings that can be carried out by the developed testing system are as follows (Chafai, M., Bentarzi, H., & Ouadi, A., 2014):

Steady State Test

Usually steady state test is used for the examination of the relay setting by injecting current and voltage at predetermined values for time longer than the setting time of relay. Then, the signal is varied slowly at a rate much smaller than resolution of relay, whichever manually turning a knob or by an automatic system (Chafai, M., Bentarzi, H., Ouadi, A., Zitouni A., & Recioui, A., 2013). This type is of less use in commissioning, due to the signal does not represent the real power system signal during faults or abnormal condition.

Dynamic State Test

Dynamic state test is inspected by simultaneously applying fundamental frequency components of voltage & current which represent power system states of pre-fault, fault and post fault (Chafai, M., Bentarzi, H., Ouadi, A., Zitouni A., & Recioui, A., 2013). Time for relay operation is measured. This type of test can be used in commissioning and troubleshooting.

Transient Test

Transient testing may be investigated by applying simultaneously fundamental and non-fundamental frequency components of voltage and current those represent real power system conditions. This type of test may be used for testing a security of the protective relay.

The signal that may be used in this type for testing may be obtained from digital fault recorders (DFR) real-time Simulator using Simulink/Matlab (Ouadi, A. Bentarzi, H., Chafai M., & Zitouni, A., 2011).

The increasing use of digital technology in devices such as protection, measurement, and control apparatus in electric power substations has created the potential for accumulating large numbers of digital recording of power system transient events. In addition to these sources of digital data, real-time system simulators may be used to generate digital data. The users of these data are faced the problem of how to convert it to real-time signal by amplifying it without distortion as the signals outputted from instrument transformers.

HARDWARE-UNDER-TEST (HUT)

The principles for testing of hardware systems cannot be directly applied to the Smart power Grid, due to the importance of hardware testing (Dufour, C., & Bélanger, J., 2014), which is why Real-Time (RT) Simulator is one of the techniques used. Real-time testing is a method of testing in which real hardware is incorporated into a simulation. This simulation can contain other hardware and software components, which can interact together as if they were physically connected. This can lead to an increase of simulation accuracy, as less hardware components have to be emulated via software (Lu, B., et al, 2007). Not all hardware can be reliably and accurately tested, in which case RT tester is the only option for its testing as part of a larger system. RT tester can be used for testing the developed of complex smart systems, such as PMU, relays and smart meters used in SG systems (Jeon, J.-H., et al, 2010). RT Simulator helps to reduce test costs while retaining the usage of simulator and does not require developing hardware emulator models for all of the power system. Certain parts of the environment which can be difficult to recreate inside a laboratory (e.g., high voltage systems) can be virtualized, while keeping other hardware in place for the simulation. RT simulator can also be used for algorithm verification on hardware and verifying that all components are capable of real-time interaction and event processing. This chapter primarily focuses on RT testing, as this kind of testing is the most relevant for overall SG testing.

The main idea is to build an affordable real-time simulation platform of power grid with hardware acceleration (FPGA) using the NI7851R Reconfigurable Card in our research laboratory. The real-time simulation could be performed using Matlab/Simulink, Labview or a simulator like Open DSS (Open Distributed Systems Simulator) which communicates with the actual HuT through the interface unit.

This real-time simulation platform will be used in our research laboratory to investigate/test new power grid related hardware and algorithms. The developed Simulink based power system model may be used for generating signals under different conditions (faults and transient disturbances) with different configurations. Besides, LabVIEW may be used for developing the graphical user interface and

Figure 2. Basic elements of a RT power grid simulator.

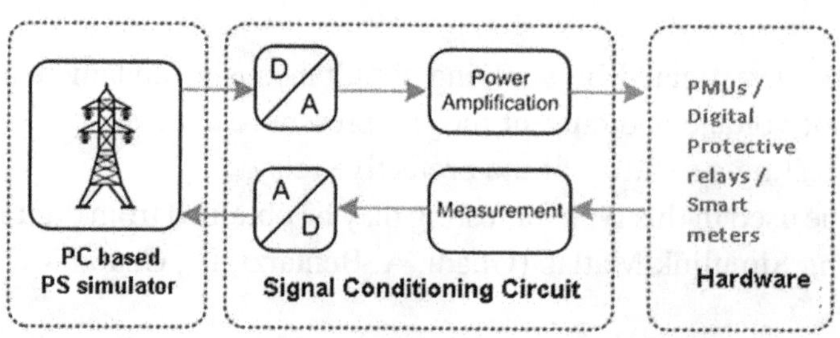

configuring the FPGA. The real-time model contains PC based power system simulator, power and measurement signal conditioning circuit as shown in Fig 2.

Portions of a system under test which can be emulated (e.g., sensors, probes) and can be subsequently controlled by the simulation platform. These emulated components can serve as an entry and measurement point for the hardware system under testing (Jeon, J.-H., et al., 2010). They must serve as a bridge between the hardware and the software simulation environment. Together the hardware, emulated components and software simulation makes up a RT simulation platform.

One of the downsides of RT simulations is that hardware systems cannot be easily cloned like software simulators; therefore, the limit on how large a simulation can be is the lab's budget. Additionally, depending on the hardware, it can be hard to separate the tested hardware from other external systems, which can increase the complexity of incorporating it into the simulation. Due to its physical nature, an experiment cannot be temporarily suspended and neither is it possible to go back in simulation time. For this reason, all RT testing system requires that the underlying testing platform is capable of real-time event processing and sending information (simulation data) to all of the simulated components (Lu, B., et al, 2007). Another major consideration point is the simulation speed: RT tester generally requires following the real wall clock (i.e., one simulation second is one real world second).

TESTER IMPLEMENTATION

FPGA based tester has been implemented using the acquisition card NI as shown in Fig.1. It consists of three main parts which are computer, NI 7851 board and amplifier.

Figure 3. Using a Host VI to Communicate with the FPGA Target

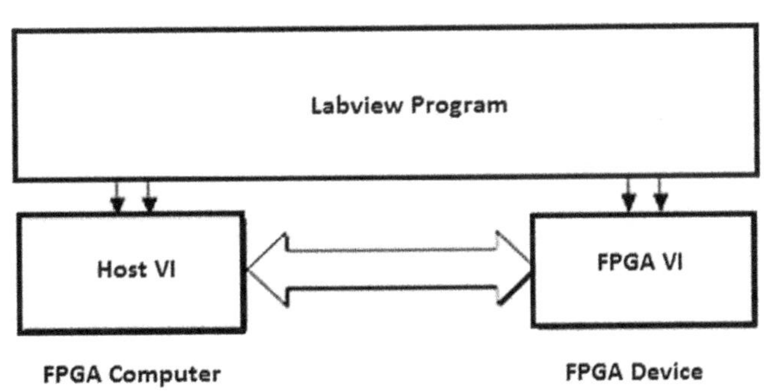

Software Part

Application development for relays tester has been investigated using SIMULINK and Labview as development tools and driver of NI7851 board (National Instruments User Manuals. 2009).

The NI LabVIEW FPGA Module allows the user to graphically implement digital circuits within an FPGA chip using LabVIEW to create VIs and functions that control the I/O, timing, and logic of the device, PCIe 7851, in our case, to generate the desired output (Bentarzi, H., Zitouni, A ., 2017). Interactive Front Panel Communication has been used to communicate directly with the FPGA board using an FPGA VI, but the communication is limited in the number of functions and operations that can be performed and lacks real times updating as shown in Fig.3. So a separate HOST VI running simultaneously that references the FPGA VI, has been used to interact with it in real time and provide the user with more control over the inputs of the FPGA, Host Vis are used to send information between the host computer and the FPGA target for the following reasons:

- Doing more data processing than you can fit on the FPGA,
- Performing operations not available on the FPGA target,
- Controlling the timing and sequencing of data transfer.

The Host VI takes advantage of a limited number of LabVIEW's inbuilt functions and relies on the FPGA interface functions like Open FPGA VI Reference, Close FPGA VI Reference to establish a connection between the FPGA VI and the Host VI and close it when needed, Read/Write Control to read a value from or write a value to a control or indicator in the FPGAVI. Using the Read/Write block. Figures 4 and 5 show the Front Panel and the Block Diagram Host VI.

Figure 4. Host VI Front Panel of Tester

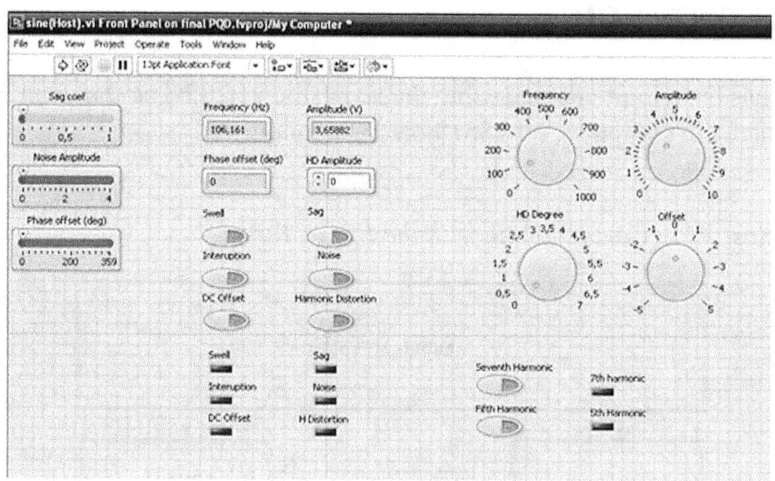

HARDWARE PART

PC outputs the signals generated by simulator via the card NI7851, where the DAC converts the data to analog form, and then to the relay under test using the amplifier (Ouadi, A. Bentarzi, H., Chafai M., & Zitouni, A., 2011).

Figure 6 shows a block diagram of the (class D) amplifier (Chafai, M., Bentarzi, H., & Ouadi, A., 2014). First the signal that will be injected to the amplifier is sampled by an analogue-to-digital converter up sampled version of the 24-bit digital input signal in the considered Bandwidth (0-5Khz). The produced signal is as input to a digital filter which drives a digital PWM. The power circuit, which can be a half-bridge converter, is followed by a low pass filter to recover the original analog signal (Texas Instruments, 2015).

Figure 5. Host VI Block Diagram of tester.

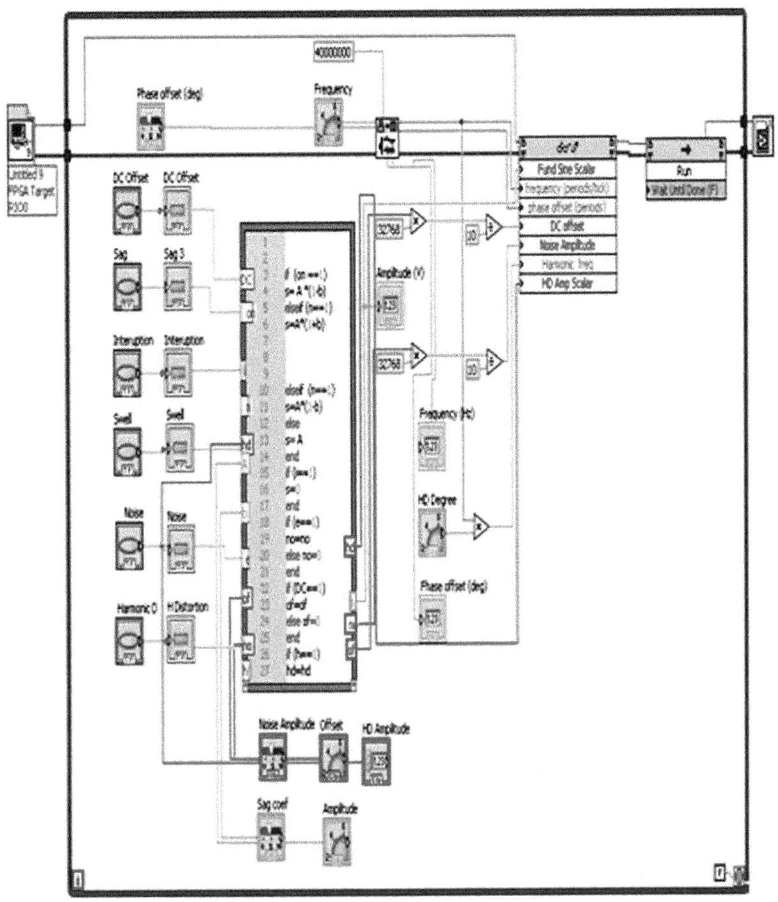

Amplifier Development

The final circuit design of this class D amplifier is shown in Figure 7. The first stage of this class D amplifier is the control circuit where the PWM signal generator is fed to the two gate drives that have as main role to create dead-time for avoiding shoot-through current that can damage the output stage. Then, the power stage which consists of H-Bridge and low pass filter is the final stage.

The PCB implementation circuit of the Class D amplifier is shown in figure 8.

This circuit first was implemented on proto board in order test it. After that it will be implemented on the PCB. The control circuit should be well isolated from the power circuit.

Figure 6. Block diagram of class D amplifier

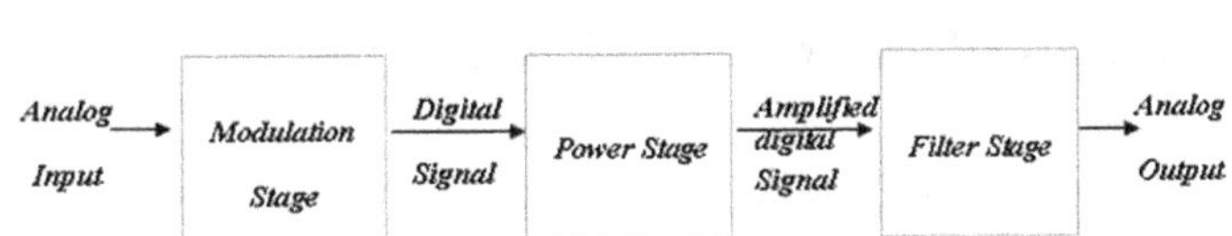

Figure 7. Class D Amplifier circuit

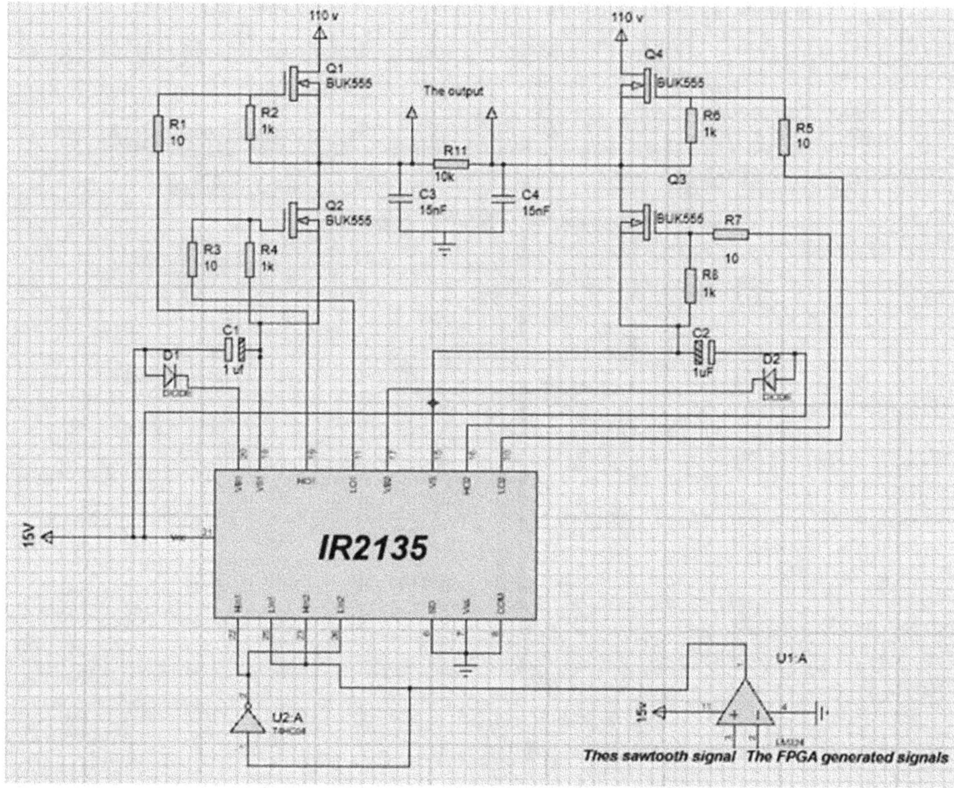

Figure 8. Class D amplifier circuit

PERFORMANCE EVALUATION OF THE TESTER

The signal of the PWM is supplied to H-bridge. The output PWM signal of the H-bridge before the using the low passive filter is shown in Fig. 9 where the H-bridge is fed with 12V.

Figure 9. H-bridge output PWM signal

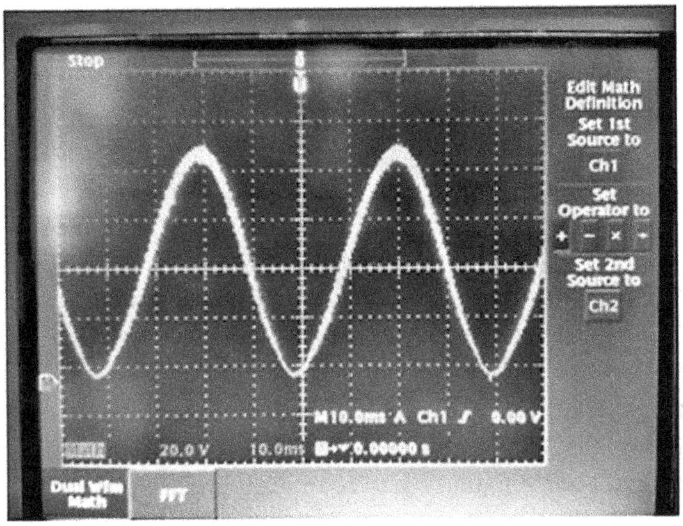

The output signal of the H-Bridge fed with 18.73V and a load of 100Ω is shown in figures 10 and 11, the input signal is a sine wave of 10V peek to peek and frequency of 50Hz.

This signal has been taken and injected to the Class D amplifier but the voltage of the H-bridge has been changed to 38.8V and the probes were changed to X10 to decrease the scale of the oscilloscope so that each division is 20V. The amplified output signal is 110V peak to peak and a frequency of 50 Hz the efficiency is approximately 100%. However, in Fig. 11 some distortions are appearing in the output amplified signal, these distortions are not due to the class D amplifier, they are part of the disturbance input signal.

Figure 10. Output signal of class D amplifier

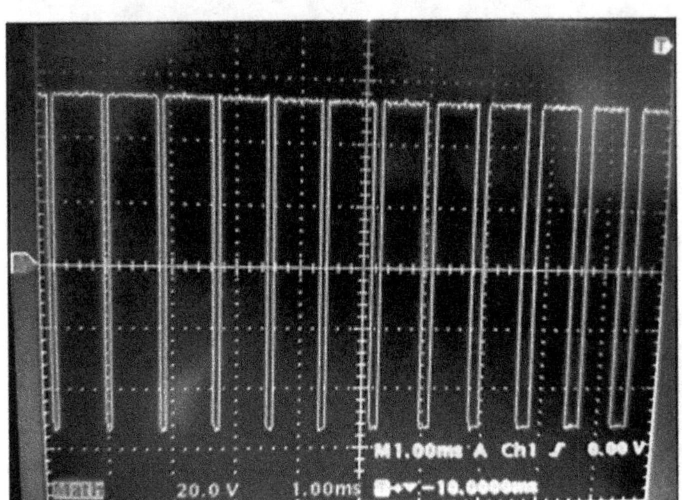

The theoretical efficiency of a Class D amplifier is 100% but this is unachievable in practice. However, this efficiency can approach significantly this value by making a good design to minimize switching losses that lead to high power loss.

The formula to calculate the efficiency is:

$$\eta = \frac{P_{OUT}}{P_{in}} * 100 \quad (1)$$

The efficiency and the gain are calculated with respect to a sine wave input of 10V and the voltage of the source that feed the H-bridge has efficiency is η=92.4%.

The total harmonic distortion (THD) is very important parameter to evaluate the performance of the Class D amplifier to have a clean signal, we don't have the proper means to do the calculation to have the exact value but it appears that the THD is less than 1%.

Finally, the proposed real-time simulator has been tested under abnormal conditions such as sag condition that can be generated by Software in the PC using Matlab or Labview. The resulted signal at the output of the power amplifier that is shown in Fig.12, can be injected to the device under test such as PMU or relay.

CONCLUSION

The FPGA based tester has been successfully designed and implemented. Simulator is used to generate three phase voltages and currents presented the actual power grid events mainly the fault and other abnormal conditions to check the relay settings. In fact, a PC has been used to generate these events signals and inject them to the relay under test via the acquisition card NI7851.

Figure 11. Amplified sine wave using Class D amplifier

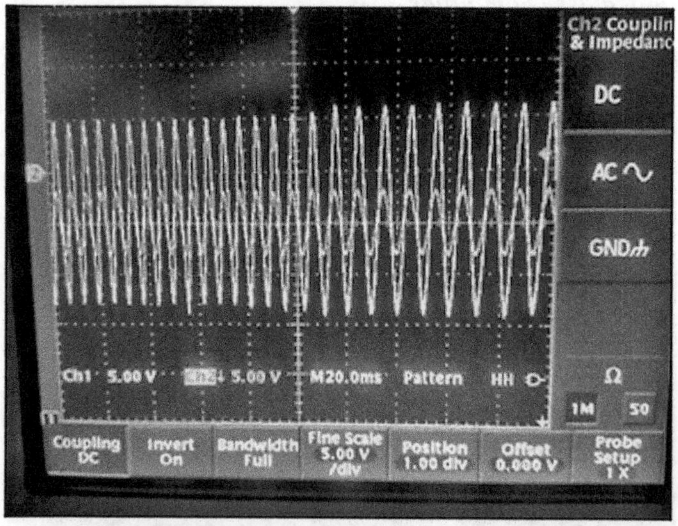

Figure 12. Voltage sag generated from Disturbances generator

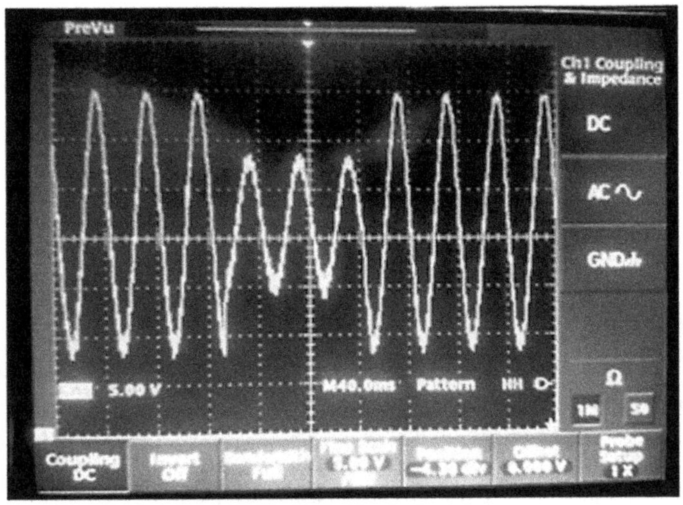

User Interface (GUI) has been developed to include some parameters of the relay, to control the generated data of waveform signals and to display the generated signals.

The Class D amplifier has been selected in this circuit to amplify the simulated disturbances signals. Design and implementation of the Class D amplifier which is the amplification part of the testing system have been investigated in this work.

The test results and performance of this tester are very satisfying. This is clearly proven by its high efficiency and very low THD. The Class D amplifier that has been used can amplify any disturbance signal with the same performance results even if the frequency of these disturbances is changing. This ability is possible because the cut-off frequency of its filter can attain 1.1 kHz.

REFERENCES

Abdelmoumene, A., Bentarzi, H., Chafai, M., & Ouadi, A. (2014). *Reliability assessment and improvement of digital protective relays. Int J Syst Assur Eng Manag.* doi:10.100713198-014-0300-z

Bentarzi, H., & Zitouni, A. (2017). Field-Programmable Gate Array Based Tester for Protective Relay, International Journal of Electrical, Computer, Energetic. *Electronic and Communication Engineering, WASET, 11*(7), 791–796.

Chafai, M., Bentarzi, H., & Ouadi, A. (2014). Real-Time Simulator for Testing Protective Relay. *International Conference on Embedded Systems and Applications (ICESA2014).*

Chafai, M., Bentarzi, H., & Ouadi, A. (2014). A New Frame Work for Protective Relay Tester. *NNGT Journal: International Journal of Signal Processing and Imaging Engineering, 1*, 60–65.

Chafai, M., Bentarzi, H., Ouadi, A., Zitouni, A., & Recioui, A. (2013). PC based Testing System for Protective Relay. *Proc. 4th International Conference on Power Engineering, Energy and Electrical Drives.* 10.1109/PowerEng.2013.6635737

Dierks, K. C. A. (2006). *Relay Specific Computer Aided Testing for Protective Relays*. Eskom, Republic of South Africa.

Dufour, C., & Bélanger, J. (2014). On the Use of Real-Time Simulation Technology in Smart Grid Research and Development. *IEEE Transactions on Industry Applications*, *50*(6), 3963–3970. doi:10.1109/TIA.2014.2315507

Faruque, M. O., & ... . (2015). Real-time simulation technologies for power systems design, testing, and analysis. *IEEE Power and Energy Technology Systems Journal*, *2*(2), 63–73. doi:10.1109/JPETS.2015.2427370

Hernandez, M. E., Ramos, G. A., Lwin, M., Siratarnsophon, P., & Santoso, S. (2018). Embedded Real-Time Simulation Platform for Power Distribution Systems. *IEEE Access: Practical Innovations, Open Solutions*, *6*, 6243–6256. doi:10.1109/ACCESS.2017.2784318

Jeon, J.-H., Kim, J.-Yu., Kim, H.-M., Kim, S.-K., Cho, C., Kim, J.-M., Ahn, J.-B., & Nam, K.-Y. (2010). Development of Hardware In-the-Loop Simulation System for Testing Operation and Control Functions of Microgrid. *IEEE Transactions on Power Electronics*, *25*(12), 2919–2929. doi:10.1109/TPEL.2010.2078518

Lu, B., Wu, X., Figueroa, H., & Monti, A. (2007). A Low Cost Real-Time Hardware-in-the-Loop Testing Approach of Power Electronics Controls. *IEEE Transactions on Industrial Electronics*, *54*(2), 919–931. doi:10.1109/TIE.2007.892253

National Instruments. (2009). *NI R Series Multifunction RIO User Manual*. NI Publication.

Ouadi, A., Bentarzi, H., Chafai, M., & Zitouni, A. (2011). Microcontroller Based Protective Relay Testing System. In *Proc. The 10th WSEAS International Conference on Circuits, Systems, Electronics, Control & Signal Processing (CSECS '11)*. Montreux University.

Phadke, A. G., Hlibka, T., & Ibrahim, M. (1976). A digital computer system for EHV Substation: Analysis and Field Tests. *IEEE Transactions on Power Apparatus and Systems*, *PAS-95*(1), 291–301. doi:10.1109/T-PAS.1976.32105

Texas Instruments. (2015). *LMx24-N, LM2902-N Low-Power, Quad-Operational Amplifiers*. Texas Instruments Press Relay testing and commissioning. Available: http://www.fecime.org/referencias/npag/chap21-370-397.pdf

Zitouni A., Ouadi A., Bentarzi H., & Chafai M. (2018). Distance relay reliability enhancement using false trip root cause analysis. *International Journal of System Assurance Engineering and Management*, 1-9.

KEY TERMS AND DEFINITIONS

Hardware-Under-Test (HuT): In power hardware in the loop (PHIL) simulations, a real-time simulated power system is interfaced to a piece of hardware, usually called hardware under test (HuT). ... Stable operation of the entire system, along with the accuracy of simulation results are the main concerns regarding a PHIL simulation.

Real-Time Simulator: Refers to a computer model of a physical system that can execute at the same rate as actual "wall clock" time.

Signal Conditioning Circuit (SCC): In electronics, signal conditioning is the manipulation of an analog signal in such a way that it meets the requirements of the next stage for further processing. In an analog-to-digital converter application, signal conditioning includes voltage or current limiting and anti-aliasing filtering.

Total Harmonic Distortion (THD): A total harmonic distortion is a measurement of the harmonic distortion presents in a signal and is defined as the ratio of the sum of the powers of all harmonic components to the power of the fundamental frequency. Distortion factor, a closely related term, is sometimes used as a synonym.

Chapter 14
Differential Protection Enhancement for Power Transformer

Hamid Bentarzi
University M'hamed Bougara of Boumerdes, Algeria

Rachid Bouderballa
Laboratory Signals and Systems, University M'hamed Bougara of Boumerdes, Algeria

ABSTRACT

A power transformer is protected against internal faults using a differential protection which is sensitive and a fast. However, during transformer magnetization (inrush current or over-excitation), the differential relay may operate unnecessarily. This phenomenon appears only when a transformer is first energized or after clearing external fault. During periodic magnetization condition due to over-excitation, the third and fifth harmonic components are largely noticed; however, during the normal apperiodic inrush conditions, the second harmonic is relatively high. In the conventional techniques, these harmonic components have been used to block differential protection to operate. However, in smart power transformer, these harmonic components are small even during the transformer magnetization; they cannot be used as block protection functions. The differential protection security has to be improved so that it can distinguish between differential current produced by magnetization and that produced by internal fault using the most advanced computer with most improved DSP algorithms.

INTRODUCTION

Power transformer is one of the most important elements in the power systems. Therefore, the protection of power transformers is crucial for the continuity of the power supply. A power transformer is mostly protected against internal faults using a differential protection which is sensitive and a fast clearing technique (Abdelmoumene, A. and Bentarzi, 2012). This technique of protection detects nonzero differential current, and then activates a circuit breaker that disconnects the power transformer. However,

DOI: 10.4018/978-1-7998-4027-5.ch014

this nonzero differential current may be produced by transformer magnetization, due to so called inrush current or over-excitation, and may cause the relay to operate unnecessarily. This magnetization current is a transient current that appears only when a transformer is first energized or after clearing external fault. Even though, it can be as great as 8 times the full load current and it contains harmonic components, however, it is harmless due to rapidity. During periodic magnetization condition due to over-excitation the third and fifth harmonic components are largely noticed; however, during the normal apperiodic inrush conditions, the second harmonic is relatively high.

Differential relays are prone to mal-operation in the presence of transformer inrush currents. Inrush currents are produced from transformer magnetic flux (Sonnemann W.K., et al, 1958). The first solution to this problem has been investigated by introducing an intentional time delay in the differential relay. Another proposal has been performed by desensitizing the relay for a given time, to overcome the inrush condition (Kennedy L. F. and Hayward C. D. 1938; Hayward C. D., 1941). Others have suggested adding a voltage signal to restrain (Guzman, A. and Zocholl, S., 2001). Much work has been conducted in making the protective relays recognize the difference between overcurrents at a fault and a magnetizing inrush current (Meckic, F., et al, 2006; Kasztenny B. and Kezunovic M., 1998).

This book chapter motivation is the presentation of the recent developed blocking techniques of differential protection during inrush conditions. This is following a number of questions that has been arisen while applying differential relays for transformer protection. Protection of large power transformers is a very challenging problem in power system relaying. Large transformers are a class of very expensive and vital components of electric power systems. Since it is very important to minimize the frequency and duration of unwanted outages, there is a high demand imposed on power transformer protective relays; this includes the requirements of dependability associated with mal-operation, security associated with no false tripping, and operating speed associated with short fault clearing time (Bouderbala R., Bentarzi, H. and Ouadi A., 2010).

This research will analyze the problem and its effect on transformer differential protection. Since the magnetizing inrush current generally contains a large second harmonic component in comparison to an internal fault, conventional transformer protection systems are designed to restrain during inrush transient phenomena by sensing this large second harmonic. However, the second harmonic component may also be generated during internal faults in the power transformer (Manana, M., et al, 2005). This may be due to current transformer (CT) saturation, presence of shunt capacitance, or the capacitance in long extra high voltage transmission lines to which the transformer may be connected. The magnitude of the second harmonic in an internal fault current can be close to or greater than that present in the magnetizing inrush current (Bouderbala R., et al, 2011). The second harmonic components in the magnetizing inrush currents tend to be relatively small in modern large power transformers because of the improvements in the power transformer core material. The commonly employed conventional differential protection technique based on the second harmonic restraint will have difficulty in distinguishing between an internal fault and the inrush current thereby threatening transformer stability (Bouderbala R., Bentarzi H., 2013). This chapter presents an approach proposed by (Bouderbala, R., Bentarzi, H., 2016) that uses two harmonics (second and fourth) for restraining or blocking a differential relay and reducing the blocking time during an internal fault. Besides, it suggests a further work for protecting a smart transformer

Three characteristics generally provide means for detecting transformer internal faults. These characteristics include an increase in phase currents, an increase in the differential current, and gas formation. When transformer internal faults occur, immediate disconnection of the faulted transformer is necessary to avoid extensive damage and preserve power system stability. Three types of protection are normally used to detect these faults: overcurrent protection for phase currents, differential protection for differential currents, and gas accumulator for arcing faults.

Overcurrent protection with fuses or relays provided the first type of transformer fault protection and is used for small transformers. Transformer differential protection is one of the most reliable and popular technique for protecting large power transformers. The percentage differential principle was applied to transformer protection to improve the security of differential protection for external faults with CT saturation.

Discrimination between an internal fault and a magnetizing inrush current has long been recognized as a challenging power transformer problem (Ong C.M., 1998). This research will analyze the problem and its effect on transformer differential protection. First, the research will review the concept of transformer differential protection and then analyze magnetizing inrush, overexcitation and current transformer saturation phenomena as possible causes of relay maloperation. Since magnetizing inrush current generally contains a large second harmonic component in comparison to an internal fault, conventional transformer protection systems are designed to restrain during inrush transient phenomena by sensing this large second harmonic. However, the second harmonic component may also be generated during internal faults in the power transformer (Brunke, J. H., Fröhlich, J.K., 2001). This may be due to CT saturation, presence of shunt capacitance, or the capacitance in long extra high voltage transmission lines to which the transformer may be connected.

Transformer overexcitation is another possible cause of power transformer relay mal-operation. The magnetic flux inside the transformer core is directly proportional to the applied voltage and inversely proportional to the system frequency (Guzmán, A. et al, 2000). Overvoltage and/or underfrequency conditions can produce flux levels that saturate the transformer core. These abnormal operating conditions can exist in any part of the power system, so any transformer may be exposed to overexcitation. Transformer overexcitation causes transformer heating and increase exciting current, noise, and vibration (Guzmán, A. et al, 2000). Though it is difficult, with differential protection, to control the amount of overexcitation that a transformer can tolerate, transformer differential protection tripping for an overexcitation condition is not desirable.

POWER TRANSFORMER ABNORMAL OPERATING CONDITIONS

Different abnormal conditions can occur in the transformer during its operation. These abnormal conditions can be divided into two classes as permanent abnormal conditions (faults) and transient abnormal conditions (magnetization current). This section deals with the basic methods that may be used to distinguish between inrush current and fault current in power transformers.

Permanent Abnormal Conditions (Faults)

Many faults can occur in the power transformer. The protection system must disconnect and isolate the power transformer from power network rapidly and correctly and hence it clears the fault as fast as possible. These faults can be classified as external and internal faults.

External Faults

They are those faults that happen outside the transformer such as shunt faults. System short circuits may produce a relatively intense rate of heating of the feeding transformers. The unit protection of the transformer should not operate for external faults. The transformer must be disconnected when such faults occur only when the faults are not cleared by other relays in pre-specified time. The transformer protection has to limit damage to the transformer during overload condition and at low frequency. The protection system has to disconnect the consumers if the voltage becomes too high.

During overload condition, an increase in copper losses may be noticed and consequently the temperature may rise. Power transformers can be temporarily overloaded. The duration of the acceptable overload period depends on the initial temperature and the cooling systems. Overloads can be carried for limited periods. The thermal time constant of naturally cooled transformers lies between 2.5-5 hours. Shorter time constants apply in the case of force-cooled transformers. Besides, the transformer core becomes overexcited when the applied voltage is too high or the applied frequency is too low. Overexcitation causes an increase in the iron loss and a substantial increase in the magnetizing current. In addition, flux is diverted from the laminated core. It is forced through surrounding steel parts such as the metal of the tank and other non-laminated parts of the transformer. In particular, the core bolts, which normally carry little flux, may be subjected to a large component of flux. Under such conditions, the bolts may be rapidly heated to a temperature that destroys their own insulation and will damage the coil insulation if the condition continues. It follows that a transformer can operate with some degree of overvoltage with a corresponding increase in frequency. Operation must not continue if the applied voltage is high and frequency is low. This reduction of system frequency has an effect with regard to flux density, similar to that of overvoltage. It follows that a transformer can operate with some degree of overvoltage with a corresponding increase in frequency, but operation must not be continued with a high voltage input at a low frequency. Operation cannot be sustained when the ratio of voltage to frequency, with these quantities given values in per unit of their rated values, exceeds unity by more than a small amount, for instance if $V/f > 1.1$. If a substantial rise in system voltage has been catered for in the design, the base of 'unit voltage' should be taken as the highest voltage for which the transformer is designed.

Short circuits produce a relatively intense heating of the feeding transformers, the copper loss increasing in proportion to the square of the per unit fault current. The typical duration of external short circuits that a transformer can sustain if the current is limited only by the self-reactance is shown in Table 1. Maximum mechanical stress on windings occurs during the first cycle of the fault. Avoidance of damage is a matter of transformer design.

Table 1. Fault withstand levels

Transformer Reactance (%)	Permitted Fault Current (Times of Current Rating)	Permitted Fault Duration (Seconds)
4	25	2
5	20	2
6	16.6	2
7	14.2	2

Internal Faults

They occur within the transformer protection zone such as incipient fault (overheating, overfluxing, overpressure) and active faults (turn-to-earth, turn-to-turn, tank fault, core fault). These internal faults can be classified into two groups.

Group I: Electrical faults that cause immediate damage but are generally detectable by unbalance of current or voltage. Amongst them are the following:

- Winding failures resulting from short circuits (turn-turn faults, phase-phase faults, phase-ground, open winding),
- Short circuit between turns of high-voltage or low-voltage windings,
- Faults to earth on a tertiary winding or short circuit between turns of a tertiary winding Statistics show that winding failures most frequently cause transformer faults (ANSI=IEEE, 1985).

Group II: These include incipient faults, which are initially minor but cause substantial damage if they are not detected and taken care of. These faults cannot be detected by monitoring currents or voltages at the terminals of the transformer. Incipient faults include the following:

- A poor electrical connection between conductors,
- A core faults (core insulation failure, shorted laminations) which causes arcing in oil,
- Terminal failures (open leads, loose connections, short circuits),
- On-load tap changer failures (mechanical, electrical, short circuit, overheating),
- Coolant failure, which causes rise of temperature,
- Bad load sharing between transformers in parallel, which can cause overheating due to circulating currents.

For a group I of faults, the transformer should be isolated as quickly as possible after the occurrence of the fault. The group II faults, though not serious in the incipient stage, may cause major faults in the course of time. Incipient faults should be cleared soon after they are detected. The IEEE Guide for Protective Relay Applications to Power Transformers (Hayward, C. D., 1941; Brunke, J. H. and Fröhlich, J.K., 2001) statistics show that winding failures most frequently cause transformer faults.

Figure 1 shows summaries of failures in those categories reported by groups of utilities in IEEE statistics USA. Winding failures, tap-changer failures and bushing failures represent more than 85% of the transformer failures. The winding failures still represent the majority of transformer failures, with tap-changers being a distant second. Two transformer characteristics causing problems for protection schemes are the low magnitude turn-to-turn faults and the high-magnitude magnetizing inrush current during energizing. Minimum internal faults can result in less than 10% of a transformer rated current. On the other hand, maximum fault current can flow for a high-side transformer bushing failure.

Winding failures, tap-changer failures and bushing failures represent more than 85% of the transformer failures. The winding failures still represent the majority of transformer failures, with tap-changers being a distant second. Two transformer characteristics causing problems for protection schemes are the low magnitude turn-to-turn faults and the high-magnitude magnetizing inrush current during energizing. Minimum internal faults can result in less than 10% of a transformer rated current. On the other hand, maximum fault current can flow for a high-side transformer bushing failure (Chafai, M., Refoufi, L., Bentarzi, H., 2016).

Figure 1. Transformer IEEE fault statistics Transient Abnormal Conditions (magnetization current)

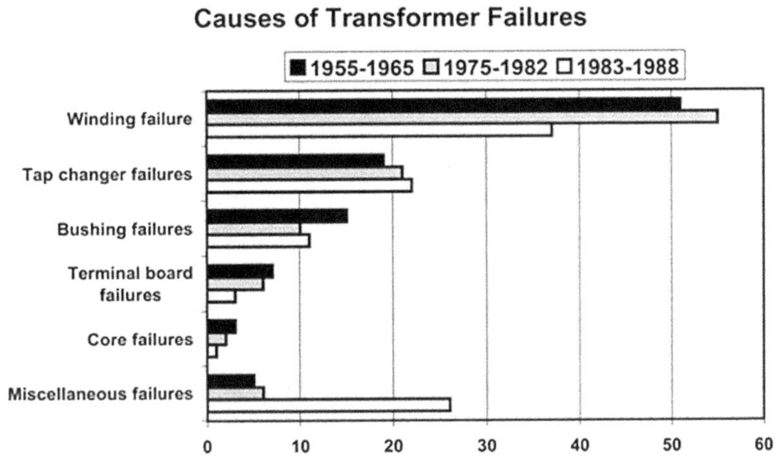

To properly set a protection function, it is necessary to have a basic understanding of the power transformer transient abnormal conditions, the function is intended to detect them. To set the inrush restraint function for transformer differential protection requires some understanding of transformer inrush currents, including the causes and characteristics of these events.

An inrush current is the surge of transient current that appears in a transformer due to inrush and over excitation conditions. The exciting voltage applied to the primary of the transformer forces the flux to build up to a maximum theoretical value of double the steady state flux plus reminisce,

$$\varphi_{MAX} = 2\varphi_M + \varphi_R \tag{1}$$

Therefore, the transformer is greatly saturated and draws more current which can be higher than the full load rating of the transformer windings.

1. Magnetizing Inrush Currents

Magnetizing inrush current in transformers results from any abrupt change of the magnetizing voltage. The magnetizing inrush currents that follow may be several times higher than the rated currents for the transformer. The waveform of transformer magnetizing current contains a proportion of harmonics that increases as the peak flux density is raised to the saturating condition. These currents are high magnitude, harmonic-rich currents generated when transformer cores are driven into saturation, and may erroneously trigger transformer overcurrent protections. The inrush current might reach 50 times the normal exciting current and few times the rated current of the power transformer. The inrush current starts very large and it decays in mill seconds to its steady state value. The inrush current (Mason, C.R., 1986; Kasztenny, B.and Kulidjian, A., 2000) is composed of harmonics of multiples of the fundamental frequency. (As given in Table 2.1).

Even though it is generally considered as a result of the transformer energizing, the magnetizing inrush current may be also caused by:

- occurrence of an external fault,
- voltage recovery after clearing an external fault,
- change of the type of a fault,
- energizing of a transformer in parallel with a transformer that is already in service.

One of the most important type of the transformer differential protection is percentage differential protection which is based on the comparison of the transformer primary and secondary currents. When these currents deviate from a predefined relationship, an internal fault is detected and the transformer is de-energized. However, during transient primary magnetizing inrush conditions, the transformer can carry very high primary current and no secondary current appears. The resulting differential current can falsely trip the circuit breaker of the transformer.

For providing a more easy-to-understand shape of inrush current, Fig. 2 shows the current waveform when the power is turned on. When the power is turned on, current begins to flow, and the initial current flow reaches the peak current value that is larger than the steady-state current value. Following this, the current value gradually decreases until it stabilizes at the steady-state current. The part during which a large current flows before reaching the steady-state current is the inrush current. If the size of the inrush current exceeds that allowed by the part in use, depending on the magnitude of the inrush current (difference between the peak current value and the steady-state current value) and length of its duration (the length of time until the peak current value converges with the steady-state current value, hereafter called the pulse width), the part used in the circuit may overheat, potentially causing the electrical device to malfunction or break down.

The magnetizing current of a transformer contains a third harmonic and progressively smaller amounts of fifth and higher harmonics. If the degree of saturation is progressively increased, not only will the harmonic content increase as a whole, but the relative proportion of fifth harmonic will increase and eventually exceed the third harmonic. At a still higher level the seventh would overtake the fifth harmonic but this involves a degree of saturation that will not be experienced with power transformers.

Figure 2. Current waveform when the device is powered on

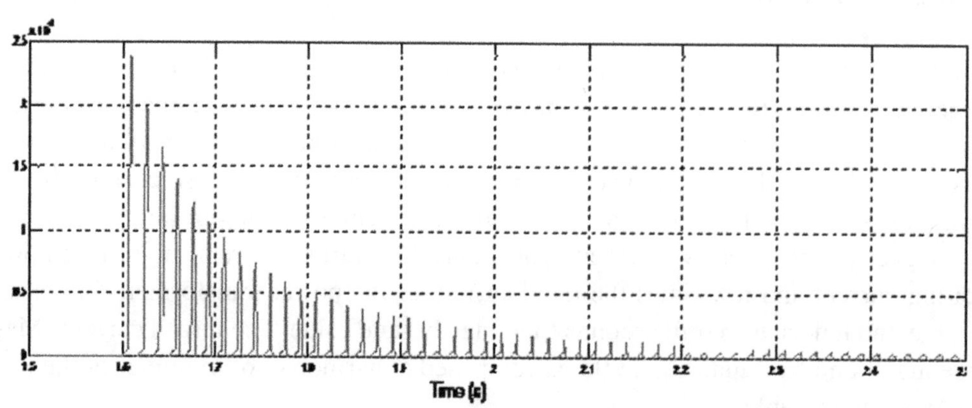

2. Transformer Over-Excitation

Over-excitation can also be caused by an increase in system voltage or a reduction in frequency. It follows, therefore, that transformers can withstand an increase in voltage with a corresponding increase in frequency but not an increase in voltage with a decrease in frequency. Operation cannot be sustained when the ratio of voltage to frequency exceeds more than a small amount. The magnetic flux inside the transformer core is directly proportional to the applied voltage and inversely proportional to the system frequency. Overvoltage and/or under-frequency conditions can produce flux levels that saturate the transformer core. These abnormal operating conditions can exist in any part of the power system, so any transformer may be exposed to over-excitation. Transformer over-excitation causes transformer heating and increases exciting current, noise, and vibration. A severely overexcited transformer should be disconnected to avoid transformer damage. Because it is difficult, with differential protection, to control the amount of over-excitation that a transformer can tolerate, transformer differential protection tripping for an over-excitation condition is not desirable. Protection against over-flux conditions does not require high-speed tripping. In fact, instantaneous tripping is undesirable, as it would cause tripping for transient system disturbances, which are not damaging to the transformer.

An alarm is triggered at a lower level than the trip setting and is used to initiate corrective action. The alarm has a definite time delay, while the trip characteristic generally has a choice of definite time delay or inverse time characteristic. Use separate transformer over-excitation protection instead, and the differential element should not trip for these conditions. One alternative is a V/Hz relay that responds to the voltage/frequency ratio.

Over-excitation of a power transformer is a typical case of ac saturation of the core that produces odd harmonics in the exciting current. Figure 3 shows the exciting current recorded during a real test of a 5 kVA, 230/115 V, single-phase laboratory transformer (Cimadevilla, R., 2013). The current corresponds to an overvoltage condition of 150 percent at nominal frequency. For comparison purposes, the peak value of the transformer nominal current is 61.5 A, and the peak value of the exciting current is 57.3 A. Table 2.3 shows the most significant harmonics of the current signal depicted in Figure 3. Harmonics are expressed as a percentage of the fundamental component. The third harmonic is the most suitable for detecting over-excitation conditions, but either the delta connection of the CTs or the delta connection compensation of the differential relay filters out this harmonic. The fifth harmonic, however, is still a reliable quantity for detecting over-excitation conditions.

Table 2. Percentage of harmonics in typical magnetizing inrush current

Harmonic Components in Magnetizing Inrush Current	Amplitude (% of Fundamental)
DC	55
2nd Harmonic	63
3rd Harmonic	26.8
4th Harmonic	5.1
5th Harmonic	4.1
6th Harmonic	3.7
7th Harmonic	2.4

Figure 3. Exciting Current of an Overexcited Transformer

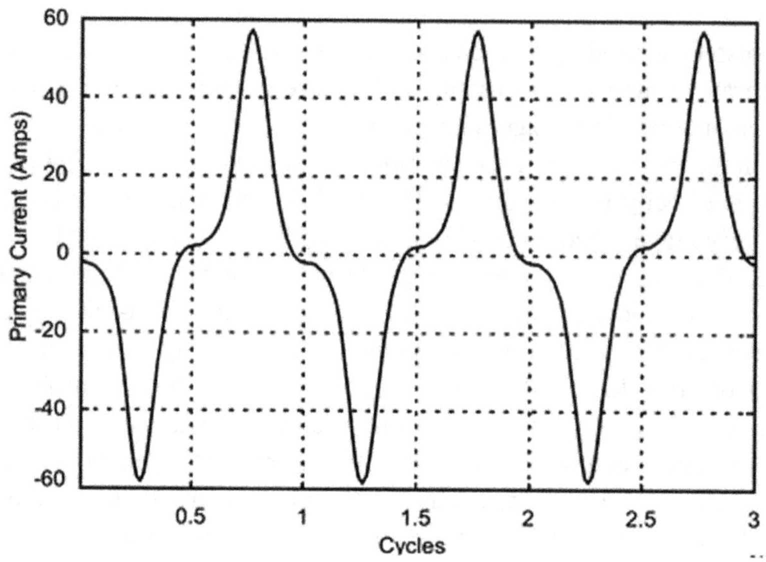

SMART TRANSFORMERS IN SMART POWER GRID

Smart transformers are solid state transformers that play a decision-making role in the electric distribution grid. They work independently to regulate the voltage while maintaining contact with the smart grid so that information and feedback about the transformer and power supply can be provided remotely. These transformers are designed to provide an optimized power supply voltage to deal with the energy needs of the utility, which is why large commercial utilities use these devices more and more today so that they can use power more efficiently and cost-effectively.

Another main advantage of smart transformers is that they respond to fluctuations within the power grid instantly acting as voltage regulators and directly reducing energy consumption to ensure that the optimized voltage is undisturbed. They immediately reduce power consumption by providing a stable and optimal power supply that supplies the electrical equipment with ideal voltage. This benefit has led them to be used in Point of Common Coupling (PCC) in microgrids to act as protecting devices for electrical equipment during power fluctuations.

Smart transformers are intelligent enough to meet the requirements of upcoming smart power systems.

This kind of transformer system is designed for applications in rural networks, which consists of a distribution transformer equipped with OLTC for automatic voltage regulation and a communication system. The power line communication channel establishes a path between distant locations where the transformer is installed and the gateway that collects data from another smart transformer installed in the region.

When smart transformers are used at the Point of Common Coupling (PCC), they control the active power exchanged between the microgrid and utility grid, depending on the state of both networks and other information communicated to the smart transformers. They prove to be powerful components that provide faster and superior voltage regulation, frequency regulation, and regulation of the harmonic behaviour of each feeder.

As shown in Figure X, a solid state transformer (SST) comprises of two or more inverter and/or converter bridges and blocks, with a medium to high-frequency transformer to provide galvanic isolated power conversion for utility and other power-delivery applications. Compared with the conventional low-frequency transformer, an SST significantly reduces the weight and volume of magnetics because of higher-frequency operation (Hengsi, Q.; Kimball, J.W., 2009). The SST introduces features that low-frequency utility transformers do not have, which include but not limit to: 1) bi-directional power flow control, 2) reactive power compensation, 3) harmonic isolation and elimination, 4) DC or multi-terminal interfacing, 5) voltage regulation and sag correction, and 6) fault isolation.

Figure 4. Single stage SST topology

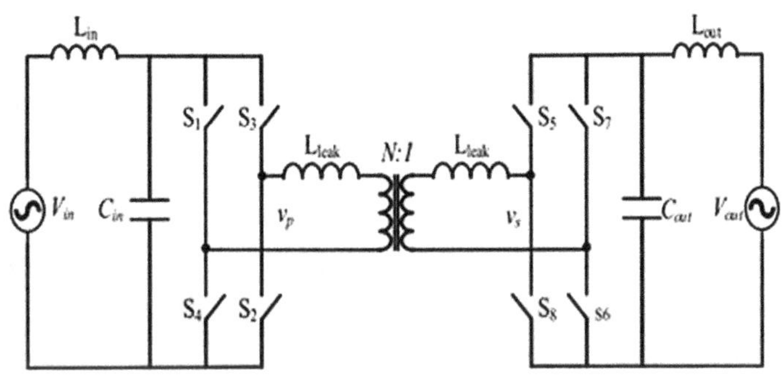

The smart transformers are excellently designed to monitor and manage power supply during fluctuations and ensure that it is voltage optimized even when new demands are required. Other benefits and functions of smart transformers include:

- Protecting the power grid from load disturbances by isolating the source from load harmonics, transients and voltage sags.
- Providing good compensation with sinusoidal currents under operation condition when non-linear loads are applied.
- Improving power quality by summarizing loads to the mains with identical phase current even for unbalanced loads.
- Minimizing grid losses and enhancing power supply reliability.

In the conventional power transformer, the harmonic components have been used to block differential protection to operate. However, in smart power transformer, these harmonic components are small even during the transformer magnetization; they cannot be used as block protection function. The research is open for developing a new differential protection scheme with high security so that it can distinguish between nonzero differential current produced by magnetization current and that produced by internal fault using the most advanced computer with most improved DSP algorithms (Guzman, A., Altuve, H., Tziouvaras, D., 2009).

POWER TRANSFORMER PROTECTION

Overview

Transformers are a critical and expensive component of the power system and should be protected properly. The type of protection for the transformers varies depending on the application and the importance of the transformer. Transformers are protected primarily against faults and overloads. The type of protection used should minimize the time of disconnection for faults within the transformer and to reduce the risk of catastrophic failure to simplify eventual repair. Any extended operation of the transformer under abnormal condition such as faults or overloads compromises the life of the transformer, which means adequate protection should be provided for quicker isolation of the transformer under such conditions. The purpose of the transformer fault protection is to:

1. remove any secondary overload faults from the transformer before it gets damaged,
2. isolate the transformer before it gets totally out of control,
3. remove the damaged transformer from the system to continue the function.

Table 3.1 summarizes the problems and the possible forms of protection that may be used. The following sections provide more detail on the individual protection methods. It is normal for a modern relay to provide all of the required protection functions in a single package, in contrast to electromechanical types that would require several relays complete with interconnections and higher overall CT burdens.

A protective system should be designed to recognize certain system abnormalities which, if undetected, could lead to damage equipments or extended loss of service. The design and specification of the system components is an important part of the protective strategies and power system are designed to withstand the usual operating contingencies that accompany load changes and line switching operation. There is several design consideration that must be weighed against cost in devising a protection strategy, the following fundamental requirements that are considered in designing the protective systems with a good performance of a relay are (Horowitz, S.H. and Phadke, A.G., 1992): reliability, selectivity, speed, sensitivity, simplicity and economy.

The relay application for protection of power system date back nearly 100 years ago. Since then, the technology employed to construct relays have improved dramatically relay size, weight, cost and functionality. Based on the technology employed for their construction, relays can be classified as follows: Electromechanical relays, Solid-state relays, Digital relays and Numerical relays.

Different Protection Types

Transformers are protected primarily against faults and overloads. The type of protection used should minimize the time of disconnection for faults within the transformer and to reduce the risk of catastrophic failure to simplify eventual repair. The electrical protection means the working principle of the protection based on the current, voltage, or frequency of that appear on the protected zone. The electrical protection of the Transformer comprises of the following: Over Current Protection, Fused Protection, Over Voltage Protection, Over Excitation Protection, and Differential Current Protection.

Table 3. Transformer faults/protection

Fault Type	Protection to Be Used
Primary winding Phase-Phase fault	Differential; Overcurrent
Primary winding Phase-Earth fault	Differential; Overcurrent
Secondary winding Phase-Phase fault	Differential
Secondary winding Phase-Earth fault	Differential; Restricted Earth Fault
Interturn fault	Differential, buchholz
Core fault	Differential, buchholz
Tank fault	Differential, buchholz; tank-earth
Overfluxing	Volts/Hz
Overloads	Thermal
External system short circuits	Time overcurrent, Instantaneous overcurrent

1. Transformer Over Current Protection

Overcurrent protection provides the first type of transformer protection and is commonly used for protection for phase and ground faults (ANSI/IEEE C37.91-1985). It's used as primary protection where differential protection is not used; also it is used to backup protection if differential protection has been used.

2. Over-Flux Protection

Transformer over-fluxing can be a result of: Overvoltage and Low system frequency.

When a transformer core is overexcited, the core is operating in a non-linear magnetic region, and creates harmonic components in the exciting current. A significant amount of current at the 5th harmonic is characteristic of overexcitation.

As the flux is proportional to the voltage, and inversely proportional to the operating frequency, the significant relaying quantity is the ratio of the per unit voltage to the per unit frequency. This is known as volts/hertz protection.

This protection is specially needed in the case of unit-connected generator transformers. If the turbine–generator is shut down with the voltage regulator in service, the volts/hertz limit of the transformers (and indeed of generators as well) could be easily exceeded. Similar conditions could also be reached by load rejection with voltage regulators disconnected, or in manual position, or with faulty instrumentation in the regulator circuits. The volts/hertz capability of transformers is specified by manufacturers. A typical capability curve is shown in Fig. 5. Many volts/hertz relays have two settings, a lower setting for alarm and a higher setting which may be used for tripping.

3. Differential Protection

Differential protection replaces overcurrent protection as the main protection for large power transformers.

A typical differential protection system is shown in Figure 6. Multiple circuits may exist, but the example is sufficient to explain the basic principle of differential protection (Einval C. H. and Linders,

J. R. 1975). It can be observed from Figure 6 that the protection zone is delimited by current transformers. Due to its very nature, differential protection does not provide backup protection to other system components. For this reason, differential protection is categorized as a unit protective scheme. The conductors bringing the current from the current transformers to the differential relay are in some situations called pilot wires.

Figure 6 shows an explanatory diagram illustrating the principle of the differential relay protection. Current transformers with similar characteristics and ratio are connected on the both sides of the transformer and a relay is connected between the two current transformers by using pilot wires. Under healthy or external fault conditions, the current distribution as shown in figure 7 (a), no current flowing in the relay. When the internal fault occurs as shown in figure 7 (b), the conditions of balance are upset and current flows in the relay to cause operation. It can be noted also that the protected zone of this differential relay is between the two current transformers. If the fault had occurred beyond, as shown in figure 7 (a), than the operation will not occur as the fault current would then flow through both current transformers thus maintaining the balance.

Figure 5. Volts/hertz capability of transformers, and relay settings

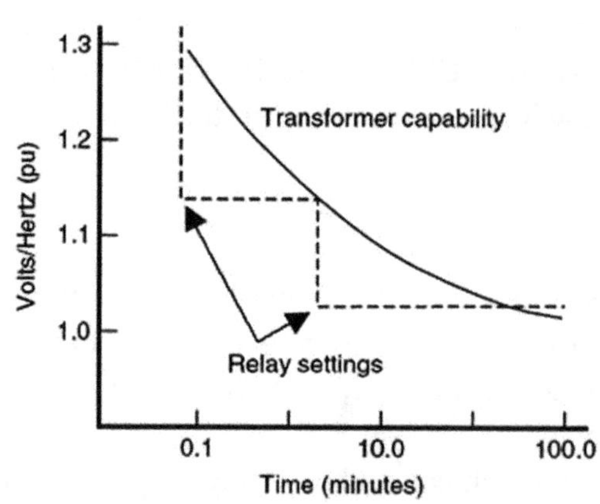

Figure 6. Typical differential power transformer protection relay

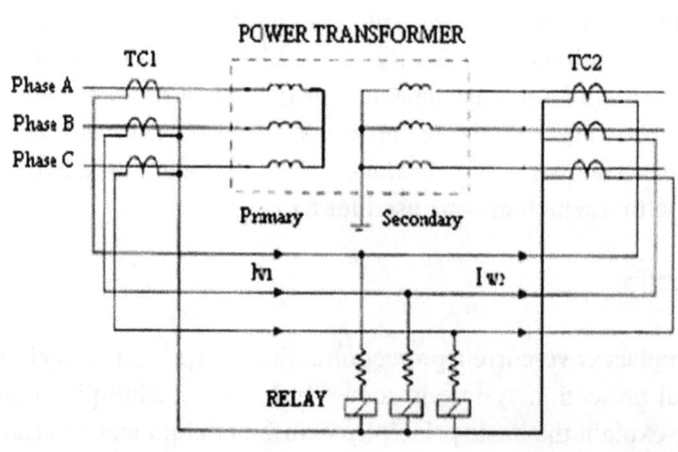

Figure 7. Current distribution: (a) Under healthy or external fault conditions, (b) during the internal fault

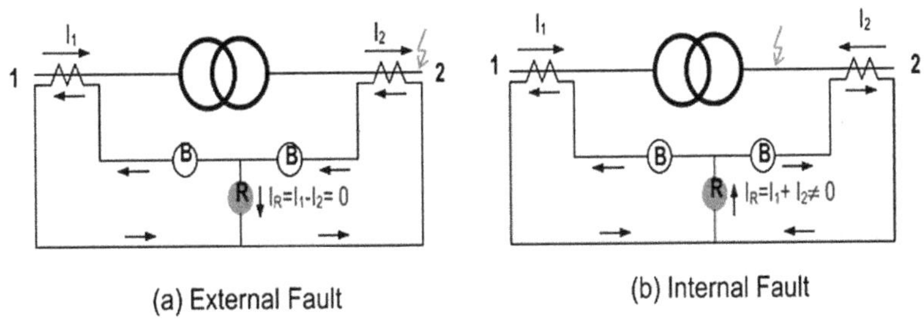

When one of the current transformers saturates, or if both current transformers saturate at different levels, false operating current appears in the differential relay and causes relay mal-operation. Some relays use the harmonics caused by the current transformer saturation for added restraint (Lewis B. J., 1998).

a. Percentage Restraint Differential Protection

Percentage restraint differential relays are more sensitive and secure than traditional differential ones (Guzman A. and Zocholl S., 2001). They have been in service for many years. Figure 8 shows a typical differential relay connection diagram. Differential elements compare an operating current with a restraining current. The operating current (also called differential current), I_{OP}, can be obtained as the phasor sum of the currents entering the protected element:

$$I_{OP} = \left|\vec{I}w_1 + \vec{I}w_2\right| \tag{2}$$

where,

I_{W1}, I_{w2} are the currents on the pilot wires of the current transformers,
I_{OP} is proportional to the fault current for internal faults and approaches zero for any other operating (ideal) conditions (Hengsi, Q.; Kimball, J.W. 2009).

Figure 8. Typical Differential Relay Connection Diagram

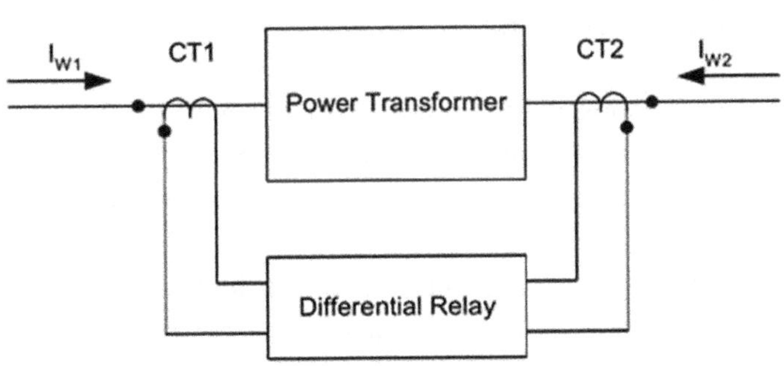

Following are the most common ways to obtain the restraining current:

$$I_{RT} = k\left|\vec{I}w_1 - \vec{I}w_2\right| \tag{3}$$

$$I_{RT} = k\left(\left|Iw_1\right| + \left|Iw_2\right|\right) \tag{4}$$

$$I_{RT} = Max\left(\left|Iw_1\right|, \left|Iw_2\right|\right) \tag{5}$$

where k is a compensation factor, usually taken as 1 or 0.5

Equations 4 and 5 offer the advantage of being applicable to differential relays with more than two restraint elements.

The differential relay generates a tripping signal if the operating current, I_{OP}, is greater than a percentage of the restraining current, I_{RT}.

Iop > SLPi.Irt (6)

where, SLPi is the slope of the ith characteristic of the differential relay

Figure 9 shows a typical differential relay operating characteristic. This characteristic consists of a straight line having a slope equal to SLP and a horizontal straight line defining the relay minimum pickup current, I_{mpu}. The relay operating region is located above the slope characteristic (Equation 6), and the restraining region is below the slope characteristic.

The current in the differential relay will not exactly zero at the normal operating condition or when external fault occur. It is normal because the tap of the transformer, the current transformer (CT) error, mismatch, and the excitation current.

The minimum pickup restraint setting, I_{mpu} adjusts the sensitivity of the relay. In non-numerical relays, the $I_{p.u(min)}$ was fixed at a typical value of 0.3 of the relay tap (Zamudio, Reference LIDV1112Av05). The characteristic consists of two slopes, SLP1 and SLP2 and a horizontal straight line defining the relay minimum pickup current, I_{mpu}. The relay operating region is located above the slope characteristic and the restraining region is below the slope characteristic (Cimadevilla, R., 2013). CT saturation is only one of the causes of false operating current in differential relays. In the case of power transformer applications other possible sources of error are:

- Mismatch between the CT ratios and the power transformer ratio,
- Variable ratio of the power transformer caused by a tap changer,
- Phase shift between the power transformer primary and secondary currents for delta-wye connections,
- Magnetizing inrush currents created by transformer transients because of energization,
- voltage recovery after the clearance of an external fault, or energization of a parallel
- transformer,
- High exciting currents caused by transformer overexcitation.

Figure 9. Percentage differential relay with dual slope characteristics

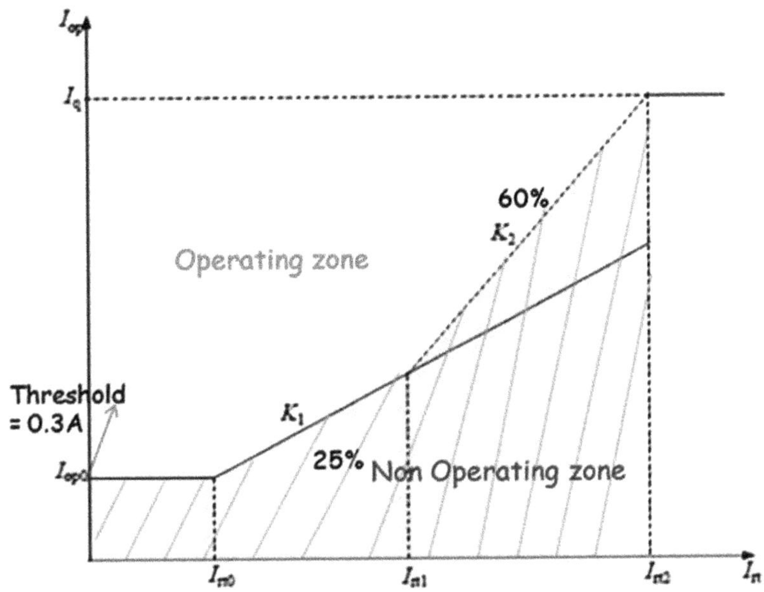

The relay percentage restraint characteristic typically solves the first two problems. A proper connection of the CTs or emulation of such a connection in a digital relay (auxiliary CTs historically provided this function) addresses the phase shift problem. A very complex problem is that of discriminating internal fault currents from the false differential currents caused by magnetizing inrush and transformer overexcitation.

b. Blocking Methods During Inrush Conditions

The transformer energization resembles the condition of an internal fault. If no inhibiting mechanism (blocking function) is provided, the differential element will trip. The magnetizing inrush currents have high component of even and odd harmonics.

The most common methods included in modern relays are based either on the measurement of the harmonic content of the differential current or on a wave shape recognition of this current. Wave shape recognition techniques represent another alternative for discriminating internal faults from inrush conditions. However, these techniques fail to identify transformer over-excitation conditions.

The magnetizing inrush current has significant 2nd harmonic content. The level of 2nd harmonic current can be used to differentiate between inrush and a fault condition. The fourth harmonic is also present in the inrush currents so it can also be used to restraint the operation. The third and fifth harmonics are normally used to detect an over-excitation condition of the power transformer. This situation occurs when the power transformer saturates with a symmetrical flux (the flux during an inrush condition was asymmetrical) because of an overvoltage or / and an under-frequency condition. The symmetrical flux originates a symmetrical magnetizing current that does not contain even harmonics but odd harmonics. The third harmonic is a good indicator for an over-excitation condition but, as it is a zero-sequence component (the three phase currents are equal), it is filtered by the delta windings or by the zero-sequence filters included in the differential relays so it will not be reliable in many transformer configurations. The fifth harmonic is normally used.

Many modern transformer differential relays use either harmonic restraint or blocking methods.

The differential relay operation condition can be expressed as;

$$I_{op} \geq SLP_i \cdot I_{rt} + k_2 I_{2h} + k_3 I_{3h} + ..., \tag{7}$$

where,

I_{op} is the fundamental component of the operating current
I_{2h}, I_{3h} are higher harmonics of the operating current
I_{rt} is the unfiltered restraint current
k_1, k_2 are the constant coefficients

A more recent set of techniques use only the second harmonic to identify currents and the fifth harmonic to avoid mal-operation for transformers due to over-excitation [23]. The basic operating equation for one phase can be expressed as follows:

$$I_{op} \geq SLP_i \cdot I_{rt} + k_2 I_{2h} + k_5 I_5 \tag{8}$$

Common harmonic restraint for three-phase transformer differential protection is a technique where the harmonic restraint quantity is proportional to the sum of the second and the fifth-harmonic components of the three relay elements. The relay operation is of the following form:

$$I_{op} \geq SLP_i \cdot I_{rt} + \sum_{n=1}^{3} \left(k_2 I_{2hn} + k_5 I_{5hm} \right) \tag{9}$$

PROPOSED BLOCKING FUNCTION FOR DIFFERENTIAL PROTECTION

Since the magnetizing inrush current generally contains a large second harmonic component in comparison to an internal fault, conventional transformer protection systems are designed to restrain during inrush transient phenomena by sensing this large second harmonic. However, the second harmonic component may also be generated during internal faults in the power transformer (Manana, M., et al 2005). This may be due to current transformer (CT) saturation, presence of shunt capacitance, or the capacitance in long extra high voltage transmission lines to which the transformer may be connected. The magnitude of the second harmonic in an internal fault current can be close to or greater than that present in the magnetizing inrush current (Bouderbala R., et al, 2011). The second harmonic components in the magnetizing inrush currents tend to be relatively small in modern large power transformers because of the improvements in the power transformer core material. The commonly employed conventional differential protection technique based on the second harmonic restraint will have difficulty in distinguishing between an internal fault and the inrush current thereby threatening transformer stability (Bouderbala R., Bentarzi

H., 2016). In this proposed approach, two harmonics (second and fourth) have used for restraining or blocking a differential relay and reducing the blocking time during an internal fault. This technique has been implemented in protection system of a three phase power transformer using Simulink/MATLAB, which ensures security for inrush conditions and provides dependability for internal faults. Then, it has been implement digital differential Protective relay for transformer: With advanced PC that operates in high speed, an acquisition card may be used for acquiring currents which will be needed by protection elements.

In a large power transformer, any switching actions can produce a large current peak due to the saturation of the transformer iron core. Owing to this core saturation, the inrush current contains, in addition to the harmonic components, a decaying dc current. Therefore, the inrush current can be modeled as follows (Degens, A. J. 1981):

$$i(t) = I_o \exp(-\lambda t) + \sum_{k=1}^{n} I_k \sin(k\omega_1 t + \theta_k) \tag{10}$$

Where k determines the order of harmonic, and $\omega 1$ is the frequency of the fundamental component. The decaying dc current can be represented by a Taylor expansion of two terms:

$$I_o \exp(-\lambda t) \approx I_o - I_o \lambda t \tag{11}$$

If it is assumed that the inrush current does not contain more than five harmonics, Eq.(10) becomes,

$$i(t) = I_o - I_o \lambda t + \sum_{k=1}^{5} I_k \cos\theta_k \sin(k\omega t) \tag{12}$$

Let X(t) denotes a stationary random process with a zero mean and suppose that one record X(t), of length T, is available. It shall be assumed that the record is sampled at equispaced intervals Δt of time tj, so that there are $n = \dfrac{T}{\Delta T}$ samples (in this case n=12). From the samples, Fourier sine and cosine coefficients X(tj) can be defined by usual relations given by:

$$S_k = \sum_{j=0}^{n-1} X(t_j) \sin(\omega_k j \Delta t) \tag{13}$$

$$C_k = \sum_{j=0}^{n-1} X(t_j) \cos(\omega_k j \Delta t) \tag{14}$$

where $\omega_k = \dfrac{2\pi k}{T}$.

If the sine and cosine terms of Eqs.(14) and (13) are replaced by their equivalent rectangular functions, then the corresponding rectangular transform term will be denoted by:

$$S'_k = \sum_{j=0}^{n-1} X(t_j) \operatorname{sgn}[\sin(\omega_k j \Delta t)] \tag{15}$$

$$C'_k = \sum_{j=0}^{n-1} X(t_j)\,\mathrm{sgn}[\cos(\omega_k j\Delta t)] \tag{16}$$

Considering that X(tj) are the last 12 differential currents with sampling frequency of 600 Hz (Thorp, J. S., Phadke, A. G., 1982). Thus, the Fourier coefficients can be obtained from the rectangular coefficients as,

$$S_k = A^{-1} S'_k \tag{17}$$

$$C_k = B^{-1} C'_k \tag{18}$$

where A and B are sparse matrices, more details about this theory are given in (Rahman, M. A., Dash, P. K., 1982). So assuming no aliasing, the Fourier coefficients can be expressed as follows:

$$S_1 = S'_1 - \left(\frac{1}{3}\right) S'_3 - \left(\frac{1}{5}\right) S'_5 \tag{19}$$

$$C_1 = C'_1 - \left(\frac{1}{3}\right) C'_3 - \left(\frac{1}{5}\right) C'_5$$

$S_2 = S'_2$, $S'_5 = S'_5$, and $C_5 = C'_5$.

In order to improve the processing speed, the quantities 1/3 and 1/5 may be generated by arithmetic shifts rather than hardware divisions. The modified formulations of the above quantities are implemented under the following form (Kaszenny, B. and Kuliidjiant, A., 2000):

$$S_1 = S'_1 - \left(\frac{1}{2} - \frac{1}{16}\right) S'_3 - \left(\frac{1}{4} - \frac{1}{32}\right) S'_5 \tag{20}$$

$$C_1 = C'_1 - \left(\frac{1}{2} - \frac{1}{16}\right) C'_3 - \left(\frac{1}{5}\right) C'_5$$

$$S_2 = \left(1 + \frac{1}{16}\right) S'_2$$

$$C_2 = \left(1 - \frac{1}{16}\right) C'_2$$

$$S_5 = \left(1 + \frac{1}{16}\right) S_3'$$

$$C_5 = \left(1 - \frac{1}{16}\right) C_5'$$

The harmonic components are found to be:

$$I_1 = \frac{2}{12}\left[S_1^2 + C_1^2\right]^2 \quad (21a)$$

$$I_2 = \frac{2}{12}\left[S_2^2 + C_2^2\right]^2 \quad (21b)$$

$$I_5 = \frac{2}{12}\left[S_5^2 + C_5^2\right]^2 \quad (21c)$$

After extraction of the fundamental, the second and the fifth harmonic components, these harmonic components will be used to produce restraining signal that may be used to block the relay. Otherwise, for internal fault case, the relay operates.

Protection Differential Relay Implementation

The above discussed approach has been implemented using Matlab/Simulink with the necessary tool box. The Matlab/Simulink which is powerful software program has been used for implementation as well as for testing and simulation.

1. Software Structure

Differential protection algorithm, which has been implemented using Simulink/ Matlab, its flow chart is shown in Fig.10. Besides, Graphical User Interface (GUI) has been developed using the same software tool, the user can select and set his wanted parameters, and makes a test by running simulation and displaying the tripping signal (see Fig.11).

2. Hardware Architecture

In protection field, current transformers are used to sense the current and provide the measured quantity as voltage signal to the input of relay. Circuit breaker is used as actuator.

The differential protection hardware whose block diagram shown in Fig.11 consists of:

- Signal transformation: current transformer (CT) transform currents of the power system to low safe values.
- Data acquisition boards: the measured values of the power system parameters fed from CTs in analog forms are passed through an anti-aliasing filter amplifier (low pass filter). Sample and hold circuits and analog multiplexed are used to sample the three different signals supplied by instrument transformers at the same time. The sampled signals are converted into digital form.
- PC: the digital signals are fed from data acquisition board to the PC where they will be processed.
- The developed differential protection has been implemented in PC associated with acquisition card AD 622 using Real time tool box with Simulink as shown in Fig.12 (Bouderbala R., Bentarzi, H. and Ouadi A., 2011).

Figure 10. Differential protection Algorithm flowchart

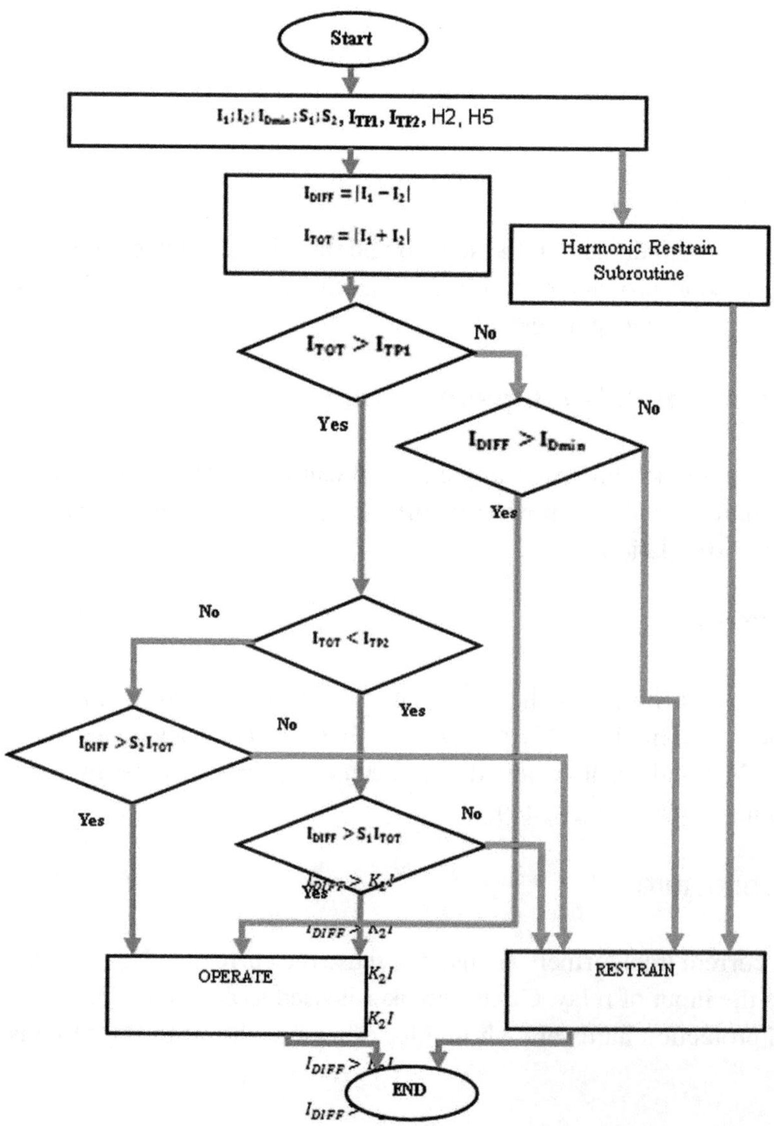

Figure 11. General block diagram of PC Based Differential relay

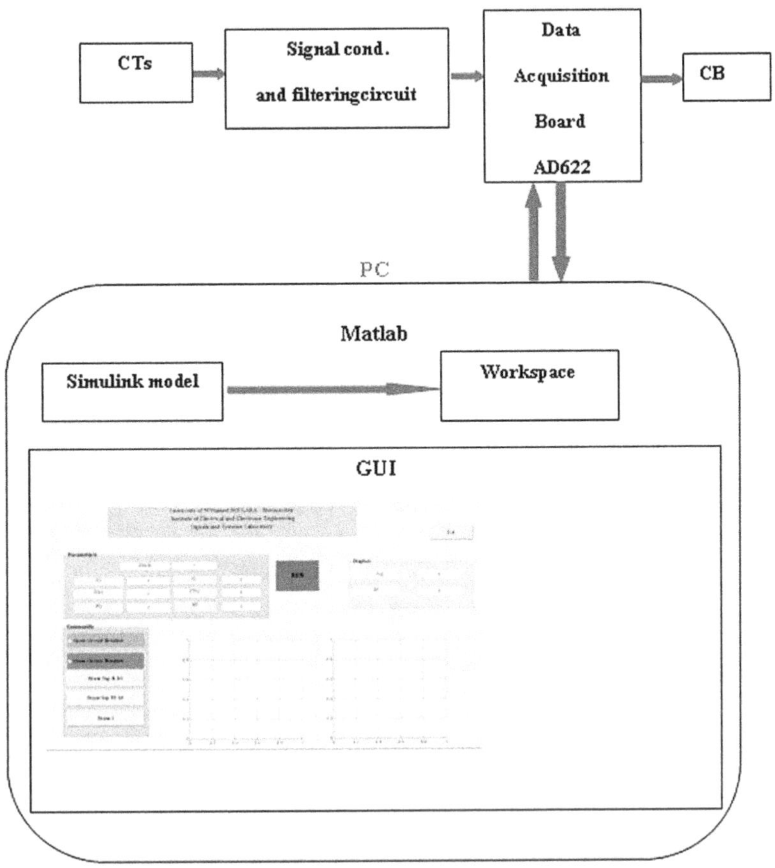

3. Differential Relay Parameters

Low impedance differential protection systems typically have 3 to 5 settings required to properly define the restraint characteristic of the relay (see Fig.12). The ensuing discussion will mainly focus on differential protection for power transformers.

Where, I_{Dmin} = minimum differential current (secondary) required to operate the relay.
I_{TP1} = turning point 1, I_{TP2} = turning point 2, S_1 = Slope 1 setting, S_2 = Slope 2 setting.
I_{RST} = Total current through the differential system, measure of system loading.
I_{DIF} = For a given value of I_{TOT}, this is the restraint current applied by the relay or alternatively the minimum differential current required to operate the relay.

The settings to be considered are I_{Dmin}, I_{TP1}, I_{TP2}, S_1, and S_2. Besides, second harmonic (H_2) and fifth harmonic (H_5) may be used when harmonic restrain technique. These are generic representations of the settings. They will differ from one manufacturer to the other.

Where there are two straight lines given with a slope of S_1=0.25 and a slope of S_2=0.6, which range from I_{rt0} to I_{rt1} and from I_{rt1} to I_{rt2}, respectively, and a horizontal straight line defining the relay minimum pickup current, I_{Dmin}=0.3A. The relay operating region is located above the slope, and the restraining region is below the slope.

The dual-slope percentage pattern adds a restraint area and avoids mal-operation caused by saturation. In comparison with a single-slope percentage scheme, the dual-slope percentage current differential protection can be regarded as a better curve fitting of transformer operational principles (Meckic, F., et al, 2006; Bouderbala, R., Bentarzi, H. and Ouadi A., 2010).

Figure 12. Differential relay Real time Simulink Model

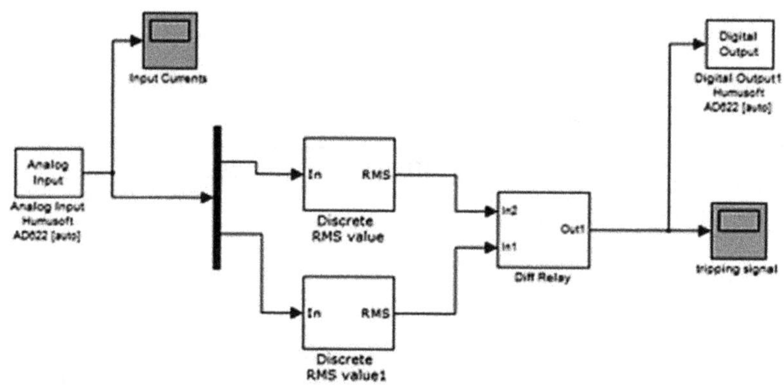

CONCLUSION

In this chapter different blocking techniques for differential protection were discussed focusing on the blocking function applied in large power transformer and smart transformer with magnetizing inrush currents tend to be relatively small. It is concluded that each of the blocking technique has its own pros and cons depending on the application involved. Also using various new and latest hardware for implementation of these techniques can deliver more reliable protective differential relay regarding the dependability and security including the rapidity.

REFERENCES

Abdelmoumene, A., & Bentarzi, H. (2012). Reliability Enhancement of Power Transformer Protection System. *Journal of Basic and Applied Scientific Research*, *2*(10), 10534–10539.

ANSI/IEEE C37.91-1985. (1985). *IEEE Guide for Protective Relay Applications to Power Transformers*. IEEE.

Bouderbala, R., & Bentarzi, H. (2013). A New Computer Based Differential Relay Framework for Power Transformer. *Lecture Notes in Electrical Engineering, 260*, 473-481. DOI: doi:10.1007/978-94-007-7262-5_54

Bouderbala, R., & Bentarzi, H. (2016). *Differential relay reliability enhancement using fourth harmonic for a large power transformer. International Journal of Syst. Assur. Engineering Mang*. doi:10.100713198-016-0475-6

Bouderbala, R., Bentarzi, H., & Ouadi, A. (2010). A New Approach Applied to Digital Differential Protection for a Large Power Transformer. *The 9th WSEAS international conference on Recent Researches in Circuits, Systems, Electronics, Control & Signal Processing Proceeding*, 202-205.

Bouderbala R., Bentarzi, H. & Ouadi A. (2011). Digital Differential Relay Reliability Enhancement of Power Transformer. *International Journal of Circuits, Systems and Signal Processing*, *5*, 263-270.

Brunke, J. H., & Fröhlich, J. K. (2001). Elimination of Transformer Inrush Currents by Controlled Switching. *IEEE Transactions on Power Delivery*, *16*(2), 276–280. doi:10.1109/61.915495

Chafai, M., Refoufi, L., & Bentarzi, H. (2016). Large power transformer reliability modeling. *Int J Syst Assur Eng Manag*, *7*(S1), 9–17. doi:10.100713198-014-0261-2

Cimadevilla, R. (2013). Application of Zero-Sequence Filter on Transformer Differential Protection. *SEAPAC*.

Cimadevilla, R., & Quintanilla, R. (2011). New protection units included in differential relays. *PAC Conference*.

Degens, A. J. (1981). Algorithm for a Digital Transformer Protection Based on a Least-Squares Curve-Fitting. *IEE Proc.*, *128*, 155.

Einval, C. H., & Linders, J. R. (1975). A Three-Phase Differential Relay for Transformer Protection. *IEEE Transactions PAS*, *94*(6), 1971–1980.

Guzman, A., Altuve, H., & Tziouvaras, D. (2009). *Power Transformer Protection Improvements with Numerical Relays*. Academic Press.

Guzmán, A., Zocholl, S., & Altuve, H. (2000). Performance analysis of traditional and Improved differential Protective transformer relays. SEL Company.

Guzman, A., Zocholl, S., Benmouyal, G., & Altuve, H. J. (2001). A current based solution for transformer differential protection – part 1: Problem Statement. *IEEE Transactions on Power Delivery*, *16*(4), 485491. doi:10.1109/61.956726

Hayward, C. D. (1941). Harmonic-Current-Restrained Relays for Transformer Differential Protection. *Transactions of the American Institute of Electrical Engineers*, *60*(6), 377–382. doi:10.1109/T-AIEE.1941.5058349

Hengsi, Q., & Kimball, J. W. (2009). AC-AC dual active bridge converter for solid state transformer. *Proceedings of the 2009 IEEE Energy Conversion Congress and Exposition*, 3039–3044. 10.1109/ECCE.2009.5316507

Horowitz, S. H., & Phadke, A. G. (1992). *Power System Relaying*. Wily & Sons.

IEEE Standard C37.91-2000. (2000). Guide for Protective Relay Applications to Power Transformers. *The Institute of Electrical and Electronics Engineers*.

Kasztenny, B., & Kezunovic, M. (1998, October). Improved Power Transformer Protection Using Numerical Relays. *IEEE Computer Applications in Power*, *11*(4), 39–45. doi:10.1109/67.721702

Kasztenny, B., & Kulidjian, A. (2000). An Improved Transformer Inrush Restraint Algorithm Increases Security While Maintaining Fault Response Performance. *53rd Annual Conference for Protective Relay Engineers.*

Kennedy, L. F., & Hayward, C. D. (1938). Harmonic-Current-Restrained Relays for Differential Protection. *Transactions of the American Institute of Electrical Engineers*, *57*(5), 262–266. doi:10.1109/T-AIEE.1938.5057808

Lewis, B. J. (1998). *Protective Relaying: Principles and Applications*. New York: Marcel Dekker.

Manana, M., Eguıluz, O. A., Dıez, G., Renedo, C., & Perez, S. (2005). Effects of magnetizing inrush current on power quality and distributed generation. *9th Spanish Portuguese Congress on Electrical Engineering.*

Mason, C. R. (1986). *The Art and Science of Protective Relaying* (2nd ed.). John Wiley.

Meckic, F., Girgis, R., & Gajic, Z. (2006). Power Transformer Characteristics and their Effects on protective Relays. *33rd Western Protective Relays Conference.*

Ong, C. M. (1998). *Dynamic Simulation of Electical Machinery using Matlab/Simulink*. Prentice Hall.

Rahman, M. A., & Dash, P. K. (1982). Fast Algorithm for Digital Protection of Power Transformer. *IEE Proc.*, *129*(C2), 79.

Sonnemann, W. K., Wagner, C. L., & Rockefeller, G. D. (1958). Magnetizing Inrush Phenomena in Transformer Banks. *Transactions of the American Institute of Electrical Engineers*, *77*, 884–892.

Thorp, J. S., & Phadke, A. G. (1982). A Microprocessor Based Three-Phase Transformer Differential Relay. *IEEE Trans.*, *PAS-101*, 426.

Chapter 15
Application of Synchro-Phasor Measurement Unit in Smart Grid Including Renewable Energy

Meriem Mahboubi
Signals and Systems Laboratory, IGEE, University M'hamed Bougara of Boumerdes, Algeria

Abderrahmane Ouadi
INELEC, Boumerdes, Algeria

ABSTRACT

Renewable energy is one of the most abundant energies in our planet. In order to satisfy the world demand of electrical energy, solar and wind energy may be used. Identical to all other types of power generation plants, the integration of these renewable energy sources in smart power grid has an impact on its operation. Thus, when the electrical power is injected into the power grid by these energy sources, the system electrical parameters must be well monitored for synchronization purpose. This can be accomplished with the aid of synchro-phasors measurement units. The phase angle of the utility is a critical parameter for the operation of power devices feeding power into the grid such as PV and wind energy inverters. There are many techniques to obtain the grid phase angle such as the zero-crossing detection and the orthogonal phase locked loop. This research work discusses the use of PMUs (Phasor Measurement Unit) for providing this important parameter to system synchronization in the case of high penetration of solar or wind energy in the power grid.

INTRODUCTION

Renewable energy is being progressively integrated with ambitious targets of renewable energy integration set at national/regional levels. Solar PV and wind power sources are the front-runners among available renewable energy sources. However, as the penetration of renewable generation increases, the impact on power system dynamics is becoming increasingly apparent, and will become a more integral part of system planning and renewable integration studies. Historically, power systems have been based

DOI: 10.4018/978-1-7998-4027-5.ch015

around large synchronous generators connected to a strongly meshed transmission network, with the dynamic characteristics of such systems being well understood and monitored. However, renewable generation stations, particularly in the form of wind and solar generation, are increasingly universally connected via power electronic interfaces, may well be connected to the distribution network, or weaker parts of the network, may offer new control capabilities, and, of course, is subject to the variability and uncertainty associated with local and regional weather patterns. The time variability and non-dispatched nature of wind generation may pose substantial challenges, particularly at higher levels of penetration, including an increase in regulation costs and incremental operating reserves, but they can also lead to increased opportunities for energy storage, demand side response, cross border interconnections, and other flexibility measures.

On the other hand, evolution of synchrophasor measurement technology, using Phasor Measurement Units (PMUs) for data measurement has opened a vast number of potential applications in the power system world, ranging from monitoring, protection, and control to enhance the smartness and efficiency of power systems. PMU-based Wide Area Monitoring Protection and Control (WAMPAC) can play a vital role in secure and stable integration of renewable energy, particularly in grid operation under high share of renewable (Phadke, A.G., 2008; Bentarzi. H., 2010). PMUs make the power system smarter and reliable has attracted many researchers to have further development in this area. Significant efforts have been dedicated to the development of efficient and precise measurement algorithms. The development of global positioning system (GPS) technology has overcome the synchronizations' difficulties and lead to the development of phasor measurement unit (PMU). PMU is a measurement device, developed in mid 1980s. It measures the electrical quantities such as voltage and current at node in the power system where it has been installed. It provides the real time phasor parameters of these quantities such as magnitude, frequency, phase angle and rate of change of frequency (ROCOF). The provided voltage and current signals by PMU from the grid may be used be to monitor completely the power system that it is observable at any moment.

Data provided by PMUs are generated at different locations, in real synchronized time. They are time stamped for synchronization by a reliable and accurate time source; the Global Positioning System (GPS) (A.G. Phadke, 2006).The objective of this chapter is to use synchrophasor measurements to solve islanding problem in a network including high penetrations of solar energy generation by performing a continuous monitoring in order to disconnect and connect the islanded generator.

LITERATURE SURVEY

In the late seventies, the oil crisis caused a drastic increase in the demand of renewable energy sources. This increase has been driven by the remarkable advantages offered by these kind of energy sources namely free resource, infinite reserve and clean conversion process. However, till today, the fuel is still the main source of energy in most power generating systems. While, global warming and energy policies have become a hot topic worldwide. Developed countries are trying to reduce their greenhouse gas emissions. For instance, the European union has committed to reduce their greenhouse gas to at least 20% levels before 1990 and to produce no less than 20% of its energy consumption from renewable sources by 2020 (European Commision, 2020). In fact, the secured and safe integration of renewable energy (RE) resources including wind and solar energy to the main grid is one of the biggest challenges for the world energy sector today (Ahmada, T., 2020).

Among all renewable energy sources, the solar energy is the fastest growing electricity source in the last decade. It is also considered as the cleanest source ever. It can be produced either off-grid or tied to the grid; grid-connected. As a matter of fact, most of the PV power generation comes from grid-connected installations, where the power is fed in the electricity network.

PV power generation is a growing business in developed countries, the first solar power concentrated commercial plant was established in 1980 (M.S. Hossain, 2016). After that, in 2016, as reported in (M.S. Hossain,2016) the top photovoltaic countries that produced energy from solar were Germany (40,988 MW), China (77,434 MW), Japan (41,600 MW), Italy (19,251 MW), United States (34,711 MW), France (6767 MW), Spain (7171 MW), UK (11,250 MW), Australia (5632 MW), Belgium (3292 MW), Thailand (2154 MW), Netherlands (1955 MW), Switzerland (1644 MW) (Hossain, M.S., 2016). The same reference is revealing that the world largest concentrated solar panel installation with 354 MW is in Mojave Desert of California. The biggest solar power stations are located in Spain: Solnova and Andasol about 150 MW both. While, the world's largest photovoltaic power station is found in the United State, the Agua Caliente Solar project with more than 250 MW production, followed by the Charanka Solar Park in India (about 221 MW) (Hossain, M.S., 2016). Further details of the solar energy use's statistical improvement throughout the world is presented in (Hossain, M.S., 2016).

As the solar energy principle imposes, the power output of the photovoltaic panels is affected by daily and seasonal weather fluctuations, resulting in intermittent electricity generation or a rapid increasing penetration of solar power in the power grid. For instance, in Germany, the high penetration of solar generation in the Electric Grid (Appen, V., 2013) has increased grid stability issues (e.g. "The 50.2 Hz Risk") which may endanger the operation of the entire interconnected grid. In order to avoid the risk of major power outages caused by events such as grid failures, the application of smart grid is a solution. It improves efficiency, reliability, cost, and sustainability of the production and distribution of electricity. A smart grid is an intelligent electrical grid that uses modern technologies, hardware, software and even practices for self-detection and self-healing. Yet, this task is complex since most utilities are facing the aging of existing infrastructure, in addition to the uncertainties and variability of the system (Klaric, M., 2018). Those combined problems can cause severe faults extending along the grid, to face this, and to ensure power grid stability, it becomes apparent that additional techniques of power system monitoring and control are required. The problems need to be identified and prepared in advance. In addition to the real time monitoring in order to prevent system instabilities at all voltage levels.

In conventional grids, the monitoring is done with a supervisory control and data acquisition (SCADA) system (Patil, G. C., 2017; Kato, D., 2014). Nevertheless, SCADA systems can only measure the magnitude of voltage and current but not the phase. Furthermore, it requires separate data acquired from oscilloscope devices installed in substations to measure information needed to analyze transient phenomena (Kato, D., 2014). A better technology to monitor the grid is phasor measurement units (PMUs) (Patil, G. C., 2017). PMUs are gaining attention as a technology that can provide operators with high-precision, time-synchronized measurements of the situation awareness they need. They have already been defined as suitable for many applications of larger renewable energy integration. Data gathered by PMUs have a short measurement cycle and are synchronized using absolute time. These advantages mean that even when a fault occurs, waveform information from multiple locations gathered at the time of the fault can be used in detailed fault analysis to create prevention measures for future faults and to minimize the consequences. This chapter emphasizes on the application of PMUs in the grid connected PV system for synchronization purpose and shows how it can be benefit in solving one of the most common problems that faces such systems; the islanding problem in a network including high penetrations of solar energy generation.

PHASOR MEASUREMENT UNIT: OVERVIEW

To understand the concept of synchro-phasors, the phasor is first defined as an analytical and time-invariant vectorial representation of an AC signal with sinusoidal waveform. Consider the pure Sinusoidal Signal:

$$x(t) = Xm\cos(\omega t + \varphi) \quad (1)$$

Where ω is the frequency of the signal. Φ is the phase angle and Xm the peak amplitude. x(t) can be represented by a unique complex number known as a phasor, given by:

x(t) = Re{ Xmej(ωt+φ) } = Re [{ej(ωt) } Xmej φ] (2)

x(t)«X = (Xm/Ö2) ej φ = (Xm/Ö2) [cos φ + jsinφ] (3)

The frequency is not explicitly stated in the phasor representation. The magnitude of the phasor is the rms value of the sinusoid, its phase angle is φ in (1), measured in a counterclockwise direction from the real axis. A sinusoidal signal and its phasor representation are illustrated in Figure 1. Then, A synchro-phasor is defined as the magnitude and angle of a fundamental frequency waveform as referenced to an absolute point of time. It is a phasor tagged with a unique, highly accurate clock time stamp; GPS (Phadke, A.G., 2008).

By referencing all PMUs to a common time base, PMU measurements become comparable over a wide area of measurement. This leads to valuable information for several electric power network-based applications (Phadke, A. G., 2008; Ouadi, A., Bentarzi, H., & Maun, J. C. 2009).

Figure 1. (a) a sinusoid signal (b) and its phasor representation
(Phadke Arun G, 2008)

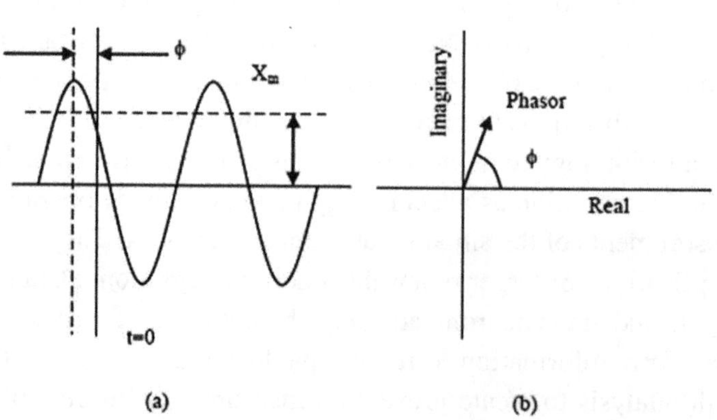

This may delay the understanding and analysis of certain power system phenomena and hence the development of certain power system applications. To remove this barrier, phasors measured across the power grid should have a common timing reference such that direct comparison is feasible.

A synchrophasor is defined according the IEEE standard C37.118 as a phasor calculated from data samples using a standard time signal as the reference for the measurement; however, synchronized phasors from remote sites have a defined common phase relationship (C37.118-2005, 2006). Thus, synchrophasors measured across an interconnected power grid will have a common timing reference and they can be compared directly as illustrated in Figure 2.

According to the same standard, a synchronizing source that provides the common timing reference may be local or global. The synchronizing signal may be distributed by broadcast or direct connection, and shall be referenced to Coordinated Universal Time (UTC). One commonly utilized synchronizing signal is the satellite signal broadcast from the Global Positioning System (GPS).

The definition of a real-time or synchronized Phasor provided in the IEEE Standard 1344-1995 (Phadke, A. G., 1993) (1344-1995, 2001) corresponds to the conventional definition described earlier, at least at rated frequency (50 Hz or 60 Hz). With real-time waveforms, it is necessary to define a time reference to measure phase angles synchronously (Power System Relaying Committee of The IEEE PES, 2006). The IEEE Standard 1344-1995 (Phadke, A. G., 1993) (1344-1995, 2001) defines the start of the second as the time reference for establishing the phasor phase angle value. The synchronized phasor measurement convention is shown in Figure 2.

The instantaneous phase angle measurement remains constant at rated frequency when using the start of the second phase reference. If the signal is at off-nominal frequency, the instantaneous phase varies with time. It can be seen later that the choice of this reference has an impact on the phasor phase angle measurement at off-nominal frequency.

Figure 2. Synchrophasor measurement conventions with respect to time (C37.118-2005, 2006).

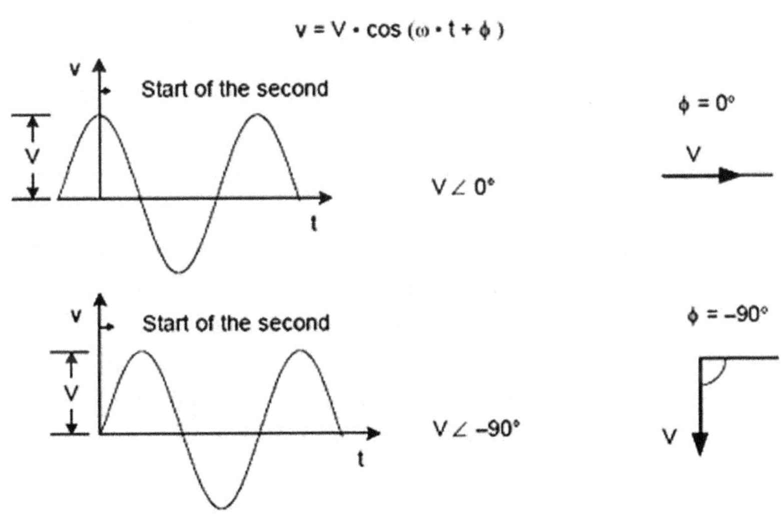

A PMU is a standalone device that measures 50/60 Hz AC voltage and current signals and provides them in phasor form with their frequency. The analog AC waveforms are digitized by an analog to digital converter for each phase and a phase-lock oscillator and a GPS reference time source, often called pulse per second (PPS) provides high-speed time synchronized sampling as illustrated in Figure 3. A PMU calculates line frequency, as well as voltage and current phasors at a high sampling rate and streams those data, along with the associated GPS time stamp, via networked communication lines. The synchrophasors can be phase or symmetrical component values.

Figure 3. PMU block diagram
(Phadke, A. G., 2008)

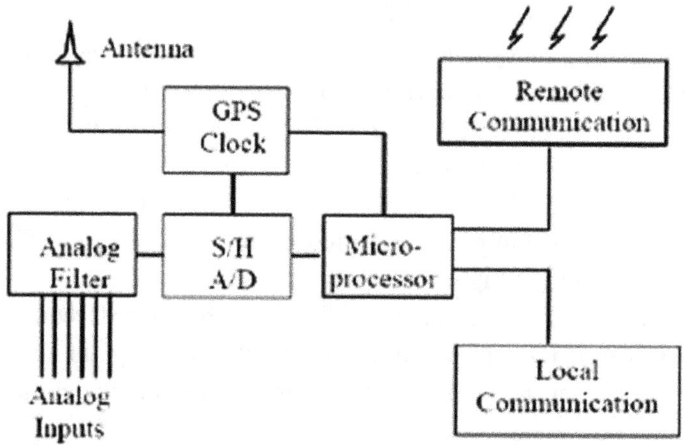

At present, phasor measurement units (PMUs) are the most accurate and advanced synchronized phasor measurement equipments. They measure 50/60 Hz sinusoidal waveforms of voltages and currents at a high sampling rate, up to 1200 samples per second and with high accuracy. From the voltage and current samples, the magnitudes and phase angles of the voltage and current signals are calculated in the phasor microprocessor of the PMU. As the PMU uses the clock signal of the Global Positioning System (GPS) to provide synchronized phase angles measurements at all the measurement points, the measured phasors are often referred to as synchrophasors. Thus, a Synchrophasor is defined as the magnitude and angle of a fundamental frequency waveform as referenced to an absolute point of time. It is a phasor tagged with a unique, highly accurate clock time stamp; GPS. By referencing all PMUs to a common time base, PMU measurements become comparable over a wide area of measurement. This leads to valuable information for several electric power network-based applications.

Figure 4 illustrates a functional block diagram of a typical PMU. The GPS receiver provides the 1 pulse-per-second (pps) signal, and a time tag consisting of the year, day, hour, minute, and second. The l-pps signal is usually divided by a phase-locked oscillator into the number of pulses per second required for the sampling of the analogue signals. The analog signals are derived from three-phase voltage and current transformers with appropriate anti-aliasing filtering. The microprocessor calculates the positive sequencevoltage and current phasors, and determines the timing message from the GPS, along with the sample number at the beginning of a window.

Figure 4. Functional block diagram of a typical PMU
(Phadke Arun G, 2008)

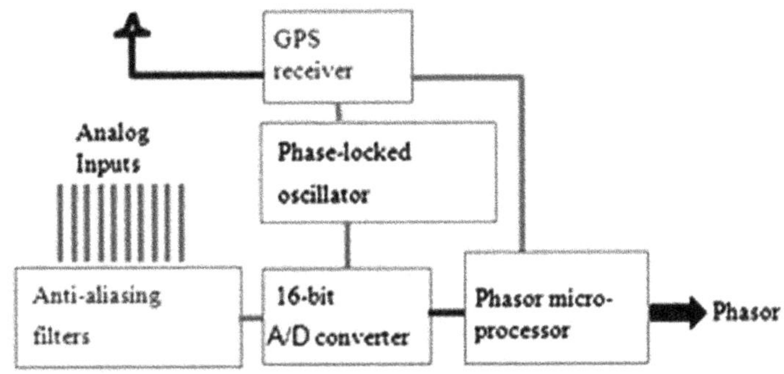

PV GENERATION SYSTEM

This section deals with the simulation of a simplified single stage single-phase grid connected Photovoltaic (PV) System. In the PV system, the chopper stage is eliminated and the output voltage is maintained by the network to which the PV system is connected.

Figure 5 shows the main parts of the PV system, which are:

1. **PV module:** This is the heart of the system, which is a set of PV cells connected either in series or in parallel to constitute PV modules and PV arrays.
2. **Single-phase full-bridge inverter with hysteresis current controller:** The hysteresis current control is used because feedback from the grid is required in the grid connected PV applications.
3. **LC Filter:** As PV arrays are integrated into the grid system, power quality issues such as Total Harmonic Distortion (THD) became an increasingly serious concern since the switching devices used to convert power introduce harmonics in the system. In order to eliminate the switching ripples efficiently, a low-pass filter at the output is required. The LC-filter is best suited to such configurations where the load impedance across the capacitor is relatively high at, and above the switching frequency. Furthermore, if a system is connected to the grid through an LC-filter, the resonance frequency varies over time as the inductance value of the grid varies (Guo, X.-Q., 2011).
4. **Controller or Microcontroller:** it may execute two tasks namely:
    a. P&O MPPT: The PV system has a non-linear current-voltage and power-voltage characteristics that continuously varies with irradiation and temperature. In order to track the continuously varying maximum power point of the solar array the maximum power point tracking (MPPT) control technique plays an important role. In this simulation the maximum power is obtained using the Perturb and Observe algorithm (P&O) in other words the output current is adjusted by P&O in the direction that increases the power at the output of the PV module.
    b. Sine wave, matching grid voltage, generation to synchronize the inverter output with the grid utility.

Thus, Maximum Power Point Tracking (MPPT), Grid synchronization, Islanding, DC-link voltage control and Current control are the major factors which are required for a grid-tied photovoltaic system to work efficiently and effectively. Phase, amplitude and frequency of the utility voltage are critical information for the operation of the grid-connected inverter system. For grid synchronization scheme may need phase, amplitude and frequency of utility voltage that is used to generate reference signal. Actuate and fast detection of these parameters is essential in order to keep grid parameters within range as mentioned in standards. This research work shows how PMU can provide these parameters and they can be used for different grid synchronization in power grid including a connected photovoltaic system.

Figure 5. PV system's main parts
(Mahboubi M., 2019)

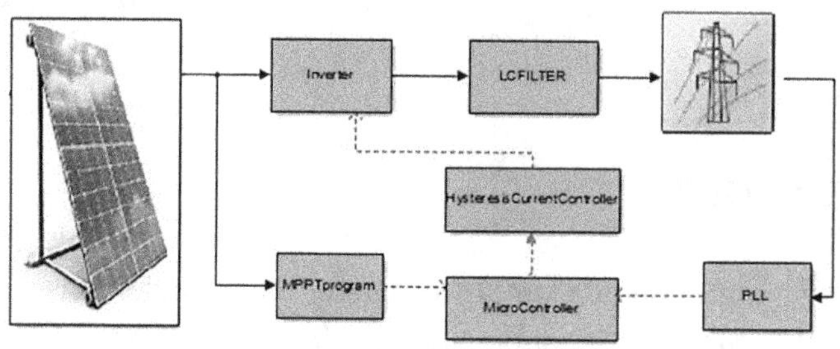

INVERTER TO GRID SYNCHRONIZATION

The phase angle of the utility is critical information for the operation of power devices feeding power into the grid like PV inverters. There are many techniques to obtain the grid phase angle like the zero-crossing detection and the orthogonal phase locked loop (Guo, X.-Q., 2011).

- **Orthogonal PLL** is a closed loop system in which an internal oscillator is controlled to keep the time and phase of an external periodical signal using a feedback loop.
- **The zero-crossing detection** is based on converting the feedback sinusoidal grid signal to an equivalent square wave signal, compare it to the ground and generate a pulse each time a rising edge is detected; it generates an equivalent periodic pulse representing its frequency.

However, when a PMU is used, it provides the phase, the amplitude and the frequency of the utility voltage. These provided data can be used to generate pulses which in turn are used to generate an equivalent reference signal i_{ref}. Then, the generated reference signal will be used in the hysteresis controller.

PMU APPLICATION IN GRID CONNECTED PV SYSTEM

Grid connected PV system requires an accurate estimate of the grid frequency to feed power synchronously to the power grid. Similar to the conventional generating plants, outputs from the solar energy source have impact on grid operation. Its rapid penetration in the grid requires real time monitoring of the system operating conditions for the secure operation of power system. This can be accomplished by using PMUs. PMU measures voltage and current at a precise time and output these quantities in phasor form. This makes the grid completely observable at any moment. Hence, utilities can drastically improve the ability to operate their system dynamics. In fact, PMUs are not only used for time monitoring, loss of mains protection is one of many real time applications that can benefit from PMU deployment. In the grid connected PV system, loss-of-mains protection is an important requirement. The role of loss-of-mains protection is to disconnect the PV system from the utility grid to prevent power islands. This islanding scenario is shown in Fig. 6. It is formed when a section of distribution network has become disconnected from utility supply, but utility customers continue to be energized by an embedded generator. This is undesirable from the point-of-view of power quality, and also poses serious hazards to workers (Rajeev A, 2016).

Figure 6. Example of islanding scenario
(Rajeev A, 2016)

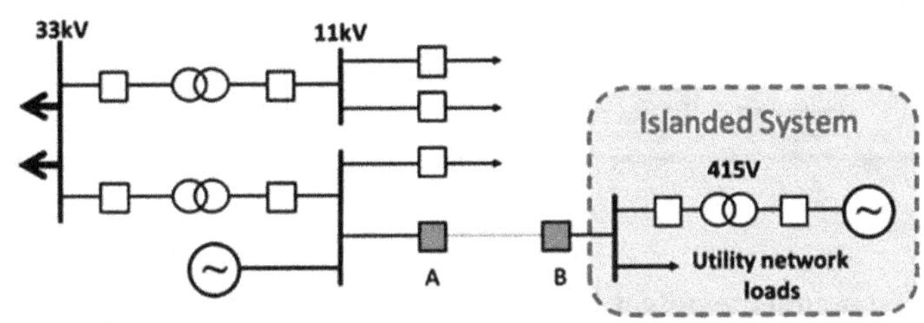

This section describes a method of loss of mains protection based on PMU technology, specifically using a PMU device of authors' design, which continuously monitors the synchronization of the PV system with the grid by means of real time data from synchro-phasors.

There are two dominant forms of traditional loss-of-mains protection, rate-of-change-of-frequency (ROCOF) and vector shift (VS). Both techniques operate on the principle that at the moment of island formation there will be an imbalance between islanded load and islanded generation. Islanding is inferred from a change in the pattern of zero-crossings of the voltage waveform (Rajeev A, 2016).

It is not possible to adequately differentiate loss-of-mains events from widespread system disturbances using traditional loss-of-mains techniques which rely only on the measurement at the point of connection of the embedded generator to the grid. A case study realized on a system that contains distributed generations is detailed in (Laverty, D.M., 2015). It shows that ROCOF method is prone to undesired operation, or nuisance tripping.

The synchrophasor data from PMU can be used to solve this problem. The proposed solution is discussed in the next section.

SYSTEM SIMULATION AND RESULTS

Case 1: Grid connected PV system synchronization using PMUs

In this research work, the PMUs technique for synchronization is used. MATLAB/Simulink software has developed in all accomplished simulations. Figure 7 shows the Simulink model of whole system that is simulated under 25°C and for different irradiances from 1000 to 800W/m^2. All results are presented here.

Figure 7. Simulink Model of a connected PV system

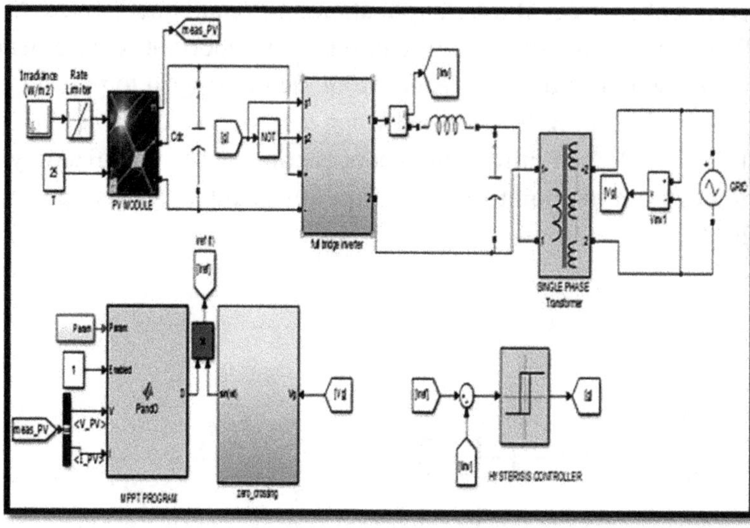

The output voltage must match the power grid voltage in phase and magnitude, regardless of the external conditions, while, the current can vary with irradiance and temperature variation as illustrated in Fig.8.

Figure 8. Inverter output voltage and current

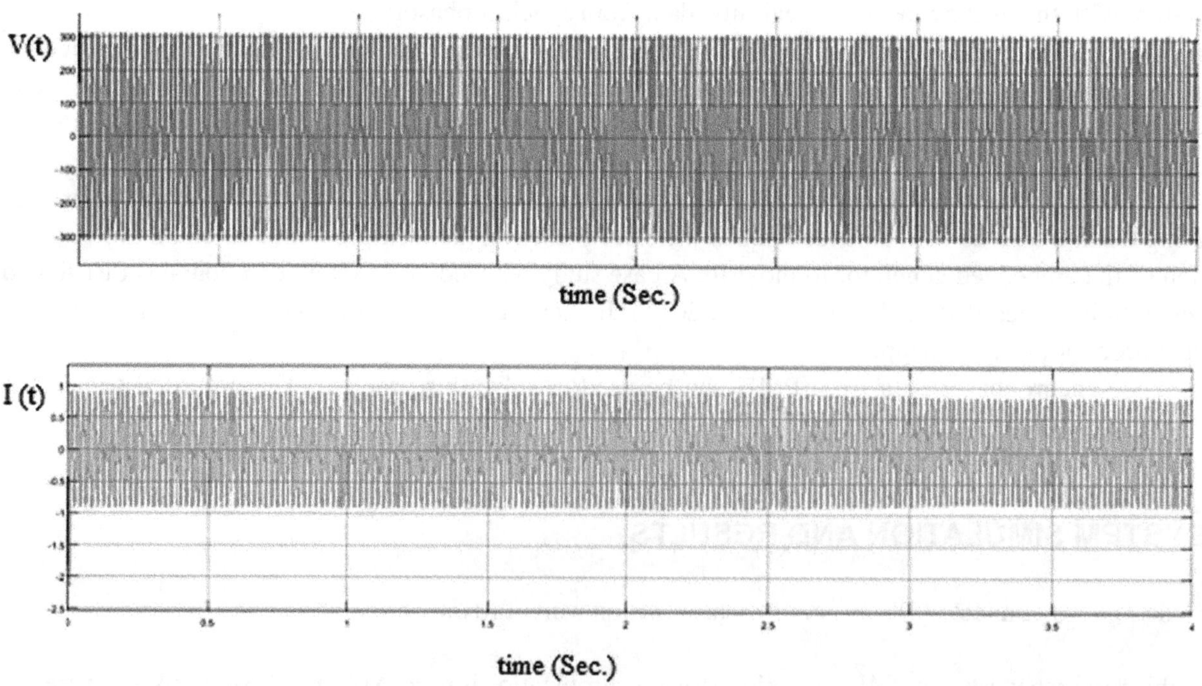

The output voltage must match the grid voltage in phase and magnitude, regardless of the external conditions, however, the current can vary with irradiance and temperature variation. When the current is in phase with the voltage, the power is positive and the inverter receives power from the grid. Since the desired functionality of this system is to generate power the current must be controlled to be out of phase by 180 degrees, resulting in negative power which means that the system supplies or injects power to the grid. To send power into grid, the system must be well synchronized. In other words, the frequency of generated current which is the key element must match with that of the grid. Different techniques are available to detect this frequency; however, in this work the PMU based frequency technique is adopted.

Figure 9 shows the output voltage to the grid versus the injected current in the case of maximum current output from the MPPT algorithm. The output power quality is good; however, the amplitude is small. This small value is due to the use of only one panel, adding to it the system losses and the small reference current.

Figure 9. Inverter output voltage and current using MPPT controller.

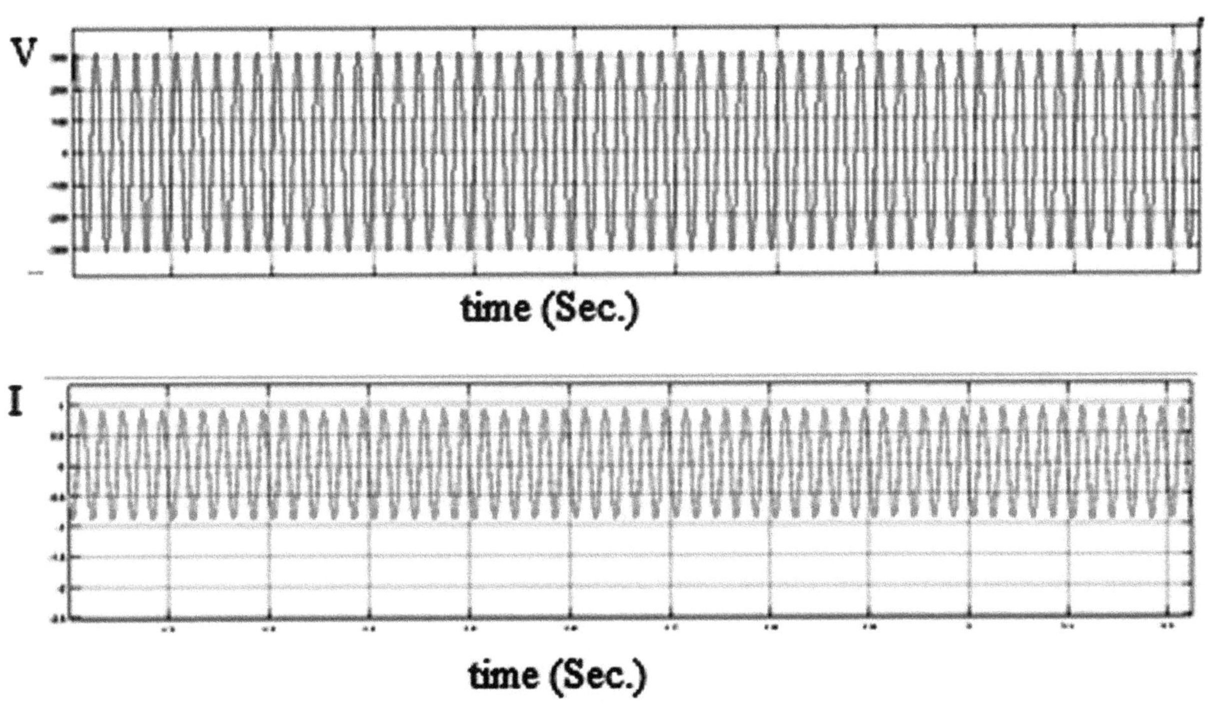

Case 2: Grid connected PV system with an islanding detection technique using PMUs

Grid connected PV system requires an accurate estimate of the grid frequency that may be used to feed power synchronously to the power grid. The synchrophasor data from PMU can be used to provide this accurate frequency. In this solution PMUs are used to compare the frequency and phase angle of the PV system with a secure point on the grid synchronously. The used algorithm and its Simulink subsystem are presented in figures 10 and 11 respectively. If the two values are different, a loss of synchronism is probable. To confirm the islanding the data from the two PMUs are compared again after a threshold time. If the result is still different the islanding is confirmed and the PV system is disconnected from the utility grid using a circuit breaker.

With this method a continuous monitoring of the PV system synchronism is the principle of operation. If synchronism is lost, then loss-of-mains is identified. Hence, the PV system is disconnected from the grid.

Figure 10. Algorithm of the proposed islanding detection technique using PMUs

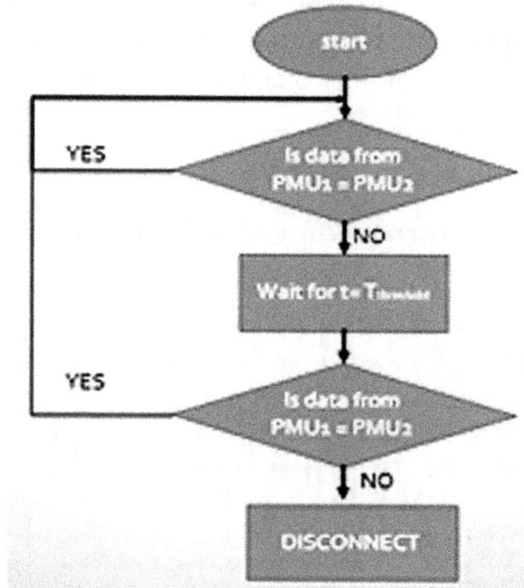

Figure 11. Islanding detection technique Simulink Subsystem using PMUs

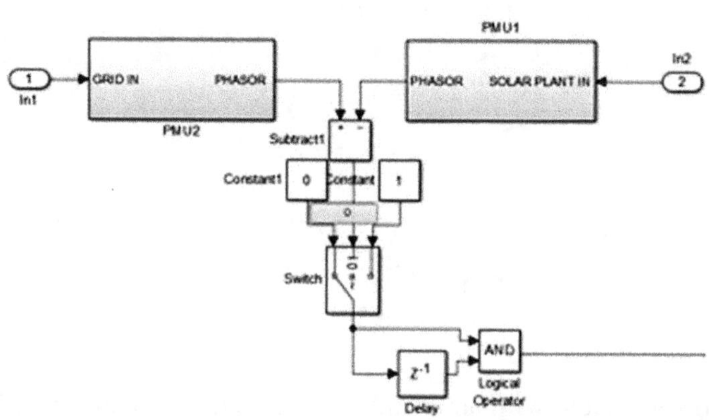

CONCLUSION

In this study, the output of a solar PV power generation system is controlled via a hysteresis current controller and used to inject a power into the utility grid. However due to the use of only one panel the inverter's output is small. Yet, for a series and parallel PV panels connection this proposed configuration can greatly satisfy the power demand, limit the use of conventional power generation techniques and also it is the only mean to tackle the future power requirements. Using solar PV power generation system saves the fossil fuels from depletion, limits global warming and keeps the environment clean and green.

This chapter presents the application of PMUs to Grid connected PV system for synchronization purpose. Then, it investigates the use of PMUs to solve islanding problem in a network including high penetrations of solar energy generation. In the first application, a PMU is used to provide the necessary data to generate pulses which in turn generate an equivalent reference signal i_{ref}. This last will be the reference signal in the hysteresis controller. While, the second application shows how the use of synchrophasor measurements makes the islanding detection easier. This is done by comparing the phasor of the PV system generator with respect to a reference from the grid. Using this method ensures a continuous monitoring of the tied system in order to disconnect the generator if islanded, and leave it connected if it is not.

ACKNOWLEDGMENT

This research was supported by DGRDT/MESRS, Algeria.

REFERENCES

Baudette, L. V., & White, A. D. (2017). Monitoring and Control of Renewable Energy Sources using Synchronized Phasor Measurements. Renewable Energy Integration (Second Edition), Practical Management of Variability, Uncertainty, and Flexibility in Power Grids, 419-434.

Bentarzi, H. (2010). Improving Monitoring, Control and Protection of Power Grid Using Wide Area Synchro-Phasor Measurements. *The 12th WSEAS Int. Conf. on Automatic control, modeling and simulation.*

European Commission. (2020). https://ec.europa.eu/clima/policies/strategies/progress_en

Guo, X.-Q., Wu, W.-Y., & Gu, H.-R. (2011). *Phase locked loop and synchronization methods for grid interfaced converters: a review*. Przegląd Elektrotechniczny.

Hossain, M. S., Madlool, N. A., Rahim, N. A., Selvaraj, J. K., Pandey, A., & Khan, A. F. (2016). Role of smart grid in renewable energy: An overview. *Renewable & Sustainable Energy Reviews*, *60*, 1168–1184. doi:10.1016/j.rser.2015.09.098

Kato, D., Horii, H., & Kawahara, T. (2014). Next-generation SCADA/EMS Designed for Large Penetration of Renewable Energy. *Hitachi Review*, *63*(4), 151.

Klari´c, M., Kuzle, I. & Ninoslav. (2018). Wind Power Monitoring and Control Based on Synchrophasor Measurement Data Mining. *Energies, MDPI, 11*(12), 1-23.

Laverty, D. M., Best, R. J., & Morrow, D. J. (2015). Loss-of-mains protection system by application of phasor measurement unit technology with experimentally assessed threshold settings. *IET Generation, Transmission & Distribution*, *9*(2), 146–153. doi:10.1049/iet-gtd.2014.0106

Mahboubi, M., Bentarzi, H., & Ouadi, A. (2019). *An application of PMU in solar energy generation Plant Connected Power Grid* [Paper presentation]. *The International Conference on Advanced Engineering in Petrochemical Industry (ICAEPI'19)*, Skikda, Algeria.

Ouadi, A., Bentarzi, H., & Maun, J. C. (2009). A new computer based phasor measurement unit framework. *Proc. 6th Int. Conf. SSD'09*, 1-6.

Patil, G. C., & Thosar, A. G. (2017). Application of synchrophasor measurements using PMU for modern power systems monitoring and control [Paper presentation]. *International Conference on Computation of Power, Energy, Information and Communication (ICCPEIC).*

Phadke, A. G. (1993). Synchronized phasor measurements in power systems. *IEEE Computer Applications in Power*, *6*(2), 10–15. doi:10.1109/67.207465

Phadke, A. G., & Thorp, J. S. (2006). *History And Applications Of Phasor Measurements*. IEEE PES Power Systems Conference and Exposition. doi:10.1109/PSCE.2006.296328

Phadke, A. G., & Thorp, S. (2008). Synchronized phasor measurement and their applications (2nd ed.). Springer. doi:10.1007/978-0-387-76537-2

Power System Relaying Committee of the IEEE PES. (2006). IEEE Standard for Sychrophasor for Power Systems. *IEEE std.1344-1995 (2001)*, 1-36.

Rajeev, A., Shanmugarajeshwaran, R., Poonurangam, T., & Sreevalsan, E. (2016). Applications of PMU in renewable energy generation [Paper presentation]. *3rd International Conference on Electrical EnergySystems (ICEES).*

Tanveer, A., Hongcai, Z., & Biao, Y. (2020). A review on renewable energy and electricity requirement forecasting models for smart grid and buildings. *Sustainable Cities and Society, 55.*

Von, A. J., Braun, M., Stetz, T., Diwold, K., & Geibel, D. (2013). Time in the sun: The challenge of high PV penetration in the German electric grid. *IEEE Power & Energy Magazine*, *11*(2), 55–64. doi:10.1109/MPE.2012.2234407

ADDITIONAL READING

Badal, F. R., Das, P., Sarker, S. K., & Das, S. K. (2019). *A survey on control issues in renewable energy integration and microgrid, Protection and Control of Modern Power Systems, Open Access*. Springer.

Bose, B. K. (2017). Power Electronics, Smart Grid and Renewable Energy Systems, *Computer Science, Engineering Proceedings of the IEEE.*

García-Vera, Y. E., Dufo-López, R., & Bernal-Agustín, J. L. (2019). Energy Management in Microgrids with Renewable Energy Sources: A Literature Review, *Applied Sciences*, 9(18).

Kumar, A., Gupta, N., & Gupta, V. (2017). A Comprehensive Review on Grid-Tied Solar Photovoltaic System. *Journal of Green Engineering*, *7*(1), 213–254. doi:10.13052/jge1904-4720.71210

Mohammed, S. & Osman A. (2017). *Grid tied systems: blackout solution,* Uinversity of Khartoum.

Purohit, D., Singh, G., & Mamodiya, U. (2017). A Review Paper on Solar Energy System, International *Journal of Engineering Research and General Science Volume 5, Issue 5, September-October, 2017 ISSN 2091-2730*

KEY TERMS AND DEFINITIONS

Phasor Measurement Unit (PMU): PMU is a measurement device, developed in mid 1980s. It measures the electrical quantities such as voltage and current at node in the power system where it has been installed. It provides the real time phasor parameters of these quantities such as magnitude, frequency, phase angle and rate of change of frequency.

Smart Power Grid: A smart grid is an electrical grid which includes a variety of operation and energy measures including smart meters, smart appliances, renewable energy resources, and different energy resources.

Zero-Crossing Detection: it is based on converting the feedback sinusoidal grid signal to an equivalent square wave signal, compare it to the ground and generate a pulse each time a rising edge is detected; it generates an equivalent periodic pulse representing its frequency.

Compilation of References

Ababei, C. (2009). Speeding Up FPGA Placement via Partitioning and Multithreading. *International Journal of Reconfigurable Computing*, *2009*, 1–9. doi:10.1155/2009/514754

Ababneh, J. I., & Bataineh, M. H. (2007). Linear phase FIR filter design using particle swarm optimization and genetic algorithms. *Digital Signal Processing*. Advance online publication. doi:10.1016/j.dsp.2007.05.011

Abbasy, N. H., & Ismail, H. M. (2009). A unified approach for the optimal PMU location for power system state estimation. *IEEE Transactions on Power Systems*, *24*(2), 806–813. doi:10.1109/TPWRS.2009.2016596

Abbod, M., Alamri, B., & Alshahrani, S. (2015). *Detection and classification of power qualityevents based on wavelet transform and artificial neural networks for smart grids, In 2015 Saudi Arabia Smart Grid*. SASG.

Abdelmoumene, A., & Bentarzi, H. (2012). Reliability Enhancement of Power Transformer Protection System. *Journal of Basic and Applied Scientific Research*, *2*(10), 10534–10539.

Abdelmoumene, A., & Bentarzi, H. (2014). A Review on Protective Relays' Developments and Trends. *Journal of Energy in Southern Africa*, *25*(2), 91–95. doi:10.17159/2413-3051/2014/v25i2a2674

Abdelmoumene, A., Bentarzi, H., Chafai, M., & Ouadi, A. (2014). *Reliability assessment and improvement of digital protective relays. Int J Syst Assur Eng Manag*. doi:10.100713198-014-0300-z

Abdoos, A. A., Pazoki, M., & Moravej, Z. (2009). Detection and classification of power quality disturbances using wavelet transform and support vector machines. *Electric Power Components and Systems*, *38*(2).

Addison, P. S. (2017). *The illustrated wavelet transform handbook: introductory theory and applications in science, engineering medicine and finance*. CRC Press.

Aggarwal, R., & Kim, C. H. (2000). Wavelet transforms in power systems, Part 1: General introduction to the wavelet transforms. *Power Engineering Journal*, *14*(2).

Aggoun, G., Boubouzel, N., Kachenoura, R., & Fellag, S. A. (2017, April). Quality of energy resulting from association converter- machine for the electric drive of a system of wind generation. *Algerian Journal of Signals and Systems*, *2*(1), 1–11.

Ahmad, K. N. E. K., Selvaraj, J., & Rahim, A. (2013). A review of the islanding detection methods in grid-connected PV inverters. *Renewable & Sustainable Energy Reviews*, *21*, 756–766. doi:10.1016/j.rser.2013.01.018

Akcayol, M. A. (2004). Application of adaptive neuro-fuzzy controller for SRM. *Advances in Engineering Software*, *35*(3–4), 129–137. doi:10.1016/j.advengsoft.2004.03.005

Al-Alawi, A., & Al-Alawi, M. (2007). Predictive control of an integrated PV-diesel water and power supply system using an artificial neural network. *Renewable Energy*, *32*(8), 1426–1439. doi:10.1016/j.renene.2006.05.003

Allegre, A. L., Bouscayrol, A., Delarue, P., Barrade, P., Chattot, E., & El-Fassi, S. (2010). Energy storage system with supercapacitor for an innovative subway. *IEEE Transactions on Industrial Electronics, 57*(12), 4001–4012. https://doi. org/ doi:10.1109/TIE.2010.2044124

Allemong, J., Radu, L., & Sasson, A. (1982). A fast and reliable state estimation algorithm for AEP's new control center. *IEEE Transactions on Power Apparatus and Systems, 4*(4), 933–944. doi:10.1109/TPAS.1982.317159

Allen, P. E., & Holberg, D. R. (2002). *CMOS Analog Circuit Design*. Oxford UniversityPress.

Almada, J. B., Leão, R. P. S., Sampaio, R. F., & Barroso, G. C. (2016). A centralized and heuristic approach for energy management of an AC microgrid. *Renewable & Sustainable Energy Reviews, 60*, 1396–1404. doi:10.1016/j.rser.2016.03.002

Alshareef, S., Talwar, S., & Morsi, W. G. (2014). A new approach based on wavelet design and machine learning for islanding detection of distributed generation. *IEEE Transactions on Smart Grid, 5*(4), 1575–1583. doi:10.1109/ TSG.2013.2296598

Aminifar, F., Khodaei, A., Fotuhi-Firuzabad, M., & Shahidehpour, M. (2010). Contingency-constrained PMU placement in power networks. *IEEE Transactions on Power Systems, 25*(1), 516–523. doi:10.1109/TPWRS.2009.2036470

Amin, M., & Stringer, J. (2008). The Electric Power Grid: Today and Tomorrow. *MRS Bulletin, 33*(4), 399–407. doi:10.1557/mrs2008.80

Amos, S. W., & James, M. (2000). *Principles of transistor circuits: introduction to the design of amplifiers, receivers, and digital circuits* (9th ed.). Newnes.

Anderson, P. A., Flemming, B., Lee, T. J., & Schweitzer, E. O. (1997). Reliability Analysis of Transmission Protection Using Fault Tree Methods. *Proceedings of the 24th Annual Western Protective Relay Conference.*

Anderson, R., & Fuloria, S. (2010), On the security economics of electricity metering. *Proc. of Workshop on Economics of Internet Security (WEIS).*

Aneke, M., & Wang, M. (2016). Energy storage technologies and real life applications – A state of the art review. *Applied Energy, 179*, 350–377. doi:10.1016/j.apenergy.2016.06.097

ANSI/IEEE C37.91-1985 (1985*). IEEE Guide for Protective Relay Applications to Power Transformers.* IEEE Publication, New York.

ANSI/IEEE C37.91-1985. (1985). *IEEE Guide for Protective Relay Applications to Power Transformers.* IEEE.

Antoniou, A. (1993). *Digital Filters: Analysis and Design* (2nd ed.). McGraw Hill.

Antoniou, A. (2006). *Digital signal processing: signals, systems and filters*. McGraw Hill.

Antoniou, A., & Lu, W.-S. (2007). *Practical Optimization: Algorithms and Engineering Applications.* Springer.

Anvari-Moghaddam, A., Rahimi-Kian, A., Mirian, M. S., & Guerrero, J. M. (2017). A multi-agent based energy management solution for integrated buildings and microgrid system. *Applied Energy, 203*, 41–56. doi:10.1016/j.apenergy.2017.06.007

Apostolov, A. (2013). Impact of IEC 61850 on the Interoperability and Reliability of Protection schemes. PES-GM.

Aravinthan, V., Karimi, B., Namboodiri, V., & Jewell, W. (2011). Wireless communication for smart grid applications at distribution level -feasibility and requirements. IEEE Power and Energy Society General Meeting, 1–8.

Asghar, M. R., D´an, G., Miorandi, D., & Chlamtac, I. (2017). *Smart Meter Data Privacy: A Survey.* IEEE.

Azzougui, Y., & Recioui, A. (2018). Hardware and software load power control in smart home applications based on Taguchi optimisation technique. *Int. J. Intelligent Systems Design and Computing*, *2*(3/4), 203–223.

Azzougui, Y., & Recioui, A. (2019). (in press). Application of the moth flame optimisation to the selective harmonic elimination in multilevel converters. *Int. J. Smart Grid and Green Communications* .

Azzougui, Y., Recioui, A., & Mansouri, A. (2019). PMU Optimal Placement in Wide Area Monitoring Systems using Grey Wolf Optimization technique. *Algerian Journal of Signals and Systems*, *4*(1), 1–7.

Azzougui, Y., Recioui, A., & Mansouri, A. (2019). PMU Optimal Placement in Wide Area Monitoring Systems using Grey Wolf Optimization Technique. Algerian. *Journal of Signals and Systems*, *4*(1), 1–7.

Baggini, A. (2008). *Handbook of power quality*. John Wiley & Sons. doi:10.1002/9780470754245

Baldwin, T. L., Mili, L., Boisen, M. B., & Adapa, R. (1993). Power System Observability with Minimal Phasor Measurement Placement. *IEEE Transactions on Power Systems*, *8*(2), 707–715. doi:10.1109/59.260810

Bansal, R., Manana, M., & Zobaa, A.F. (2011). *Power quality: Monitoring, analysis and enhancement*. Academic Press.

Barros, J., & Diego, R. I. (2006). Application of the wavelet-packet transform to the estimation of harmonic groups in current and voltage waveforms. *IEEE Transactions on Power Delivery*, *21*(1), 533–535. doi:10.1109/TPWRD.2005.848437

Batra, N., Kelly, J., Parson, O., Dutta, H., Knottenbelt, W., Rogers, A., Singh, A., & Srivastava, M. (2014), NILMTK: An Open Source Toolkit for Non-intrusive Load Monitoring. *Proc. of ACM Intl. Conf. on Future Energy Systems (e-Energy).*

Battacharyya, S. S., Murthy, P. K., & Lee, E. A. (1996). *Software synthesis from dataflow graphs*. Kluwer Academic Publishers. doi:10.1007/978-1-4613-1389-2

Baudette, L. V., & White, A. D. (2017). Monitoring and Control of Renewable Energy Sources using Synchronized Phasor Measurements. Renewable Energy Integration (Second Edition), Practical Management of Variability, Uncertainty, and Flexibility in Power Grids, 419-434.

Baumeister, T. (2016). *Literature review on smart grid cyber security*. https://csdl-techreports.googlecode.com/svn/trunk/techreports/2010/10-11/10-11.pdf

Bayindir, R., Colak, I., Fulli, G., & Demirtas, K. (2016). Smart grid technologies and applications. *Renewable & Sustainable Energy Reviews*, *66*, 499–516. doi:10.1016/j.rser.2016.08.002

Begovic, M. M., Djuric, P. M., Dunlap, S., & Phadke, A. G. (1993). Frequency tracking in power networks in the presence of harmonics. *IEEE Transactions on Power Delivery*, *8*(2), 480–486. doi:10.1109/61.216849

Behzadi, M. S., & Niasati, M. (2015). Comparative performance analysis of a hybrid PV/FC/battery stand-alone system using different power management strategies and sizing approaches. *International Journal of Hydrogen Energy*, *40*(1), 538–548. doi:10.1016/j.ijhydene.2014.10.097

Bei, X., Yoon, Y. J., & Abur, A. (2008). *Optimal placement and utilization of phasor measurements for state estimation*. Texas A&M University.

Belaidi, H., Belkalem, J., Abed, M. A., & Bentarzi, H. (2018). IoT Path Planning Approach for Mobile Robots. *2018 International Conference on Applied Smart Systems (ICASS)*, 1-8. 10.1109/ICASS.2018.8652050

Belaidi, H., Hentout, A., & Bentarzi, H. (2019). Human–Robot Shared Control for Path Generation and Execution. *International Journal of Social Robotics*, *11*(4), 609–620. doi:10.100712369-019-00520-3

Benmouyal, G. (1995). Removal of DC-offset in Current Waveforms Using Digital Mimic Filtering. *IEEE Transactions on Power Delivery, 10*(2), 621–630. doi:10.1109/61.400869

Bennett, C., & Highfill, D. (2008). Networking AMI smart meters. *Proc. IEEE Energy 2030 Conference*, 1 – 8.

Bentarzi, H. (2010). Improving Monitoring, Control and Protection of Power Grid Using Wide Area Synchro-Phasor Measurements. *The 12th WSEAS Int. Conf. on Automatic control, modeling and simulation.*

Bentarzi, H., Ouadi, A., & Maun, J. C. (2014). A New PMU Based Power Swing Detector to EnhanceReliability of Distance Relay. *Proceedings of CIGRE Session 45.*

Bentarzi, H., & Zitouni, A. (2017). Field-Programmable Gate Array Based Tester for Protective Relay, International Journal of Electrical, Computer, Energetic. *Electronic and Communication Engineering, WASET, 11*(7), 791–796.

Bera, S., Gupta, P., & Misra, S. (2015). D2S: Dynamic demand scheduling in smart grid using optimal portfolio selection strategy. *IEEE Transactions on Smart Grid, 6*(3), 1434–1442. doi:10.1109/TSG.2014.2386325

Berkowitz, J., Elkan, C., & Lipton, Z. C. A. (2015). *Critical review of recurrent neural networks forsequence learning.* arXiv:1506.00019

Bevrani, H., Watanabe, M., & Mitani, Y. (2014). *Power System Monitoring and Control.* John Wiley & Sons, Inc. doi:10.1002/9781118852422

Bhimte, A., Mathew, R. K., & Kumaravel, S. (2015). Development of smart energy meter in LabVIEW for power distribution systems. *2015 Annual IEEE India Conference (INDICON)*, 1–6. 10.1109/INDICON.2015.7443547

Bicker, R., & Jabar, A. A. A. (2014). Simulation of Non-stationary Signal Analysis Using Wavelet Transform Based on LabVIEW and Matlab. In *2014 European Modelling Symposium* (pp. 138-144). IEEE.

Blume, D., Stephanblome, T., & Schlabbach, J. (2001). *Voltage quality in electrical power systems.* IET.

Blume, S. W. (2016). *Electric power system basics for the nonelectrical professional.* John Wiley & Sons.

Boait, P. (2009). Smart metering of renewable micro generators by output pattern recognition. *Proc. International Conference and Exhibition on Electricity Distribution*, 1 – 4. 10.1049/cp.2009.0541

Bocklisch, T. (2015). Hybrid energy storage systems for renewable energy applications. *Energy Procedia, 73*, 103–111. doi:10.1016/j.egypro.2015.07.582

Bollen, M. H., Watson, N. R., & Arrillaga, J. O. (2000). Power quality following deregulation. *Proceedings of the IEEE, 88*(2).

Borges, F. (2017). *On privacy-preserving protocols for smart metering systems.* Springer. doi:10.1007/978-3-319-40718-0

Borlase, S. (2018). *Smart Grids: Advanced Technologies and solutions.* CRC Press.

Bouderbala R., Bentarzi, H. & Ouadi A. (2011). Digital Differential Relay Reliability Enhancement of Power Transformer. *International Journal of Circuits, Systems and Signal Processing, 5*, 263-270.

Bouderbala, R., & Bentarzi, H. (2013). A New Computer Based Differential Relay Framework for Power Transformer. *Lecture Notes in Electrical Engineering, 260*, 473-481. Doi:10.1007/978-94-007-7262-5_54

Bouderbala, R., Bentarzi, H., & Ouadi A. (2011). Digital Differential Relay Reliability Enhancement of Power Transformer. *International Journal of Circuits, Systems and Signal Processing, 5*, 263-270.

Bouderbala, R., Bentarzi, H., & Ouadi, A. (2010). A New Approach Applied to Digital Differential Protection for a Large Power Transformer. *The 9th WSEAS international conference on Recent Researches in Circuits, Systems, Electronics, Control & Signal Processing Proceeding*, 202-205.

Bouderbala, R., & Bentarzi, H. (2016). *Differential relay reliability enhancement using fourth harmonic for a large power transformer. International Journal of Syst. Assur. Engineering Mang*. doi:10.100713198-016-0475-6

Bouhafs, F., Mackay, M., & Merabti, M. (2012). Links to the future: Communication requirements and challenges in the smart grid. *Power and Energy Magazine, IEEE, 10*(1), 24–32. doi:10.1109/MPE.2011.943134

Boutora, S., Bentarzi, H., & Ouadi, A. (2011). Analysis of the Disturbances in Distribution Networks using Matlab and ATP. *International Journal of Exergy*, *5*(1), 9–16.

Boxwell, M. (2017). *Solar electricity Handbook*. Academic Press.

Brka, A., Kothapalli, G., & Al-abdeli, Y. M. (2015). ScienceDirect Predictive power management strategies for stand- alone hydrogen systems : Lab-scale validation. *International Journal of Hydrogen Energy*, *40*(32), 9907–9916. doi:10.1016/j.ijhydene.2015.06.081

Brunke, J. H., & Fröhlich, J. K. (2001). Elimination of Transformer Inrush Currents by Controlled Switching. *IEEE Transactions on Power Delivery*, *16*(2), 276–280. doi:10.1109/61.915495

Cano, A., Jurado, F., Sánchez, H., Fernández, L. M., & Castañeda, M. (2014). Optimal sizing of stand-alone hybrid systems based on PV/WT/FC by using several methodologies. *Journal of the Energy Institute*, *87*(4), 330–340. doi:10.1016/j.joei.2014.03.028

Capitan, J., Spaan, M. T. J., Merino, L., & Ollero, A. (2012). Decentralized Multi-Robot Cooperation with Auctioned POMDPs. *Proceedings of the Twenty-Fourth International Conference on Automated Planning and Scheduling, 2012 IEEE International Conference on Robotics and Automation*, 14-18.

Cappers, P., MacDonald, J., Page, J., Potter, J., & Stewart, E. (2016). *Future opportunities and challenges with using demand response as a resource in distribution system operation and planning activities*. Lawrence Berkeley Nat. Lab., Berkeley, CA, USA, Tech. Rep. LBNL-1003951.

Cecilia, A. A., & Sudarsanan, K. (2016). A survey on smart grid. *2016 International Conference on Emerging Trends in Engineering, Technology and Science (ICETETS)*, 1–7.

Chafai, M., Bentarzi, H., Ouadi, A., Zitouni, A., & Recioui, A. (2013). *PC based testing system for protective relay*. 4th International Conference on Power Engineering, Energy and Electrical Drives, POWENG"2013, Istanbul, Turkey.

Chafai, M., Bentarzi, H., & Ouadi, A. (2014). A New Frame Work for Protective Relay Tester. *NNGT Journal: International Journal of Signal Processing and Imaging Engineering*, *1*, 60–65.

Chafai, M., Bentarzi, H., & Ouadi, A. (2014). Real-Time Simulator for Testing Protective Relay. *International Conference on Embedded Systems and Applications (ICESA2014)*.

Chafai, M., Bentarzi, H., Ouadi, A., Zitouni, A., & Recioui, A. (2013), PC based testing system for protective relay. *4th International Conference on Power Engineering, Energy and Electrical Drives, POWENG"2013*. 10.1109/PowerEng.2013.6635737

Chafai, M., Refoufi, L., & Bentarzi, H. (2016). Large power transformer reliability modeling. *Int J Syst Assur Eng Manag*, *7*(S1), 9–17. doi:10.100713198-014-0261-2

Chakrabarti, S., & Kyriakides, E. (2007). Optimal placement of phasor measurement units for state estimation. *Proc. 7th IASTED Int. Conf. Power and Energy Systems*, 1–6.

Chakrabarti, S., & Kyriakides, E. (2008). Optimal placement of phasor measurement units for power system observability. *IEEE Transactions on Power Systems*, *23*(3), 1433–1440. doi:10.1109/TPWRS.2008.922621

Chakravarthi, K. R., & Ortmeyer, T. M. (1985). *The effects of power system harmonics on power system equipment and loads. IEEE Trans. Power Appar. Syst.*

Chandra, P., & Pahuja, P. (2015). Power Quality Monitoring Using LABVIEW. *International Journal Of Electrical Electronics & Computer Engineering*, *4*(2), 59.

Chao, L., Wu, X., Wu, J., Li, P., Han, Y., & Li, L. (2012), Implementations and Experiences of Wide-area HVDC Damping Control in China Southern Power Grid. *IEEE PES General Meeting.* doi: 10.1109/PESGM.2012.6345363

Charu, S., & Barjeev, T. (2011). An approach for optimal PMU placement using binary particle swarm optimization with conventional measurements. *International Journal of Engineering Science and Technology*, *3*(3), 56–63.

Chaudhuri, N., Ray, S., Majumder, R., & Chaudhuri, B. (2010), A New Approach to Continuous Latency Compensation With Adaptive Phasor Power Oscillation Damping Controller (POD). *IEEE Transactions on Power Systems*, *25*(2), 939–946. doi:10.1109/TPWRS.2009.2031908

Chellali, F. B. M., Recioui, A., Yaiche, M. R., & Bentarzi, H. (2014). A hybrid wind/solar/diesel stand alone system optimisation for remote areas in Algeria. *International Journal of Renewable Energy Technology*, *5*(1), 12–24. doi:10.1504/IJRET.2014.059658

Chen, J. (1994). *Accurate frequency estimation with phasor angles* (Master thesis). Virginia Polytechnic Institute and State University.

Chena, J., Thorpa, J. S., & Dobsonb, I. (2005). Cascading dynamics and mitigation assessment in power system disturbances via a hidden failure model. *Electrical Power and Energy Systems*, *27*(4), 318–326. doi:10.1016/j.ijepes.2004.12.003

Chenine, M., & Nordström, L. (2011), Modeling and Simulation of Wide-Area Communication for Centralized PMU-Based Applications. *IEEE Transactions on Power Delivery*, *26*(3), 1372 –1380. doi:10.1109/TPWRD.2011.2106805

Chenine, M., Al Khatib, I., Ivanovski, J., Maden, V., & Nordström, L. (2010), PMU traffic shaping in IP-based Wide Area communication. *5th International Conference on Critical Infrastructure (CRIS)*, 1–6, 10.1109/CRIS.2010.5617564

Chilukuri, M. V., & Li, J. (2014). Power supply quality nalysis using S-transform and SVM classifier. *Journal of Power and Energy Engineering*, *2*(04), 438–447. doi:10.4236/jpee.2014.24059

Chong, L. W., Wong, Y. W., Rajkumar, R. K., Rajkumar, R. K., & Isa, D. (2016). Hybrid energy storage systems and control strategies for stand-alone renewable energy power systems. *Renewable & Sustainable Energy Reviews*, *66*, 174–189. doi:10.1016/j.rser.2016.07.059

Chouhan, N. S., & Ferdowsi, M. (2009). Review of energy storage systems. *North American Power Symposium (NAPS)*, 1–5. https://doi.org/10.1109/NAPS.2009.5484062

Cimadevilla, R. (2013). Application of Zero-Sequence Filter on Transformer Differential Protection. *SEAPAC.*

Cimadevilla, R., & Quintanilla, R. (2011). New protection units included in differential relays. *PAC Conference.*

Cleveland, F. M. (2008). Cyber security issues for advanced metering infrastructure. *Proc. IEEE Power and Energy Society General Meeting - Conversion and Delivery of Electrical Energy*, 1 – 5.

Cooper, A., & Shuster, M. (2019). *Report on Electric Company Smart Meter Deployments: Foundation for a Smart Grid.* https://www.edisonfoundation.net/iei/publications/Documents/IEI_Smart%20Meter%20Report_2019_FINAL.pdf

Cortelazzo, G., & Lightner, M. R. (1984). Simultaneous design in both magnitude and group-delay of IIR and FIR filters based on multiple criterion optimization. *IEEE Transactions on Circuits and Systems*, *32*(5), 949–967.

Crossley, D. (2007). Smart Metering, Load Control and Energy-using Behaviour. *Workshop on Smart Metering to Use Less Energy*, 10.

Cuijpers, C., & Koops, B.-J. (2013). Smart metering and privacy in Europe: Lessons from the Dutch case. *In European Data Protection. Eds. Springer*, *2013*, 269–293.

Cui, Y., Li, A., & Liu, Z. (2015). A classification method for complex power quality disturbances using EEMD and rank wavelet SVM. *IEEE Transactions on Smart Grid*, *6*(4), 1678–1685. doi:10.1109/TSG.2015.2397431

Cyr, C., & Kamwa, I. (2010), WACS design at Hydro-Quebec. *PMU Tutorial, IEEE PES General Meeting.*

Dabra, V., Paliwal, K. K., Sharma, P., & Kumar, N. (2017). Optimization of photovoltaic power system: A comparative study. *Protection and Control of Modern Power Systems*, *2*(1), 3. Advance online publication. doi:10.118641601-017-0036-2

Daki, H., El Hannani, A., Aqqal, A., Haidine, A., & Dahbi, A. (2017). Big Data management in smart grid: Concepts, requirements and implementation. *Journal of Big Data*, *4*(13), 13. Advance online publication. doi:10.118640537-017-0070-y

Dambhare, S., Dua, D., Gajbhiye, R. K., & Soman, S. A. (2008). *Optimal Zero: An ILP Approach.* Indian Institute of Technology.

Daponte, P., D'apuzzo, M., Testa, A., & Angrisani, L. (1998). A measurement method based on the wavelet transform for power quality analysis. *IEEE Transactions on Power Delivery*, *13*(4).

Dash, P. K., Padhee, M., & Panigrahi, T. K. (2012). A hybrid time–frequency approach based fuzzy logic system for power island detection in grid connected distributed generation. *International Journal of Electrical Power & Energy Systems*, *42*(1), 453–464. doi:10.1016/j.ijepes.2012.04.003

Daubechies, I. (1992). *Ten lectures on wavelets*. Siam. doi:10.1137/1.9781611970104

Davim, P., & Ozel, T. (2009). *Intelligent maching*. Wiley-IEEE Press.

Decker, I. C., Silva, A. S., Agostini, M. N., Prioste, F. B., Mayer, B. T., & Dotta, D. (2011). Experience and applications of phasor measurements to the Brazilian interconnected power system. *European Transactions on Electrical Power*, *21*(4), 1557–1573. doi:10.1002/etep.537

Degens, A. J. (1981). Algorithm for a Digital Transformer Protection Based on a Least-Squares Curve-Fitting. *IEE Proc.*, *128*, 155.

Dekhandji, F. Z. (2017). Signal processing deployment in power quality disturbance detection and classification. Acta Physica Polonica A, 132(3), 415-419.

Dekhandji, F. Z. (2018), Optimal placement of power factor correction capacitors using a discrete teaching learning based optimization. WSEAS Journal of Power Systems, 13, 226-234.

Dekhandji, F. Z., Azzougui, Y., & Recioui, A. (2019). PMU Optimal Placement for Fault Deetction in Wide Area Monitoring Systems based on the Dron Suadron Optimization. *Algerian Journal of Signals and Systems*, *4*(1), 25–31.

Dekhandji, F. Z., & Bentarzi, H. (2019, December). Theoretical and Experimental Investigation of the Supply Voltage Unbalance Effects on Squirrel Cage Induction Motors. *Algerian Journal of Signals and Systems*, *4*(2), 51–60.

Dekhandji, F. Z., Bouaraki, M., Zaidi, N., & Recioui, A. (2019). Demand Response and Optimal Sizing of Hybrid Renewable Energy Systems. *International Journal of Control Energy and Electrical Engineering*, *7*, 35–40.

Dekhandji, F. Z., Douche, M., & Zebidi, N. (2017). DVR and D-STATCOM Mitigation Techniques of Power Quality Effects on Induction Motors. *Algerian Journal of Signals and Systems*, *2*(2), 110–129.

Dekhandji, F. Z., Kassah, M. T., & Bensalah, A. (2020). Harmonic Reduction and Elimination in Three Phase PWM Inverters using a Spiral-Inspired Optimization Technique, *Algerian J. Env. Sc. Technology*, *6*(3), 1436–1447.

Dekhandji, F. Z., Talhaoui, S., & Arkab, Y. (2019, December). Power Quality Detection, Classification and Monitoring Using LABVIEW. *Algerian Journal of Signals and Systems*, *4*(2), 101–111.

Deng, Y., Lin, H., Phadke, A., Sandeep, S., & Thorp, J. M. L. (2012), Communication network modeling and simulation for wide area measurement applications. IEEE PES ISGT. doi:10.1109/ISGT.2012.6175664

Depuru, S. S. S. R., Wang, L., Devabhaktuni, V., & Gudi, N. (2011), Smart meters for power grid—Challenges, issues, advantages and status. 2011 IEEE/PES Power Systems Conference and Exposition, 1–7.

Depuru, S. S. S. R., Wang, L., Devabhaktuni, V., & Gudi, N. (2011, March). *Smart European Commission (2006) European SmartGrids Technology Platform: Vision and Strategy for Europe's Electricity*. http://ec.europa.eu/research/energy/pdf/smartgrids_en.pdf

Diego, R. I., De Apràiz, M., & Barros, J. (2012), Applications of wavelets in electric powerquality: Voltage events. Electric Power Systems Research, 130-136.

Dierks, K. C. A. (2006). *Relay Specific Computer Aided Testing for Protective Relays*. Eskom, Republic of South Africa.

Dolezilek, D., Fischer, N., & Schloss, R. (2012). *Improvements in Synchronous Wide-Area Data Acquisition Design and Deployment for Telecontrol and Teleprotection*. https://www.selinc.com/WorkArea/DownloadAsset.aspx?id=99365

Dong, Y., & Kezunovic, M. (2011). Communication infrastructure for emerging transmission-level smart grid applications. IEEE Power and Energy Society General Meeting, 1–7. doi:10.1109/PES.2011.6039640

Dua, D., Dambhare, S., Gajbhiye, R. K., & Soman, S. A. (2008). Optimal multistage scheduling of PMU placement: An ILP approach. *IEEE Transactions on Power Delivery*, *23*(4), 1812–1820. doi:10.1109/TPWRD.2008.919046

Dufour, C., & Bélanger, J. (2014). On the Use of Real-Time Simulation Technology in Smart Grid Research and Development. *IEEE Transactions on Industry Applications*, *50*(6), 3963–3970. doi:10.1109/TIA.2014.2315507

Dulout, J., Jammes, B., Seguier, L., & Alonso, C. (2015). Control and design of a hybrid energy storage system. *2015 17th European Conference on Power Electronics and Applications, EPE-ECCE Europe 2015*. https://doi.org/10.1109/EPE.2015.7309468

Dumitrescu, B., & Niemisto, R. (2004). Multistage IIR filter design using convex stability domains defined by positive realness. *IEEE Transactions on Signal Processing*, *52*(4), 962–974. doi:10.1109/TSP.2004.823497

Durgadevi, S., Sundari, K. T., Raaghavi, D., & Akshaya, R. S. (2019). Comparative study of Controller Optimisation for CSTR using Particle Swarm Optimization Technique. *5th International Conference on Electrical Energy Systems, ICEES 2019*. https://doi.org/10.1109/ICEES.2019.8719302

Ebling, C. E. (2000). *Introduction to Reliability and Maintainability Engineering*. McGraw Hill Publishing House.

Eid, C., Reneses Guillén, J., Frías Marín, P., & Hakvoort, R. (2014). The economic effect of electricity net-metering with solar PV: Consequences for network cost recovery, cross subsidies and policy objectives. *Energy Policy*, *75*, 244–254. doi:10.1016/j.enpol.2014.09.011

Einan, M., Torkaman, H., & Pourgholi, M. (2017). Optimized Fuzzy-Cuckoo Controller for Active Power Control of Battery Energy Storage System, Photovoltaic, Fuel Cell and Wind Turbine in an Isolated Micro-Grid. *Batteries*, *3*(3), 23. doi:10.3390/batteries3030023

Einval, C. H., & Linders, J. R. (1975). A Three-Phase Differential Relay for Transformer Protection. *IEEE Transactions PAS*, *94*(6), 1971–1980.

Ekanayake, J., Jenkins, N., Liyanage, K., Wu, J., & Yokoyama, A. (2012). *Smart grid: technology and applications*. Wiley.

Ekici, S. (2008), A Pattern Recognition System For Power disturbance classification. *The 5th International Conference on Electrical and Computer Systems (EECS'08).*

Ekren, O., & Ekren, B. Y. (2010). Size optimization of a PV/wind hybrid energy conversion system with battery storage using simulated annealing. *Applied Energy*, *87*(2), 592–598. doi:10.1016/j.apenergy.2009.05.022

Ekwue A, Aggarwal R Warwick K (1997), Artificial Intelligence Techniques in Power Systems (Energy Engineering). *The Institution of Engineering and Technology.*

El-Arroudi, K., Joos, G., Kamwa, I., & McGillis, D. T. (2007). Intelligent based approach to islanding detection in distributed generation. *IEEE Transactions on Power Delivery*, *22*(2), 828–835. doi:10.1109/TPWRD.2007.893592

Environnement, V. (2010). Smart grids. *Sci. Chron. Mag*, *18*, 1–7.

Ericsson, G. N. (2010). Cyber Security and Power System Communication - Essential Parts of a Smart Grid Infrastructure. *IEEE Transactions on Power Delivery*, *25*(3), 1501–1507. doi:10.1109/TPWRD.2010.2046654

European Commission. (2020). https://ec.europa.eu/clima/policies/strategies/progress_en

Faccio, M., Gamberi, M., Bortolini, M., & Nedaei, M. (2018). State-of-art review of the optimization methods to design the configuration of hybrid renewable energy systems (HRESs). *Frontiers in Energy*, *12*(4), 591–622. doi:10.100711708-018-0567-x

Fais, M., Ghani, A., Abidin, A. F., & Hannoon, N. S. (2017). Real time detection and classification of single and multiple power quality disturbance based on embedded S-transform algorithm in labview. *Proceedings of the 2017 international symposium on Industrial Engineering and Operations Management (IEOM)*, 280–289.

Fang, X., Misra, S., Xue, G., & Yang, D. (2012). Smart Grid — The New and Improved Power Grid: A Survey. *IEEE Communications Surveys and Tutorials*, *14*(4), 944–980. doi:10.1109/SURV.2011.101911.00087

Fan, Z., Kulkarni, P., Gormus, S., Efthymiou, C., Kalogridis, G., Sooriyabandara, M., Zhu, Z., Lambotharan, S., & Chin, W. (2012). Smart grid communications: Overview of research challenges, solutions, and standardization activities. *IEEE Communications Surveys and Tutorials*, *99*, 1–8.

Farhangi, H. (2010). The path of the smart grid. *IEEE Power & Energy Magazine, 8*(1), 18-28.

Farrokhabadi, M., Canizares, C. A., Simpson-Porco, J. W., Nasr, E., Fan, L., Mendoza-Araya, P. A., Tonkoski, R., Tamrakar, U., Hatziargyriou, N., Lagos, D., Wles, R. W., Paolone, M., Liserre, M., Meegahapola, L., Kabalan, M., Hajimiragha, A. H., Peralta, D., Elizondo, M. A., Schneider, K. P., ... Reilly, J. (2020). Microgrid Stability definitions, Analysis &examples. *IEEE Transactions on Power Systems*, *35*(1), 13–29. doi:10.1109/TPWRS.2019.2925703

Faruque, M. O., & … . (2015). Real-time simulation technologies for power systems design, testing, and analysis. *IEEE Power and Energy Technology Systems Journal*, *2*(2), 63–73. doi:10.1109/JPETS.2015.2427370

Farzin, H., Fotuhi-Firuzabad, M., & Moeini-Aghtaie, M. (2017). Stochastic Energy Management of Microgrids during Unscheduled Islanding Period. *IEEE Transactions on Industrial Informatics, Volume*, *13*(3), 1–9. doi:10.1109/TII.2016.2646721

Fathima, A. H., & Palanisamy, K. (2015). Optimization in microgrids with hybrid energy systems - A review. *Renewable & Sustainable Energy Reviews*, *45*, 431–446. doi:10.1016/j.rser.2015.01.059

Fausett, L. V. (1994). *Fundamentals of neural networks: architectures, algorithms, and applications*. Prentice-Hall.

Floyd, T. L. (2005). *Electronic devices*. Pearson.

Frank, K. D., Rich, C., & Longcore, T. (2006). Effects of artificial night lighting on moths. *Journal of Ecological consequences of artificial night lighting. Hindawi Publishing Corporation*, *15*(1), 305–344.

Freitas, W., Xu, W., Affonso, C. M., & Huang, Z. (2005). Comparative analysis between ROCOF and vector surge relays for distributed generation applications. *IEEE Transactions on Power Delivery*, *20*(2), 1315–1324. doi:10.1109/TPWRD.2004.834869

Fritz, W., & Gupta, G. (2016). *Voltage unbalance for power systems and mitigation techniques a survey.* In *2016 IEEE 1st International Conference on Power Electronics, Intelligent Control and Energy Systems (ICPEICES)*. IEEE.

Galli, S., Scaglione, A., & Wang, Z. (2010). Power line communications and the smart grid. *IEEE International Conference Smart Grid Communications (SmartGridComm)*, 303–308. 10.1109/SMARTGRID.2010.5622060

García, P., Torreglosa, J. P., Fernández, L. M., & Jurado, F. (2013). Optimal energy management system for stand-alone wind turbine/photovoltaic/ hydrogen/battery hybrid system with supervisory control based on fuzzy logic. *International Journal of Hydrogen Energy*, *38*(33), 14146–14158. doi:10.1016/j.ijhydene.2013.08.106

Gaston, K., Bennie, J., Davies, T., & Hopkins, J. (2013). The ecological impacts of nighttime light pollution: A mechanistic appraisal. *Biological Reviews of the Cambridge Philosophical Society*, *88*(4), 912–927. doi:10.1111/brv.12036 PMID:23565807

Gaviano, A., Weber, K., & Dirmeier, C. (2012). Challenges and Integration of PV and Wind Energy Facilities from a Smart Grid Point of View. *Energy Procedia*, *25*, 118–125. doi:10.1016/j.egypro.2012.07.016

Gawre, S. K., Kumar, T., & Kumar, V. (2014). Power quality analysis using wavelet transform: a review. *International Journal of Innovative Research in Science, Engineering and Technology*, *3*(3).

Gee, A. M., Robinson, F. V. P., & Dunn, R. W. (2013). Analysis of battery lifetime extension in a small-scale wind-energy system using supercapacitors. *IEEE Transactions on Energy Conversion*, *28*(1), 24–33. doi:10.1109/TEC.2012.2228195

Geetha, A., Usha, S., Thamizhthentral, T. M., Ganguly, U., Kumar, A., & Narnoli, P. K. (2020). *Design of a Solar Charge Controller Using Fuzzy Logic MPPT Technique For A Solar Panel*. Academic Press.

Gellings, C. W. (2009). *The Smart Grid: Enabling Energy Efficiency and Demand IEA Demand Side Management Programme*. http://www.ieadsm.org

Geramian, S. S., Abyane, H. A., & Mazlumi, K. (2008). Determination of Optimal PMU Placement for Fault Location Using Genetic Algorithm. *13th International Conference on Harmonics and Quality of Power*. 10.1109/ICHQP.2008.4668810

Gezer, C., & Buratti, C. (2011), A ZigBee smart energy implementation for energy efficient buildings. *Proc. IEEE 73rd Veh. Technol. Conf. (VTC Spring)*, 1–5.

Ghaviha, N., Campillo, J., Bohlin, M., & Dahlquist, E. (2017). Review of Application of Energy Storage Devices in Railway Transportation. *Energy Procedia*, *105*, 4561–4568. doi:10.1016/j.egypro.2017.03.980

Ghazvini, M., Abbaspour-Tehrani-fard, A., & Fotuhi-Firuzabad, M. (2015). A particle swarm optimization-based approach to achieve optimal design and operation strategy of standalone hybrid energy systems. *Turkish Journal of Electrical Engineering and Computer Sciences*, *23*(2), 335–353. doi:10.3906/elk-1212-11

Ghosh P., Maiti A. (2012). Instantaneous Power Measurement using Hall Sensor. *Int. J. Sci. Res. Publ.*, *2*(8).

Giray, M. M., & Sachdev, M. S. (1989). Off-normal frequency measurements in electric power systems. *IEEE Transactions on Power Delivery*, *4*(3), 1573–1578. doi:10.1109/61.32645

Girgis, A.A., & Brown, R.G. (1981). Application of Kalman filter in computer relaying. *IEEE Trans. on PAS*, *100*(7), 3387-3395.

Girgis, A. A., & Brown, R. G. (1981). Application of Kalman filter in computer relaying. *IEEE Trans. on PAS*, *PAS-100*(7), 3387–3395. doi:10.1109/TPAS.1981.316681

Girgis, A. A., & Brown, R. G. (1982). A new Kalman filtering-Based Digital Distance Relay. *IEEE Trans. PAS*, *101*(9), 3471–3480. doi:10.1109/TPAS.1982.317574

Girgis, A. A., & Ham, F. M. (1982). A new FFT-based digital frequency relay for load shedding. *IEEE Trans. on PAS*, *101*(2), 433–439. doi:10.1109/TPAS.1982.317125

Girgis, A. A., & Hwang, T. L. (1984). Optimal estimation of voltage phasors and frequency deviation using linear and non-linear Kalman filtering: Theory and limitations. *IEEE Trans. on PAS*, *103*(10), 2943–2951. doi:10.1109/TPAS.1984.318297

Glavin, M. E. E., Chan, P. K. W., Armstrong, S., & Hurley, W. G. G. (2008). A stand-alone photovoltaic supercapacitor battery hybrid energy storage system. *2008 13th International Power Electronics and Motion Control Conference, EPE-PEMC 2008*, 1688–1695. https://doi.org/10.1109/EPEPEMC.2008.4635510

Glavin, M. E., & Hurley, W. G. (2012). Optimisation of a photovoltaic battery ultracapacitor hybrid energy storage system. *Solar Energy*, *86*(10), 3009–3020. doi:10.1016/j.solener.2012.07.005

Gou, B. (2008). Generalized integer linear programming formulation for optimal PMU placement. *IEEE Transactions on Power Systems*, *23*(3), 1099–1104. doi:10.1109/TPWRS.2008.926475

Graditi, G., Ippolito, M. G., Telaretti, E., & Zizzo, G. (2016). Technical and economical assessment of distributed electrochemical storages for load shifting applications: An Italian case study. *Renewable & Sustainable Energy Reviews*, *57*, 515–523. doi:10.1016/j.rser.2015.12.195

Grantham, A., Pudney, P., Ward, L. A., Whaley, D., & Boland, J. (2017). The viability of electrical energy storage for low-energy households. *Solar Energy*, *155*, 1216–1224. doi:10.1016/j.solener.2017.07.063

Greveler, U., Justus, B., & Loehr, D. (2012). *Multimedia content identification through smart meter power usage profiles*. Computers, Privacy and Data Protection.

Guerrero, J., Blaabjerg, F., Zhelev, T., Hemmes, K., Monmasson, E., Jemei, S., Comech, M., Granadino, R., & Frau, J. (2010). Distributed Generation: Toward a New Energy Paradigm. *IEEE Industrial Electronics Magazine*, *4*(1), 52–64. doi:10.1109/MIE.2010.935862

Gu, J.-C., & Yu, S.-Li. (2003). Removal of DC offset in current and voltage signals a novel Fourier Algorithm. *IEEE Transactions on Power Delivery*, *15*(1), 73–79.

Gu, J., Shen, K., Yu, S., & Yu, C. (2006). Removal of dc offset and subsynchronous resonance in current signals for series compensated transmission lines using novel Fourier filter algorithm. *Electric Power Systems Research, 76*(5), 327–335. doi:10.1016/j.epsr.2005.08.002

Guney, M. S., & Tepe, Y. (2017). Classification and assessment of energy storage systems. *Renewable and Sustainable Energy Reviews, 75*(November), 1187–1197. https://doi.org/ doi:10.1016/j.rser.2016.11.102

Gungor, V. C., Sahin, D., Kocak, T., & Ergüt, S. (2011). *Smart grid communications and networking.* Türk Telekom, Tech. Rep. 11316-01.

Gungor, V. C., Sahin, D., Kocak, T., Ergut, S., Buccella, C., Cecati, C., & Hancke, G. P. (2011). Smart grid technologies: Communication technologies and standards. *IEEE Transactions on Industrial Informatics, 7*(4), 529–539. doi:10.1109/TII.2011.2166794

Guo, T., Liu, Y., Zhao, J., Zhu, Y., & Liu, J. (2020). A dynamic wavelet-based robust wind power smoothing approach using hybrid energy storage system. *International Journal of Electrical Power and Energy Systems, 116*(October), 105579. https://doi.org/ doi:10.1016/j.ijepes.2019.105579

Guo, X.-Q., Wu, W.-Y., & Gu, H.-R. (2011). *Phase locked loop and synchronization methods for grid interfaced converters: a review.* Przegląd Elektrotechniczny.

Guo, Y., Kezunovic, M., & Chen, D. (2003). Simplified Algorithm for Removal of the Effect of Exponentially Decaying DC-offset on the Fourier Algorithm. *IEEE Transactions on Power Delivery, 18*(3), 711–717. doi:10.1109/TPWRD.2003.813894

Gurlin, S. L. (2011). Smart Grid and its Development Prospects In the Asia-Pacific Region. *Journal of Emerging Trends in Computing and Information Sciences.*

Gurusinghe, D. R., Rajapakse, A. D., & Narendra, K. (2012). Valuation of Steady-State and Dynamic Performance of a Synchronized Phasor Measurement Unit. *IEEE elect. Pow. And Ene. Conf. 2012.*

Gurusinghe, D. R., Rajapakse, A. D., & Narendra, K. (2012). Valuation of Steady-State and Dynamic Performance of a Synchronized Phasor Measurement Unit. In *Proceedings IEEE elect. Pow. And Ene. Conf. 2012.* IEEE. 10.1109/EPEC.2012.6474980

Guzman, A., Altuve, H., & Tziouvaras, D. (2009). *Power Transformer Protection Improvements with Numerical Relays.* Academic Press.

Guzmán, A., Zocholl, S., & Altuve, H. (2000). Performance analysis of traditional and Improved differential Protective transformer relays. SEL Company.

Guzman, A., Zocholl, S., Benmouyal, G., & Altuve, H. J. (2001). A current based solution for transformer differential protection – part 1: Problem Statement. *IEEE Transactions on Power Delivery, 16*(4), 485–491. doi:10.1109/61.956726

Hahn, S. L. (1996). *Hilbert transforms in signal processing.* Artech House.

Haider, R., Kim, C. H., Ghanbari, T., & Bukhari, S. B. A. (2018). Harmonic-signature-based islanding detection in grid-connected distributed generation systems using Kalman filter. *IET Renewable Power Generation, 12*(15), 1813–1822. doi:10.1049/iet-rpg.2018.5381

Hajian, M. A., Ranjbar, M., Amraee, T., & Shirani, A. R. (2007). Optimal Placement of Phasor Measurement Units: Particle Swarm Optimization Approach. *Proc. Of the 14th International Conference on Intelligent System Applications to Power Systems.*

Hajian, M., Ranjbar, A. M., Amraee, T., & Mozafari, B. (2011). Optimal placement of PMUs to maintain network observability using a modified BPSO algorithm. *Int. J. Electr. Power*, *33*(1), 28–34. doi:10.1016/j.ijepes.2010.08.007

Hamzeh, M., & Mokhtari, H. (2009). Power quality comparison of active islanding detection methods in a single phase PV grid connected inverter. *Proceedings of IEEE International Symposium on Industrial Electronics*. 10.1109/ISIE.2009.5219766

Han, J., Choi, C.-S., Park, W.-K., & Lee, I. (2011). Green home energy management system through comparison of energy usage between the same kinds of home appliances. *2011 IEEE 15th International Symposium on Consumer Electronics (ISCE)*, 1–4.

Hart, G. (1992). Nonintrusive appliance load monitoring. *Proceedings of the IEEE*, *80*(12), 1870–1891. doi:10.1109/5.192069

Hashemi, F., Ghadimi, N., & Sobhani, B. (2013). Islanding detection for inverter-based DG coupled with using an adaptive neuro-fuzzy inference system. *International Journal of Electrical Power & Energy Systems*, *45*(1), 443–455. doi:10.1016/j.ijepes.2012.09.008

Hashemi, F., Mohammadi, M., & Kargarian, A. (2017). Islanding detection method for microgrid based on extracted features from differential transient rate of change of frequency. *IET Generation, Transmission & Distribution*, *11*(4), 891–904. doi:10.1049/iet-gtd.2016.0795

Hassan, S. Z., Li, H., Kamal, T., Arifoğlu, U., Mumtaz, S., & Khan, L. (2017). Neuro-fuzzy wavelet based adaptive MPPT algorithm for photovoltaic systems. *Energies*, *10*(3), 1–16. doi:10.3390/en10030394

Hauser, C. H., Bakken, D. E., & Bose, A. (2005). A failure to communicate: Next generation communication requirements, technologies, and architecture for the electric power grid. *IEEE Power & Energy Magazine*, *3*(2), 47–55. doi:10.1109/MPAE.2005.1405870

Haykin, S. (1994). *Neural Networks: a comprehensive foundation* (Vol. 2). Prentice Hall PTR.

Hayward, C. D. (1941). Harmonic-Current-Restrained Relays for Transformer Differential Protection. *Transactions of the American Institute of Electrical Engineers*, *60*(6), 377–382. doi:10.1109/T-AIEE.1941.5058349

He, J., Lu, C., Wu, X., Li, P., & Wu, J. (2009). Design and experiment of wide area HVDC supplementary damping controller considering time delay in China southern power grid. *Generation, Transmission Distribution, IET*, *3*(1), 17–25. doi:10.1049/iet-gtd:20080129

Heidari, M., Seifossadat, G., & Razaz, M. (2013). Application of decision tree and discrete wavelet transform for an optimized intelligent-based islanding detection method in distributed systems with distributed generations. *Renewable & Sustainable Energy Reviews*, *27*, 525–532. doi:10.1016/j.rser.2013.06.047

Hemmati, R., & Saboori, H. (2016). Emergence of hybrid energy storage systems in renewable energy and transport applications – A review. *Renewable & Sustainable Energy Reviews*, *65*, 11–23. doi:10.1016/j.rser.2016.06.029

Hengsi, Q., & Kimball, J. W. (2009). AC-AC dual active bridge converter for solid state transformer. *Proceedings of the 2009 IEEE Energy Conversion Congress and Exposition*, 3039–3044. 10.1109/ECCE.2009.5316507

Hernandez, M. E., Ramos, G. A., Lwin, M., Siratarnsophon, P., & Santoso, S. (2018). Embedded Real-Time Simulation Platform for Power Distribution Systems. *IEEE Access: Practical Innovations, Open Solutions*, *6*, 6243–6256. doi:10.1109/ACCESS.2017.2784318

Hersch, P., & Zweibel, K. (2019), Basic Photovoltaic Principles and Methods. *Solar Energy Research Institute (SERI)*. Available: https://www.nrel.gov/docs/legosti/old/1448.pdf

Ho, C. Y.-F., Ling, B. W.-K., Chi, Z.-W., Shikh-Bahaei, M., Liu, Y.-Q., & Teo, K.-L. (2008). Design of near allpass strictly stable minimal-phase real valued rational IIR filters. *IEEE Transactions on Circuits and Wystems. II, Express Briefs*, *55*(8), 781–785. doi:10.1109/TCSII.2008.922417

Hongbin, S., & Qinglai, G. (2019). Review of Challenges and Research Opportunities for Voltage Control in Smart Grids. *IEEE Transactions on Power Systems*, *34*(4), 2790–2801. doi:10.1109/TPWRS.2019.2897948

Horowitz, S. H., & Phadke, A. G. (1992). *Power System Relaying*. Wily & Sons.

Hossain, M., Amanullah, M. T. O., & Shawkat Ali, A. B. M. (2013). Smart Grid: Opportunities, developments and trends. Springer-Verlag.

Hossain, M. S., Madlool, N. A., Rahim, N. A., Selvaraj, J. K., Pandey, A., & Khan, A. F. (2016). Role of smart grid in renewable energy: An overview. *Renewable & Sustainable Energy Reviews*, *60*, 1168–1184. doi:10.1016/j.rser.2015.09.098

Hosseini, S. S. S., & Gandomi, A. H. (2010). Discussion of "economic load dispatch - A comparative study on heuristic optimization techniques with an improved coordinated aggregation-based PSO.". *IEEE Transactions on Power Systems*, *25*(1), 590. doi:10.1109/TPWRS.2009.2037531

Hossen, A. (2001). A new fast approximate hilbert transform with different applications. IIUM Engineering Journal, 2(2).

Huang, S., Jan, R., & Yang, W. (2006). A ring-based information collection architecture in wireless sensor networks. *International Journal of Sensor Networks*, *1*(3/4), 190–199. doi:10.1504/IJSNET.2006.012034

Hubert, T., & Grijalva, S. (2012). Modeling for residential electricity optimization in dynamic pricing environments. *IEEE Transactions on Smart Grid*, *3*(4), 2224–2231. doi:10.1109/TSG.2012.2220385

Huihui, Q., Wei, Z., Pengfei, W., & Chunling, C. (2012). Transient power quality signal generatorand detector platform. *Energy Procedia*, *16*, 1380–1385. doi:10.1016/j.egypro.2012.01.219

Hu, Q., & Li, F. (2013). Hardware design of smart home energy management system with dynamic price response. *IEEE Transactions on Smart Grid*, *4*(4), 1878–1887. doi:10.1109/TSG.2013.2258181

Huu, D. N. (2016). An adaptive control of hybrid battery-supercapacitor storage for integration of wind and solar. *2016 IEEE International Conference on Sustainable Energy Technologies (ICSET)*, 157–162. https://doi.org/10.1109/ICSET.2016.7811774

Ibe, A., & Cory, B. (1986). A travelling wave-based fault locator for two-and three-terminal networks. *IEEE Transactions on Power Delivery*, *1*(2), 283–288. doi:10.1109/TPWRD.1986.4307961

IEEE Application Guide for IEEE Std 1547(TM) (2009). IEEE Standard for Interconnecting Distributed Resources with Electric Power Systems. *IEEE Std 1547.2-2008*, 15, 1-217. doi:10.1109/IEEESTD.2008.4816078

IEEE Guides and Standards. (1991). *IEEE Guides and Standards for Protective Relaying Systems*. IEEE Publication.

IEEE Power System Relay Committee Working Group I17 Report. (2005). *Transmission Relay System Performance Comparison*. IEEE Publication.

IEEE Standard C37.118.1-2011. (2011). *IEEE standard for synchrophasor measurements for power systems*. IEEE Publication.

IEEE Standard C37.91-2000. (2000). Guide for Protective Relay Applications to Power Transformers. *The Institute of Electrical and Electronics Engineers*.

IEEE Std. C37.118-2005. (2006). *IEEE Standard for Synchrophasors for Power Systems*, IEEE Publication.

IEEE. (2006). C37.118-2005. *IEEE Standard for Synchrophasors for Power Systems.*

IEEE. (2011). IEEE Standard for Synchrophasor Measurements for Power Systems. *IEEE Std C37.118.1-2011 (Revision of IEEE Std C37.118-2005)*, 1–61. doi:. doi:10.1109/IEEESTD.2011.6111219

Iong, S., & Kim, K. H. (2004). An islanding detection method for distributed generations using voltage unbalance and total harmonic distortion of current. *IEEE Transactions on Power Delivery*, *19*(2), 745–752. doi:10.1109/TPWRD.2003.822964

Isaksson, A. (1988). Digital Protective Relaying through Recursive Least-Squares Identification. *IEE Proceedings. Part C. Generation, Transmission and Distribution*, *135*(5), 441–449. doi:10.1049/ip-c.1988.0057

Iskin, I., Daim, T., Kayakutlu, G., & Altuntas, M. (2012). Exploring renewable energy pricing with analytic network process— Comparing a developed and a developing economy. *Energy Econ*, *34*(4), 882–891. doi:10.1016/j.eneco.2012.04.005

Ismail, M. S., Moghavvemi, M., & Mahlia, T. M. I. (2014). Genetic algorithm based optimization on modeling and design of hybrid renewable energy systems. *Energy Conversion and Management*, *85*, 120–130. doi:10.1016/j.enconman.2014.05.064

Jahan, M. S., & Amjady, N. (2013). Solution of large-scale security constrained optimal power flow by a new bi-level optimization approach based on enhanced gravitational search algorithm. *IET Generation, Transmission & Distribution*, *7*(12), 1481–1491. doi:10.1049/iet-gtd.2012.0697

Jazayeri, P., Schellenberg, A., Rosehart, W. D., Doudna, J., Widergren, S., Lawrence, D., Mickey, J., & Jones, S. (2005). A survey of load control programs for price and system stability. *IEEE Transactions on Power Systems*, *20*(3), 1504–1509. doi:10.1109/TPWRS.2005.852147

Jenkins, N., Long, C., & Wu, J. (2015). An Overview of the Smart Grid in Great Britain. *Engineering*, *1*(4), 413–421. doi:10.15302/J-ENG-2015112

Jeon, J.-H., Kim, J.-Yu., Kim, H.-M., Kim, S.-K., Cho, C., Kim, J.-M., Ahn, J.-B., & Nam, K.-Y. (2010). Development of Hardware In-the-Loop Simulation System for Testing Operation and Control Functions of Microgrid. *IEEE Transactions on Power Electronics*, *25*(12), 2919–2929. doi:10.1109/TPEL.2010.2078518

Jiang, W., Hu, R., Chen, W., Wang, X., Li, Z., Yang, Y., & Wang, J. (2014). Research of multipurpose optimization control in hybrid energy storage system. *2014 IEEE Energy Conversion Congress and Exposition (ECCE)*, 2550–2554. https://doi.org/ doi:10.1109/ECCE.2014.6953741

Jiang, A., & Kwan, H. K. (2009). IIR digital filter design with new stability constraint based on argument principle. *IEEE Transactions on Circuits and Systems. I, Regular Papers*, *56*(3), 583–593. doi:10.1109/TCSI.2008.2002116

Jiang, A., & Kwan, H. K. (2010). Minimax design of IIR digital filters using iterative SOCP. *IEEE Transactions on Circuits and Systems. I, Regular Papers*, *57*(6), 1326–1337. doi:10.1109/TCSI.2009.2028749

Jiangnon, P., Yuanzhang S., & Wang H. F., Optimal Placement for full Network Observability using Tabu Search Algorithm. *Int. J. Elect. Power Energy Syst.*, *23*, 1525-1526.

Jiang, S., Ji, Z., & Shen, Y. (2014). A novel hybrid particle swarm optimization and gravitational search algorithm for solving economic emission load dispatch problems with various practical constraints. *International Journal of Electrical Power & Energy Systems*, *55*, 628–644. doi:10.1016/j.ijepes.2013.10.006

Jiao, S., Bo, Z., Liu, W., & Yang, Q. (2001). New Principles to Detect Faults During Power Swing. *Seventh IEE Int. Conf. on Dev. in Power Syst. Protect.*, 515 – 518.

Ji, B., Yuan, X., Li, X., Huang, Y., & Li, W. (2014). Application of quantum-inspired binary gravitational search algorithm for thermal unit commitment with wind power integration. *Energy Conversion and Management*, *87*, 589–598. doi:10.1016/j.enconman.2014.07.060

Jingcheng, G., Xiao, Y., Liu, J., Liang, W., & Chen, C. L. (2012). A survey of communication/networking in Smart Grids. *Future Generation Computer Systems*, *28*(2), 391–404. doi:10.1016/j.future.2011.04.014

Johns, A. T., & Salman, S. K. (1995). Digital protection for power system. IEE power series N°15.

Johns.A.T., & salman S.K. (1995). Digital protection for power system. *IEE Power Series,* 15.

Johnston, R., Hauser, C., Gjermundrod, K., & Bakken, D. (2006). Distributing Time Synchronous Phasor Measurement Data Using the Grid Stat Communication Infrastructure. *In System Sciences, 2006. HICSS*, *06*, 245b–245b. doi:10.1109/HICSS.2006.127

Jonsson, M., & Daalder, J. (2001). A New Protection Scheme to Prevent Mal-Trips Due to Power Swings. *IEEE/PES Transmission and Distrib. Conf.*, *2*, 724 – 729. 10.1109/TDC.2001.971327

Jovanovic-Dolecek, G. (2006). Design of IIR Notch filters with maximally flat or equiripple magnitude characteristics. *14th Europeon Signal Processing Conference (EUSIPCO 2006)*, Florence, Italy.

Justo, J. J., Mwasilu, F., Lee, J., & Jung, J. W. (2013). AC-microgrid versus DC-microgrid with distributed energy resources: A review. *Renewable & Sustainable Energy Reviews*, *24*, 387–405. doi:10.1016/j.rser.2013.03.067

Kabalci, Y. (2016). A survey on smart metering and smart grid communication. *Renewable & Sustainable Energy Reviews*, *57*, 302–318. doi:10.1016/j.rser.2015.12.114

Kalogridis, G., Cepeda, R., Denic, S., Lewis, T., & Efthymiou, C. (2011). Elecprivacy: Evaluating the privacy protection of electricity management algorithms. *IEEE Transactions on Smart Grid*, *2*(4), 750–758. doi:10.1109/TSG.2011.2160975

Kamwa, I., Dobrescu, M., Heniche, A., Cyr, C., & Cadieux, P. (2011). A fundamental study of wide-area damping controllers with application to fuzzy-logic based PSS design for dynamic shunt compansators. *Power Systems Computation Conference (PSCC).*

Kane, P. O., & Fox, B. (1997). Loss of mains detection for embedded generation by system impedance monitoring. *Proceedings of 6th International Conference on Development of Power System Protection.*10.1049/cp:19970037

Kansal, P., & Bose, A. (2012). Bandwidth and latency reuirements for smart transmission grid applications. *Smart Grid, IEEE Transaction*, *3*(3), 1344–1352. doi:10.1109/TSG.2012.2197229

Kappagantu, R., & Arul Daniel, S. (2018). Challenges and Issues of Smart Grid Implementation: A Case of Indian Scenario. *Journal of Electrical Systems and Information Technology*, *5*(3), 453–467. doi:10.1016/j.jesit.2018.01.002

Karaboga, N. (2004). Design of minimum phase digital IIR filters by using Genetic Algorithm. *Proceedings of the 6th Nordic signal Processing Symposium-NORSIG.*

Karavas, C. S., Kyriakarakos, G., Arvanitis, K. G., & Papadakis, G. (2015). A multi-agent decentralized energy management system based on distributed intelligence for the design and control of autonomous polygeneration microgrids. *Energy Conversion and Management*, *103*, 166–179. doi:10.1016/j.enconman.2015.06.021

Karimi, H., Yazdami, A., & Iravni, R. (2008). Negative-sequence current injection for fast islanding detection of a distributed resource unit. *IEEE Transactions on Power Electronics*, *23*(1), 298–307. doi:10.1109/TPEL.2007.911774

Kasztenny, B., & Kezunovic, M. (1998). Improved Power Transformer Protection UsingNumerical Relays. *IEEE Computer Applications in Power*, *11*(4), 39–45. doi:10.1109/67.721702

Kasztenny, B., & Kulidjian, A. (2000). An Improved Transformer Inrush Restraint Algorithm Increases Security While Maintaining Fault Response Performance. *53rd Annual Conference for Protective Relay Engineers.*

Kato, D., Horii, H., & Kawahara, T. (2014). Next-generation SCADA/EMS Designed for Large Penetration of Renewable Energy. *Hitachi Review*, *63*(4), 151.

Katrel S. R., & Rojatkar D. V. (2017). Home Automation: Past, Present And Future. *International Research Journal of Engineering and Technology, 4*(10).

Katsanevakis, M., Stewart, R. A., & Lu, J. (2017). Aggregated applications and benefits of energy storage systems with application-specific control methods: A review. *Renewable and Sustainable Energy Reviews, 75*(August), 719–741. https://doi.org/ doi:10.1016/j.rser.2016.11.050

Kaur, R., Dhillon, J. S., & Singh, D. (2010). Multicriterion Optimization approach to the design of stable recursive Digital Filter. *International Journal of Engineering Science and Technology*, *2*(8), 3515–3524.

Kaźmierkowski, M. P., Krishnan, R., & Blaabjerg, F. (2002). Control in power electronics: selected problems. Academic Press.

Kennedy, B. W. (2000). *Power quality primer*. McGraw Hill Professional.

Kennedy, J., & Eberhart, R. (1995). Particle swarm optimization. *Neural Networks, 1995. Proceedings., IEEE International Conference*, 1942-1948.

Kennedy, L. F., & Hayward, C. D. (1938). Harmonic-Current-Restrained Relays for Differential Protection. *Transactions of the American Institute of Electrical Engineers*, *57*(5), 262–266. doi:10.1109/T-AIEE.1938.5057808

Kerdphol, T., Qudaih, Y., & Mitani, Y. (2015). Optimal Battery Energy Storage Size Using Particle Swarm Optimization for Microgrid System. *International Review of Electrical Engineering*, *10*(2), 277–285. doi:10.15866/iree.v10i2.5350

Kester, W. (2009). *ADC Architectures III: Sigma-Delta ADC Basics, Analogue Devices, MT-022 Tutorial, 2009*. http://www.analog.com/static/imported-files/tutorials/MT-022.pdf

Kezunovic, M., & Kasztenny, B. (2000). Design Optimization and Performance Evaluation of the Relaying Algorithms, Relays and Protective Systems Using Advanced testing Tools. *IEEE Transactions on Power Delivery*, *15*(4), 1129–1135. doi:10.1109/61.891493

Khamis, A., Shareef, H., Mohamed, A., & Bizkevelci, E. (2015). Islanding detection in a distributed generation integrated power system using phase space technique and probabilistic neural network. *Neurocomputing*, *148*, 587–599. doi:10.1016/j.neucom.2014.07.004

Khan, R. H., & Khan, J. Y. (2013). A comprehensive review of the application characteristics and traffic requirements of a smart grid communications network. *Computer Networks*, *57*(3), 825–845. doi:10.1016/j.comnet.2012.11.002

Khan, R., McLaughlin, K., Laverty, D., & Sezer, S. (2016). Analysis of IEEE C37.118 and IEC 61850-90-5 Synchrophasor. *Proceedings of Power and Energy Society General Meeting (PESGM)*. 10.1109/PESGM.2016.7741343

Khodaparastan, M., Vahedi, H., Khazaeli, F., & Oraee, H. (2017). A hybrid islanding detection method for inverter based DGs using SFS and ROCOF. *IEEE Transactions on Power Delivery*, *32*(5), 2162–2170. doi:10.1109/TPWRD.2015.2406577

Kian, A., & Keyhani, A. (2001). Stochastic price modeling of electricity in deregulated energy markets. *Proceedings of the 34th Annual Hawaii International Conference on System Sciences*, 7–12. 10.1109/HICSS.2001.926292

Kim, J., Filali, F., & Young-Bae, K. (2015). Trends and Potentials of the Smart Grid Infrastructure: From ICT Sub-System to SDN-Enabled Smart Grid Architecture. *Applied Sciences (Basel, Switzerland)*, *5*(4), 706–727. doi:10.3390/app5040706

Kim, M. S., Haider, R., Cho, G. J., Kim, C. H., Won, C. Y., & Chai, J. S. (2019). Comprehensive review of islanding detection merhod for distributed generation systems. *Energies*, *12*(837), 1–21. doi:10.3390/en12050837

Klari´c, M., Kuzle, I. & Ninoslav. (2018). Wind Power Monitoring and Control Based on Synchrophasor Measurement Data Mining. *Energies, MDPI, 11*(12), 1-23.

Kolter, J. Z., & Johnson, M. J. (2011). Redd A public data set for energy disaggregation research. *Proc. of ACM KDD Workshop on Data Mining Applications in Sustainability (SustKDD).*

Komninos, N., Philippou, E., & Pitsillides, A. (2014). Survey in smart grid and smart home security: Issues, challenges and countermeasures. *IEEE Communications Surveys and Tutorials*, *16*(4), 1933–1954. doi:10.1109/COMST.2014.2320093

Koponen, P. (2008). *Definition of Smart Metering and Applications and Identification of Benefits Project: European Smart Metering Alliance*. VTT Technical Research Centre of Finland.

Korkali, M., & Abur, A. (2009). Reliable measurement design against loss of PMUs with variable number of channels. *North American Power Symposium (NAPS).*

Kouzou, A. L., Bentarzi, H., Mohammedi, R. D., & Laoumer, M. (2019). Optimal Placement of Phasor Measurement Unit in Power System using Meta-Heuristic Algorithms. *Journal Electrotehnica, Electronica. Automatica*, *67*(2).

Koykul, W. (2011). Current status and action plan of utility sector on smart grid and smart community in Thailand. *Thailand-Japan Workshop on Smart Community in Thailand: Provincial Electricity Authority.*

Kumar A. (2014). *Hall-Effect Current Sensors for Power Electronic Applications: Design and Performance Validation.* Indian Institute of Science Bangalore.

Kumar, A., Srikanth, P., Nayyar, A., Sharma, G., Krishnamurthi, R., & Alazab, M. (2020). A Novel Simulated-Annealing Based Electric Bus System Design, Simulation, and Analysis for Dehradun Smart City. *IEEE Access: Practical Innovations, Open Solutions*, *8*, 89395–89424. doi:10.1109/ACCESS.2020.2990190

Kushiro, N., Suzuki, S., Nakata, M., Takahara, H., & Inoue, M. (2003). Integrated residential gateway controller for home energy management system. *IEEE Transactions on Consumer Electronics*, *49*(3), 629–636. doi:10.1109/TCE.2003.1233787

Kuzlu, M., Pipattanasomporn, M., & Rahman, S. (2014). *Communication network requirements for major smart grid applications in HAN, NAN and WAN. In Computer Networks* (Vol. 67). Elsevier.

Lam, H. K. (2012). *Computational intelligence and its applications: evolutionary comutaion, fuzzy logic, neural network and support vector machine techniques*. World Scientific.

Lang, M. C. (2000). Least-Squares Design of IIR Filters with Prescribed Magnitude and Phase Response and a Pole Radius Constraint. *IEEE Transactions on Signal Processing*, *48*(11), 3109–3121. doi:10.1109/78.875468

Larsen, R. W. (2011). *LABVIEW for engineers*. Prentice Hall/Pearson.

Laverty, D. M., Best, R. J., & Morrow, D. J. (2015). Loss-of-mains protection system by application of phasor measurement unit technology with experimentally assessed threshold settings. *IET Generation, Transmission & Distribution*, *9*(2), 146–153. doi:10.1049/iet-gtd.2014.0106

Lee, P. K., & Lai, L. L. (2009). Smart Metering in Micro-Grid Applications. *Proc. IEEE Power & Energy Society General Meeting*, 1 – 5.

Lewis, B. J. (1998). *Protective Relaying: Principles and Applications*. New York: Marcel Dekker.

Lewis, R. P., Igic, P., & Zhongfu, Z. (2009). Assessment of communication methods for smart electricity metering in the U.K. *Proc. IEEE PES/IAS Conf. Sustainable Alternative Energy (SAE)*, 1-4. 10.1109/SAE.2009.5534884

Li, N., Chen, L., & Low, S. H. (2011). Optimal demand response based on utility maximization in power networks. 2011 IEEE power and energy society general meeting, 1–8.

Li, C., Cao, C., Cao, Y., Kuang, Y., Zeng, L., & Fang, B. (2014). A review of islanding detection methods for microgrid. *Renewable & Sustainable Energy Reviews*, *35*, 211–220. doi:10.1016/j.rser.2014.04.026

Lien, K.-P., Liu, C.-W., Yu, C.-S., & Jiang, J.-A. (2006). Transmission Network Fault Loca-tion Observability With Minimal PMU Placement. *IEEE Transactions on Power Delivery*, *21*(3), 1128–1136. doi:10.1109/TPWRD.2005.858806

Lin, X., Gao, Y., & Liu, P. (2008). A Fast Unblocking Scheme for Distance Protection to Identify Symmetrical Fault Occurring During Power Swings. *IEEE Transactions on Power Delivery*, *1*, 73–77. doi:10.1109/TPWRD.2007.911154

Lin, Y.-H., Liu, C.-W., & Chen, C.-S. (2004). A new PMU-based fault detection/location technique for transmission lines with consideration of arcing fault discrimination-part I: Theory and algorithms. *IEEE Transactions on Power Delivery*, *19*(4), 1587–1593. doi:10.1109/TPWRD.2004.832407

Lin, Y.-H., Liu, C.-W., & Chen, C.-S. (2004). A new PMU-based fault detection/location technique for transmission lines with consideration of arcing fault discrimination-partII: Performance evaluation. *IEEE Transactions on Power Delivery*, *19*(4), 1594–1601. doi:10.1109/TPWRD.2004.835276

Liserre, M., Sauter, T., & Hung, J. Y. (2010). Future energy systems: Integrating renewable energy sources into the smart power grid through industrial electronics. *IEEE Ind. Electron*, *4*(1), 18–37. doi:10.1109/MIE.2010.935861

Liu, F., Kang, Y., Zhang, Y., Duan, S., & Lin, X. (2010). Improved SMS islanding detction method for grid connected converters. *IET Renewable Power Generation*, *4*(1), 36–42. doi:10.1049/iet-rpg.2009.0019

Liu, L., Kong, F., Liu, X., Peng, Y., & Wang, Q. (2015). A review on electric vehicles interacting with renewable energy in smart grid. *Renewable & Sustainable Energy Reviews*, *51*, 648–661. doi:10.1016/j.rser.2015.06.036

Li, W., Logenthiran, T., Phan, V., & Woo, W. L. (2016). *Intelligent Multi-Agent System for Power Grid Communication. In 2016 IEEE Region 10 Conference*. TENCON.

Lobaccaro, G., Carlucci, S., & Löfström, E. (2016). A Review of Systems and Technologies for Smart Homes and Smart Grids. *Energies*, *9*(5), 348. doi:10.3390/en9050348

Lobos, T., & Janic, P. (2006). Automated classification of power-quality disturbances using SVM and RBF networks. *IEEE Transactions on Power Delivery*, *21*(3), 1663–1669. doi:10.1109/TPWRD.2006.874114

Locke, G., & Gallagher, P. D. (2010). *Nist Framework and Roadmap for Smart Grid Interoperability Standards, Release 1.0; Smart Grid technologies: Communication technologies and standards*. National Institute of Standards and Technology.

Logenthiran, T., Srinivasan, D., Khambadkone, A. M., & Aung, H. N. (2010). Multi-Agent System (MAS) for short-term generation scheduling of a microgrid. *2010 IEEE International Conference on Sustainable Energy Technologies, ICSET 2010*. https://doi.org/10.1109/ICSET.2010.5684943

Lo, K. L., Zakaria, Z., & Sohod, M. H. (2005). Determination of Consumers' Load Profiles based on Two-stage Fuzzy C-Means, Power System Research Group, Department of Electronics & Engineering. *Proceedings of the 5th WSEAS Int. Conf. on Power Systems and Electromagnetic Compatibility*, 212-217.

Lopes, L. A. C., & Sun, H. (2006). Performance assessment of active frequency drifting islanding detection methods. *IEEE Transactions on Energy Conversion*, *21*(1), 171–180. doi:10.1109/TEC.2005.859981

Lu, B., Wu, X., Figueroa, H., & Monti, A. (2007). A Low Cost Real-Time Hardware-in-the-Loop Testing Approach of Power Electronics Controls. *IEEE Transactions on Industrial Electronics*, *54*(2), 919–931. doi:10.1109/TIE.2007.892253

Luo, X., Wang, J., Dooner, M., & Clarke, J. (2015). Overview of current development in electrical energy storage technologies and the application potential in power system operation. *Applied Energy*, *137*, 511–536. doi:10.1016/j.apenergy.2014.09.081

Luque, A., & Hegedus, S. (2003). Handbook of photovoltaic science and engineering. Hoboken, NJ: Wiley.

Lutovac, M. D., & Milic, L. D. (1997). Design of computationally efficient elliptic IIR filters with a reduced number of shift-and-add operations in multipliers. *IEEE Transactions on Signal Processing*, *45*(10), 2422–2430. doi:10.1109/78.640708

Lu, X., Wang, W., & Ma, J. (2012). Authentication and integrity in the smart grid: An empirical study in substation automation systems. *International Journal of Distributed Sensor Networks*, *8*(6), 175262. doi:10.1155/2012/175262

Mahari, A., & Heresh, S. (2013). Optimal PMU placement for power system observability using BICA considering measurement redundancy. *Electric Power Systems Research*, *103*, 78–85. doi:10.1016/j.epsr.2013.04.013

Mahat, P., Chen, Z., & Bak-Jensen, B. (2009). A hybrid islanding detection technique using average rate of voltage change and real power shift. *IEEE Transactions on Power Delivery*, *24*(2), 764–771. doi:10.1109/TPWRD.2009.2013376

Mahboubi, M., Bentarzi, H., & Ouadi, A. (2019). *An application of PMU in solar energy generation Plant Connected Power Grid* [Paper presentation]. *The International Conference on Advanced Engineering in Petrochemical Industry (ICAEPI'19)*, Skikda, Algeria.

Mahmood, A., Aamir, M., & Anis, M. I. (2008). Design and implementation of AMR smart grid system. *Proc. IEEE Canada Electric Power Conference*, 6 - 7. 10.1109/EPC.2008.4763340

Mahmood, A., Khan, I., Razzaq, S., Najam, Z., Khan, N. A., Rehman, M. A., & Javaid, N. (2014). Home appliances coordination scheme for energy management (HACS4EM) using wireless sensor networks in smart grids. *Procedia Computer Science*, *32*, 469–476. doi:10.1016/j.procs.2014.05.449

Mahmood, D., Javaid, N., Alrajeh, N., Khan, Z., Qasim, U., Ahmed, I., & Ilahi, M. (2016, March 15). Realistic scheduling mechanism for smart homes. *Energies*, *9*(3), 202. doi:10.3390/en9030202

Majumder, R., Chaudhuri, B., & Pal, B. (2007). Implementation and test results of a wide-area measurement-based controller for damping interarea oscillations considering signal-transmission delay. *Generation, Transmission Distribution, IET*, *1*(1), 1–7. doi:10.1049/iet-gtd:20050493

Manana, M., Eguıluz, O. A., Dıez, G., Renedo, C., & Perez, S. (2005). Effects of magnetizing inrush current on power quality and distributed generation. *9th Spanish Portuguese Congress on Electrical Engineering*.

Mao, J., Mohiuddin, K. M., & Jain, A. K. (1996). Artificial neural networks: A tutorial. *Computer*, *1*(3).

Marijic, Z., Ilic, Z., & Bazant, A. (2010). Fixed-data-rate power minimization algorithm for ofdm-based power-line communication networks. *IEEE Transactions on Power Delivery*, *25*(1), 141–149. doi:10.1109/TPWRD.2009.2035286

Martens, J., Hinton, G. E., & Sutskever, I. (2011). Generating text with recurrent neural networks. *Proceedings of the 28th International Conference on Machine Learning (ICML-11)*, 1017-1024.

Martin, K. E., Hamai, D., Adamiak, M. G., Anderson, S., Begovic, M., Benmouyal, G., Brunello, G., Burger, J., Cai, J. Y., Dickerson, B., Gharpure, V., Kennedy, B., Karlsson, D., Phadke, A. G., Salj, J., Skendzic, V., Sperr, J., Song, Y., Huntley, C., ... Price, E. (2008). Exploring the IEEE Standard C37.118-2005 Synchrophasors for Power Systems. *IEEE Transactions on Power Delivery*, *23*(4), 1805–1811. doi:10.1109/TPWRD.2007.916092

Mason, C. R. (1986). *The Art and Science of Protective Relaying* (2nd ed.). John Wiley.

Masoum, M. A., & Fuchs, E. (2011). *Power quality in power systems and electrical machines.* Academic Press.

Massoud Amin, S., & Wollenberg, B. F. (2005). Towards a SmartGrid: Power delivery for the 21st Century. *IEEE Power & Energy Magazine*, *3*(5), 34–41. doi:10.1109/MPAE.2005.1507024

Math Works Inc. (2002). *Power System Blockset User's guide*. Author.

McClenny & Tokash, W. (2017). *2Q17 Hybrid Advanced Battery Markets report.* Navigent Research. Retrieved from www.naviantresearch.com

McGranaghan, M. F., Beaty, Hw., Santos, S., & Dugan, R. C. (1996). *Electrical power systems quality* (2nd ed.). McGraw-Hill.

Meckic, F., Girgis, R., & Gajic, Z. (2006). Power Transformer Characteristics and their Effects on protective Relays. *33rd Western Protective Relays Conference.*

Medina, J., Muller, N., & Roytelman, I. (2016). Demand response and distribution grid operations: Opportunities and challenges. IEEE Trans. Smart Grid, 1(2), 193–198.

Menon, V., & Nehrir, M. H. (2007). A hybrid islanding detection technique using voltage unbalance and frequency set point. *IEEE Transactions on Power Systems*, *22*(1), 442–448. doi:10.1109/TPWRS.2006.887892

Metke, A., & Ekl, R. (2010). Smart grid security technology. In *Innovative Smart Grid Technologies* (pp. 1–7). ISGT.

Miceli, R. (2013). Energy Management and Smart Grids. *Energies*, *6*(4), 2262–2290. doi:10.3390/en6042262

Millman, J., & Grabel, A. (1999). *Microelectronics* (2nd ed.). McGraw-Hill Book Company.

Milosevic, B., & Begovic, M. (2003). Nondominated sorting genetic algorithm for optimal phasor measurement placement. *IEEE Transactions on Power Systems*, *18*(1), 69–75. doi:10.1109/TPWRS.2002.807064

Mirjalili, S. (2015). Moth-flame optimization algorithm: A novel nature-inspired heuristic paradigm. *Knowledge-Based Systems*, *89*, 228–249. doi:10.1016/j.knosys.2015.07.006

Mirjalili, S., & Lewis, A. (2014). Grey Wolf Optimizer. *Advances in Engineering Software*, *69*, 46–61. doi:10.1016/j.advengsoft.2013.12.007

Mishra, S., Salama, M. A., Liew, A. C., & Dash, P. K. (2000). Classification of power system disturbances using a fuzzy expert system and a Fourier linear combiner. *IEEE Transactions on Power Delivery*, *15*(2).

Misra, S., Mondal, A., Banik, S., Khatua, M., Bera, S., & Obaidat, M. S. (2013). Residential Energy Management in Smart Grid: A Markov Decision Process-Based Approach. *2013 IEEE International Conference on Green Computing and Communications and IEEE Internet of Things and IEEE Cyber, Physical and Social Computing*, 1152-1157. 10.1109/GreenCom-iThings-CPSCom.2013.200

Mitra, M., Sengupta, S., & Chattopadhyay, S. (2011). *Electric power quality*. Springer.

Mohammadi-Ivatloo, B. (2009). *Optimal placement of PMUs for power system observability using topology based formulated algorithms. Sahrif University of Technology.*

Momoh, J. A. (2012). *Smart grid: fundamentals of design and analysis*. Wiley. doi:10.1002/9781118156117

Morcos, M. M., & Ibrahim, W. A. (2002). Artificial intelligence and advanced mathematical tools for power quality applications: A survey. *IEEE Transactions on Power Delivery*, *17*(2).

Mosavi, M. R., Akhyani, A. A., & Rahmati, A. (2012). *A PMU Placement Optimal Method in Power Systems using Modified ACO Algorithm and GPS Timing*. Iran University of Science and Technology.

Moxley, R. (2010). *Analyse relay fault data to improve service reliability*. Schweitzer Engineering Laboratories, Inc.

Mo, Y., Kim, T.-H., Brancik, K., Dickinson, D., Lee, H., Perrig, A., & Sinopoli, B. (2012). Cyber-physical security of a smart grid infrastructure. *Proceedings of the IEEE, 100*(1), 195–209. doi:10.1109/JPROC.2011.2161428

Müller, H. H., & Castro, C. A. (2016). Genetic algorithm-based phasor measurement unit placement method considering observability and security criteria. *IET Generation, Transmission & Distribution, 10*(1), 270–280. doi:10.1049/iet-gtd.2015.1005

Mumtaz, S., & Khan, L. (2017). Indirect adaptive neurofuzzy Hermite wavelet based control of PV in a grid-connected hybrid power system. *Turkish Journal of Electrical Engineering and Computer Sciences, 25*, 4341–4353. doi:10.3906/elk-1702-175

Munasinghe, M., & Warfor, J. J. (1982). Electricity Pricing Theory and Case Studies. Academic Press.

Muneta L. M. & Rey G. R. (2011). *Electrical generation and distribution systems and power quality disturbances*. Academic Press.

Nagaraj, S. B., Nongpiur, R., & Antoniou, A. (2010). Design of Nearly Linear-Phase Recursive Digital Filters by Using Unconstrained Least-pth MinimaxOptimization. *Proceedings of the 4th International Symposium on Communications, Control and Signal Processing, ISCCSP 2010*.

Nagaraj, S. B. R. N., & Antonious, A. (2010). Design of Nearly Linear-Phase Recursive Digital Filters by Using Unconstrained Least-pthMinimax Optimization. *Proceedings of the 4th International Symposium on Communications, Control and Signal Processing, ISCCSP 2010*. 10.1109/ISCCSP.2010.5463429

Nashelsky, L., & Boylestad, R. L. (1978). *Electronic devices and circuit theory*. Prentice/Hall International.

National Instruments. (2009). *NI R Series Multifunction RIO User Manual*. NI Publication.

NETL. (2007). *A systems view of the modern grid*. NETL.

NETL. (2009). *West Virginia Smart Grid implementation Plan*. NETL.

Nguyen, C. T., & Srinivasan, K. (1984). A new technique for rapid tracking of frequency deviations based on level crossing. *IEEE Trans. on PAS, 103*(8), 2230–2236. doi:10.1109/TPAS.1984.318536

Niyato, D., & Ping, W. (2012). Cooperative transmission for meter data collection in smart grid. *Communications Magazine, IEEE, 50*(4), 90–97. doi:10.1109/MCOM.2012.6178839

Noelia, U. P. (2016). State of the Art and Trends Review of Smart Metering in Electricity Grids. *Applied Sciences, MDPI, 6*(68). Advance online publication. doi:10.3390/app6030068

Nongpiur, R.C., Shpak, D.J., & Antonious, A. (2013). *Improved Design Method for Nearly linear-phase filters using Constrained Optimization*. Academic Press.

Nuqui, R. F., & Phadke, A. G. (2005). Phasor Measurement Unit placement techniques for complete and incomplete observability. *IEEE Transactions on Power Delivery, 20*(4), 2381–2388. doi:10.1109/TPWRD.2005.855457

Olivares, D. E., Mehrizi-Sani, A., Etemadi, A. H., Cañizares, C. A., Iravani, R., Kazerani, M., Hajimiragha, A. H., Gomis-Bellmunt, O., Saeedifard, M., Palma-Behnke, R., Jimenez-Estevez, G. A., & Hatziargyriou, N. D. (2014). Trends in microgrid control. *IEEE Transactions on Smart Grid, 5*(4), 1905–1919. doi:10.1109/TSG.2013.2295514

Ong, C. M. (1998). *Dynamic Simulation of Electical Machinery using Matlab/Simulink*. Prentice Hall.

Osman, A., & Malik, O. (2004). Protection of parallel transmission lines using wavelet transform. *IEEE Transactions on Power Delivery*, *19*(1), 49–55. doi:10.1109/TPWRD.2003.820419

Ouadi, A., Bentarzi, H., & Maun, J. C. (2009). A new computer based phasor measurement unit framework. *Proc. 6th Int. Conf. SSD'09*, 1-6.

Ouadi, A., Bentarzi, H., Chafai, M., & Zitouni, A. (2011). Microcontroller Based Protective Relay Testing System. In *Proc. The 10th WSEAS International Conference on Circuits, Systems, Electronics, Control & Signal Processing (CSECS '11)*. Montreux University.

Ouadi, A., Bentarzi, H., & Chafai, M. (2016). A new PMU based power swing detector to prevent mal-operation of distance relay. *Russian Electrical Engineering*, *87*(10), 572–578. doi:10.3103/S1068371216100059

Ouadi, A., Bentarzi, H., & Maun, J. C. (2009). A new computer based phasor measurement unit framework. *Proc. 6th Int. Conf. SSD'09*, 1-6. 10.1109/SSD.2009.4956773

Ouadi, A., Bentarzi, H., & Maun, J. C. (2010). Improvement of Phasor Measurement Unit Performance. *Proc. 9th WSEAS International Conference on CIRCUITS, SYSTEMS, ELECTRONICS, CONTROL & SIGNAL PROCESSING (CSECS '10)*, 40-44.

Ouadi, A., Bentarzi, H., & Recioui, A. (2013). Optimal multiobjective design of digital filters using spiral optimization technique. *SpringerPlus*, *2*(1), 461. doi:10.1186/2193-1801-2-461 PMID:24083108

Ouadi, A., Bentarzi, H., & Recioui, A. (2014). Optimal Multi-Objective Design of Digital Filters Using Taguchi Optimization Technique. *Journal of Electrical Engineering*, *65*(1), 21–29. doi:10.2478/jee-2014-0003

Ouadi, A., Bentarzi, H., & Zitouni, A. (2018). Spiral Search Optimization Algorithm Applied to (IIR) Digital Filter Design. *International Conference On Engineering Technologies*.

Ould Amrouche, S., Rekioua, D., Rekioua, T., & Bacha, S. (2016). Overview of energy storage in renewable energy systems. *International Journal of Hydrogen Energy*, *41*(45), 20914–20927. doi:10.1016/j.ijhydene.2016.06.243

Ozturk, Y., Senthilkumar, D., Kumar, S., & Lee, G. (2013). An intelligent home energy management system to improve demand response. *IEEE Transactions on Smart Grid*, *4*(2), 694–701. doi:10.1109/TSG.2012.2235088

Pai, F. S., & Huang, S. J. (2001). A detection algorithm for islanding prevention of dispersed consumer owned storage and generation units. *IEEE Transactions on Energy Conversion*, *16*(4), 346–351. doi:10.1109/60.969474

Palaniappan, S. (2015). Home Automation Systems - A Study. *International Journal of Computer Applications*, *116*(11).

Palensky, P., & Dietrich, D. (2011). Demand Side Management: Demand Response, Intelligent Energy Systems, and Smart Loads. *IEEE Transactions on Industrial Informatics*, *7*(3), 381–388. doi:10.1109/TII.2011.2158841

Pan, Z., Sun, H., & Guo, Q. (2013). TOU-based optimal energy management for smart home. IEEE PES ISGT Europe, 1–5.

Park, N., & Kim, M. (2014). Implementation of load management application system using smart grid privacy policy in energy management service environment. *Cluster Computing*, *17*(3), 653–664. doi:10.100710586-014-0367-y

Parsons, J. D. (2000). Mobile |Radio Propagation Chanel (2nd ed.). John Wiley & Sons, Inc.

Paruchuri, V., Durresi, A., & Ramesh, M. (2008). Securing powerline communications. *Proc. IEEE Int. Symp. Power Line Commun. Appl., (ISPLC)*, 64–69.

Patil, G. C., & Thosar, A. G. (2017). Application of synchrophasor measurements using PMU for modern power systems monitoring and control [Paper presentation]. *International Conference on Computation of Power, Energy, Information and Communication (ICCPEIC).*

Patrao, I., Figueres, E., Garcerá, G., & González-Medina, R. (2015). Microgrid architectures for low voltage distributed generation. *Renewable & Sustainable Energy Reviews, 43*, 415–424. doi:10.1016/j.rser.2014.11.054

Peizhong, Y., Iwayemi, A., & Zhou, C. (2011). Developing ZigBee deployment guideline under WiFi interference for smart grid applications. *IEEE Transactions on Smart Grid, 2*(1), 110–120. doi:10.1109/TSG.2010.2091655

Peretto, L. (2010). The role of measurements in the smart grid era. *IEEE Instrumentation & Measurement Magazine, 13*(3), 22–25. doi:10.1109/MIM.2010.5475163

Phadke, A. G., Thorp, J. S., & Adamiak, M. G. (1983). A New Measurement Technique for Tracking Voltage Phasors, Local System Frequency, and Rate of Change Frequency. *IEEE Transactions on Power Apparatus and Systems, 102*(5), 1025–1038. doi:10.1109/TPAS.1983.318043

Phadke, A. G. (1993). Synchronized phasor measurements in power systems. *IEEE Computer Applications in Power, 6*(2), 10–15. doi:10.1109/67.207465

Phadke, A. G., Hlibka, T., Adamiak, M. G., Ibrahim, M., & Thorp, J. S. (1981). A Microcomputer Based Ultra High Speed Distance Relay: Field Tests. *IEEE Trans. On PAS, 100*(4), 2026–2036. doi:10.1109/TPAS.1981.316480

Phadke, A. G., Hlibka, T., & Ibrahim, M. (1976). A digital computer system for EHV Substation: Analysis and Field Tests. *IEEE Transactions on Power Apparatus and Systems, PAS-95*(1), 291–301. doi:10.1109/T-PAS.1976.32105

Phadke, A. G., Hlibka, T., Ibrahim, M., & Adamiak, M. G. (1979). A Microcomputer Based Symmetrical Component Distance Relay. *Proceedings of PICA*, 47-55. 10.1109/PICA.1979.720045

Phadke, A. G., Ibrahim, M., & Hlibka, T. (1977). Fundamental basis for Distance Relaying with Symmetrical Components. *IEEE Trans. on PAS, 96*(2), 635–646. doi:10.1109/T-PAS.1977.32375

Phadke, A. G., & Thorp, J. S. (1988). *Computer Relaying for Power Systems.* John Wiley and Sons.

Phadke, A. G., & Thorp, J. S. (1988). *Computer relaying for Power Systems.* Research Studies Press Ltd. John Wiley & Sons Inc.

Phadke, A. G., & Thorp, J. S. (2006). *History And Applications Of Phasor Measurements.* IEEE PES Power Systems Conference and Exposition. doi:10.1109/PSCE.2006.296328

Phadke, A. G., & Thorp, J. S. (2008). *Synchronized Phasor Measurements and Their Applications.* doi:10.1007/978-0-387-76537-2

Phuangpornpitak, N., & Tia, S. (2013). Opportunities and Challenges of Integrating Renewable Energy in Smart Grid System. *Energy Procedia, 34*, 282–290. doi:10.1016/j.egypro.2013.06.756

Pipattanasomporn, M., Kuzlu, M., & Rahman, S. (2012). Demand response implementation in a home area network: a conceptual hardware architecture. IEEE Innovative Smart Grid Technologies (ISGT) Conference, 1–8. doi:10.1109/ISGT.2012.6175675

Planas, E., Gil-de-Muro, A., Andreu, J., Kortabarria, I., & Martínez de Alegría, I. (2013). General aspects, hierarchical controls and droop methods in microgrids: A review. *Renewable & Sustainable Energy Reviews, 17*, 147–159. doi:10.1016/j.rser.2012.09.032

Poularikas, A. D. (2010). *Trasforms and applications handbook.* CRC press. doi:10.1201/9781420066531

Poullikkas, A., Kourtis, G., & Hadjipaschalis, I. (2013). *A review of net metering mechanism for electricity renewable energy sources*. Electricity Authority of Cyprus.

Power System Relaying Committee of the IEEE PES. (2006). IEEE Standard for Sychrophasor for Power Systems. *IEEE std.1344-1995 (2001),* 1-36.

Prabhash, N., Panigrahi, C. K., & Dasgupta, A. (2015, December). Energy Management System in Smart Grid: An Overview. *International Journal of Research and Scientific Innovation, II*(XII), 31–48.

Prasad, R. (2004). *OFDM For Wireless Communications Systems*. Artech House.

Preethi, V., & Harish, G. (2016). Design and implementation of smart energy meter. *2016 International Conference on Inventive Computation Technologies (ICICT)*, 1, 1–5. 10.1109/INVENTIVE.2016.7823225

Qdr, Q. (2006). *Beneðts of demand response in electricity markets and recommendations for achieving them. United States Dept. Energy, Washington*. Tech. Rep.

Quinn, E. (2009). *Privacy and the new energy infrastructure*. Available at SSRN http://ssrn.com/abstract=1370731

Radhika, P., Subasri, V., Kanagavalli, R., & Devaraj, D. (2006). Power Quality monitoring usingwavelet transform and artificial Neural Networks. *2006 India International Conference on Power Electronics, IEEE*, 425-430.

Radosavljević, J., Jevtić, M., Arsić, N., & Klimenta, D. (2014). Optimal power flow for distribution networks using gravitational search algorithm. *Electrical Engineering, 96*(4), 335–345. doi:10.100700202-014-0302-5

Ragib, H., Rakesh, B., & Himanshu, K. (2006). Analysis NASPInet Data Flows. IEEE PES Power System Conference and Exhibition. doi:10.1109/PSCE.2009.4840100

Rahimi, F., & Ipakchi, A. (2010). Demand response as a market resource under the smart grid paradigm. *IEEE Transactions on Smart Grid, 1*(1), 82–88. doi:10.1109/TSG.2010.2045906

Rahman, M. A., & Dash, P. K. (1982). Fast Algorithm for Digital Protection of Power Transformer. *IEE Proc., 129*(C2), 79.

Rajeev, A., Shanmugarajeshwaran, R., Poonurangam, T., & Sreevalsan, E. (2016). Applications of PMU in renewable energy generation [Paper presentation]. *3rd International Conference on Electrical EnergySystems (ICEES).*

Raju, L., Milton, R. S., & Mahadevan, S. (2016). Multi Agent Systems based Distributed Control and Automation of Micro-grid using MACSimJX. *2016 10th International Conference on Intelligent Systems and Control (ISCO),* 7-8.

Raju, L., Milton, R. S., & Mahadevan, S. (2016). Multiagent Systems Based Modeling and Implementation of Dynamic Energy Management of Smart Microgrid Using MACSimJX. Hindawi Publishing Corporation. doi:10.1155/2016/9858101

Raju, L., & Morais, A. A. (2020). *Multi-Agent Systems Based Advanced Energy Management of Smart Micro-grid. In Multi Agent Systems - Strategies and Applications*. IntechOpen.

Ram, G. T., Laxmi, A. J., & Sushama, M. (2010). Detection of power quality disturbances using wavelet transforms. *International Journal of the Computer, the Internet and Management, 18*(1).

Rao, K.U., Jade, S., & Sridhar, S. (2016). Detection and classification of power quality disturbances inthe supply to induction motor using wavelet transform and neural networks. *Balkan Journal of Electrical and Computer Engineering, 4*(1).

Rao, J. G., & Pradhan, A. K. (2012). Differential Power-Based Symmetrical Fault Detection During Power Swing. *IEEE Transactions on Power Delivery, 27*(3), 1557–1564. doi:10.1109/TPWRD.2012.2196527

Rao, R. V., & Patel, V. K. (2012). Multi-objective optimization of combined Brayton and inverse Brayton cycles using advanced optimization algorithms. *Engineering Optimization*. Advance online publication. doi:10.1080/0305215X.2011.624183

Rao, R. V., Savsani, V. J., & Vakharia, D. P. (2011). Teaching-learning-based optimization: A novel method for constrained mechanical design optimization problems. *Computer Aided Design*, *43*(3), 303–315. doi:10.1016/j.cad.2010.12.015

Rao, R. V., Savsani, V. J., & Vakharia, D. P. (2012). Teaching-learning-based optimization: An optimization method for continuous non-linear large scale problems. *Inform Sci*, *183*(1), 1–15. doi:10.1016/j.ins.2011.08.006

Rao, V. J., Savsani, J. B., & Balic, J. (2012). Teaching learning based optimization algorithm for constrained and unconstrained real parameter optimization problems. *Engineering Optimization*, *44*(12), 1447–1462. Advance online publication. doi:10.1080/0305215X.2011.652103

Rasheed, M. B., Javaid, N., Ahmad, A., Awais, M., Khan, Z. A., Qasim, U., & Alrajeh, N. (2016). Priority and delay constrained demand side management in real-time price environment with renewable energy source. *International Journal of Energy Research*, *40*(14), 2002–2021. doi:10.1002/er.3588

Rasheed, M. B., Javaid, N., Ahmad, A., Khan, Z. A., Qasim, U., & Alrajeh, N. (2015). An efðcient power scheduling scheme for residential load management in smart homes. *Applied Sciences (Basel, Switzerland)*, *5*(4), 1134–1163. doi:10.3390/app5041134

Ray, P. K., Kishor, N., & Mohanty, S. R. (2012). Islanding and power quality disturbance detection in grid-connected hybrid power system using wavelet and S-transform. *IEEE Transactions on Smart Grid*, *3*(3), 1082–1094. doi:10.1109/TSG.2012.2197642

Recioui, A., Bentarzi, H., & Ouadi, A. (2015). Application of a binary teaching learning-based algorithm to the optimal placement of phasor measurement units. *Progress in Clean Energy*, *1*, 817–830.

Recioui, A., Bentarzi, H., & Tsebia, M. (2016). PMU Deployment in Power System Oscillation Monitoring. *Sustainable Aviation*, 312-322.

Recioui, A., Bentarzi, H., & Tsebia, M. (2016). PMU deployment in power system oscillation monitoring. *Sustainable Aviation*, 313-322.

Recioui, A. (2018). *Application of Teaching Learning-Based Optimization to the Optimal Placement of Phasor Measurement Units. In Handbook of Research on Emergent Applications of Optimization Algorithms*. IGI Global.

Recioui, A. (2018). *Optimal Placement of Power Factor Correction Capacitors Using Taguchi Optimization Method. In Handbook of Research on Emergent Applications of Optimization Algorithms*. IGI Global.

Recioui, A., Bentarzi, H., & Ouadi, A. (2015). Application of a binary teaching learning-based algorithm to the optimal placement of phasor measurement units. *Progress in Clean Energy*, *1*, 817–830. doi:10.1007/978-3-319-16709-1_60

Recioui, A., & Dassa, K. (2017, June). Design of Standalone Micro-Grid Systems Using Teaching Learning Based Optimization. *Algerian Journal of Signals and Systems*, *2*(2), 75–85.

Recioui, A., Tazibt, M., & Benhenni, H. (2017, September). Fault Detection in Wide Area Monitoring Systems via PMU Optimal Placement. *Algerian Journal of Signals and Systems*, *2*(3), 149–161.

Redfern, M. A., Usta, O., & Fielding, G. (1993). Protection against loss of utility grid supply for a dispersed storage and generation unit. *IEEE Transactions on Power Delivery*, *8*(3), 948–954. doi:10.1109/61.252622

Reis, M. V. G., Barros, T. A. S., Moreira, A. B., Nascimento, F. E. R., & Villalava, M. G. (2015). Analysis of the Sandia Frequency Shift (SFS) islanding detection method with a single phase photovoltaic distributed generation system. *Proceedings of IEEE PES Innovative Smart Grid Technologies Latin America (ISGT LATAM)*, 10.1109/ISGT-LA.2015.7381140

Renner, S., & … . (2011). European smart metering landscape report. *Imprint*, *2*, 1–168.

Reston Filho, J., Affonso, C., & Oliveira, R. (2009). Pricing analysis in the Brazilian energy market: A decision tree approach. 2009 IEEE Bucharest Power Tech, 1–6.

Robinson, R. E., & Stuber, M. G. (n.d.). *Itron white paper: advanced metering infrastructure risk analysis for advanced metering*. www.itron.com/asset.asp?path=support/whitepaper/pdf/itr_ 016898.pdf

Rohit, A. K., Devi, K. P., & Rangnekar, S. (2017). An overview of energy storage and its importance in Indian renewable energy sector: Part I – Technologies and Comparison. *Journal of Energy Storage*, *13*, 10–23. doi:10.1016/j.est.2017.06.005

Romero-Troncoso,, R.J., Osornio-Rios,, R.A., Garcia-Perez,, A., & Cabal-Yepez,, E., & Granados-Lieberman, D. (2011). Techniques and methodologies for power quality analysis and disturbances classification in power systems: A review. *IET Generation, Transmission & Distribution*, *5*(4).

Roop, M. E., Bagovic, M., Rohtgi, A., Kern, G. A., Bonn, R. H., & Gonzalez, S. (2000). Determining the relative effectiveness of islanding detection methods using phase criteria and nondection zones. *IEEE Transactions on Energy Conversion*, *15*(3), 290–296. doi:10.1109/60.875495

Ropp, M. E., Begovic, M., & Rohatgi, A. (1999). Analysis and performance assessment of the active frequency drift method of islanding prevention. *IEEE Transactions on Energy Conversion*, *14*(3), 810–816. doi:10.1109/60.790956

Roy Choudhury, D., & Shail, B. J. (2010). Linear Integrated Circuits. New Age International Publishers.

Russell, S., & Norvig, P. (2009). *Artificial intelligence: A modern approach. Publisher: Pearson*. Prentice Hall Pa.

Sachdev, M. S., & Baribeau, M. A. (1979). A New Algorithm for Digital Impedance Relay. *IEEE Transactions on Power Apparatus and Systems*, *98*(6), 2232–2240. doi:10.1109/TPAS.1979.319422

Sachdev, M. S., & Giray, M. M. (1985). A least error squares technique for determining power system frequency. *IEEE Trans. on PAS*, *104*(2), 437–443. doi:10.1109/TPAS.1985.319059

Sadeghi, Naghshbandy, & Bahramara. (2020). Optimal sizing of hybrid renewable energy systems in presence of electric vehicles using multi-objective particle swarm optimization. *Energy*, *209*, 118471. https://doi.org/https://doi.org/10.1016/j.energy.2020.118471

Sai, V. O., Shieh, C. S., Lin, Y. C., Horng, M. F., Nguyen, T. T., Le, Q. D., & Jiang, J. Y. (2017). Comparative Study on Recent Development of Heuristic Optimization Methods. *Proceedings - 2016 3rd International Conference on Computing Measurement Control and Sensor Network, CMCSN 2016*, 68–71. https://doi.org/10.1109/CMCSN.2016.29

Saikong, W., & Kulworawanichpong, T. (2017). Comparative Study of Energy Consumption for Electric Vehicles with Various On-board Energy Storage Systems. *Energy Procedia*, *138*, 81–86. doi:10.1016/j.egypro.2017.10.060

Samanta, C. K., & Mishra, D. P. (2013). Power Quality Monitoring by Disturbance Detection using Hilbert Phase Shifting. *Int. Conf. on Recent Trends in Power, Control and Instrumentation Engineering*.

Samantaray, S. R., Babu, B. C., & Dash, P. K. (2011). Probabilistic neural network based islanding detection in distributed generation. *Electric Power Components and Systems*, *39*(3), 191–203. doi:10.1080/15325008.2010.526986

Samantaray, S. R., Samui, A., & Chitti Babu, B. (2011). Time-frequency transform-based islanding detection in distributed generation. *IET Renewable Power Generation*, *5*(6), 431–438. doi:10.1049/iet-rpg.2010.0166

Samerchur, S., Premrudeepreechacharn, S., Kumsuwun, Y., & Higuchi, K. (2011). Power control of single-phase voltage source inverter for grid-connected photovoltaic systems. *2011 IEEE/PES Power Systems Conference and Exposition*, 1–6. https://doi.org/10.1109/PSCE.2011.5772504

Sankaran, C. (2001). *Power quality*. CRC Press.

Saoud, A., & Recioui, A. (2017). A review on Data communication in smart grids. *Algerian Journal of Signals and Systems*, *2*(3), 162–179.

Saoud, A., & Recioui, A. (2017, September). A review on Data communication in smart grids. *Algerian Journal of Signals and Systems*, *2*(3), 162–179.

Sarma, D. S., & Venkatesh, C. (2010). Techniques used for Detection of Power Quality Events-a Comparative Study. *16th National Power Systems Conference.*

Sazli, M. H. (2006). A brief review of feed-forward neural networks. University of Ankara. doi:10.1501/0003168

Segovia, R., & Sánchez, M. (2011). *Set of common functional requirements of the Smart Meter*. Jt. Contrib. DG ENER DG INFSO.

Shadkam, E., & Bijari, M. (2014). Evaluation the Efficiency of Cuckoo Optimization Algorithm. *International Journal on Computational Science & Applications*, *4*(2), 39–47. doi:10.5121/ijcsa.2014.4205

Shahid, M. (2005). *(MIMO) Systems with Multicarrier Time Delay Diversity Modulation*. Naval Postgraduate School.

Shahraeini, M., & Javidi, M. H. (2011). *A Survey on Topological Observability of Power Systems*. IEEE Press. doi:10.1109/PEAM.2011.6135115

Sharma, K., & Saini, L. M. (2017). Performance analysis of smart metering for smart grid: An overview. *Renewable & Sustainable Energy Reviews*, *49*, 720–735. doi:10.1016/j.rser.2015.04.170

Sharmeela, C., Rajeshwari, V., Keerthana, B., & Murugesan, P. (2017). Localizing Power Quality Disturbances Using Hilbert Transform In labview. *National Conference On Innovations And Research In Electrical & Electronics Engineering*.

Shehory, O. (1999). *Architectural properties of multi-agent systems*. Technical Report CMU-RI-TR-98-28, The Robotics Institute, Carnegie Mellon University.

Shim, J. W., Cho, Y., Kim, S. J., Min, S. W., & Hur, K. (2013). Synergistic control of SMES and battery energy storage for enabling dispatchability of renewable energy sources. *IEEE Transactions on Applied Superconductivity*, *23*(3). Advance online publication. doi:10.1109/TASC.2013.2241385

Shirazi, E., & Jadida, S. (2019). A multiagent design for self-healing in electric power distribution systems. *Electric Power Systems Research*, *171*, 230–239. doi:10.1016/j.epsr.2019.02.025

Silva, D. (n.d.). *New 'smart' electrical meters raise privacy issues*. Available: http://www.physorg.com/news176703307.html

Singam, B., & Hui, L. Y. (2006). Assessing SMS and PJD schemes of anti islanding with varying quality factor. *Proceedings of the IEEE International Power and Energy Conference*.10.1109/PECON.2006.346645

Singh, P., & Lather, J. S. (2020). Accurate power-sharing, voltage regulation, and SOC regulation for LVDC microgrid with hybrid energy storage system using artificial neural network. *International Journal of Green Energy*, *17*(12), 756–769. Advance online publication. doi:10.1080/15435075.2020.1798767

Skvarenina, T. L. (2018). *The power electronics Handbook*. CRC Press. doi:10.1201/9781420037067

Smart Metering Implementation Programme: Statement of Design Requirements, Ofgem E-Serve. (2010). http://www.decc.gov.uk/assets/decc/consultations/smartmeter-imp-prospectus/225-smart-metering-imp-programme-design.pdf

Sodhi, R., Srivastava, S. C., & Singh, S. N. (2009). *Optimal PMU placement to ensure system observability under contingencies. In Power & Energy Society General Meeting, 2009 (PES '09)*. IEEE.

Sonalgaz Company internal report (2013), Algiers, Algeria

Song, P., Lei, Y., & Fu, Y. (2020). Multi-objective optimization and matching of power source for PhEV based on genetic algorithm. *Energies*, *13*(5), 1127. Advance online publication. doi:10.3390/en13051127

Sonnemann, W. K., Wagner, C. L., & Rockefeller, G. D. (1958). Magnetizing Inrush Phenomena in Transformer Banks. *Transactions of the American Institute of Electrical Engineers*, *77*, 884–892.

Sonnenenergie, D. G. (2007). Planning and Installing. Infor.

Souran, D. M., Safa, H. H., Moghadam, B. G., Ghasempour, M., & Heravi, P. T. (2016). Smart Grid Technology in Power Systems. *Soft. Computer Applications (Nottingham)*, *357*, 1367–1381.

Stevens, J., Bonn, R., Ginn, J., Gonzalez, S., & Kern, G. (2000). *Development and testing of an approach to anti-islanding in utility interconnected photovoltaic systems*. Sandia National Laboratories. doi:10.2172/759506

Stouffer, K., Falco, J., Scarfone, K., Stouffer, K., Falco, J., & Scarfone, K. (2006). Guide to supervisory control and data acquisition (scada) and industrial control systems security. *SPIN*.

Su, B., Dong, X. Z., Bo, Z. Q., Sun, Y. Z., Caunce, B. R., Tholomier, D. J., & Apostolov, A. (2005). Fast Detector of Symmetrical Fault During Power Swing for Distance Relay. *IEEE Power Eng. Soc. Gen. Meet.*, *2*, 1836 – 1841.

Sullivan, J. L., & John, W. A. (1998). PCLS IIR digital filters with simultaneous frequency response magnitude and group delay specifications. *IEEE Transactions on Signal Processing*, *46*(11), 2853–2861. doi:10.1109/78.726800

Suresh, M., & Meenakumari, R. (2019). An improved genetic algorithm-based optimal sizing of solar photovoltaic/wind turbine generator/diesel generator/battery connected hybrid energy systems for standalone applications. *International Journal of Ambient Energy*, *0*(0), 1–8. doi:10.1080/01430750.2019.1587720

Syal, M. M., & Ofei-Amoh, K. (2013). Smart-grid technologies in housing. *Cityscape (Washington, D.C.)*, 283–288.

Takagi, T., Yamakoshi, Y., Yamaura, M., Kondow, R., & Matsushima, T. (1982). Development of a new type fault locator using the one-terminal voltage and current data. *IEEE Transactions on Power Apparatus and Systems*, *8*(8), 2892–2898. doi:10.1109/TPAS.1982.317615

Tamura, K., & Yasuda, K. (2011). Primary study of spiral dynamics inspired optimization. *IEEE Transactions on Electrical and Electronic Engineering*, *6*(S1), 98–100.

Tang & Man. (1998). Design and Optimization of IIR Filter Structure Using Hierarchical Genetic Algorithms. *IEEE Transactions on Industrial Electronics*, *45*(3).

Tankasala, G. R., Sanisetty, S., & Vala, V. K. (2012). Optimal Placement of Phasor Measurement Units for State Estimation using Artificial Intelligence Techniques. *International Journal of Scientific and Engineering Research*, *3*(2).

Tan, S., Debraj, D., Wen-Zhan, S., Yang, J., & Das Sajal, K. (2016). *Survey of Security Advances in Smart Grid: A Data Driven Approach*. IEEE COMST.

Tanveer, A., Hongcai, Z., & Biao, Y. (2020). A review on renewable energy and electricity requirement forecasting models for smart grid and buildings. *Sustainable Cities and Society, 55*.

Tan, X., Li, Q., & Wang, H. (2013). Advances and trends of energy storage technology in Microgrid. *International Journal of Electrical Power & Energy Systems*, *44*(1), 179–191. doi:10.1016/j.ijepes.2012.07.015

Tariq, F., & Dooley, L. S. (2013). Smart Grid Communication and Networking Technologies: Recent Developments and Future Challenges. In Smart Grids (pp. 199–213). Springer London.

Tarroja, B., Zhang, L., Wifvat, V., Shaffer, B., & Samuelsen, S. (2016). Assessing the stationary energy storage equivalency of vehicle-to-grid charging battery electric vehicles. *Energy*, *106*, 673–690. doi:10.1016/j.energy.2016.03.094

Taskovski, D., & Markovska, M. (2018). The effectiveness of wavelet based features on power quality disturbances classification in noisy environment. *18th International Conference on Harmonics and Quality of Power (ICHQP)*, 1-6.

Texas Instruments. (2015). *LMx24-N, LM2902-N Low-Power, Quad-Operational Amplifiers*. Texas Instruments Press Relay testing and commissioning. Available: http://www.fecime.org/referencias/npag/chap21-370-397.pdf

Theodorakatos, N. P., Manousakis, N. M., & Korres, G. N. (2015). Optimal Placement of Phasor Measurement Units with Linear and Non-linear Models. *Electric Power Components and Systems*, *43*(4), 357–373. doi:10.1080/15325008.2014.981319

Thiaw, L., Sow, G., & Fall, S. (2014). Application of Neural Networks Technique in Renewable Energy Systems. *2014 First International Conference on Systems Informatics, Modelling and Simulation Application*, 6–12. https://doi.org/10.1109/SIMS.2014.12

Thorp, J. S., & Phadke, A. G. (1982). A Microprocessor Based Three-Phase Transformer Differential Relay. *IEEE Trans.*, *PAS-101*, 426.

Timothy, M. H., & Edwin, K. P. C. (2018). A Partially Observable Markov Decision Process Approach to Residential Home Energy Management. *IEEE Transactions on Smart Grid, Volume*, *9*(2), 1271–1281. doi:10.1109/TSG.2016.2582701

Torrent-Fontbona, F., & López, B. (2015). Power re-allocation for reducing contracted electric power costs. *Energy and Building*, *89*, 112–122. doi:10.1016/j.enbuild.2014.12.030

TrongNghia, L., Wen-Long, C., Dang Khoa, T., & Tran Hiep, N. (2016). Advanced Metering Infrastructure Based on Smart Meters in Smart Grid. *Intechopen, 2016*. Advance online publication. doi:10.5772/63631

TrongNghia, L., WenLong, C., DangKhoa, T., & Tran, H. N. (2016). *Advanced Metering Infrastructure Based on Smart Meters in Smart Grid*. Intech Open.

Trujillo, C. L., Velasco, D., Figueres, E., & Garcera, G. (2010). Analysis of active islanding detection methods for grid connected microconverter for renewable energy processing. *Applied Energy*, *87*(11), 3591–3605. doi:10.1016/j.apenergy.2010.05.014

Tsikalakis, A. G., Member, S., Hatziargyriou, N. D., & Member, S. (2011). Centralized Control for Optimizing Microgrids Operation. *IEEE Transfer Energy Converters*, *23*(1), 1–8. doi:10.1109/PES.2011.6039737

Tziouvaras, D. (2007). *Relay performance during major system disturbances*. Schweitzer Engineering Laboratories, Inc. doi:10.1109/CPRE.2007.359905

Ucar, A., Demir, Y., & Eristi, H. (2010). Wavelet-based feature extraction and selection for classification of power system disturbances using support vector machines. Electric Power Systems Research, 80(7), 743-752.

Ur Rahman, Ali, Ullah, Ur Rahman, Iqbal, Ahmad, Zeb, Ali, Shahzad, & Beenish. (2012). Advancement in Wide Area Monitoring Protection and Control Using PMU's Model in MATLAB/SIMULINK Taj. *Smart Grid and Renewable Energy*, (3), 294–307.

Valenti, M. C., & Feliachi, A. (2002). *Communication Delays in Wide Area Measurement Systems.* IEEE.

Vapnic, V. (1998). The support vector method of function estimation. In *Nonlinear Modeling* (pp. 55–85). Springer. doi:10.1007/978-1-4615-5703-6_3

Vera Yimy, E. G., Dufo-López, R., & Bernal-Agustín, J. L. (2019). Energy Management in Microgrids with Renewable Energy Sources: A Literature Review. *Applied Sciences (Basel, Switzerland)*, *9*(3854), 1–28.

Vivekanandan, V., & Wong, V. W. S. (2006). Ordinal MDS-based localisation for wireless sensor networks. *International Journal of Sensor Networks*, *1*(3/4), 169–178. doi:10.1504/IJSNET.2006.012032

Vlot, A., & Gunnink, J. W. (2011). *Fibre metal laminates: an introduction.* Springer Science & Business Media.

Von, A. J., Braun, M., Stetz, T., Diwold, K., & Geibel, D. (2013). Time in the sun: The challenge of high PV penetration in the German electric grid. *IEEE Power & Energy Magazine*, *11*(2), 55–64. doi:10.1109/MPE.2012.2234407

Vytelingum, P., Voice, T. D., Ramchurn, S. D., Rogers, A., & Jennings, N. R. (2010). Agent-based micro-storage management for the smart grid. In *Proceedings of the 9th International Conference on Autonomous Agents and Multiagent Systems.* International Foundation for Autonomous Agents and Multiagent Systems.

Waghmode, L. Y., & Patil, R. B. (2013). An Overview of Fault Tree Analysis (FTA) Method for Reliability Analysis. *Journal of Engineering Research and Studies*, *4*(1), 6–8.

Wang, R., Ziad, M., & Lee, Y. (2001). Data Quality. The Kluwer International Series on Advances in Database Systems. Springer.

Wang, S., Gao, W., Wang, J., & Lin, J. (2012). Synchronized Sampling Technology-Based Compensation for Network Effects in WAMS Communication. *IEEE Transactions on Smart Grid*, *3*(2), 837–845. doi:10.1109/TSG.2012.2183902

Wang, Y., Chen, Q., Kang, C., Zhang, M., Wang, K., & Zhao, Y. (2015). Load Profiling and Its Application to Demand Response: A Review. *Science and Technology, 20*(2), 117-129.

Wang, B., Xu, J., Cao, B., & Ning, B. (2017). Adaptive mode switch strategy based on simulated annealing optimization of a multi-mode hybrid energy storage system for electric vehicles. *Applied Energy*, *194*, 596–608. doi:10.1016/j.apenergy.2016.05.030

Wang, G., Gao, F., Liu, J., Li, Q., & Zhao, Y. (2020). Design consideration and performance analysis of hybrid islanding detection method combining voltage unbalance/ toal harmonic distortion and bilateral reactive power variation. *CPSS Transactions on Power Electronics and Applications*, *5*(1), 86–100. doi:10.24295/CPSSTPEA.2020.00008

Wang, W., Xu, Y., & Khanna, M. (2011). A survey on the communication architectures in smart grid. *Computer Networks*, *55*(15), 3604–3629. doi:10.1016/j.comnet.2011.07.010

Webster, J. G. (1998). *The measurement, instrumentation and sensors handbook.* CRC press.

Webster, J. G. (1998). *The Measurement, Instrumentation and Sensors Handbook.* CRC Press.

Weixiao, M., Ruofei, M., & Hsiao-Hwa, C. (2014). Smart Grid Neighborhood Area Networks: A Survey. *IEEE Network*, *28*(1), 24–32. doi:10.1109/MNET.2014.6724103

Weranga, K. S. K., Kumarawadu, S., & Chandima, D. P. (2014). *Smart Metering Design and Applications.* Springer Singapore. doi:10.1007/978-981-4451-82-6

Widraw, B. (1975). Adaptive noise cancelling; principles and applications. Proc. IEEE, 1692.

Wietfeld, C., Georg, H., Groening, S., Lewandowski, C., Mueller, C., & Schmutzler, J. (2011). Wireless M2M communication networks for smart grid applications. Wireless Conference 2011 – Sustainable Wireless Technologies (European Wireless), 1–7.

Wood, G., & Newborough, M. (2003). Dynamic energy-consumption indicators for domestic appliances: Environment, behaviour and design. *Energy and Building*, *35*(8), 821–841. doi:10.1016/S0378-7788(02)00241-4

Wu, C. H., Lin, C. H., Cheng, F. S., & Lin, W. M. (2008). Detection and classification of multiple power- quality disturbances with wavelet multiclass SVM. *IEEE Transactions on Power Delivery*, *23*(4), 2575–2582. doi:10.1109/TPWRD.2008.923463

Wu, H., Tsakalis, K., & Heydt, G. (2004). Evaluation of time delay effects to wide-area power system stabilizer design. Power Systems. *IEEE Transaction*, *19*(4), 1935–1940. doi:10.1109/TPWRS.2004.836272

Wysocki, L. C. (2006). Complex Orthogonal Space Time Processing in Wireless Communications. Academic Press.

Xu, B. (2006). *Optimal Monitoring And Visualization Of Steady State Power System Operation* (Ph.D Dissertation). Texas A&M University.

Xu, B., & Abur, A., (2005). *Optimal placement of phasor measurement units for state estimation.* Final Project Report PSERC.

Xu, B., & Abur, A. (2004). Observability analysis and measurement placement for systems with PMUs. *IEEE PES Power Syst. Conf. Expo.*, *2*, 943–946.

Xu, W., Zhang, G., Li, C., Wang, W., Wang, G., & Kliber, J. (2007). A power line signaling based technique for anti-islanding protection of distributed generators part I scheme & analysis. *IEEE Transactions on Power Delivery*, *22*(3), 1758–1766. doi:10.1109/TPWRD.2007.899618

Yalcintas, M., Hagen, W. T., & Kaya, A. (2015). An analysis of load reduction and load shifting techniques in commercial and industrial buildings under dynamic electricity pricing schedules. *Energy and Building*, *88*, 15–24. doi:10.1016/j.enbuild.2014.11.069

Yalla, M. V. V. S. (1992). A digital multifunction protective relays. *IEEE Transactions on Power Delivery*, *7*(1), 193–201. doi:10.1109/61.108907

Yan, Y., Qian, Y., Sharif, H., & Tipper, D. (2013). A Survey on Smart Grid Communication Infrastructures: Motivations, Requirements and Challenges. IEEE Communications Surveys & Tutorials, 15(1), 5 – 20.

Yang, Y., Xu, T., Cao, Y., Wang, X., & Chen, C. L. (2009). Power quality disturbances detection based on hilbert phase-shifting. 2009 Asia-Pacific Power and Energy Engineering Conference, 1-4.

Yang, H., Zhou, W., Lu, L., & Fang, Z. (2008). Optimal sizing method for stand-alone hybrid solar-wind system with LPSP technology by using genetic algorithm. *Solar Energy*, *82*(4), 354–367. doi:10.1016/j.solener.2007.08.005

Yan, Y., Qian, Y., Sharif, H., & Tipper, D. (2013). A Survey on Smart Grid Communication Infrastructures: Motivations, Requirements and Challenges. *IEEE Communications Surveys and Tutorials*, *15*(1), 5–20. doi:10.1109/SURV.2012.021312.00034

Yao, W., Jiang, L., Wu, Q., Wen, J., & Cheng, S. J. (2010). Design of wide-area damping controllers based on networked predictive control considering communication delays. *IEEE PES General Meeting*, 1–8. doi: 10.1109/PES.2010.5589556

Yao, L., Yang, B., Cui, H., Zhuang, J., Ye, J., & Xue, J. (2016). Challenges and progresses of energy storage technology and its application in power systems. *Journal of Modern Power Systems and Clean Energy*, *4*(4), 519–528. doi:10.100740565-016-0248-x

Yarali, A. (2008). Wireless mesh networking technology for commercial and industrial customers. Proc. Elect. Comput. Eng., CCECE, 47–52.

Ye, Z., Kolwalkar, A., Zhang, Y., Du, P., & Walling, R. (2004). Evaluation of anti-islanding schemes based on nondetection zone concept. *IEEE Transactions on Power Electronics*, *19*(5), 1171–1176. doi:10.1109/TPEL.2004.833436

Yin, H., & Fan, L. (2010). PMU data-based fault location techniques. In *Int North American Power Symposium (NAPS)*. IEEE Press.

Young, S., & Stanic, R. (2009). *Smart Meter to HAN Communications.* Smart Grid Australia Intelligent Networking Working Group. http://smartgridaustralia.com.au/SGA/Documents/IN_Work_Group_SmartMeter_HAN_Comms.pdf

Yu, C.-S. (2006). A Discrete Fourier Transform-based Adaptive MIMIC phasor Estimator for Distance Relaying Applications. *IEEE Transactions on Power Delivery*, *21*(4), 1836–1846. doi:10.1109/TPWRD.2006.874609

Yu, S. L., & Gu, J. C. (2001). Removal of Decaying DC in Current and Voltage Using a Modified Fourier Filter Algorithm. *IEEE Transactions on Power Delivery*, *15*(4).

Zachar, M., & Daoutidis, P. (2018). Energy management and load shaping for commercial microgrids coupled with flexible building environment control. *Journal of Energy Storage*, *16*, 61–75. doi:10.1016/j.est.2017.12.017

Zafar, A., Shafique, A., Nazir, Z., & Zia, M. F. (2018). A Comparison of Optimization Techniques for Energy Scheduling of Hybrid Power Generation System. *Proceedings of the 21st International Multi Topic Conference, INMIC 2018*, (June), 1–6. https://doi.org/10.1109/INMIC.2018.8595665

Zhang, X. (2008). Design of maximally flat IIR filters with flat group delay responses. *Signal Processing*, *88*(7), 1792–1800. doi:10.1016/j.sigpro.2008.01.016

Zheng, J., Gao, D. W., & Lin, L. (2013). Smart Meters in Smart Grid: An Overview. Academic Press.

Zhou, B., Li, W., Chan, K. W., Cao, Y., Kuang, Y., Liu, X., & Wang, X. (2016). Smart home energy management systems: Concept, configurations, and scheduling strategies. *Renewable & Sustainable Energy Reviews*, *61*, 30–40. doi:10.1016/j.rser.2016.03.047

Zhu, K., Al-Hammouri, A., & Nordström, L. (2012). To concentrate or not to concentrate, Performance Analysis of ICT system with Data Concentrations for Wide-area Monitoring and Control Systems. *IEEE PES General Meeting*. doi:10.1109/PESGM.2012.6344977

Zhu, K., Chenine, M., Nordström, L., Holmström, S., & Ericsson, G. (2013). An Empirical Study of Synchrophasor Communication Delay in a Utility TCP/IP Network. *International Journal of Emerging Electric Power Systems*, *14*(4), 341–350. doi:10.1515/ijeeps-2013-0062

Zhu, K., Chenine, M., Nordström, L., Holmström, S., & Ericsson, G. (2013). Design Requirements for Wide-Area Damping Systems - Using Empirical Data from a Utility IP Network. *IEEE Transactions on Smart Grid Revision.*

Zin, A. M., Mokhtar, A. S., Ismail, N. M., Zaren, N., & Khokhar, S. (2013). *Automatic classification of power quality disturbances: A review. In 2013 IEEE Student Conference Research and Developemen*. IEEE.

Zitouni A., Ouadi A., Bentarzi H., & Chafai M. (2018). Distance relay reliability enhancement using false trip root cause analysis. *International Journal of System Assurance Engineering and Management*, 1-9.

Zitouni, A., Bentarzi, H., & Recioui, A. (2019). BTLBO Based SPMUs Placement Optimization for Fault Observation in Power Grid. *Algerian Journal of Signals and Systems*, *4*(1).

About the Contributors

Abdelmadjid Recioui is a Professor at the Institute of Electrical Engineering and Electronics University of Boumerdes, Algeria. He obtained a PhD degree in electrical and electronic engineering option telecommunications from the Institute of Electrical Engineering and Electronics, University of Boumerdes in 2011. He holds also Master (Magister) Electronic System Engineering degree which has been achieved at the Institute of Electrical Engineering and Electronics, University of Boumerdes in 2006. In June 2002, he finished his engineering studies at the institute of Electrical Engineering and Electronics, University of Boumerdes. He is a research Assistant at the laboratory signals and systems from January 2008 to Present in Laboratory: signals and systems, Inst. of Electrical Engineering and electronics, University of Boumerdes. His research interests include: Antennas, Wireless Communication Systems, antenna array synthesis and design, capacity enhancement, system optimization, smart antennas, power system protection, power system optimization, power system communications.

* * *

Hadjira Belaidi is an assistant professor of Electronics at Institute of Electrical and Electronic Engineering since 2014. She holds an engineer degree in electrical and electronic engineering (2006) from the Institute of Electrical and Electronic Engineering, University M'hamed Bougara of Boumerdès and a magister degree in electrical drives and control (2009) from the same institute. Then, she received her Doctorate degree in electronics (2015) from Institute of Electrical and Electronic Engineering, University M'hamed Bougara of Boumerdès. Currently, she occupies the position in the sub-director in charge of post-graduation, scientific research and external relations of the Institute of Electrical and Electronic Engineering, University M'hamed Bougara of Boumerdès since 2018. Her research interests include and not limited to all-terrain mobile robots, environment modeling and manipulator robots. She served as reviewer in several journals and a scientific member and chair man in several international conferences.

Hamid Bentarzi was born in Legata, Boumerdes, Algeria. He received both Electrical Engineering and Magister Degrees with honors from "Institut National d'Electricite et d'Electronique" (INELEC), Boumerdes, Algeria, in 1989 and 1992 respectively and Ph.D in Microelectronic systems from "Ecole Nationale Polytechnique" (ENP), Algiers, Algeria, in 2004. Hamid was a lecturer at INELEC, Boumerdes, Algeria, since 1991. After 1999,he has been a faculty member at the Department of Electrical and Electronic Engineering, Faculty of Engineering, University of Boumerdes, Algeria. Since 2001, he has been head of research team working in developing microelectronic systems applied to power systems in the Signal and System Laboratory, Boumerdes, Algeria. Prof. Bentarzi interests are in the fields of microelectronics, electrical protection systems and systems reliability. He has authored and co-authored over 70 technical papers. Besides, he has been a member of IEEE as well as organizing and technical committee of several conferences including WSEAS group.

Fatma Zohra Dekhandji is an associate professor at the institute of electrical and electronic engineering, University of Boumerdes, Algeria. She holds a PhD from the university of Boumerdes. Her research interests include power quality, smart grids, industrial electronics.

Maheswari M. is currently working as Lecturer, Engineering Department in Nizwa College of Technology, Sultanate of Oman. She completed her B.E. Electrical and Electronics Engineering degree in K.S. Rangasamy College of Technology Thiruchengode with first class with distinction, M.E. Power Systems in Government College of Technology, Coimbatore and with first class with distinction and Ph.D. at Anna University, Chennai in the year 2015. She has 18 years of teaching experience and served in various administrative positions. She has published her research work in 15 various international journals, 18 International Conferences, 4 book chapters and 2 patents. She is also a reviewer of leading journals.

Ahmed Maache is an Assistant Professor at the Institute of Electrical and Electronics Engineering, University of M'Hamed Bougara of Boumerdes. He received his PhD in Electrical and Electronics Engineering from the University of Southampton, UK. His current research interests include hardware acceleration using FPGAs, Real-time simulation and Hardware-In-the-Loop, Smart Grids.

Meriem Mahboub was born in Algiers, Algeria, in 1992. She received her Bachelor and master degree from the institute (IGEE), University M'hamed Bougara, Boumerdes, Algeria, in 2014 and 2016 respectively, in electrical and electronics engineering; control option. She is currently working toward the Ph.D. degree in the Department of Electrical Engineering, at IGEE, University M'hamed Bougara, Boumerdes, Algeria. Her research interests include phasor measurement unit and its applications.

Loganathan Nachimuthu was born in India, in 1970. He received B.E. electrical and electronics engineering degree in Government College of Engineering, Salem, India in 1999, M.E degree in power and energy system engineering in the University Vishveshwarya College of Engineering, Bangalore, India and Ph.D degree from the Anna University Chennai, India in 2013. Currently he is working as a Lecturer in Engineering Department electrical engineering section at Nizwa College of Technology, Nizwa, Sultanate of Oman. His research interest includes in High voltage engineering, Smart power system protection, power system Engineering.

Abderrahmane Ouadi is a Lecturer from the Institute of Electrical Engineering and Electronics University of Boumerdes, Algeria. He received his Magister from the Institut National d'Electricit et Electronique (INELEC), Boumerdes, in 1996 and PhD in Electrical Engineering from the Institute of Electrical Engineering and Electronics University of Boumerdes in 2015. He is an Associate Professor at IGEE and member of Signals and Systems Laboratory. His current research interests include power systems measurement, protection, the control systems, microelectronics and telecommunications. His current research interests include power system protection and wide area measurement systems improvement using the recent developed technologies of communication and programmable devices. He has been author and co-author of many technical papers.

Muhammad Asif Rabbani currently works as a lecturer in Electrical and Electronics Engineering department, Cyprus Science University. He has been working as a business development and technical manager in various engineering companies related to power and renewable energy sector before coming in research field. As he bears a good technical on field experience so this helps him to do research which can be beneficial for the industry as well. Currently doing research on the fabrication of super capacitors and Thyristor switched capacitors and hybrid energy storage systems for future smart grids.

Mahiraj Singh Rawat received the B.Tech. degree in Electrical & Electronics Engineering from Uttar Pradesh Technical University, Lucknow, Uttar Pradesh, India, M.Tech. degree in Electrical Engineering with specialization in Power System from National Institute of Technology, Hamirpur, Himanchal Pradesh, India and Ph.D. in Electrical Engineering from National Institute of Technology, Kurukshetra, Haryana, India in 2007, 2011 and 2019 respectively. He is currently working as Assistant Professor in Department of Electrical Engineering, National Institute of Technology, Uttarakhand. His current research interest are stability and renewable energy integration issues of power system, integration of FACTS devices, and distribution system analysis. He is also the member of IEEE and IEEE Power and Energy Society.

Shelly Vadhera is currently working as an Associate Professor in National Institute of Technology, Kurukshetra, Haryana, India. Her current research area focuses on Power Systems and Renewable Energy.

N. Suthanthira Vanitha, currently working as Professor/EEE and Dean - Quality Assurance, Muthayammal Engineering College,Rasipuram, Namakkal Dt. She received B.E.,(EEE), K.S.R. College of Technology, Madras University, M.E.,(Applied Electronics),Mohamed Sathak Engineering College, Madurai Kamaraj University and Ph.D., Biomedical Instrumentation & Embedded Systems, Anna University, Chennai. Her Research interest lies in the area of Biomedical, Embedded Systems, Robotics and Renewable Energy Sources etc. Published 159 Research Publications in National, International Journals and Conferences.

Abdelkader Zitouni was born in August 14th, 1965 in Hussein-Dey, Algiers. He received His Ingéniorat degree in Electronic Instrumentation from University Houari Boumediene of Algiers (USTHB) in 1989 and Magister degrees in Applied Electronics from Institut of Electrical and Electronic Engineering (IGEE), University M'Hamed Bougara of Boumerdes in 1999. He achieved his Doctorat d'Etat "PhD" degree at IGEE in 2019. Currently he is an Associate Professor at IGEE and member of Signals and Systems Laboratory, with activities centered upon the control and the protection systems, microelectronics and telecommunications. His current research interests include power system protection and wide area measurement systems improvement using the recent developed technologies of communication and programmable devices. He has been author and co-author many technical papers.

Index

H

I

L

M

N

O

P

R

S

T

W

Z

www.ingramcontent.com/pod-product-compliance
Lightning Source LLC
LaVergne TN
LVHW062114300926
849161LV00010B/2458